Quantum Signatures of Chaos

Springer
Berlin
Heidelberg
New York
Barcelona
Hong Kong
London
Milan
Paris
Singapore
Tokyo

Physics and Astronomy ONLINE LIBRARY

http://www.springer.de/phys/

Springer Series in Synergetics

http://www.springer.de/phys/books/sssyn

An ever increasing number of scientific disciplines deal with complex systems. These are systems that are composed of many parts which interact with one another in a more or less complicated manner. One of the most striking features of many such systems is their ability to spontaneously form spatial or temporal structures. A great variety of these structures are found, in both the inanimate and the living world. In the inanimate world of physics and chemistry, examples include the growth of crystals, coherent oscillations of laser light, and the spiral structures formed in fluids and chemical reactions. In biology we encounter the growth of plants and animals (morphogenesis) and the evolution of species. In medicine we observe, for instance, the electromagnetic activity of the brain with its pronounced spatio-temporal structures. Psychology deals with characteristic features of human behavior ranging from simple pattern recognition tasks to complex patterns of social behavior. Examples from sociology include the formation of public opinion and cooperation or competition between social groups.

In recent decades, it has become increasingly evident that all these seemingly quite different kinds of structure formation have a number of important features in common. The task of studying analogies as well as differences between structure formation in these different fields has proved to be an ambitious but highly rewarding endeavor. The Springer Series in Synergetics provides a forum for interdisciplinary research and discussions on this fascinating new scientific challenge. It deals with both experimental and theoretical aspects. The scientific community and the interested layman are becoming ever more conscious of concepts such as self-organization, instabilities, deterministic chaos, nonlinearity, dynamical systems, stochastic processes, and complexity. All of these concepts are facets of a field that tackles complex systems, namely synergetics. Students, research workers, university teachers, and interested laymen can find the details and latest developments in the Springer Series in Synergetics, which publishes textbooks, monographs and, occasionally, proceedings. As witnessed by the previously published volumes, this series has always been at the forefront of modern research in the above mentioned fields. It includes textbooks on all aspects of this rapidly growing field, books which provide a sound basis for the study of complex systems.

Series Editor

Hermann Haken

Institut für Theoretische Physik
und Synergetik
der Universität Stuttgart
70550 Stuttgart, Germany

and
Center for Complex Systems
Florida Atlantic University
Boca Raton, FL 33431, USA

Advisory Board

Åke Andersson

Royal Institute of Technology
Department of Infrastructure
and Planning (RP)
10044 Stockholm, Sweden

Luigi Lugiato

Dipartimento di Fisica
Universitá degli Studi di Milano
Via Celoria 16
20133 Milan, Italy

Bernold Fiedler

Freie Universität Berlin
Institut für Math I
Arnimallee 2–6
14195 Berlin, Germany

Jürgen Parisi

Fachbereich Physik, Abt. Energie-
und Halbleiterforschung
Universität Oldenburg
26111 Oldenburg, Germany

Yoshiki Kuramoto

Department of Physics
Graduate School of Sciences
Kyoto University
Kyoto 606-8592, Japan

Manuel G. Velarde

Instituto Pluridisciplinar (USM)
Paseo Juan XXIII, No. 1
28040 Madrid, Spain

Fritz Haake

Quantum Signatures of Chaos

Second Revised and Enlarged Edition
With 66 Figures

 Springer

Professor Dr. Fritz Haake

Institut für Theoretische Physik, Fachbereich 7, Physik
University of Essen
Postfach 10 37 46
45117 Essen, Germany

The first edition appeared as Springer Series in Synergetics, Volume 54, under the title:
F. Haake, *Quantum Signatures of Chaos*

Library of Congress Cataloging-in-Publication Data

Haake, Fritz.
    Quantum signatures of chaos / Fritz Haake.-- 2nd ed.
      p. cm. -- (Springer series in synergetics, ISSN 0172-7389)
    The first edition appeared as Springer series in synergetics, volume 54 under the title:
F. Haake, Quantum signatures of chaos--T.p. verso.
    Includes bibliographical references and index.
    ISBN 3540677232 (alk. paper)
      1. Quantum chaos. I. Title. II. Series.

  Q172.5.C45 H33 2000
  003'.857--dc21

                                                          00-063786

ISSN 0172-7389

ISBN 3-540-67723-2 2nd Edition Springer-Verlag Berlin Heidelberg New York

ISBN 3-540-53144-0 1st Edition Springer-Verlag Berlin Heidelberg New York

Springer-Verlag Berlin Heidelberg New York
a member of BertelsmannSpringer Science+Business Media GmbH

© Springer-Verlag Berlin Heidelberg 1992, 2001
Printed in Germany

The use of general descriptive names, registered names, trademarks, etc. in this publication does not imply, even in the absence of a specific statement, that such names are exempt from the relevant protective laws and regulations and therefore free for general use.

Typesetting by the author using a Springer TeX macro package.
Final typesetting: LE-TeX Jelonek, Schmidt & Vöckler GbR, Leipzig
Cover design: *design & production*, Heidelberg

Printed on acid-free paper   SPIN: 10663193   55/3141/mf - 5 4 3 2 1 0

Für Gitta und Julia

Preface

The warm reception of the first edition, as well as the tumultuous development of the field of quantum chaos have tempted me to rewrite this book and include some of the important progress made during the past decade.

Now we know that quantum signatures of chaos are paralleled by wave signatures. Whatever is undergoing wavy space-time variations, be it sound, electromagnetism, or quantum amplitudes, each shows exactly the same manifestations of chaos. The common origin is nonseparability of the pertinent wave equation; that latter "definition" of chaos, incidentally, also applies to classical mechanics if we see the Hamilton–Jacobi equation as the limiting case of a wave equation. At any rate, drums, concert halls, oscillating quartz blocks, microwave and optical oscillators, electrons moving ballistically or with impurity scattering through mesoscopic devices all provide evidence and data for wave or quantum chaos. All of these systems have deep analogies with billiards, much as the latter may have appeared of no more than academic interest only a decade ago. Of course, molecular, atomic, and nuclear spectroscopy also remain witnesses of chaos, while the chromodynamic innards of nucleons are beginning to attract interest as methods of treatment become available.

Of the considerable theoretical progress lately achieved, the book focuses on the deeper statistical exploitation of level dynamics, improved control of semiclassical periodic-orbit expansions, and superanalytic techniques for dealing with various types of random matrices. These three fields are beginning, independently and in conjunction, to generate an understanding of why certain spectral fluctuations in classically nonintegrable systems are universal and why there are exceptions.

Only the rudiments of periodic-orbit theory and superanalysis appeared in the first edition. More could not have been included here had I not enjoyed the privilege of individual instruction on periodic-orbit theory by Jon Keating and on superanalysis by Hans-Jürgen Sommers and Yan Fyodorov. Hans-Jürgen and Yan have even provided their lecture notes on the subject. While giving full credit and expressing my deep gratitude to these three colleagues, I must bear all blame for blunders.

Reasonable limits of time and space had to be respected and have forced me to leave out much interesting material such as chaotic scattering and the semiclassical art of getting spectra for systems with mixed phase spaces. Equally regrettably, no justice could be done here to the wealth of experiments that have now been performed, but I am happy to see that gap filled by my much more competent colleague Hans-Jürgen Stöckmann.

Incomplete as the book must be, it now contains more material than fits into a single course in quantum chaos theory. In some technical respects, it digs deeper than general introductory courses would go. I have held on to my original intention though, to provide a self-contained presentation that might help students and researchers to enter the field or parts thereof.

The number of co-workers and colleagues from whose knowledge and work I could draw has increased considerably over the years. Having already mentioned Yan Fyodorov, Jon Keating, and Hans-Jürgen Sommers, I must also express special gratitude to my partner and friend Marek Kuś whose continuing help was equally crucial. My thanks for their invaluable influence go to Sergio Albeverio, Daniel Braun, Peter Braun, Eugene Bogomolny, Chang-qi Cao, Dominique Delande, Bruno Eckhardt, Pierre Gaspard, Sven Gnutzmann, Peter Goetsch, Siegfried Grossmann, Martin Gutzwiller, Gregor Hackenbroich, Alan Huckleberry, Micha Kolobov, Pavel Kurasov, Robert Littlejohn, Nils Lehmann, Jörg Main, Alexander Mirlin, Jan Mostowski, Alfredo Ozorio de Almeida, Pjotr Peplowski, Ravi Puri, Jonathan Robbins, Kazik Rzą_żewski, Henning Schomerus, Carsten Seeger, Thomas Seligmann, Frank Steiner, Hans-Jürgen Stöckmann, Jürgen Vollmer, Joachim Weber, Harald Wiedemann, Christian Wiele, Günter Wunner, Dmitri Zaitsev, Kuba Zakrzewski, Martin Zirnbauer, Marek Zukowski, Wojtek Zurek, and, last but not at all least, Karol Życzkowski.

In part this book is an account of research done within the Sonderforschungsbereich "Unordnung und Große Fluktuationen" of the Deutsche Forschungsgemeinschaft. This fact needs to be gratefully acknowledged, since coherent longterm research of a large team of physicists and mathematicians could not be maintained without the generous funding we have enjoyed over the years through our Sonderforschungsbereich.

Times do change. Like many present-day science authors I chose to pick up LaTeX and key all changes and extensions into my little machine myself. As usually happens when learning a new language, the beginning is all effort, but one eventually begins to enjoy the new mode of expressing oneself. I must thank Peter Gerwinski, Heike Haschke, and Rüdiger Oberhage for their infinite patience in getting me going.

Essen, July 2000   *F. Haake*

Foreword to the First Edition

The interdisciplinary field of synergetics grew out of the desire to find general principles that govern the spontaneous formation of ordered structures out of microscopic chaos. Indeed, large classes of classical and quantum systems have been found in which the emergence of ordered structures is governed by just a few degrees of freedom, the so-called order parameters. But then a surprise came with the observation that a few degrees of freedom may cause complicated behavior, nowadays generally subsumed under the title "deterministic chaos" (not to be confused with microscopic chaos, where many degrees of freedom are involved). One of the fundamental problems of chaos theory is the question of whether deterministic chaos can be exhibited by quantum systems, which, at first sight, seem to show no deterministic behavior at all because of the quantization rules. To be more precise, one can formulate the question as follows: How does the transition occur from quantum mechanical properties to classical properties showing deterministic chaos?

Fritz Haake is one of the leading scientists investigating this field and he has contributed a number of important papers. I am therefore particularly happy that he agreed to write a book on this fascinating field of quantum chaos. I very much enjoyed reading the manuscript of this book, which is written in a highly lively style, and I am sure the book will appeal to many graduate students, teachers, and researchers in the field of physics. This book is an important addition to the Springer Series in Synergetics.

Stuttgart, February 1991   *H. Haken*

Preface to the First Edition

More than sixty years after its inception, quantum mechanics is still exerting fascination on every new generation of physicists. What began as the scandal of non-commuting observables and complex probability amplitudes has turned out to be the universal description of the micro-world. At no scale of energies accessible to observation have any findings emerged that suggest violation of quantum mechanics.

Lingering doubts that some people have held about the universality of quantum mechanics have recently been resolved, at least in part. We have witnessed the serious blow dealt to competing hidden-variable theories by experiments on correlations of photon pairs. Such correlations were found to be in conflict with any local deterministic theory as expressed rigorously by Bell's inequalities. – Doubts concerning the accommodation of dissipation in quantum mechanics have also been eased, in much the same way as in classical mechanics. Quantum observables can display effectively irreversible behavior when they are coupled to an appropriate environmental system containing many degrees of freedom. Even in closed quantum systems with relatively few degrees of freedom, behavior resembling damping is possible, provided the system displays chaotic motion in the classical limit.

It has become clear that the relative phases of macroscopically distinguishable states tend, in the presence of damping, to become randomized in exceedingly short times; that remains true even when the damping is so weak that it is hardly noticeable for quantities with a well-defined classical limit. Consequently, a superposition (in the quantum sense) of different readings of a macroscopic measuring device would, even if one could be prepared momentarily, escape observation due to its practically instantaneous decay. While this behavior was conjectured early in the history of quantum mechanics it is only recently that we have been able to see it explicitly in rigorous solutions for specific model systems.

There are many intricacies of the classical limit of quantum mechanics. They are by no means confined to abrupt decay processes or infinitely rapid oscillations of probability amplitudes. The classical distinction between regular and chaotic motion, for instance, makes itself felt in the semiclassical regime that is typically associated with high degrees of excitation. In that regime quantum effects like the discreteness of energy levels and interference phenomena are still discernible while the correspondence principle suggests the onset of validity of classical mechanics.

The semiclassical world, which is intermediate between the microscopic and the macroscopic, is the topic of this book. It will deal with certain universal modes of behavior, both dynamical and spectral, which indicate whether their classical counterparts are regular or chaotic. Conservative as well as dissipative systems will be treated.

The area under consideration often carries the label "quantum chaos". It is a rapidly expanding one and therefore does not yet allow for a definite treatment. The material presented reflects subjective selections. Random-matrix theory will enjoy special emphasis. A possible alternative would have been to make current developments in periodic-orbit theory the backbone of the text. Much as I admire the latter theory for its beauty and its appeal to classical intuition, I do not understand it sufficiently well that I can trust myself to do it justice. With more learning, I might yet catch up and find out how to relate spectral fluctuations on an energy scale of a typical level spacing to classical properties. There are other regrettable omissions. Most notable among these may be the ionization of hydrogen atoms by microwaves, for which convergence of theory and experiment has been achieved recently. Also too late for inclusion is the quantum aspect of chaotic scattering, which has seen such fine progress in the months between the completion of the manuscript and the appearance of this book.

This book grew out of lectures given at the universities of Essen and Bochum. Most of the problems listed at the end of each chapter have been solved by students attending those lectures. The level aimed at was typical of a course on advanced quantum mechanics. The book accordingly assumes the reader to have a good command of the elements of quantum mechanics and statistical mechanics, as well as some background knowledge of classical mechanics. A little acquaintance with classical nonlinear dynamics would not do any harm either.

I could not have gone through with this project without the help of many colleagues and coworkers. They have posed many of the questions dealt with here and provided most of the answers. Perhaps more importantly, they have, within the theory group in Essen, sustained an atmosphere of dedication and curiosity, from which I keep drawing knowledge and stimulus. I can only hope that my young coworkers share my own experience of receiving more than one is able to give. I am especially indebted to Michael Berry, Oriol Bohigas, Giulio Casati, Boris Chirikov, Barbara Dietz, Thomas Dittrich, Mario Feingold, Shmuel Fishman, Dieter Forster, Robert Graham, Rainer Grobe, Italo Guarneri, Klaus-Dieter Harms, Michael Höhnerbach, Ralf Hübner, Felix Israilev, Marek Kuś, Georg Lenz, Maciej Lewenstein, Madan Lal Mehta, Jan Mostowski, Akhilesh Pandey, Dirk Saher, Rainer Scharf, Petr Šeba, Dima Shepelyansky, Uzy Smilansky, Hans-Jürgen Sommers, Dan Walls, and Karol Życzkowski.

Angela Lahee has obliged me by smoothening out some clumsy Teutonisms and by her careful editing of the manuscript. My secretary, Barbara Sacha, deserves a big thank you for keying version upon version of the manuscript into her computer.

My friend and untiring critic Roy Glauber has followed this work from a distance and provided invaluable advice. – I am grateful to Hermann Haken for his invitation to contribute this book to his series in synergetics, and I am all the more honored since it can fill but a tiny corner of Haken's immense field. However, at least Chap. 8 does bear a strong relation to several other books in the series inasmuch as it touches upon adiabatic-elimination techniques and quantum stochastic processes. Moreover, that chapter represents variations on themes I learned as a young student in Stuttgart, as part of the set of ideas which has meanwhile grown to span the range of this series. The love of quantum mechanics was instilled in me by Hermann Haken and his younger colleagues, most notably Wolfgang Weidlich, as they were developing their quantum theory of the laser and thus making the first steps towards synergetics.

Essen, January 1991   *F. Haake*

Contents

1. Introduction

Long as it may have taken to realize, we are now certain that there are two radically different types of motion in classical Hamiltonian mechanics: the *regular motion* of *integrable* systems and the *chaotic motion* of *nonintegrable* systems. The harmonic oscillator and the Kepler problem show regular motion, while systems as simple as a periodically driven pendulum or an autonomous conservative double pendulum can display chaotic dynamics. To identify the type of motion for a given system, one may look at a bundle of trajectories originating from a narrow cloud of points in phase space. The distance between any two such trajectories grows exponentially with time in the chaotic case; the growth rate is the so-called Lyapunov exponent. For regular motion, on the other hand, the distance in question may increase like a power of time but never exponentially; the corresponding Lyapunov exponent can thus be said to vanish. I should add that neither subexponential nor exponential separation can prevail indefinitely. Limits are set by Poincaré recurrences or, in some cases, the accessible volume of phase space. Usually, however, such limits become effective on time scales way beyond those on which regular or chaotic behavior is manifest.

When quantum effects are important for a physical system under study, the notion of a phase-space trajectory loses its meaning and so does the notion of a Lyapunov exponent measuring the separation between trajectories. In cases with a discrete energy spectrum, exponential separation is strictly excluded even for expectation values of observables, on time scales on which level spacings are resolvable. The dynamics is then characterized by quasi-periodicity, i.e., recurrences rather than chaos in the classical sense. The quasi-period, i.e., a typical recurrence time, is inversely proportional to a typical level spacing and must of course tend to infinity in the classical limit (formally, $\hbar \to 0$).

Having lost the classical distinction between regular motion and chaos when turning to quantum mechanics, one naturally wonders whether there are other, genuinely quantum mechanical criteria allowing one to distinguish two types of quantum dynamics. Such a distinction, if at all possible, should parallel the classical case: As $\hbar \to 0$, one group should become regular and the other chaotic. It has become clear during the last decade that intrinsically quantum mechanical distinction criteria do exist. Some are based on the energy spectrum; others on the energy eigenvectors, or on the temporal evolution of suitable expectation values.

A surprising lesson was taught to us by experiments with classical waves, electromagnetic and sound. Classical fields show much the same signatures of chaos as quantum probability amplitudes. Quantum or wave chaos arises whenever the pertinent wave equation is nonseparable for given boundary conditions. The character of the field is quite immaterial; it may be real, complex, or vector; the wave equation may be linear like Schrödinger's or Maxwell's or nonlinear like the Gross–Pitaevski equation for Bose–Einstein condensates. In all cases, classical trajectories or rays arise in the limit of short wavelengths, and these trajectories are chaotic in the sense of positive Lyapunov exponents whenever the underlying wave problem is nonseparable. *Nonseparability is shared by the wave problem and the classical Hamilton–Jacobi equation ensuing in the short-wave limit and may indeed be seen as the deepest characterization of chaos.*

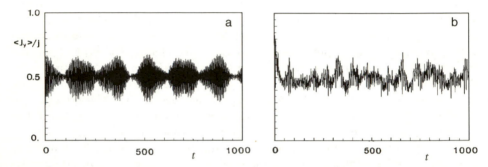

Fig. 1.1. Quasi-periodic behavior of a quantum mean value (angular momentum component for a periodically kicked top) under conditions of classically regular motion (**a**) and classical chaos (**b**). For details see Sect. 7.6

I shall discuss quantum (and wave) distinctions between "regular" and "irregular" motions at some length in the chapters to follow. It should be instructive to jump ahead a little and infer from Fig. 1.1 that the temporal quasi-periodicity in quantum systems with discrete spectra can manifest itself in a way that tells us whether the classical limit will reveal regular or chaotic behavior. The figure displays the time evolution of the expectation value of a typical observable of a periodically kicked top, i.e., a quantum spin J with fixed square, $J^2 = j(j + 1) = $ const. In the classical limit, $j \rightarrow \infty$, the classical angular momentum J/j is capable of regular or chaotic motion depending on the values of control parameters and on the initial orientation. Figure 1.1 pertains to values of the control parameters for which chaos and regular motion coexist in the classical phase space. Parts a and b of Fig. 1.1 refer to initial quantum states localized entirely within classically regular and classically chaotic parts of phase space, respectively. In both cases quasi-periodicity is manifest in the form of recurrences. Obviously, however, the quasi-periodicity in Fig. 1.1a is a nearly perfect periodicity ("collapse and revival") while Fig. 1.1b shows an erratic sequence of recurrences. I shall show later how one can characterize this qualitative difference in quantitative terms. In any event, the difference in question is intrinsically quantum mechanical; the mean temporal separation of recurrences of either type is proportional to j and thus diverges in the classical limit. In contrast to classical chaos, quantum mechanically irregular motion cannot be characterized by extreme sensitivity to tiny changes of initial data: Due to the unitarity of quantum dynamics, the overlap of two wave functions remains time-independent, $|\langle\phi(t)|\psi(t)\rangle|^2 = |\langle\phi(0)|\psi(0)\rangle|^2$, provided the time-dependences of $\phi(t)$ and $\psi(t)$ are generated by the same Hamiltonian. However, an alternative characterization of classical chaos, extreme sensitivity to slight changes of the dynamics, does carry over into quantum mechanics, as illustrated in Fig. 1.2. This figure refers to the same dynamical system as in Fig. 1.1 and shows the time-dependent overlap of two wave functions. For each curve, the two states involved originate from one and the same initial state but evolve with slightly different values (relative difference 10^{-4}) of one control parameter; that (tiny!) difference apart, all control parameters are set as in Fig. 1.1, as are the two initial states used. The time-dependent overlap remains close to unity at all times, if the initial state is located in a classically regular part of the phase space. For the initial state residing in the classically chaotic region, however, the overlap falls exponentially, down to a level of order $1/j$. Such sensitivity of the overlap to changes of a control parameter is quite striking and may indeed serve as a quantum criterion of irregular motion.

Somewhat more widely known are the following two possibilities for the statistics of energy levels (or quasi-energy levels for periodically driven systems). Generic classically integrable systems with two or more degrees of freedom have quantum levels that tend to cluster and are not prohibited from crossing when a parameter in the Hamiltonian is varied [1]. The typical distribution of the spacings of neighboring levels is exponential, $P(S) = \exp(-S)$, just as if the levels arose as the uncorrelated events in a Poissonian random process. Classically nonintegrable systems with their phase spaces dominated by chaos, on the other hand, enjoy the privilege of levels that are correlated such that crossings are strongly resisted [2–7]. There are three universal degrees of level repulsion: linear, quadratic, and quartic [$P(S) \sim S^\beta$ for $S \to 0$ with $\beta = 1, 2$, or 4]. To which universality class a given nonintegrable system can belong is determined by the set of its symmetries. As will be explained in detail later, anti-unitary symmetries such as time-reversal invariance play an especially important role. Broadly speaking, in systems without any antiunitary symmetry generically, $\beta = 2$; in the presence of an antiunitary symmetry and with sufficiently high geometric invariances, one typically finds linear level repulsion; the strongest resistance to level crossings, $\beta = 4$, is characteristic of time-reversal invariant systems possessing Kramers' degeneracy but no geometric symmetry at all.

Clearly, the alternative between level clustering and level repulsion belongs to the worlds of quanta and waves. It parallels, however, the classical distinction between predominantly regular and predominantly chaotic motion. Figure 1.3 illustrates the four generic possibilities mentioned, obtained from the numerically determined quasi-energies of various types of periodically kicked tops; the corresponding portraits of trajectories in the respective classical phase spaces (Fig. 7.3) show the correlation between the quantum and the classical distinction of the two types of dynamics.

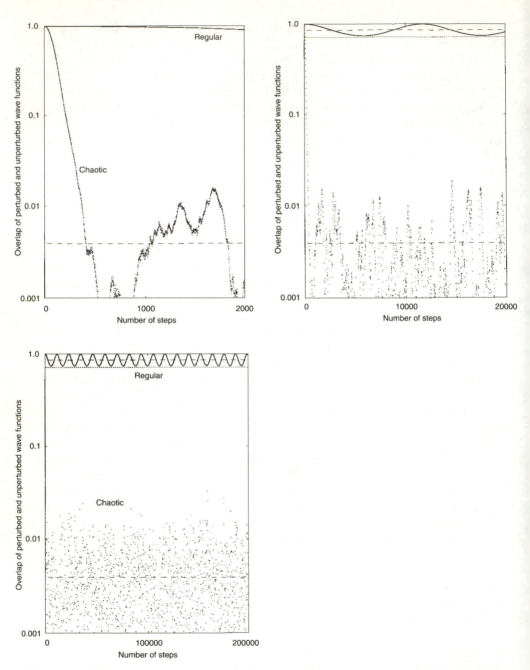

Fig. 1.2. Time dependence of the overlap of two wave vectors of the kicked top, both originating from the same initial state but evolving with slightly different values of one control parameter. Upper and lower curves refer to conditions of classically regular and classically chaotic motion, respectively. For details see Sect. 7.6. Courtesy of A. Peres [233]

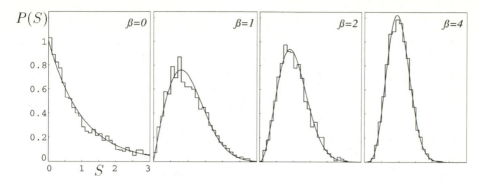

Fig. 1.3. Level spacing distributions for kicked tops under conditions of classically regular motion $\beta = 0$ and classical chaos $\beta = 1, 2, 4$. The latter three curves pertain to tops from different universality classes and display linear $\beta = 1$, quadratic $\beta = 2$, and quartic level repulsion $\beta = 4$

An experimentally determined spectrum of nuclear, atomic, or molecular levels will in general, if taken at face value, display statistics close to Poissonian, even when there is no reason to suppose that the corresponding classical many-body problem is integrable or at least nearly integrable. To uncover spectral correlations, the complete set of levels must be separated into subsets, each of which has fixed values of the quantum numbers related to the symmetries of the system. Subsets that are sufficiently large to allow for a statistical analysis generally reveal level repulsion of the degree expected on grounds of symmetry.

Of the three universal degrees of level repulsion, only the linear and quadratic varieties have been observed experimentally to date. As for linear repulsion, the first data came from nuclear physics in the 1960s [5–10]; mostly much later, confirmation came from microwaves [11–14], molecular [16] and atomic [17] spectroscopy, and sound waves [18, 19]. Figure 1.4 displays such results from various fields and provides evidence for the kinship of quantum and wave chaos.

As regards quadratic level repulsion, it would be hopeless to look in nuclei; even though the weak interaction does break time-reversal invariance, that breaking is far too feeble to become visible in level-spacing distributions. Equally unwieldy are normal-size atoms with a magnetic field to break conventional time-reversal invariance. A homogeneous field would not change the linear degree of repulsion since it preserves antiunitary symmetry (generalized time-reversal); such a symmetry survives even in nonaligned electric and magnetic fields if both fields are homogeneous. Rydberg atoms exposed to strong and sufficiently inhomogeneous magnetic fields do possess quadratically repelling levels and will eventually reveal that property to spectroscopists. Promise also lies in systems with half-integer j that are invariant under conventional time reversal. One then confronts Kramers' degeneracy. If all geometric symmetries could be broken, quartic repulsion would be expected, while sufficiently low geometric symmetries would still allow for the quadratic case. To reduce geometric symmetries, appropriately oriented magnetic and electric fields could be used. Needless to say, experimental data of such a quantum kind would be highly welcome. In view of the obstinacy

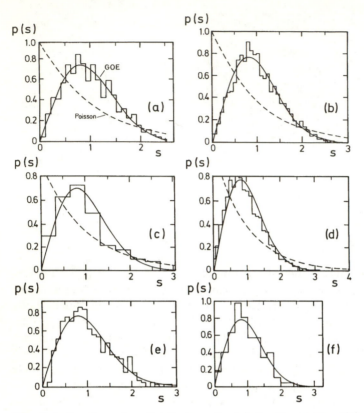

Fig. 1.4. Level spacing distributions for (a) the Sinai billiard[7], (b) a hydrogen atom in a strong magnetic field [20], (c) an NO_2 molecule [16], (d) a vibrating quartz block shaped like a three dimensional Sinai billiard [21], (e) the microwave spectrum of a three-dimensional chaotic cavity [22], (f) a vibrating elastic disc shaped like a quarter stadium [23]. Courtesy of Stöckmann

of single atoms and molecules, the recent first observation of quadratic-level repulsion in microwave experiments [24, 25] was a most welcome achievement, all the more so since it once more underscored that wave chaos is not an exclusive privilege of quanta.

As already emphasized, alternatives such as nearly periodic versus erratic quasi-periodicity and level clustering versus repulsion are quantum or wave concepts without meaning in the strict classical or ray limit. Conversely, the notion of the Lyapunov exponent as the logarithmic rate of separation between phase space trajectories is inapplicable when quantum effects are important. Such complementarity notwithstanding, the quantum distinctions of what becomes regular or chaotic motion classically are most pronounced in semiclassical situations, i.e., when the systems studied are highly excited and have action scales far exceeding Planck's constant. Wave functions on smaller action scales are insensitive to the hierarchy of phase-space structures that distinguish regular and chaotic behavior. In fact, for few-body systems, the ground state and its first few neighbors in energy can often be calculated quite well by variational techniques; however, the classical trajectories of correspondingly low energy are accessible, if chaotic, only through considerable numerical effort.

At high energies, Schrödinger's equation for integrable systems yields to "torus quantization". This beautiful semiclassical approximation scheme is based on the fact that the phase-space trajectory of an integrable system winds around a torus fixed by the set of constant actions. By expressing the Hamiltonian in terms of the actions and allowing each action to be an integer multiple of Planck's constant, one usually obtains good approximations to the high-lying energy levels. As already mentioned, the spectrum so derived has an exponential spacing distribution (provided there are at least two degrees of freedom). A short exposition of the underlying phenomenon of level clustering and its explanation through torus quantization will be the object of Chap. 5.

Similarly as unwieldy as Newton's equations to nonnumerical techniques is Schrödinger's equation with respect to highly excited states of nonintegrable systems, on the other hand. Torus quantization may still work reasonably in near integrable cases where slight perturbations of integrable limiting cases have only modified but not destroyed the vast majority of the original tori. When chaos is fully developed, however, Schrödinger's equation can only be solved numerically.

In numerical treatments of systems with global chaos, an interesting predictability paradox arises. As indicated by a positive Lyapunov exponent, small but inevitable numerical inaccuracies will amplify exponentially and thus render impossible long-range predictions of classical trajectories. Much more reliable by comparison can be the corresponding quantum predictions for mean values, at least in the case of discrete levels; quasi-periodic time evolution, even if very many sinusoidal terms contribute, is incapable of exponential runaway. A stupefying illustration of quantum predictions outdoing classical ones is given by the kinetic energy of the periodically kicked rotator in Fig. 1.5. The quantum mean (dotted) originates from the momentum eigenstate with a vanishing eigenvalue while the classical curve (full) represents an average over 250 000 trajectories starting from a cloud of points with zero momentum and equipartition for the conjugate angle. The classical average displays diffusive growth of the kinetic energy – which, as will be shown later, is tantamount to chaos – until time is reversed after 500 kicks. Instead of retracing its path back to the initial state, the classical mean soon turns again to diffusive growth. The quantum mean, on the other hand, is symmetric around $t = 500$ without noticeable error. Evidently, such a result suggests that a little caution is necessary with respect to our often too naive notions of classical determinism and quantum indeterminacy.

Nuclear physicists have known for quite some time that mean-field theories not only give a rather satisfactory account of the low-energy excitations of nuclei, but also give reasonable approximations for the average level density and other average properties at high excitation energies. An average is meant as one over many excited states with neighboring energies. Such averages may be, and in general are, energy-dependent but they constitute sensible concepts only if they vary but little over energy scales of the order of a typical level spacing. In practice, an average level density can be defined for all but the lightest nuclei, at energies

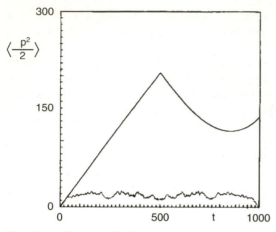

Fig. 1.5. Classical (*full curve*) and quantum (*dotted*) mean kinetic energy of the periodically kicked rotator, determined by numerical iteration of the respective maps (see Chap. 7 for details). The quantum mean follows the classical diffusion for times up to some "break" time and then begins to display quasi-periodic fluctuations. After 500 kicks the direction of time was reserved; while the quantum mean accurately retraces its history, the classical mean reverts to diffusive growth, thus revealing the extreme sensitivity of chaotic systems to tiny perturbations (here round-off errors). For details see Chaps. 7, 9. Courtesy of Graham and Dittrich [352]

of about a few MeV and above. Relative to the average level density of a nucleus, the precise locations in energy of the individual levels appear, from a statistical point of view, to be determined by fluctuations; their calculation is beyond the scope of mean-field theories such as those based on the shell model. Theory would in fact be at a loss here, were it not for the fortunate fact that some statistical characteristics of the fluctuations turn out to be universal. The level-spacing distribution already discussed is among those characteristics that appear insensitive to details of the nuclear structure and are instead determined by no more than symmetries and some kind of maximum-disorder principle. The situation just described for nuclei is also found, either experimentally or from numerical work, in highly excited atoms and molecules and in fact, also in almost all systems displaying strong chaos in the classical limit.

Following a conjecture of *Bohigas, Giannoni*, and *Schmit* [7], certain universal features of spectral fluctuations in classically chaotic systems have been found to be well described by random-matrix theory [26, 27]. Based on a suggestion by *Wigner*, this theory takes random Hermitian matrices as models of Hamiltonians H of autonomous systems. Unitary random matrices are employed similarly for periodically driven systems, as models of the unitary "Floquet" operators F describing the change of the quantum state during one cycle of the driving; powers F^n of the Floquet operator with $n = 1, 2, 3, \ldots$ yield a stroboscopic description of the driven dynamics under consideration. There are four important classes for both Hermitian and unitary random matrices, one related to classically integrable systems and three to nonintegrable ones. The four classes in question illustrate the quantum distinction between regular and irregular dynamics discussed above: their eigenvalues (energies in the Hermitian case and quasi-energies, i.e., eigenphases in the unitary case) cluster and repel according to one of three universal degrees, respectively. The matrix ensembles with level repulsion can be defined by the requirement of maximum statistical independence of all matrix elements within the constraints imposed by symmetries.

I shall present a brief review of random-matrix theory in Chap. 4. Since several excellent texts on this subject are available [26–28], I shall attempt neither completeness nor rigor but keep to an introductory style and intuitive arguments. When slightly fancy machinery is employed for some more up-to-date issues, it is patiently developed rather than assumed as a prerequisite.

Considerable emphasis will be given to an interesting reinterpretation of random-matrix theory as "level dynamics" in Chap. 6. Such a reinterpretation was already sought by *Dyson* but could be implemented only recently on the basis of a discovery of *Pechukas* [29]. The fate of the N eigenvalues and eigenvectors of an $N \times N$ Hermitian matrix, $H = H_0 + \lambda V$, is in one-to-one correspondence with the classical Hamiltonian dynamics of a particular one-dimensional N-particle system upon changing the weight λ of a perturbation V [29–31]. This fictitious system, now often called Pechukas–Yukawa gas, has λ as a time, the eigenvalues E_n of H as coordinates, the diagonal elements V_{nn} of the perturbation V in the H representation as momenta, and the off-diagonal elements V_{nm} related to certain angular momenta. As I shall explain in Chap. 6, random-matrix theory now emerges as a result of standard equilibrium statistical mechanics for the fictitious N-particle system. Moreover, the universality of spectral fluctuations of chaotic dynamics will find a natural explanation on that basis.

The fictitious N-particle model also sheds light on another important problem. A given Hamiltonian $H_0 + \lambda V$ may have H_0 integrable and V breaking integrability. The level dynamics will then in general display a transition from level clustering to level repulsion. We shall see that such transitions can be understood as equilibration processes for certain observables in the fictitious N-particle model.

A related phenomenon is the transition from one universality class to another displayed by a Hamiltonian $H(\lambda) = H_0 + \lambda V$ when H_0 and V have different (antiunitary) symmetries. An ad hoc description of such transitions could previously be given in terms of Dyson's Brownian-motion model, a certain "dynamic" generalization of random-matrix theory. As will become clear below, that venerable model is in fact rigorously implied by level dynamics, i.e., the λ-dependence of $H_0 + \lambda V$, provided the energy scale is reset in a suitable λ-dependent manner.

A certain class of periodically driven systems of which the kicked rotator is a prototype displays an interesting quantum anomaly. Like all periodically driven systems, those in question are generically nonintegrable classically. Even under the conditions of fully developed classical chaos, however, the kicked rotator does not display level repulsion. The reason for this anomaly can be inferred from Fig. 1.5. The kinetic energy, and thus the quantum mechanical momentum uncertainty, does not follow the classical diffusive growth indefinitely. After a certain break time the quantum mean enters a regime of quasi-periodic behavior. Clearly, the eigenvectors of the corresponding Floquet operator then have an upper limit to their width in the momentum representation. They are, in fact, exponentially localized on this basis, and the width is interpretable as a "localization length". Two eigenvectors much further apart in momentum than a localization length have no overlap and thus no matrix elements of noticeable magnitude with respect to any observable; they have no reason, therefore, to stay apart in quasi-energy and will display Poissonian level statistics. A theoretical understanding of the phenomenon has been generated by *Grempel, Fishman,* and *Prange* [32] who were able to map the Schrödinger equation for the kicked rotator onto that for Anderson's one-dimensional tight-binding model of a particle in a random potential. Actually, the potential arrived at in the map is pseudo-random, but that restriction does not prevent exponential localization, which is rigorously established only for the strictly random case. It is amusing to find that number-theoretical considerations are relevant in a quantum context here. Quantum localization occurs only in the case where a certain dimensionless version of Planck's constant is an irrational number. Otherwise, the equivalent tight-binding model has a periodic potential and thus extended eigenvectors of the Bloch type and eigenvalues forming continuous bands. I shall present a description of this fascinating situation in Chap. 7 on quantum localization.

A separate chapter will be devoted to *Gutzwiller*'s periodic-orbit theory. The basic trace formulas for maps and autonomous flows will be derived and discussed. As an application, I shall present the current (still unsatisfactory) status of attempts to explain the universality of spectral fluctuations of fully chaotic systems with the help of periodic orbits.

Next, I shall turn to dissipative systems. Instead of Schrödinger's equation, one now faces irreversible master equations. Real energies (or quasi-energies) are replaced with complex eigenvalues of generators of infinitesimal time translations (or nonunitary generalizations of the Floquet operator in the case of periodically driven systems) while the density operator or a suitable representative, like the Wigner function, takes over the role of the wave function. One would again like to identify genuinely quantum mechanical criteria to distinguish two types of motion, one becoming regular and the other chaotic in the classical limit. There is a general qualitative argument, however, suggesting that such quantum distinctions will be a lot harder to establish for dissipative than for Hamiltonian systems. The difference between, say, a complicated limit cycle and a strange attractor becomes apparent when phase-space structures are considered over several orders of magnitude of action scales. The density matrix or the Wigner function will of course reflect phase-space structures on action scales upward of Planck's constant and will thus indicate, with reasonable certainty, the difference between a strange and a simple attractor. But for any representative of the density operator to reveal this difference in terms of genuinely quantum mechanical features without classical meaning, it would have to embody coherences with respect to states distinct on action scales that are large compared to Planck's constant. Such coherences between or superpositions of "macroscopically distinct" states are often metaphorized as Schrödinger cat states. In the presence of even weak damping, such superpositions tend to decohere so rapidly in time to mixtures that observing them becomes difficult if not impossible.

An illustration of the dissipative death of quantum coherences is presented in Fig. 1.6. As in Fig. 1.1a, column a in Fig. 1.6 shows quantum recurrences for an angular momentum component of a periodically kicked top without damping. The control parameters and the initial state are kept constant for all curves in the column and are chosen such that the classical limit would yield regular behavior. Proceeding down the column, the spin quantum number j is increased to demonstrate the rough proportionality of the quasi-period to j (which may be thought of as inversely proportional to Planck's constant). The curves in column b of Fig. 1.6 correspond to their neighbors in column a in all respects except for the effect of weak dissipation. The damping mechanism is designed so as to leave j a good quantum number, and the damping constant is so small that classical trajectories would not be influenced noticeably during the times over which the plots extend. The sequences of quantum mechanical "collapses and revivals" is seen to be strongly altered by the dissipation, even though the quasi-period is in all cases smaller than the classical decay time. I shall argue below that the lifetime of the quantum coherences is shorter than the classical decay time and also shorter than the time constant for quantum observables not sensitive to coherences between "macroscopically" distinct states by a factor of the order of $1/j \sim \hbar$. Similarly dramatic is the effect of dissipation on the erratic recurrences found under conditions of classical chaos. Obviously, dissipation will then tend to wipe out the distinction between regular and erratic recurrences.

The quantum localization in the periodically kicked rotator also involves coherences between states that are distinct on action scales far exceeding Planck's constant. Indeed, if expressed in action units, the localization length is large compared to $\hbar$. One must therefore expect localization to be destroyed by dissipation. Figure 1.7 confirms that expectation. As a particular damping constant is increased, the time dependence of the kinetic energy of the rotator tends to resemble indefinite classical diffusion ever more closely.

The influence of dissipation on level statistics is highly interesting. As long as the widths assigned to the individual levels by a damping mechanism are smaller than the spacings between neighboring levels, there is no noticeable change of the spacing distribution; in particular, the distinction between clustering and repulsion is still possible. The original concept of a spacing loses its meaning, however, when the typical level width exceeds a typical spacing in the absence of dissipation. A possible extension of this concept is the Euclidean distance in the complex plane between the eigenvalues of the relevant master equation or generalized Floquet operator. Numerical evidence for a damped version of the periodically kicked top suggests that these spacings display linear and cubic repulsion under conditions of classically regular and chaotic motion, respectively. Linear repulsion, incidentally, is a rigorous property of Poissonian random processes in the plane. Cubic repulsion, on the other hand, will turn out to be characteristic of random matrices unrestricted by either Hermiticity or unitarity.

The final chapter is devoted to superanalysis and its application to random matrices and disordered systems. This is in respectful reference to the effec-

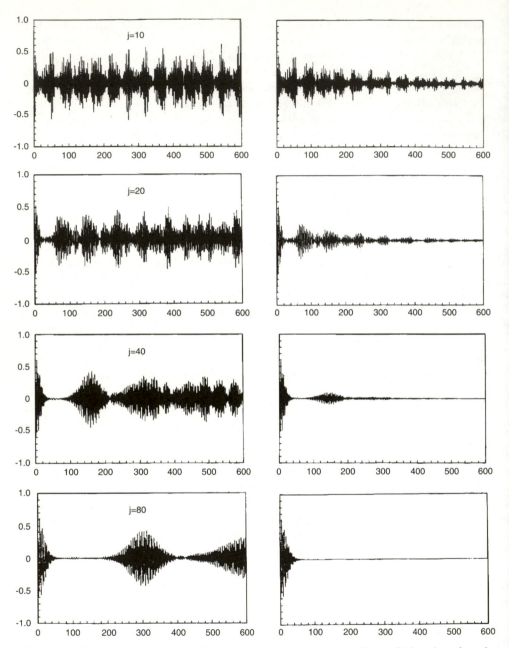

Fig. 1.6. Quantum recurrences for periodically kicked top without (*left column*) and with (*right column*) damping under conditions of classically regular motion. For details see Chap. 9

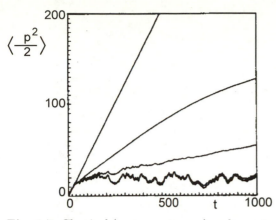

Fig. 1.7. Classical (*uppermost curve*) and quantum mean kinetic energy of periodically kicked rotator with damping. The stronger the damping, the steeper are the curves. In all cases, the damping is so weak that the classical curve is still indistinguishable from that without dissipation. Courtesy of Graham and Dittrich [352]

tive merger of the fields of quantum chaos and disordered systems that we have witnessed during the past decade. The so-called supersymmetry technique has brought about many insights into various ensembles of random matrices, including banded and sparse matrices, and promises further progress. While several monographs on results and the method are available, a pedagogically oriented introduction seems missing and the present text aims to fill just that gap. Readers willing to carefully study Chap. 10 should end up motivated and equipped to carry on the game toward new applications.

2. Time Reversal and Unitary Symmetries

2.1 Autonomous Classical Flows

A classical Hamiltonian system is called time-reversal invariant if from any given solution $\boldsymbol{x}(t)$, $\boldsymbol{p}(t)$ of Hamilton's equations an independent solution $\boldsymbol{x}'(t')$, $\boldsymbol{p}'(t')$, is obtained with $t' = -t$ and some operation relating $\boldsymbol{x}'$ and $\boldsymbol{p}'$ to the original coordinates $\boldsymbol{x}$ and momenta $\boldsymbol{p}$. The simplest such invariance, to be referred to as conventional, holds when the Hamiltonian is an even function of all momenta,

$$t \to -t , \quad \boldsymbol{x} \to \boldsymbol{x} , \quad \boldsymbol{p} \to -\boldsymbol{p} , \quad H(\boldsymbol{x}, \boldsymbol{p}) = H(\boldsymbol{x}, -\boldsymbol{p}) . \tag{2.1.1}$$

This is obviously not a canonical transformation since the Poisson brackets $\{p_i, x_j\} = \delta_{ij}$ are not left intact. The change of sign brought about for the Poisson brackets is often acknowledged by calling classical time reversal anticanonical. We should keep in mind that the angular momentum vector of a particle is bilinear in $\boldsymbol{x}$ and $\boldsymbol{p}$ and thus odd under conventional time reversal.

Systems with an external magnetic field are not invariant under conventional time reversal since the minimal-coupling term in the Hamiltonian of a charged particle, $(\boldsymbol{p} - (e/c)\boldsymbol{A})^2/2m$, is not even in $\boldsymbol{p}$. Such systems may nonetheless have some other, nonconventional time-reversal invariance, to be explained in Sect. 2.9.

Hamiltonian systems with no time-reversal invariance must not be confused with dissipative systems. The differences between Hamiltonian and dissipative dynamics are drastic and well known. Most importantly from a theoretical point of view, all Hamiltonian motions conserve phase-space volumes according to Liouville's theorem, while for dissipative processes such volumes contract in time. The difference between Hamiltonian systems with and without time-reversal invariance, on the other hand, is subtle and has never attracted much attention in the realm of classical physics. It will become clear below, however, that the latter difference plays an important role in the world of quanta [33–35].

2.2 Spinless Quanta

The Schrödinger equation

$$i\hbar\dot{\psi}(\boldsymbol{x}, t) = H\psi(\boldsymbol{x}, t) \tag{2.2.1}$$

is time-reversal invariant if, for any given solution $\psi(\boldsymbol{x}, t)$, there is another one, $\psi'(\boldsymbol{x}, t')$, with $t' = -t$ and ψ' uniquely related to ψ. The simplest such invariance, again termed conventional, arises for a spinless particle with the real Hamiltonian

$$H(\boldsymbol{x}, \boldsymbol{p}) = \frac{\boldsymbol{p}^2}{2m} + V(\boldsymbol{x}) \,, \quad V(\boldsymbol{x}) = V^*(\boldsymbol{x}) \,, \tag{2.2.2}$$

where the asterisk denotes complex conjugation. The conventional reversal is

$$t \rightarrow -t \,, \quad \boldsymbol{x} \rightarrow \boldsymbol{x} \,, \quad \boldsymbol{p} \rightarrow -\boldsymbol{p} \,,$$
$$\psi(\boldsymbol{x}) \rightarrow \psi^*(\boldsymbol{x}) = K\psi(\boldsymbol{x}) \,. \tag{2.2.3}$$

In other words, if $\psi(\boldsymbol{x}, t)$ solves (2.2.1) so does $\psi'(\boldsymbol{x}, t) = K\psi(\boldsymbol{x}, -t)$. The operator K of complex conjugation obviously fulfills

$$K^2 = 1 \,, \tag{2.2.4}$$

i.e., it equals its inverse, $K = K^{-1}$. Its definition also implies

$$K\left(c_1\psi_1(\boldsymbol{x}) + c_2\psi_2(\boldsymbol{x})\right) = c_1^* K\psi_1(\boldsymbol{x}) + c_2^* K\psi_2(\boldsymbol{x}) \,, \tag{2.2.5}$$

a property commonly called antilinearity. The transformation $\psi(\boldsymbol{x}) \rightarrow K\psi(\boldsymbol{x})$ does not change the modulus of the overlap of two wave functions,

$$\left|\langle K\psi|K\phi\rangle\right|^2 = \left|\langle\psi|\phi\rangle\right|^2 \,, \tag{2.2.6}$$

while the overlap itself is transformed into its complex conjugate,

$$\langle K\psi|K\phi\rangle = \langle\psi|\phi\rangle^* = \langle\phi|\psi\rangle \,. \tag{2.2.7}$$

The identity (2.2.7) defines the property of antiunitarity which implies antilinearity [33] (Problem 2.4).

It is appropriate to emphasize that I have defined the operator K with respect to the position representation. Dirac's notation makes this distinction of K especially obvious. If some state vector $|\psi\rangle$ is expanded in terms of position eigenvectors $|x\rangle$,

$$|\psi\rangle = \int d\boldsymbol{x}\psi(\boldsymbol{x})|\boldsymbol{x}\rangle \,, \tag{2.2.8}$$

the operator K acts as

$$K|\psi\rangle = \int d\boldsymbol{x}\psi^*(\boldsymbol{x})|\boldsymbol{x}\rangle \,, \tag{2.2.9}$$

i.e., as $K|\boldsymbol{x}\rangle = |\boldsymbol{x}\rangle$. A complex conjugation operator K' can of course be defined with respect to any representation. It is illustrative to consider a discrete basis and introduce

$$K'|\nu\rangle = K'\sum_\nu \psi_\nu|\nu\rangle = \sum_\nu \psi_\nu^*|\nu\rangle \,. \tag{2.2.10}$$

Conventional time reversal, i.e., complex conjugation in the position representation, can then be expressed as

$$K = UK'$$ (2.2.11)

with a certain symmetric unitary matrix U, the calculation of which is left to the reader as problem 2.5. Unless otherwise stated, the symbol K will be reserved for complex conjugation in the coordinate representation, as far as orbital wave functions are concerned. Moreover, antiunitary time-reversal operators will, for the most part, be denoted by T. Only the conventional time-reversal for spinless particles has the simple form $T = K$.

2.3 Spin-1/2 Quanta

All time-reversal operators T must be antiunitary

$$\langle T\psi | T\phi \rangle = \langle \phi | \psi \rangle \,,$$ (2.3.1)

because of (i) the explicit factor i in Schrödinger's equation and (ii) since they should leave the modulus of the overlap of two wave vectors invariant. It follows from the definition (2.3.1) of antiunitarity that the product of two antiunitary operators is unitary. Consequently, any time-reversal operator T can be given the so-called standard form

$$T = UK \,,$$ (2.3.2)

where U is a suitable unitary operator and K the complex conjugation with respect to a standard representation (often chosen to be the position representation for the orbital part of wave functions).

Another physically reasonable requirement for every time-reversal operator T is that any wave function should be reproduced, at least to within a phase factor, when acted upon twice by T,

$$T^2 = \alpha \,, \quad |\alpha| = 1 \,.$$ (2.3.3)

Inserting the standard form (2.3.2) in (2.3.3) yields[1] $UKUK = UU^*K^2 = UU^* = \alpha$, i.e., $U^* = \alpha U^{-1} = \alpha U^\dagger = \alpha \tilde{U}^*$. The latter identity once iterated gives $U^* = \alpha^2 U^*$, i.e., $\alpha^2 = 1$ or

$$T^2 = \pm 1 \,.$$ (2.3.4)

The positive sign holds for conventional time reversal with spinless particles. It will become clear presently that $T^2 = -1$ in the case of a spin-1/2 particle.

---

[1] Matrix transposition will always be represented by a tilde, while the dagger † will denote Hermitian conjugation.

A useful time-reversal operation for a spin-1/2 results from requiring that

$$TJT^{-1} = -J \tag{2.3.5}$$

holds not only for the orbital angular momentum but likewise for the spin. With respect to the spin, however, T cannot simply be the complex conjugation operation since all purely imaginary Hermitian 2×2 matrices commute with one another. The more general structure (2.3.2) must therefore be considered. Just as a matter of convenience, I shall choose K as the complex conjugation in the standard representation where the spin operator S takes the form

$$S = \tfrac{\hbar}{2}\boldsymbol{\sigma} \, ,$$

$$\sigma_x = \begin{pmatrix} 0 & 1 \\ 1 & 0 \end{pmatrix} \, , \quad \sigma_y = \begin{pmatrix} 0 & -i \\ i & 0 \end{pmatrix} \, , \quad \sigma_z = \begin{pmatrix} 1 & 0 \\ 0 & -1 \end{pmatrix} \, . \tag{2.3.6}$$

The matrix U is then constrained by (2.3.5) to obey

$$T\sigma_x T^{-1} = UK\sigma_x K U^{-1} = U\sigma_x U^{-1} = -\sigma_x$$

$$T\sigma_y T^{-1} = UK\sigma_y K U^{-1} = -U\sigma_y U^{-1} = -\sigma_y \tag{2.3.7}$$

$$T\sigma_z T^{-1} = UK\sigma_z K U^{-1} = U\sigma_z U^{-1} = -\sigma_z \, ,$$

i.e., U must commute with σ_y and anticommute with σ_x and σ_z. Because any 2×2 matrix U can be represented as a sum of Pauli matrices, we can write

$$U = \alpha\sigma_x + \beta\sigma_y + \gamma\sigma_z + \delta \, . \tag{2.3.8}$$

The first of the equations (2.3.7) immediately gives $\alpha = \delta = 0$; the second yields $\gamma = 0$, whereas β remains unrestricted by (2.3.7). However, since U is unitary, β must have unit modulus. It is thus possible to choose $\beta = i$ whereupon the time-reversal operation reads

$$T = i\sigma_y K = e^{i\pi\sigma_y/2}K \, . \tag{2.3.9}$$

This may be taken to include, if necessary, the time reversal for the orbital part of wave vectors by interpreting K as complex conjugation both in the position representation and in the standard spin representation. In this sense I shall refer to (2.3.9) as conventional time reversal for spin-1/2 quanta.

The operation (2.3.9) squares to minus unity, in contrast to conventional time reversal for spinless particles. Indeed, $T^2 = i\sigma_y K i\sigma_y K = (i\sigma_y)^2 = -1$.

If one is dealing with N particles with spin 1/2, the matrix U must obviously be taken as the direct product of N single-particle matrices,

$$\begin{aligned} T &= i\sigma_{1y}i\sigma_{2y} \ldots i\sigma_{Ny}K \\ &= \exp\left[i\frac{\pi}{2}\left(\sigma_{1y} + \sigma_{2y} + \ldots + \sigma_{Ny}\right)\right]K = \exp\left(i\pi\frac{S_y}{\hbar}\right)K \, , \end{aligned} \tag{2.3.10}$$

where S_y now is the y-component of the total spin $S = \hbar(\boldsymbol{\sigma}_1 + \boldsymbol{\sigma}_2 + \ldots + \boldsymbol{\sigma}_N)/2$. The square of T depends on the number of particles according to

$$T^2 = \begin{cases} +1 & N \text{ even} \\ -1 & N \text{ odd} \, . \end{cases} \tag{2.3.11}$$

2.4 Hamiltonians Without T Invariance

All Hamiltonians can be represented by Hermitian matrices. Before proceeding to identify the subclasses of Hermitian matrices to which time-reversal invariant Hamiltonians belong, it is appropriate to pause and make a few remarks about Hamiltonians unrestricted by antiunitary symmetries.

An arbitrary Hamiltonian becomes real in its eigenrepresentation,

$$H = \text{diag} \left(E_1, E_2, \ldots \right) = H^* . \tag{2.4.1}$$

Under a unitary transformation U,

$$H_{\mu\nu} = U_{\mu\lambda} E_\lambda U^\dagger_{\lambda\nu} = U_{\mu\lambda} E_\lambda U^*_{\nu\lambda} , \tag{2.4.2}$$

H preserves Hermiticity, $(H_{\mu\nu})^* = \tilde{H}_{\mu\nu} = H_{\nu\mu}$, but ceases, in general, to be real.

Now, I propose to construct the class of "canonical transformations" that change a Hamiltonian matrix without destroying its Hermiticity and without altering its eigenvalues. To this end it is important to look at each irreducible part of the matrix H separately, i.e., to think of good quantum numbers related to a complete set of mutually commuting conserved observables (other than H itself) as fixed. Eigenvalues are preserved under a similarity transformation with an arbitrary nonsingular matrix A. To show that $H' = A\,HA^{-1}$ has the same eigenvalues as H, it suffices to write out H' in the H representation,

$$H'_{\mu\nu} = \sum_\lambda A_{\mu\lambda} E_\lambda \left(A^{-1} \right)_{\lambda\nu} , \tag{2.4.3}$$

and to multiply from the right by A. The columns of A are then recognized as eigenvectors of H', and the eigenvalues of H turn out to be those of H' as well.

For A to qualify as a canonical transformation, H' must also be Hermitian,

$$\left(A\,HA^{-1} \right)^\dagger = A\,HA^{-1} \Leftrightarrow [H, A^\dagger A] = 0 . \tag{2.4.4}$$

Excluding the trivial solution where $A^\dagger A$ is a function of H and recalling that all other mutually commuting conserved observables are multiples of the unit matrix in the space considered, one concludes that $A^\dagger A$ must be the unit matrix, at least to within a positive factor. That factor must itself be unity if A is subjected to the additional constraint that it should preserve the normalization of vectors,

$$A^\dagger A = \mathbb{1} . \tag{2.4.5}$$

The class of canonical transformations of Hamiltonians unrestricted by antiunitary symmetries is thus constituted by unitary matrices. Obviously, for an N-dimensional Hilbert space that class is the group $U(N)$.

It is noteworthy that Hamiltonian matrices unrestricted by antiunitary symmetries are in general complex. They can, of course, be given real representations, but any such representation will become complex under a general canonical (i.e., unitary) transformation.

2.5 T Invariant Hamiltonians, $T^2 = 1$

When we have an antiunitary operator T with

$$[H, T] = 0 , \quad T^2 = 1 \tag{2.5.1}$$

the Hamiltonian H can always be given a real matrix representation and such a representation can be found without diagonalizing H.

As a first step toward proving the above statement, I demonstrate that, with the help of an antiunitary T squaring to plus unity, T invariant basis vectors ψ_ν can be constructed. Take any vector ϕ_1 and a complex number a_1. The vector

$$\psi_1 = a_1 \phi_1 + T a_1 \phi_1 \tag{2.5.2}$$

is then T invariant, $T\psi_1 = \psi_1$. Next, take any vector ϕ_2 orthogonal to ψ_1 and a complex number a_2. The combination

$$\psi_2 = a_2 \phi_2 + T a_2 \phi_2 \tag{2.5.3}$$

is again T invariant. Moreover, ψ_2 is orthogonal to ψ_1 since

$$
\begin{aligned}
\langle \psi_2 | \psi_1 \rangle &= a_2^* \langle \phi_2 | \psi_1 \rangle + a_2 \langle T\phi_2 | \psi_1 \rangle \\
&= a_2 \left\langle T^2 \phi_2 | \psi_1 \right\rangle^* = a_2 \langle \phi_2 | \psi_1 \rangle^* = 0 .
\end{aligned} \tag{2.5.4}
$$

By proceeding in this way, we eventually arrive at a complete set of mutually orthogonal vectors. If desired, the numbers a_ν can be chosen so as to obtain the normalization $\langle \psi_\mu | \psi_\nu \rangle = \delta_{\mu\nu}$.

With respect to a T invariant basis, the Hamiltonian $H = THT$ is real,

$$
\begin{aligned}
H_{\mu\nu} &= \langle \psi_\mu | H \psi_\nu \rangle = \langle T\psi_\mu | TH\psi_\nu \rangle^* \\
&= \langle \psi_\mu | THT^2 \psi_\nu \rangle^* = \langle \psi_\mu | THT\psi_\nu \rangle^* = H_{\mu\nu}^* .
\end{aligned} \tag{2.5.5}
$$

Note that the Hamiltonians in question can be made real without being diagonalized first. It is therefore quite legitimate to say that they are generically real matrices. The canonical transformations that are admissible now form the group $SO(N)$ of real orthogonal matrices O, $O\tilde{O} = 1$. Beyond preserving eigenvalues and Hermiticity, an orthogonal transformation also transforms a real matrix H into another real matrix $H' = OH\tilde{O}$. The orthogonal group is obviously a subgroup of the unitary group considered in the last section.

It may be worthwhile to look back at Sect. 2.2 where it was shown that the Schrödinger equation of a spinless particle is time-reversal invariant provided the Hamiltonian is a real operator in the position representation. The present section generalizes that previous statement.

2.6 Kramers' Degeneracy

For any Hamiltonian invariant under a time reversal T,

$$[H, T] = 0 , \tag{2.6.1}$$

i.e., if ψ is an eigenfunction with eigenvalue E, so is $T\psi$. As shown above, we may choose the equality $T\psi = \psi$ without loss of generality if $T^2 = +1$. Here, I propose to consider time-reversal operators squaring to minus unity,

$$T^2 = -1 \, . \tag{2.6.2}$$

In this case, ψ and $T\psi$ are orthogonal,

$$\langle \psi | T\psi \rangle = \langle T\psi | T^2\psi \rangle^* = -\langle T\psi | \psi \rangle^* = -\langle \psi | T\psi \rangle = 0 \, , \tag{2.6.3}$$

and therefore all eigenvalues of H are doubly degenerate. This is Kramers' degeneracy. It follows that the dimension of the Hilbert space must, if finite, be even. This fits with the result of Sect. 2.3 that $T^2 = -1$ is possible only if the number of spin-1/2 particles in the system is odd; the total-spin quantum number s is then a half-integer and $2s + 1$ is even.

In the next two sections, I shall discuss the structure of Hamiltonian matrices with Kramers' degeneracy, first for the case with additional geometric symmetries and then for the case in which T is the only invariance.

2.7 Kramers' Degeneracy and Geometric Symmetries

As an example of geometric symmetry, let us consider a parity such that

$$[R_x, H] = 0 \, , \quad [R_x, T] = 0 \, , \quad R_x^2 = -1 \, . \tag{2.7.1}$$

This could be realized, for example, by a rotation through π about, say, the x-axis, $R_x = \exp(i\pi J_x/\hbar)$; note that since $T^2 = -1$ only half-integer values of the total angular momentum quantum number are admitted.

To reveal the structure of the matrix H, it is convenient to employ a basis ordered by parity,

$$R_x |n\pm\rangle = \pm i |n\pm\rangle \, . \tag{2.7.2}$$

Moreover, since T changes the parity,

$$R_x T |n\pm\rangle = T R_x |n\pm\rangle = \mp i T |n\pm\rangle \, , \tag{2.7.3}$$

the basis can be organized such that

$$T |n\pm\rangle = \pm |n\mp\rangle \, . \tag{2.7.4}$$

For the sake of simplicity, let us assume a finite dimension $2N$. The matrix H then falls into four $N \times N$ blocks

$$H = \begin{pmatrix} H^+ & 0 \\ 0 & H^- \end{pmatrix} \tag{2.7.5}$$

two of which are zero since H has vanishing matrix elements between states of different parity. Indeed, $\langle m+|R_x H R_x^{-1}|n-\rangle$ is equal to $+\langle m+|H|n-\rangle$ due to the invariance of H under R_x and equal to $-\langle m+|H|n-\rangle$ because of (2.7.2). The T invariance relates the two blocks $H^\pm$:

$$
\begin{aligned}
\langle m+|H|n+\rangle &= \langle m+|THT^{-1}|n+\rangle \\
&= -\langle m+|TH|n-\rangle = -\langle T(m+)|T^2H|n-\rangle^* \\
&= +\langle m-|H|n-\rangle^* = \langle n-|H|m-\rangle \ . \tag{2.7.6}
\end{aligned}
$$

At this point Kramers' degeneracy emerges: Since they are the transposes of one another, H^+ and H^- have the same eigenvalues. Moreover, they are in general complex and thus have $U(N)$ as their group of canonical transformations.

Further restrictions on the matrices $H^\pm$ arise from additional symmetries. It is illustrative to admit one further parity R_y with

$$
[R_y, H] = 0 \ , \quad [R_y, T] = 0 \ , \quad R_x R_y + R_y R_x = 0 \ , \quad R_y^2 = -1 \tag{2.7.7}
$$

which might be realized as $R_y = \exp\left(i\pi J_y/\hbar\right)$. The anticommutativity of R_x and R_y immediately tells us that R_y changes the R_x parity, just as T does, $R_x R_y|m\pm\rangle = \mp iR_y|m\pm\rangle$. The basis may thus be chosen according to

$$
R_y|n\pm\rangle = \pm|n\mp\rangle \tag{2.7.8}
$$

which is indeed the same as (2.7.4) but with R_y instead of T. Despite this similarity, the R_y invariance imposes a restriction on H that goes beyond those achieved by T precisely because R_y is unitary while T is antiunitary. The R_y invariance implies, together with (2.7.8), that

$$
\begin{aligned}
\langle m+|H|n+\rangle &= \langle m+|R_yHR_y^{-1}|n+\rangle \\
&= \langle m-|H|n-\rangle \ , \tag{2.7.9}
\end{aligned}
$$

i.e., $H^+ = H^-$ while the T invariance had given $H^+ = (H^-)^*$ [see (2.7.6)]. Thus we have the result that H^+ and H^- are identical real matrices. Their group of canonical transformation is reduced by the new parity from $U(N)$ to $O(N)$.

As a final illustration of the cooperation of time-reversal invariance with geometrical symmetries, the case of full isotropy, $[H, \boldsymbol{J}] = 0$, deserves mention. The appropriate basis to use here is $|\alpha jm\rangle$ where $\hbar m$ and $\hbar^2 j(j+1)$ are the eigenvalues of J_z and $\boldsymbol{J}^2$, respectively. The Hamiltonian matrix then falls into blocks given by

$$
\langle \alpha jm|H|\beta j'm'\rangle = \delta_{jj'}\delta_{mm'}\left\langle \alpha \left|H^{(j,m)}\right| \beta \right\rangle \ ,
$$
$$
j = \tfrac{1}{2}, \tfrac{3}{2}, \tfrac{5}{2}, \dots \ , \quad m = \pm\tfrac{1}{2}, \pm\tfrac{3}{2}, \dots, \pm j \ . \tag{2.7.10}
$$

It is left to the reader as Problem 2.9 to show that

(i) due to T invariance, for any fixed value of j, the two blocks with differing signs of m are transposes of one another,

$$
H^{(j,m)} = \tilde{H}^{(j,-m)} \ , \tag{2.7.11}
$$

and thus have identical eigenvalues (Kramers' degeneracy!), and

(ii) invariance of H under rotations about the y-axis makes the two blocks equal.

The two statements above imply that the blocks $H^{(j,m)}$ are all real and thus have the orthogonal transformations as their canonical transformations.

To summarize, unitary transformations are canonical both when there is no time-reversal invariance and when a time-reversal invariance with Kramers' degeneracy ($T^2 = -1$) is combined with one parity. Orthogonal transformations constitute the canonical group when time-reversal invariance holds, either with or without Kramers' degeneracy, in the first case, however, only in the presence of certain geometric symmetries. An altogether different group of canonical transformations will be encountered in Sect. 2.8.

2.8 Kramers' Degeneracy Without Geometric Symmetries

When a time reversal with $T^2 = -1$ is the only symmetry of H, it is convenient to adopt a basis of the form

$$|1\rangle, T|1\rangle, |2\rangle, T|2\rangle, \ldots |N\rangle, T|N\rangle . \tag{2.8.1}$$

[Note that in (2.7.5) another ordering of states $|n+\rangle$ and $T|n+\rangle = |n-\rangle$ was chosen.] Sometimes I shall write $|Tn\rangle$ for $T|n\rangle$ and $\langle Tn|$ for the corresponding Dirac bra. For the sake of simplicity, the Hilbert space is again assumed to have the finite dimension $2N$.

If the complex conjugation operation K is defined relative to the basis (2.8.1), the unitary matrix U in $T = UK$ takes a simple form which is easily found by letting T act on an arbitrary state vector

$$
\begin{aligned}
|\psi\rangle &= \sum_m \left(\psi_{m+}|m\rangle + \psi_{m-}|Tm\rangle\right) , \\
T|\psi\rangle &= \sum_m \left(\psi_{m+}^*|Tm\rangle - \psi_{m-}^*|m\rangle\right) .
\end{aligned}
\tag{2.8.2}
$$

Clearly, in each of the two-dimensional subspaces spanned by $|m\rangle$ and $|Tm\rangle$, the matrix U, to be called Z from now on, takes the form

$$Z_{mm} = \begin{pmatrix} 0 & -1 \\ 1 & 0 \end{pmatrix} \equiv \tau_2 \tag{2.8.3}$$

while different such subspaces are unconnected,

$$Z_{mn} = 0 \quad \text{for} \quad m \neq n . \tag{2.8.4}$$

The $2N \times 2N$ matrix Z is obviously block diagonal with the 2×2 blocks (2.8.3) along the diagonal. In fact it will be convenient to consider Z as a diagonal $N \times N$ matrix, whose nonzero elements are themselves 2×2 matrices given by (2.8.3).

Similarly, the two pairs of states $|m\rangle, T|m\rangle$, and $|n\rangle, T|n\rangle$ give a 2×2 submatrix of the Hamiltonian

$$
\begin{pmatrix}
\langle m|H|n\rangle & \langle m|H|Tn\rangle \\
\langle Tm|H|n\rangle & \langle Tm|H|Tn\rangle
\end{pmatrix}
\equiv h_{mn} .
\tag{2.8.5}
$$

The full $2N \times 2N$ matrix H may be considered as an $N \times N$ matrix each element of which is itself a 2×2 block h_{mn}. The reason for the pairwise ordering of the basis (2.8.1) is, as will become clear presently, that the restriction imposed on H by time-reversal invariance can be expressed as a simple property of h_{mn}.

As is the case for any 2×2 matrix, the block h_{mn} can be represented as a linear combination of four independent matrices. Unity and the three Pauli matrices $\boldsymbol{\sigma}$ may come to mind first, but in fact the condition of time-reversal invariance will take a slightly nicer form if the anti-Hermitian matrices $\boldsymbol{\tau} = -i\boldsymbol{\sigma}$ are employed,

$$\tau_1 = \begin{pmatrix} 0 & -i \\ -i & 0 \end{pmatrix}, \quad \tau_2 = \begin{pmatrix} 0 & -1 \\ 1 & 0 \end{pmatrix}, \quad \tau_3 = \begin{pmatrix} -i & 0 \\ 0 & +i \end{pmatrix} \tag{2.8.6}$$
$$\tau_i \tau_j = \varepsilon_{ijk} \tau_k, \quad \tau_i \tau_j + \tau_j \tau_i = -2\delta_{ij}.$$

Four coefficients $h_{mn}^{(\mu)}$, $\mu = 0, 1, 2, 3$, characterize the block h_{mn},

$$h_{mn} = h_{mn}^{(0)} 1 + \boldsymbol{h}_{mn} \cdot \boldsymbol{\tau}. \tag{2.8.7}$$

Now, time-reversal invariance gives

$$
\begin{aligned}
h_{mn} &= \left(THT^{-1} \right)_{mn} \\
&= \left(ZKHKZ^{-1} \right)_{mn} \\
&= \left(ZH^* Z^{-1} \right)_{mn} \\
&= -\tau_2 h_{mn}^* \tau_2 \\
&= -\tau_2 \left\{ h_{mn}^{(0)*} 1 + \boldsymbol{h}_{mn}^* \cdot \boldsymbol{\tau}^* \right\} \tau_2 \\
&= h_{mn}^{(0)*} 1 + \boldsymbol{h}_{mn}^* \cdot \boldsymbol{\tau},
\end{aligned} \tag{2.8.8}
$$

which simply means that the four amplitudes $h_{mn}^{(\mu)}$ are all real:

$$h_{mn}^{(\mu)} = h_{mn}^{(\mu)*}. \tag{2.8.9}$$

For historical reasons, matrices with the property (2.8.9) are called "quaternion real". Note that this property does look nicer than the one that would have been obtained if we had used Pauli's triple $\boldsymbol{\sigma}$ instead of the anti-Hermitian $\boldsymbol{\tau}$.

The Hermiticity of H implies the relation

$$h_{mn} = h_{nm}^\dagger, \tag{2.8.10}$$

which in turn means that the four real amplitudes $h_{mn}^{(\mu)}$ obey

$$
\begin{aligned}
h_{mn}^{(0)} &= h_{nm}^{(0)} \\
h_{mn}^{(k)} &= -h_{nm}^{(k)}, \quad k = 1, 2, 3.
\end{aligned} \tag{2.8.11}
$$

It follows that the $2N \times 2N$ matrix H is determined by $N(2N-1)$ independent real parameters.

With the structure of the Hamiltonian now clarified, it remains to identify the canonical transformations that leave this structure intact. To that end we must find the subgroup of unitary matrices that preserve the form $T = ZK$ of the time-reversal operator. In other words, the question is to what extent there is freedom in choosing a basis with the properties (2.8.1). The allowable unitary basis transformations S have to obey

$$T = STS^{-1} = SZKS^{-1} = SZ\tilde{S}K ,$$

i.e.,

$$SZ\tilde{S} = Z . \tag{2.8.12}$$

The requirement (2.8.12) defines the so-called symplectic group $Sp(N)$. I leave it to the reader as Problem 2.11 to show that the solutions S of (2.8.12) do indeed form a group.

The symplectic transformations just found are in fact the relevant canonical transformations since they leave a quaternion real Hamiltonian quaternion real. To prove that statement, we shall demonstrate that if H is T invariant, then so is SHS^{-1}.

With the help of $ZS^* = SZ$ and $\tilde{S}Z = ZS^{-1}$ and that both of these identities are reformulations of (2.8.12),

$$
\begin{aligned}
TSHS^{-1}T^{-1} &= ZKSHS^{-1}KZ^{-1} = ZS^*KHK\left(S^{-1}\right)^* Z^{-1} \\
&= SZKHK\tilde{S}(-Z) = SZKHK(-Z)S^{-1} \\
&= STHT^{-1}S^{-1} = SHS^{-1} .
\end{aligned} \tag{2.8.13}
$$

Now, the list of canonical transformations for Hamiltonians of various symmetry properties is complete.

2.9 Nonconventional Time Reversal

We have defined conventional time reversal by

$$
\begin{aligned}
T\boldsymbol{x}T^{-1} &= \boldsymbol{x} \\
T\boldsymbol{p}T^{-1} &= -\boldsymbol{p} \\
T\boldsymbol{J}T^{-1} &= -\boldsymbol{J}
\end{aligned} \tag{2.9.1}
$$

and, for any pair of states,

$$
\begin{aligned}
\langle T\phi|T\psi\rangle &= \langle\psi|\phi\rangle \\
T^2 &= \pm 1 .
\end{aligned} \tag{2.9.2}
$$

The motivation for this definition is that many Hamiltonians of practical importance are invariant under conventional time reversal, $[H, T] = 0$. An atom and a molecule in an isotropic environment, for instance, have Hamiltonians of that symmetry. But, as already mentioned in Sect. 2.1, conventional time reversal is broken by an external magnetic field.

In identifying the canonical transformations of Hamiltonians from their symmetries in Sects. 2.5–2.8, extensive use was made of (2.9.2) but none, as the reader is invited to review, of (2.9.1). In fact, and indeed fortunately, the validity of (2.9.1) is not at all necessary for the above classification of Hamiltonians according to their group of canonical transformations.

Interesting and experimentally realizable systems often have Hamiltonians that commute with some antiunitary operator obeying (2.9.2) but not (2.9.1). There is nothing strange or false about such a "nonconventional" time-reversal invariance: it associates another, independent solution, $\psi'(t) = T\psi(-t)$, with any solution $\psi(t)$ of the Schrödinger equation, and is thus as good a time-reversal symmetry as the conventional one.

An important example is the hydrogen atom in a constant magnetic field [36, 37]. Choosing that field as $\boldsymbol{B} = (0, 0, B)$ and the vector potential as $\boldsymbol{A} = \boldsymbol{B} \times \boldsymbol{x}/2$ and including spin-orbit interaction, one obtains the Hamiltonian

$$H = \frac{\boldsymbol{p}^2}{2m} - \frac{e^2}{r} - \frac{eB}{2mc}(L_z + gS_z) + \frac{e^2 B^2}{8mc^2}(x^2 + y^2) + f(r)\boldsymbol{LS}. \qquad (2.9.3)$$

Here $\boldsymbol{L}$ and $\boldsymbol{S}$ denote orbital angular momentum and spin, respectively, while the total angular momentum is $\boldsymbol{J} = \boldsymbol{L} + \boldsymbol{S}$. This Hamiltonian is not invariant under conventional time reversal, T_0, but instead under

$$T = e^{i\pi J_x/\hbar}T_0. \qquad (2.9.4)$$

If spin is absent, $T^2 = 1$, whereas $T^2 = -1$ with spin. In the subspaces of constant J_z and $\boldsymbol{J}^2$, one has the orthogonal transformations as the canonical group in the first case (Sect. 2.2) and the unitary transformations in the second case (Sect. 2.7 and Problem 2.10).

When a homogeneous electric field $\boldsymbol{E}$ is present in addition to the magnetic field, the operation T in (2.9.4) ceases to be a symmetry of H since it changes the electric-dipole perturbation $-e\boldsymbol{x} \cdot \boldsymbol{E}$. But $T = RT_0$ is an antiunitary symmetry where the unitary operator R represents a reflection in the plane spanned by $\boldsymbol{B}$ and $\boldsymbol{E}$. Note that the component of the angular momentum lying in this plane changes sign under that reflection since the angular momentum is a pseudovector. While the Zeeman term in H changes sign under both conventional time reversal and under the reflection in question, it is left invariant under the combined operation. The electric-dipole term as well as all remaining terms in H are symmetric with respect to both T_0 and R such that $[H, RT_0] = 0$ indeed results.

As another example *Seligman* and *Verbaarschot* [38] proposed two coupled oscillators with the Hamiltonian

$$\begin{aligned} H &= \tfrac{1}{2}\left(p_1 - a\,x_2^3\right)^2 + \tfrac{1}{2}\left(p_2 + a\,x_1^3\right)^2 \\ &\quad + \alpha_1 x_1^6 + \alpha_2 x_2^6 - \alpha_{12}(x_1 - x_2)^6. \end{aligned} \qquad (2.9.5)$$

Here, too, T_0 invariance is violated if $a \neq 0$. As long as $\alpha_{12} = 0$, however, H is invariant under

$$T = e^{i\pi L_2/\hbar}T_0 \qquad (2.9.6)$$

and thus representable by a real matrix. The geometric symmetry TT_0^{-1} acts as $(x_1, p_1) \to (-x_1, -p_1)$ and $(x_2, p_2) \to (x_2, p_2)$ and may be visualized as a rotation through π about the 2-axis if the two-dimensional space spanned by x_1 and x_2 is imagined embedded in a three-dimensional Cartesian space. However, when $a \neq 0$ and $\alpha_{12} \neq 0$, the Hamiltonian (2.9.5) has no antiunitary symmetry left and therefore is a complex matrix. (Note that H is a complex operator in the position representation anyway.)

From an experimental point of view, it is not an easy matter to realize situations where the unitary group is canonical. Nuclei, atoms, and molecules in isotropic environments have isotropic and T_0-invariant Hamiltonians. Homogeneous magnetic fields merely replace T_0 by a nonconventional time-reversal invariance. To make magnetic fields substantially inhomogeneous across a nucleus is certainly impossible for the human experimenter; the situation may look a little better for the most highly excited atomic states, the so-called Rydberg states. Systems with half-integer j and Kramers' degeneracy might be realizable and could belong to any of the three universality classes, as follows from Sects. 2.7 and 2.8.

2.10 Stroboscopic Maps for Periodically Driven Systems

Time-dependent perturbations, especially periodic ones, are characteristic of many situations of experimental interest. They are also appreciated by theorists inasmuch as they provide the simplest examples of classical nonintegrability: Systems with a single degree of freedom are classically integrable, if autonomous, but may be nonintegrable if subjected to periodic driving.

Quantum mechanically, one must tackle a Schrödinger equation with an explicit time dependence in the Hamiltonian,

$$i\hbar\dot{\psi}(t) = H(t)\psi(t) . \tag{2.10.1}$$

The solution at $t > 0$ can be written with the help of a time-ordered exponential

$$U(t) = \left\{ \exp\left[\frac{-i}{\hbar} \int_0^t dt' H(t') \right] \right\}_+ \tag{2.10.2}$$

where the "positive" time ordering requires

$$[A(t)B(t')]_+ = \begin{cases} A(t)B(t') & \text{if } t > t' \\ B(t')A(t) & \text{if } t < t' \end{cases} . \tag{2.10.3}$$

Of special interest are cases with periodic driving,

$$H(t + n\tau) = H(t) , \quad n = 0, \pm 1, \pm 2, \dots . \tag{2.10.4}$$

The evolution operator referring to one period τ, the so-called Floquet operator

$$U(\tau) \equiv F , \tag{2.10.5}$$

is worthy of consideration since it yields a stroboscopic view of the dynamics,

$$\psi(n\tau) = F^n \psi(0) .$$ (2.10.6)

Equivalently, F may be looked upon as defining a quantum map,

$$\psi([n+1]\tau) = F\psi(n\tau) .$$ (2.10.7)

Such discrete-time maps are as important in quantum mechanics as their Newtonian analogues have proven in classical nonlinear dynamics.

The Floquet operator, being unitary, has unimodular eigenvalues (involving eigenphases alias quasi-energies) and mutually orthogonal eigenvectors,

$$F\Phi_\nu = e^{-i\phi_\nu}\Phi_\nu ,$$
$$\langle\Phi_\mu|\Phi_\nu\rangle = \delta_{\mu\nu} .$$ (2.10.8)

I shall in fact be concerned only with normalizable eigenvectors. With the eigenvalue problem solved, the stroboscopic dynamics can be written out explicitly,

$$\psi(n\tau) = \sum_\nu e^{-in\phi_\nu} \langle\Phi_\nu|\psi(0)\rangle \Phi_\nu .$$ (2.10.9)

Monochromatic perturbations are relatively easy to realize experimentally. Much easier to analyse are perturbations for which the temporal modulation takes the form of a periodic train of delta kicks,

$$H(t) = H_0 + \lambda V \sum_{n=-\infty}^{+\infty} \delta(t - n\tau) .$$ (2.10.10)

The weight of the perturbation V in $H(t)$ is measured by the parameter λ, which will be referred to as the kick strength. The Floquet operator transporting the state vector from immediately after one kick to immediately after the next reads

$$F = e^{-i\lambda V/\hbar}e^{-iH_0\tau/\hbar} .$$ (2.10.11)

The simple product form arises from the fact that only H_0 is on between kicks, while H_0 is ineffective "during" the infinitely intense delta kick.

It may be well to conclude this section with a few examples. Of great interest with respect to ongoing experiments is the hydrogen atom exposed to a monochromatic electromagnetic field. Even the simplest Hamiltonian,

$$H = \frac{p^2}{2m} - \frac{e^2}{r} - Ez\cos\omega t ,$$ (2.10.12)

defies exact solution. The classical motion is known to be strongly chaotic for sufficiently large values of the electric field E: A state that is initially bound (with respect to $H_0 = p^2/2m - e^2/r$) then suffers rapid ionization. The quantum modifications of this chaos-enhanced ionization have been the subject of intense discussion. See Ref. [39] for the early efforts, and for a brief sketch of the present situation, see Sect. 7.1.

A fairly complete understanding has been achieved for both the classical and quantum behavior of the kicked rotator [40], a system of quite some relevance for microwave ionization of hydrogen atoms. The Hamiltonian reads

$$H(t) = \frac{1}{2I}p^2 + \lambda \cos\phi \sum_{n=-\infty}^{+\infty} \delta(t - n\tau) \, . \tag{2.10.13}$$

The classical kick-to-kick description is *Chirikov*'s standard map [41]. Most of the chapter on quantum localization will be devoted to that prototypical system.

Somewhat richer in their behavior are the kicked tops for which H_0 and V in (2.10.10, 11) are polynomials in the components of an angular momentum $\mathbf{J}$. Due to the conservation of $\mathbf{J}^2 = \hbar^2 j(j+1)$, $j = \frac{1}{2}, 1, \frac{3}{2}, 2, \ldots$, kicked tops enjoy the privilege of a finite-dimensional Hilbert space. The special case

$$H_0 \propto J_x \, , \quad V \propto J_z^2 \tag{2.10.14}$$

can be realized experimentally in various ways to be discussed in Sect. 9.15.

2.11 Time Reversal for Maps

It is easy to find the condition which the Hamiltonian $H(t)$ must satisfy so that a given solution $\psi(t)$ of the Schrödinger equation

$$i\hbar\dot{\psi}(t) = H(t)\psi(t) \tag{2.11.1}$$

yields an independent solution,

$$\tilde{\psi}(t) = T\psi(-t) \, , \tag{2.11.2}$$

where T is some antiunitary operator. By letting T act on both sides of the Schrödinger equation,

$$-i\hbar\frac{\partial}{\partial t}T\psi(t) = TH(t)T^{-1}T\psi(t) \tag{2.11.3}$$

or, with $t \to -t$,

$$i\hbar\frac{\partial}{\partial t}T\psi(-t) = TH(-t)T^{-1}T\psi(-t) \, . \tag{2.11.4}$$

For (2.11.4) to be identical to the original Schrödinger equation, $H(t)$ must obey

$$H(t) = TH(-t)T^{-1} \, , \tag{2.11.5}$$

a condition which reduces to that studied previously for time-dependent Hamiltonians.

For periodically driven systems it is convenient to express the time-reversal symmetry (2.11.5) as a property of the Floquet operator. As a first step in searching for that property, we again employ the formal solution of (2.11.1). Distinguishing now between positive and negative times,

$$\psi(t) = \begin{cases} U_+(t)\psi(0), & t > 0 \\ U_-(t)\psi(0), & t < 0 \end{cases} \tag{2.11.6}$$

where $U_+(t)$ is positively time-ordered as explained in (2.10.2, 3) while the negative time order embodied in $U_-(t)$ is simply the opposite of the positive one. Now, I assume $t > 0$ and propose to consider

$$\tilde{\psi}(t) = T\psi(-t) = TU_-(-t)\psi(0) = TU_-(-t)T^{-1}T\psi(0) . \tag{2.11.7}$$

If $H(t)$ is time-reversal invariant in the sense of (2.11.5), this $\tilde{\psi}(t)$ must solve the original Schrödinger equation (2.11.1) such that

$$U_+(t) = TU_-(-t)T^{-1} . \tag{2.11.8}$$

The latter identity is in fact equivalent to (2.11.5). The following discussion will be confined to τ-periodic driving, and we shall take the condition (2.11.8) for $t = \tau$. The backward Floquet operator $U_-(-\tau)$ is then simply related to the forward one. To uncover that relation, we represent $U_-(-\tau)$ as a product of time evolution operators, each factor referring to a small time increment,

$$\begin{aligned} U_-(-\tau) &= \left\{ \exp\left[-\mathrm{i} \int_0^{-\tau} dt' H(t') \right] \right\}_- \\ &= \mathrm{e}^{\mathrm{i}\Delta tH(-t_n)/\hbar} \mathrm{e}^{\mathrm{i}\Delta tH(-t_{n-1})/\hbar} \cdots \\ &\quad \cdots \mathrm{e}^{\mathrm{i}\Delta tH(-t_2)/\hbar} \mathrm{e}^{\mathrm{i}\Delta tH(-t_1)/\hbar} . \end{aligned} \tag{2.11.9}$$

As illustrated in Fig. 2.1, we choose equidistant intermediate times between $t_{-n} = -\tau$ and $t_n = \tau$ with the positive spacing $t_{i+1} - t_i = \Delta t = \tau/n$.

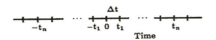

Fig. 2.1. Discretization of the time used to evaluate the time-ordered exponential in (2.11.9)

The intervals $t_{i+1} - t_i$ are assumed to be so small that the Hamiltonian can be taken to be constant within each of them. Note that the positive sign appears in each of the n exponents in the second line of (2.11.9) since Δt is defined to be positive while the time integral in the negatively time-ordered exponential runs toward the left on the time axis. Now, we invoke the assumed periodicity of the Hamiltonian, $H(-t_{n-i}) = H(t_i)$, to rewrite $U_-(-\tau)$ as

$$\begin{aligned} U_-(-\tau) &= \mathrm{e}^{\mathrm{i}\Delta tH(0)/\hbar} \mathrm{e}^{\mathrm{i}\Delta tH(t_1)/\hbar} \cdots \\ &\quad \cdots \mathrm{e}^{\mathrm{i}\Delta tH(t_{n-2})/\hbar} \mathrm{e}^{\mathrm{i}\Delta tH(t_{n-1})/\hbar} = U_+(\tau)^\dagger \end{aligned} \tag{2.11.10}$$

which is indeed the Hermitian adjoint of the forward Floquet operator. Now, we can revert to a simpler notation, $U_+(\tau) = F$, and write the time-reversal property (2.11.8) for $t = \tau$ together with (2.11.10) as a time-reversal "covariance" of the Floquet operator

$$TFT^{-1} = F^\dagger = F^{-1} \ . \tag{2.11.11}$$

This is a very intuitive result indeed: The time-reversed Floquet operator TFT^{-1} is just the inverse of F.

An interesting general statement can be made about periodically kicked systems with Hamiltonians of the structure (2.10.10). If H_0 and V are both invariant under some time reversal T_0 (conventional or not), the Hamiltonian $H(t)$ is T_0-covariant in the sense of (2.11.5) provided the zero of time is chosen halfway between two successive kicks. The Floquet operator (2.10.11) is then not covariant in the sense (2.11.11) with respect to T_0, but it is covariant with respect to

$$T = \mathrm{e}^{\mathrm{i}H_0\tau/\hbar}T_0 \ . \tag{2.11.12}$$

The reader is invited, in Problem 2.12, to show that the Floquet operator defined so as to transport the wave vector by one period starting at a point halfway between two successive kicks is T_0-covariant.

The above statement implies that for the periodically kicked rotator defined by (2.10.13), F has an antiunitary symmetry of the type (2.11.12). Similarly, the Floquet operator of the kicked top (2.10.14) is covariant with respect to

$$T = \mathrm{e}^{\mathrm{i}H_0\tau/\hbar}K \tag{2.11.13}$$

where K is complex conjugation in the standard representation of angular momenta in which J_x and J_z are real and J_y is imaginary.

2.12 Canonical Transformations for Floquet Operators

The arguments presented in Sects. 2.4, 2.5, 2.7, and 2.8 for Hermitian Hamiltonians carry over immediately to unitary Floquet operators. We shall assume a finite number of dimensions throughout.

First, irreducible $N \times N$ Floquet matrices without any T covariance have $U(N)$ as their group of canonical transformations. Indeed, any transformation from that group preserves eigenvalues, unitarity, and normalization of vectors. The proof is analogous to that sketched in Sect. 2.4.

Next, when F is T covariant with $T^2 = 1$, one can find a T invariant basis in which F is symmetric. In analogy with the reasoning in (2.5.5), one takes matrix elements in $TF^\dagger T^{-1} = F$ with respect to T invariant basis states,

$$
\begin{aligned}
F_{\mu\nu} &= \langle\psi_\mu|TF^\dagger T^{-1}|\psi_\nu\rangle = \langle T\psi_\mu|T^2 F^\dagger T|\psi_\nu\rangle^* \\
&= \langle\psi_\mu|F^\dagger T|\psi_\nu\rangle^* = F_{\nu\mu} \ .
\end{aligned}
\tag{2.12.1}
$$

It is worth recalling that time-reversal invariant Hamiltonians were also found, in Sect. 2.5, to be symmetric if $T^2 = 1$. Of course, for unitary matrices $F = \tilde{F}$ does not imply reality. The canonical group is now $O(N)$, as was the case for $[H, T] = 0$, $T^2 = 1$. To see this, we assume that $F = \tilde{F}$ and that O is unitary, and require $OFO^\dagger$ to be symmetric,

$$OFO^\dagger = \tilde{O}^\dagger F \tilde{O} . \tag{2.12.2}$$

Multiplication from the left with $\tilde{O}$ and from the right with O gives

$$\tilde{O}OF = F\tilde{O}O . \tag{2.12.3}$$

Since F must be assumed irreducible, the product $O\tilde{O}$ must be unity, i.e., O must be an orthogonal matrix.

Finally, if F is T covariant with $T^2 = -1$, there is again Kramers' degeneracy. To prove this, let

$$\begin{aligned} F|\phi_\nu\rangle &= \mathrm{e}^{-\mathrm{i}\phi_\nu}|\phi_\nu\rangle , \\ F^\dagger|\phi_\nu\rangle &= \mathrm{e}^{\mathrm{i}\phi_\nu}|\phi_\nu\rangle \end{aligned} \tag{2.12.4}$$

and let T act on the latter equation:

$$\mathrm{e}^{-\mathrm{i}\phi_\nu}T|\phi_\nu\rangle = TF^\dagger T^{-1}T|\phi_\nu\rangle = FT|\phi_\nu\rangle . \tag{2.12.5}$$

The orthogonality of $|\phi_\nu\rangle$ and $T|\phi_\nu\rangle$, which follows from $T^2 = -1$, has already been demonstrated in (2.6.3). The dimension of the Hilbert space must again be even.

Which group of transformations is canonical depends, as for time-independent Hamiltonians, on whether or not F has geometric invariances. Barring any such invariances for the moment, again we employ the basis (2.8.1), thus giving the time-reversal operator the structure

$$T = ZK . \tag{2.12.6}$$

The restriction imposed on F by T covariance can be found by considering the 2×2 block

$$\begin{aligned} F_{mn} &= \begin{pmatrix} \langle m|F|n\rangle & \langle m|F|Tn\rangle \\ \langle Tm|F|n\rangle & \langle Tm|F|Tn\rangle \end{pmatrix} \\ &= f_{mn}^{(0)}\mathbf{1} + \boldsymbol{f}_{mn}\cdot\boldsymbol{\tau} \end{aligned} \tag{2.12.7}$$

which must equal the corresponding block of $TF^\dagger T^{-1}$. In analogy with (2.8.8),

$$\begin{aligned} f_{mn} &= \left(TF^\dagger T^{-1}\right)_{mn} = \left(ZKF^\dagger KZ^{-1}\right)_{mn} \\ &= -\left(Z\tilde{F}Z\right)_{mn} = -\tau_2 \tilde{f}_{nm}\tau_2 \\ &= -\tau_2\left\{ f_{nm}^{(0)}\mathbf{1} + \boldsymbol{f}_{nm}\cdot\tilde{\boldsymbol{\tau}} \right\}\tau_2 = f_{nm}^{(0)}\mathbf{1} - \boldsymbol{f}_{nm}\cdot\boldsymbol{\tau} . \end{aligned} \tag{2.12.8}$$

Now, the restrictions in question can be read off as

$$f_{mn}^{(0)} = f_{nm}^{(0)}$$

$$\boldsymbol{f}_{mn} = -\boldsymbol{f}_{nm} \ . \tag{2.12.9}$$

They are identical in appearance to (2.8.11) but, in contrast to the amplitudes $h_{mn}^{(\mu)}$, the $f_{mn}^{(\mu)}$ are in general complex numbers.

The pertinent group of canonical transformation is the symplectic group defined by (2.8.12) since SFS^{-1} is T covariant if F is. Indeed, reasoning in parallel to (2.8.13),

$$
\begin{aligned}
TSFS^{-1}T^{-1} &= ZKSFS^{-1}KZ^{-1} = ZS^*KFK\tilde{S}Z^{-1} \\
&= SZKFKZ^{-1}S^{-1} = STFT^{-1}S^{-1} \\
&= SF^\dagger S^{-1} = \left(SFS^{-1}\right)^\dagger \ .
\end{aligned}
\tag{2.12.10}
$$

To complete the classification of Floquet operators by their groups of canonical transformations, it remains to allow for geometric symmetries in addition to Kramers' degeneracy. Since there is no difficulty in transcribing the considerations of Sect. 2.7, one can state without proof that the group in question is $U(N)$ when there is one parity R_x with $[T, R_x] = 0$, $R_x^2 = -1$, while additional geometric symmetries may reduce the group to $O(N)$, where the convention for N is the same as in Sect. 2.7.

We conclude this section with a few examples of Floquet operators from different universality classes, all for kicked tops. These operators are functions of the angular momentum components J_x, J_y, J_z and thus entail the conservation law $\boldsymbol{J}^2 = \hbar^2 j(j+1)$ with integer or half-integer j. The latter quantum number also defines the dimension of the matrix representation of F as $(2j+1)$.

The simplest top capable of classical chaos, already mentioned in (2.10.14), has the Floquet operator [42, 43]

$$F = e^{-i\lambda J_z^2/2j\hbar^2} e^{-ipJ_x/\hbar} \ . \tag{2.12.11}$$

Its dimensionless coupling constants p and λ may be said to describe a linear rotation and a nonlinear torsion. (For a more detailed discussion, see Sect. 7.6.) The quantum number j appears in the first unitary factor in (2.12.11) to give to the exponents of the two factors the same weight in the semiclassical limit $j \gg 1$. This simplest top belongs to the orthogonal universality class: Its F operator is covariant with respect to generalized time reversal

$$T = e^{ipJ_x/\hbar} e^{i\pi J_y/\hbar} T_0 \tag{2.12.12}$$

where T_0 is the conventional time reversal. By diagonalizing F, the level spacing distribution has been shown [42, 43] to obey the dictate of the latter symmetry, i.e., to display linear level repulsion (Chap. 3) under conditions of classical chaos.

An example of the unitary universality class is provided by

$$F = e^{-i\lambda' J_y^2/2j\hbar^2} e^{-i\lambda J_z^2/2j\hbar^2} e^{-ipJ_x/\hbar} \ . \tag{2.12.13}$$

Indeed, the quadratic level repulsion characteristic of this class (Chap. 3) is obvious from Fig. 1.2c that was obtained [44, 45] by diagonalizing F for $j = 500$, $p = 1.7$, $\lambda = 10$, $\lambda' = 0.5$.

Finally, the Floquet operator

$$F = \mathrm{e}^{-\mathrm{i}V}\mathrm{e}^{-\mathrm{i}H_0} \, , \qquad (2.12.14)$$

with

$$H_0 = \lambda_0 J_z^2 / j\hbar^2 \, ,$$

$$V = \lambda_1 J_z^4 / j^3 \hbar^4 + \lambda_2 \left(J_x J_z + J_z J_x \right) / \hbar^2 + \lambda_3 \left(J_x J_y + J_y J_x \right) / \hbar^2 \, ,$$

is designed so as to have no conserved quantity beyond $\boldsymbol{J}^2$ (i.e., in particular, no geometric symmetry) but a time-reversal covariance with respect to $T = \mathrm{e}^{-\mathrm{i}H_0}T_0$. Now, since $T^2 = +1$ and $T^2 = -1$ for j integer and half integer, respectively, the top in question may belong to either the orthogonal or the symplectic class. These alternatives are most strikingly displayed in Fig. 1.3b, d. Both graphs were obtained [44, 45] for $\lambda_0 = \lambda_1 = 2.5$, $\lambda_2 = 5$, $\lambda_3 = 7.5$, values which correspond to global classical chaos. The only parameter that differs in the two cases is the angular momentum quantum number: $j = 500$ (orthogonal class) for graph b and $j = 499.5$ (symplectic class) for d. The difference in the degree of level repulsion is obvious (Chap. 3). Such a strong reaction of the degree of level repulsion to a change as small as one part per thousand is really rather remarkable. No quantity with a well-defined classical limit could respond so dramatically.

2.13 Experimental Distinction Between Universality Classes

Nuclei, atoms, and molecules in isotropic environments enjoy conventional time-reversal invariance, the tiny and hard to detect electroweak effects apart. Homogeneous magnetic fields merely replace T_0 by a nonconventional time-reversal invariance. To break all time-reversal symmetries by making magnetic fields substantially inhomogeneous across a nucleus is certainly impossible to the human experimenter; the situation may look a little better for the most highly excited Rydberg states of atoms or molecules whose diameters are sufficiently large to probe realizable magnetic-field inhomogeneities.

Atoms and molecules with half-integer electronic angular momentum and Kramers' degeneracy could and indeed should be checked spectroscopically for their universality class. That class may be orthogonal, unitary, or symplectic depending on the geometric symmetries present, as follows from Sects. 2.7 and 2.8.

The electrons involved in forming solids see an electrostatic potential constituted by the spatial array of atomic cores (nuclei surrounded by inner-shell electrons). Even for perfect crystals, a homogeneous magnetic field $\boldsymbol{B}$ breaks time-reversal invariance, provided only no rotation through π about any axis perpendicular to $\boldsymbol{B}$ leaves the aforementioned crystal field invariant. Disordered solids possess no geometric symmetry in their "crystal fields" and therefore cannot be time-reversal invariant in a magnetic field.

The foregoing considerations do not specify any measurable quantity capable of revealing the universality class or even the presence or absence of time-reversal invariance. I withhold fulfilling the promise made in the title of the present section till Sect. 3.5.

2.14 Beyond Dyson's Threefold Way

We have been concerned with the orthogonal, unitary, and symplectic universality classes of Hamiltonians or Floquet operators. This classification goes back to *Dyson* [46] who deduced it from group theoretical arguments about complex Hermitian (or unitary) matrices. Much of the present book builds on Dyson's classification.

More recently, *Zirnbauer* [47] has argued that further symmetry classes exist and pointed to lots of realizations in solid state as well as particle physics. That possibility arises for systems of many bosons or fermions whose indistinguishability provides the respective Hilbert spaces with further structure: The notions of real, complex, and quaternionic vectors must be refined to accommodate Zirnbauer's suggestion of an "tenfold way".

2.15 Problems

2.1. Consider a particle with the Hamiltonian $H = (\boldsymbol{p} - (e/c)\boldsymbol{A})^2/2m + V(|\boldsymbol{x}|)$ where the vector potential $\boldsymbol{A}$ represents a magnetic field $\boldsymbol{B}$ constant in space and time. Show that the motion is invariant under a nonconventional time reversal which is the product of conventional time reversal with a rotation of π about an axis perpendicular to $\boldsymbol{B}$.

2.2. Generalize the statement in Problem 2.1 to N particles with isotropic pair interactions.

2.3. Show that $K\boldsymbol{x}K^{-1} = \boldsymbol{x}$, $K\boldsymbol{p}K^{-1} = -\boldsymbol{p}$, and $K\boldsymbol{L}K^{-1} = -\boldsymbol{L}$, where $\boldsymbol{L} = \boldsymbol{x} \times \boldsymbol{p}$ is the orbital angular momentum and K the complex conjugation defined with respect to the position representation.

2.4. (a) Show that antiunitary implies antilinearity. (b) Show that antilinearity and $|\langle K\psi|K\phi\rangle|^2 = |\langle\psi|\phi\rangle|^2$ together imply the antiunitarity of K.

2.5. Show that $U_{\mu\nu} = U_{\nu\mu} = \int dx \langle\mu|x\rangle \langle\nu|x\rangle$, $U^\dagger = U^{-1}$, for $K = U\tilde{K}$ where K and $\tilde{K}$ are the complex conjugation operations in the continuous basis $|x\rangle$ and the discrete basis $|\mu\rangle$, respectively.

2.6. Show that for spin-1 particles, time reversal can be simply complex conjugation.

2.7. Show that the unitary matrix U in $T = UK$ is symmetric or antisymmetric when T squares to unity or minus unity, respectively.

2.8. Show that $\langle \phi | \psi \rangle = \langle T\phi | T\psi \rangle^*$ for $T = UK$ with $U^\dagger = U^{-1}$.

2.9. Show that time-reversal invariance with $T^2 = -1$ together with full isotropy implies that the canonical transformations are given by the orthogonal transformations.

2.10. Find the group of canonical transformations for a Hamiltonian obeying $[T, H] = 0$, $T^2 = -1$ and having cylindrical symmetry.

2.11. Show that the symplectic matrices S defined by $SZ\tilde{S} = Z$ form a group.

2.12. Let H_0 and V commute with an antiunitary operator T. Show that $TFT^{-1} = F^\dagger$ with $F = e^{-H_0\tau/2\hbar} e^{-ikV/\hbar} e^{-iH_0\tau/2\hbar}$.

2.13. What would be the analogue of $H(t) = TH(-t)T^{-1}$ if the Floquet operator were to commute with some T_0?

2.14. Show that the eigenvectors of unitary operators are mutually orthogonal.

2.15. Show that $U(N)$ is canonical for Floquet operators without any T covariance.

2.16. Show that $U(N) \otimes U(N)$ is canonical for Floquet operators with $TFT^{-1} = F^\dagger$, $T^2 = -1$, $[R_x, F] = 0$, $[T, R_x] = 0$, $R_x^2 = -1$.

2.17. Show that $O(N) \otimes O(N)$ is canonical if, in addition to the symmetries in Problem 2.16, there is another parity R_y commuting with F and T but anticommuting with R_x.

2.18. Give the group of canonical transformations for Floquet operators in situations of full isotropy.

3. Level Repulsion

3.1 Preliminaries

In the previous chapter, I classified Hamiltonians H and Floquet operators F by their groups of canonical transformations. Now I propose to show that orthogonal, unitary, and symplectic canonical transformations correspond to level repulsion of, respectively, linear, quadratic, and quartic degree [48, 49]. The different canonical groups are thus interesting not only from a mathematical point of view but also have distinct measurable consequences. It is a fascinating feature of quantum mechanics that different behavior under time reversal actually becomes observable experimentally. The origin of this phenomenon lies in the antilinearity of the quantum mechanical time-reversal operator – a property alien to classical mechanics.

Resistance of levels to crossings is a generic property of Hamiltonians and Floquet operators just as nonintegrability is typical for classical Hamiltonian systems with more than one degree of freedom. (In fact, as will become clear in Chap. 5, Hamiltonians with integrable classical limits and more than one degree of freedom do not, as a rule, display level repulsion.) Therefore, knowing that the degree of level repulsion is 1, 2, or 4 for some system implies nothing more than (1) some information about the symmetries and (2) that the system is classically nonintegrable. Conversely, the great universality of spectral fluctuations calls for explanation.

To reveal the phenomenon of repulsion, the levels of H or F must be divided into multiplets; each such multiplet has fixed values for all observables except

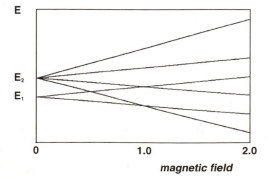

E

E_2

E_1

0   1.0   2.0

magnetic field

Fig. 3.1. Level crossings between different Zeeman multiplets

H or F from the corresponding complete set of conserved quantities. Levels belonging to different multiplets have no inhibition to cross when a parameter in H or F is varied. Such intermultiplet crossings are in fact well known, e.g., from the Zeeman levels pertaining to multiplets of different values of the total angular momentum quantum number j (Fig. 3.1). The levels within one multiplet typically avoid crossings, as illustrated in Fig. 3.2 for the quasi-energy of a kicked top.

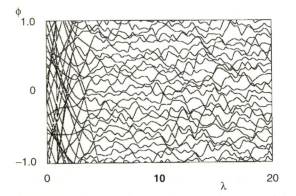

Fig. 3.2. Dependence of the quasi-energies of a kicked top on some control parameter λ. For $\lambda \lesssim 3$, the classical motion is predominantly regular, whereas classical chaos prevails for $\lambda > 3$. Note that revel repulsion is much more pronounced in the classically chaotic range

A simple explanation of level repulsion was given by *von Neumann* and *Wigner* in 1929 [50]. In fact, the present chapter will expound and generalize what is nowadays often referred to as the von Neumann–Wigner theorem.

3.2 Symmetric Versus Nonsymmetric H or F

When two levels undergo a close encounter upon variation of a parameter in a Hamiltonian, their fate can be studied by nearly degenerate perturbation theory. Assuming that each of the two levels is nondegenerate, one may deal with a two-dimensional Hilbert space spanned by approximants $|1\rangle$, $|2\rangle$ to the corresponding eigenvectors. By diagonalizing the 2×2 matrix

$$H = \begin{pmatrix} H_{11} & H_{12} \\ H_{12}^* & H_{22} \end{pmatrix} , \tag{3.2.1}$$

one obtains the approximate eigenvalues

$$E_{\pm} = \tfrac{1}{2}\left(H_{11} + H_{22}\right) \pm \sqrt{\tfrac{1}{4}\left(H_{11} - H_{22}\right)^2 + |H_{12}|^2} . \tag{3.2.2}$$

An important difference between Hamiltonians with unitary and with orthogonal canonical transformations becomes manifest at this point. In the first case, the Hamiltonian (3.2.1) is complex. The discriminant in (3.2.2),

$$D = \tfrac{1}{4}\left(H_{11} - H_{22}\right)^2 + \left(\mathrm{Re}\left\{H_{12}\right\}\right)^2 + \left(\mathrm{Im}\left\{H_{12}\right\}\right)^2 ,   \tag{3.2.3}$$

is thus the sum of three nonnegative terms. When orthogonal transformations are canonical, on the other hand, the matrix (3.2.1) is real symmetric, $\mathrm{Im}\left\{H_{12}\right\} = 0$, whereupon the discriminant D has only two nonnegative contributions. Clearly, by varying a single parameter in the Hamiltonian, the discriminant and thus the level spacing $|E_+ - E_-|$ can in general be minimized but not made to vanish (Fig. 3.3). To make a level crossing, $E_+ = E_-$, a generic possibility rather than an unlikely exception, three parameters must be controllable when unitary transformations apply, whereas two suffice in the orthogonal case. The resistance of levels to crossings should therefore be greater for complex Hermitian Hamiltonians than for real symmetric ones.

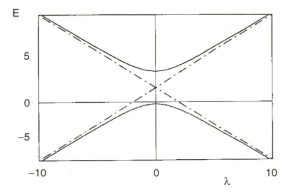

Fig. 3.3. Hyperbolic form of a typical avoided crossing of two levels

The above argument can likewise be applied to Floquet operators. The most general unitary 2×2 matrix reads

$$F = \begin{pmatrix} \alpha + i\beta & \gamma + i\delta \\ -\gamma + i\delta & \alpha - i\beta \end{pmatrix} , \quad \text{with}$$

$$\alpha^2 + \beta^2 + \gamma^2 + \delta^2 = 1 .   \tag{3.2.4}$$

Apart from a common phase factor, which can be set equal to unity by a proper choice of the reference point of the eigenphases of F, the four parameters α, β, γ, δ are real. Three of these parameters, say β, γ, and δ, may be taken as independent. The unimodular eigenvalues of F are

$$e_\pm = \alpha \pm i\sqrt{\beta^2 + \gamma^2 + \delta^2} .   \tag{3.2.5}$$

Again, the number of nonnegative additive contributions to the discriminant is generally three, unless the additional condition that F be symmetric, $\gamma = 0$, reduces that number to two.

One might argue (correctly!) that due to the unitarity condition, $\alpha^2 + \beta^2 + \gamma^2 + \delta^2 = 1$, it suffices to control the single parameter $1 - \alpha^2$. However, control over $\beta^2 + \gamma^2 + \delta^2$ allows one to achieve $E_+ = E_-$ in the Hamiltonian case as well. What counts in this argument is that the discriminants in (3.2.2) and (3.2.5) can be controlled by the same number of independent parameters. I shall refer to this number as the *codimension* of a level crossing.

To summarize, both for Hamiltonian and Floquet operators, the number n of controllable parameters necessary to enforce a level crossing is three or two depending on whether H and F are unrestricted (beyond Hermiticity and unitarity, respectively) or are symmetric matrices. These two possibilities correspond to the alternatives of unitary and orthogonal canonical transformation groups:

$$n = \begin{cases} 2 & \text{orthogonal} \\ 3 & \text{unitary} \end{cases} . \tag{3.2.6}$$

It remains to be shown that the codimension n of a level crossing yields the degree of level repulsion as $n - 1$ (Sect. 3.4).

3.3 Kramers' Degeneracy

When there is an invariance of H or a covariance of F,

$$THT^{-1} = H \quad \text{or} \quad TFT^{-1} = F^\dagger , \tag{3.3.1}$$

under some antiunitary time-reversal transformation that squares to minus unity,

$$T^2 = -1 , \tag{3.3.2}$$

each level is doubly degenerate. The close encounter of two levels must therefore be discussed for a four-dimensional Hilbert space [51]. Four approximants to the eigenvectors can be chosen and ordered as in (2.8.1), i.e., as $|1\rangle, T|1\rangle, |2\rangle, T|2\rangle$. The 4×4 matrix H or F can then be represented so that it consists of four 2×2 blocks h_{mn} or f_{mn}; see (2.8.5) and (2.12.7). The symmetry (3.3.1, 3.3.2) yields the restrictions (2.8.11) and (2.12.9) for the amplitudes $h_{mn}^{(\mu)}$ and $f_{mn}^{/(\mu)}$, respectively. Barring, for the moment, any geometric symmetry, there are no further restrictions. In a slightly less belligerent notation, the corresponding 4×4 Hamiltonian reads

$$H = \begin{pmatrix} \alpha + \beta & 0 & \gamma - i\sigma & -\varepsilon - i\delta \\ 0 & \alpha + \beta & \varepsilon - i\delta & \gamma + i\sigma \\ \gamma + i\sigma & \varepsilon + i\delta & \alpha - \beta & 0 \\ -\varepsilon + i\delta & \gamma - i\sigma & 0 & \alpha - \beta \end{pmatrix} \tag{3.3.3}$$

with six real parameters α, β, γ, δ, ε, σ. Rather than invoking (2.8.11), it is of course also possible to verify the structure (3.3.3) directly by using $T^2 = -1$ and the ordering of the four basis vectors given above. For instance,

$$\begin{aligned} \langle 1|H|T1\rangle &= \langle 1|THT^{-1}|T1\rangle = \langle 1|TH|1\rangle \\ &= -\langle T1|H|1\rangle^* = -\langle 1|H|T1\rangle = 0 . \end{aligned} \tag{3.3.4}$$

Since $T^2 = -1$ the quartic secular equation of (3.3.3) is biquadratic and yields the two double roots

$$E_\pm = \alpha \pm \sqrt{\beta^2 + \gamma^2 + \delta^2 + \varepsilon^2 + \sigma^2} . \tag{3.3.5}$$

Similarly, the 4×4 Floquet matrix F also has the structure (3.3.3), but by virtue of its unitarity, the five parameters $i\beta$, $i\gamma$, $i\delta$, $i\varepsilon$, $i\sigma$ become real, and α remains real; actually, α may be considered to be given in terms of the other five parameters since unitarity requires

$$\alpha^2 + (i\beta)^2 + (i\gamma)^2 + (i\delta)^2 + (i\varepsilon)^2 + (i\sigma)^2 = 1 . \tag{3.3.6}$$

As in (3.2.4), I have chosen a phase factor common to all elements of F so as to make the two unimodular eigenvalues

$$e_\pm = \alpha \pm i\sqrt{(i\beta)^2 + (i\gamma)^2 + (i\delta)^2 + (i\varepsilon)^2 + (i\sigma)^2} \tag{3.3.7}$$

complex conjugates of one another.

For both H and F discriminants appear in the eigenvalues (3.3.5) and (3.3.7) which have five nonnegative additive contributions. The codimension of a level crossing is thus

$$n = 5 \qquad \text{symplectic} . \tag{3.3.8}$$

As for the orthogonal and unitary cases, it will become clear presently that the value $n = 5$ is larger by one than the degree of the level repulsion characteristic of Hamiltonian and Floquet operators whose group of canonical transformations is symplectic.

It is instructive to check that the number n reduces to 3 and 2 when geometric symmetries are imposed so as to break H or F down to block diagonal form with, respectively, $U(2)$ and $O(2)$ as canonical transformations of the 2×2 blocks on the diagonal. For instance, by assuming a parity R_x with the properties (2.7.1), one may choose the four basis vectors as parity eigenstates such that

$$R_x|1\rangle = -i|1\rangle , \quad R_x|T2\rangle = -i|T2\rangle$$
$$R_x|2\rangle = i|2\rangle , \quad R_x|T1\rangle = i|T1\rangle . \tag{3.3.9}$$

It follows that

$$\begin{aligned} \langle 1|H|2\rangle &= \langle 1|R_x H R_x^{-1}|2\rangle = \langle 1|R_x H R_x|2\rangle \\ &= -\langle 1|H|2\rangle = 0 , \end{aligned}$$

i.e.,

$$\gamma = \sigma = 0 ,$$

but there are no further restrictions and so $n = 3$. Alternatively, by rearranging the order of the basis states to $|1\rangle$, $|T2\rangle$, $|2\rangle$, $|T1\rangle$, the matrix H indeed becomes block diagonal

$$H = \begin{pmatrix} \alpha + \beta & -\varepsilon - i\delta & 0 & 0 \\ -\varepsilon + i\delta & \alpha - \beta & 0 & 0 \\ 0 & 0 & \alpha - \beta & \varepsilon + i\delta \\ 0 & 0 & \varepsilon - i\delta & \alpha + \beta \end{pmatrix} . \tag{3.3.10}$$

Both of the nonzero blocks have fixed parity. Time reversal connects the two blocks but has no consequences within a block. Clearly, then, $U(2)$ is the canonical group, and $n = 3$ for each block.

Finally, an additional parity R_y obeying (2.7.7) allows one to endow the basis with

$$R_y|1\rangle = |T1\rangle , \quad R_y|T1\rangle = -|1\rangle$$
$$R_y|2\rangle = |T2\rangle , \quad R_y|T2\rangle = -|2\rangle \tag{3.3.11}$$

whereupon $\langle 1|H|T2\rangle = -\langle T1|H|2\rangle$, i.e., $\delta = 0$. The matrix H becomes real, the canonical group of the 2×2 blocks is $O(2)$, and the index n is reduced to 2.

3.4 Universality Classes of Level Repulsion

It is intuitively clear that an accidental level crossing becomes progressively less likely, the larger the number of parameters in H or F necessary to enforce a degeneracy. This number of parameters, or the "codimension of a level crossing", it was found in the preceding sections, takes on values characteristic of the group of canonical transformations,

$$n = \begin{cases} 2 & \text{orthogonal} \\ 3 & \text{unitary} \\ 5 & \text{symplectic .} \end{cases} \tag{3.4.1}$$

Now I shall show that the degree of level repulsion expressed by the level spacing distribution $P(S)$ is determined by n according to

$$P(S) \propto S^{n-1} = S^\beta \quad \text{for} \quad S \to 0 . \tag{3.4.2}$$

Linear, quadratic, and quartic level repulsion is thus typical for Hamiltonians and Floquet operators with, respectively, orthogonal, unitary, and symplectic canonical transformations. Following common practice, I have introduced the exponent of level repulsion as $\beta = n - 1$.

The limiting behavior (3.4.2) can be derived from the following elementary argument [66, 67]. The level spacing distribution for a given spectrum can be defined as

$$P(S) = \langle \delta(S - \Delta E) \rangle \tag{3.4.3}$$

where ΔE stands for a distance between neighboring levels and the angular brackets $\langle \ldots \rangle$ mean an average over all ΔE. For $S \to 0$, i.e., for S smaller than the average separation, the level pairs contributing to $P(S)$ correspond to encounters sufficiently close for ΔE to be representable by nearly degenerate perturbation theory. In a unified notation, such spacings can be written as

$$\Delta E = \sqrt{\boldsymbol{x}^2} = \sqrt{x_1^2 + \ldots + x_n^2} . \tag{3.4.4}$$

Here ΔE may refer to energies, as in (3.2.2) and (3.3.5), or to quasi-energies. Of course, the level pairs contributing to $P(S)$ for a fixed small value of S will generally possess different values of the n parameters $\boldsymbol{x}$. The average over the level spectrum can therefore be understood as an average over $\boldsymbol{x}$ with a suitable weight $W(\boldsymbol{x})$. The latter function may not be easy to obtain, but fortunately its precise form does not matter for the asymptotic behavior of

$$P(S) = \int d^n x W(\boldsymbol{x}) \delta \left(S - \sqrt{\boldsymbol{x}^2} \right) . \tag{3.4.5}$$

By simply rescaling the n integration variables $\boldsymbol{x} \to S\boldsymbol{x}$,

$$P(S) = S^\beta \int d^n x W(S\boldsymbol{x}) \delta \left(1 - \sqrt{\boldsymbol{x}^2} \right) . \tag{3.4.6}$$

The power law (3.4.2) thus applies, provided that the weight $W(\boldsymbol{x})$ is neither zero nor infinite at $\boldsymbol{x} = 0$. A finite nonvanishing value of $W(0)$ must, however, be considered generic since the dictates of symmetries are already accounted for in the codimension $n = \beta + 1$ of a level crossing.

The full probability distribution $P(S)$ in $0 \leq S < \infty$ is not as easily found as its asymptotic form for small S. It will be shown below that $P(S)$ can be constructed on the basis of the theory of random matrices or by using methods from equilibrium statistical mechanics.

Trying to establish the spacing distribution $P(S)$ and the degree of level repulsion for some dynamical system is a sensible undertaking only if the number of levels is large: Only in that "semiclassical" situation can a smooth histogram be built up even for small S where $P(S)$ itself is small; but then indeed $P(S)$ becomes "self-averaging" for a single spectrum.

3.5 Experimental Observation of Level Repulsion

Systematic statistical analyses of complex energy spectra first became popular among nuclear physicists half a century ago. Upon being sorted into histograms, the spacings between highly excited neighboring levels of nuclei revealed linear level repulsion and thus time-reversal invariance of the strong interaction mainly responsible for nuclear structure [48, 49, 52]. While the electroweak interaction does not enjoy that symmetry, it is far too weak to be detectable in spectral fluctuations on the scale of a mean level spacing in complex nuclei.

There are rather fewer experimental verifications of linear level repulsion in electronic spectra. Evidence was found for NO_2 molecules [53] and Rb atoms [54].

While the nuclear, atomic, and molecular data just mentioned are now generally accepted as quantum manifestations of chaos, the somewhat less difficult to observe linear repulsion between eigenfrequences of microwave resonators with sufficiently irregular shape [55–60] requires explanation in the framework of classical wave theory rather than quantum mechanics. In fact, the Helmholtz equation for any of the components of the electromagnetic field within a resonator involves the very same differential operator, $\nabla^2 + k^2$ with k denoting the wave vector, as does Schrödinger's equation for a free particle in a container; as long as the boundary conditions at the walls do not mix different components of the electromagnetic field, there is a complete mathematical equivalence between the quantum and the classical wave problem. The chaos of which one sees the quantum or wave signatures is that of a "billiard" with the shape of the "box" in question, at least if the boundary conditions express specular reflection of the point particle that idealizes the billiard ball.

The first of the microwave experiments just mentioned [55] was actually meant to simulate sound waves of air in a concert hall and was done in complete innocence of the jargon of present-day chaology; the simulation of sound by microwaves was preferred to "listening" since concert halls tend to have overlapping rather than separated resonances and are thus not too well suited for picking up spacing statistics. It is amusing to realize, though, that audio engineers do go for wave chaos when designing the boundaries of concert halls.

Meanwhile, acoustic chaos has also been ascertained experimentally as linear repulsion of the elastomechanical eigenfrequencies of irregularly shaped quartz blocks [61, 62]. Inasmuch as the vibrating crystal is anisotropic and supports longitudinal as well as transverse sound waves, these experiments suggest that the distinction between chaotic and regular waves arises quite independently of the character of the medium and the detailed form of the pertinent wave equation. *The common origin of quantum and wave chaos may be seen in the nonseparability of the wave equation, just as we may attribute chaos in classical mechanics to the nonseparability of the Hamilton–Jacobi equation. If we see the Hamilton–Jacobi equation as the short-wave limit of wave theories, we can consider nonseparability as the universal chaos criterion.*

Neither quadratic nor quartic level repulsion has been observed to date in nuclei, atoms, and molecules. As already explained in Sect. 2.13, it might be possible to break time-reversal invariance for Rydberg atoms with strongly inhomogeneous magnetic fields and thus realize the quadratic case. In the absence of such experiments on quantum systems, the recent observation of quadratic repulsion in microwave resonators with broken time-reversal invariance was a most welcome achievement [63, 64].

For a comprehensive review of experimentally observed quantum manifestations of chaos, the reader is referred to *Stöckmann*'s recent book [65].

3.6 Problems

3.1. Show that a unitary 4×4 matrix with Kramers' degeneracy can be given in the form

$$
F = \begin{pmatrix}
\alpha + i\beta & 0 & \sigma + i\gamma & \delta - i\varepsilon \\
0 & \alpha + i\beta & \delta + i\varepsilon & -\sigma + i\gamma \\
-\sigma + i\gamma & -\delta + i\varepsilon & \alpha - i\beta & 0 \\
-\delta - i\varepsilon & \sigma + i\gamma & 0 & \alpha - i\beta
\end{pmatrix}
$$

with the six real parameters α, β, ... obeying $\alpha^2 + \beta^2 + \gamma^2 + \delta^2 + \varepsilon^2 + \sigma^2 = 1$. Use basis vectors $|1\rangle$, $|T1\rangle$, $|2\rangle$, $|T2\rangle$, and $T^2 = -1$.

3.2. Show that the matrix F from Problem 3.1 has $n = 3$ if it is invariant under a parity R_x with $[T, R_x] = 0$, $R_x^2 = -1$. Moreover, show that $n = 2$ if a second parity holds with $[T, R_y] = 0$, $R_x R_y + R_y R_x = 0$, $R_y^2 = -1$. What is the structure of F if full isotropy holds?

3.3. Verify that nearly degenerate perturbation theory typically gives a hyperbolic form to a level crossing.

3.4. In the orthogonal case, adjacent energy levels can be steered to a crossing in a two-dimensional parameter space. Show that the two energy surfaces are connected at the degeneracy like the two sheets of a double cone (this is *Berry's* diabolo! [66, 67]).

4. Random-Matrix Theory

4.1 Preliminaries

A wealth of empirical and numerical evidence suggests universality for local fluctuations in quantum energy or quasi-energy spectra of systems that display global chaos in their classical phase spaces. Exceptions apart, all such Hamiltonian matrices of sufficiently large dimension yield the same spectral fluctuations provided they have the same group of canonical transformations (see Chap. 2). In particular, the level spacing distribution $P(S)$ generally takes the form characteristic of the universality class defined by the canonical group. Most notable among the exceptions barred by the (mathematically hard to substantiate) term "untypical" are systems with "localization" that will be discussed in Chap. 7. Conversely, "generic" classically integrable systems with at least two degrees of freedom tend to display universal local fluctuations of yet another type, to be considered in Chap. 5.

The aforementioned universality is the starting point for the theory of random matrices. After early success in reproducing universal features in spectra of highly excited nuclei, that theory was boosted into even higher esteem when the connection of "generical integrable" and "chaotic" with different types of universal spectral fluctuations was spelled out by *Berry* and *Tabor* [68] and *Bohigas, Giannoni,* and *Schmit* [69], with important hints due to *McDonald* and *Kaufman* [70], *Casati, Valz-Gris,* and *Guarneri* [71], and *Berry* [72]. In its by now classic version, random-matrix theory deals with three "Gaussian" ensembles of Hermitian matrices, one for each group of canonical transformations. Any member of an ensemble can serve as a model of a Hamiltonian. Similarly, there are three ensembles of random unitary matrices to represent Floquet or scattering matrices. "Poissonian" ensembles of diagonal matrices with independent, random, diagonal elements are often used to model integrable Hamiltonians. Even systems with localization have recently been accommodated in their own universality class of banded random matrices that is to be touched upon in Chap. 10.

Spectral fluctuations such as those expressed in the level spacing distribution are represented by suitable ensemble averages. Then, the extent to which an individual Hamiltonian or Floquet operator can be expected to be faithful to ensemble averages is open to discussion. A partial answer to this question is provided by a certain ergodicity property of the various ensembles. Different approaches to understanding the success of random-matrix theory will be presented in Chap. 6 (level dynamics) and Chap. 9 (semiclassical approximations).

This chapter is mostly an account of classic material but also includes a brief introduction to and some applications of Grassmann analysis. The latter thread will be spun further toward "superanalysis" in Chap. 10. For more extensive treatments and greater mathematical rigor, the reader may consult Refs. [73, 74]. An up-to-date review can be found in [75].

For the sake of notational convenience, I set $\hbar = 1$ throughout this chapter.

4.2 Gaussian Ensembles of Hermitian Matrices

The construction of the Gaussian ensembles will be illustrated by considering real symmetric 2×2 matrices with $O(2)$ as their group of canonical transformations (If reflections are to be excluded, the group would be $SO(2)$). What we are seeking is a probability density $P(H)$ for the three independent matrix elements H_{11}, H_{22}, H_{12} normalized as

$$\int_{-\infty}^{+\infty} dH_{11} dH_{22} dH_{12} P(H) = 1 . \tag{4.2.1}$$

Two requirements suffice to determine $P(H)$. First, $P(H)$ must be invariant under any canonical, i.e., orthogonal transformation of the two-dimensional basis,

$$P(H) = P(H') , \quad H' = OH\tilde{O} , \quad \tilde{O} = O^{-1} . \tag{4.2.2}$$

Second, the three independent matrix elements must be uncorrelated. The function $P(H)$ must therefore be the product of three densities, one for each element,

$$P(H) = P_{11}(H_{11}) P_{22}(H_{22}) P_{12}(H_{12}) . \tag{4.2.3}$$

The latter assumption can be reinterpreted as one of minimum-knowledge input or of maximum disorder.

To exploit (4.2.2,3), it suffices to consider an infinitesimal orthogonal change of basis,

$$O = \begin{pmatrix} 1 & -\Theta \\ \Theta & 1 \end{pmatrix} , \tag{4.2.4}$$

for which $H' = OH\tilde{O}$ gives

$$\begin{array}{rcl} H'_{11} & = & H_{11} - 2\Theta H_{12} \\ H'_{22} & = & H_{22} + 2\Theta H_{12} \\ H'_{12} & = & H_{12} + \Theta(H_{11} - H_{22}) . \end{array} \tag{4.2.5}$$

Factorization and the invariance of $P(H)$ yield

$$P(H) = P(H) \left\{ 1 - \Theta \left[2H_{12} \frac{d \ln P_{11}}{dH_{11}} - 2H_{12} \frac{d \ln P_{22}}{dH_{22}} \right. \right.$$
$$\left. \left. - (H_{11} - H_{22}) \frac{d \ln P_{12}}{dH_{12}} \right] \right\} . \tag{4.2.6}$$

Since the infinitesimal angle Θ is arbitrary, its coefficient in (4.2.6) must vanish,

$$\frac{1}{H_{12}} \frac{d \ln P_{12}}{dH_{12}} - \frac{2}{H_{11} - H_{22}} \left(\frac{d \ln P_{11}}{dH_{11}} - \frac{d \ln P_{22}}{dH_{22}} \right) = 0 . \tag{4.2.7}$$

This gives three differential equations, one for each of the three independent functions $P_{ij}(H_{ij})$ since each P_{ij} has its own exclusive argument H_{ij}. The solutions are Gaussians and have the product

$$P(H) = C \exp \left[-A \left(H_{11}^2 + H_{22}^2 + 2H_{12}^2 \right) - B \left(H_{11} + H_{22} \right) \right] . \tag{4.2.8}$$

Of the three integration constants, B can be made to vanish by appropriately choosing the zero of energy, A fixes the unit of energy, and C is determined by normalization. Without loss of generality, then, $P(H)$ can be written as

$$P(H) = Ce^{-A \operatorname{Tr} H^2} . \tag{4.2.9}$$

The discussion of complex Hermitian Hamiltonians with $U(2)$ as the group of canonical transformations proceeds analogously. There are four real parameters $H_{11}, H_{22}, \operatorname{Re}\{H_{12}\}, \operatorname{Im}\{H_{12}\}$ to be dealt with now, and they are all assumed statistically independent. The probability density $P(H)$ thus factorizes as in (4.2.3) but with P_{12} as a function of $\operatorname{Re}\{H_{12}\}, \operatorname{Im}\{H_{12}\}$ or, equivalently, of H_{12} and H_{12}^*. The normalization (4.2.1) is modified such that P_{12} is integrated over the whole complex H_{12} plane,

$$\int_{-\infty}^{+\infty} dH_{11} dH_{22} d^2 H_{12} P(H) = 1 \tag{4.2.10}$$

where $d^2 H_{12} = d\operatorname{Re} H_{12} d\operatorname{Im} H_{12}$. The three functions $P_{11}(H_{11})$, $P_{22}(H_{22})$, and $P_{12}(H_{12}, H_{12}^*)$ are determined by demanding that $P(H)$ is invariant under unitary transformations of the matrix H. Up to an inconsequential phase factor, the general infinitesimal change of basis is represented by the 2×2 matrix

$$U = 1 - i\boldsymbol{\varepsilon} \cdot \boldsymbol{\sigma} \tag{4.2.11}$$

with $\boldsymbol{\varepsilon}$ an infinitesimal vector and $\boldsymbol{\sigma}$ the triple of Pauli matrices. This U shifts the Hamiltonian by

$$dH = -i \left[\boldsymbol{\varepsilon} \cdot \boldsymbol{\sigma}, H \right] . \tag{4.2.12}$$

Now, the analogue of (4.2.7) reads

$$(\varepsilon_x + i\varepsilon_y) \left[H_{12} \left(-\frac{d \ln P_{11}}{dH_{11}} + \frac{d \ln P_{22}}{dH_{22}} \right) + (H_{11} - H_{22}) \frac{\partial \ln P_{12}}{\partial H_{12}^*} \right]$$
$$+ 2\varepsilon_z H_{12} \frac{\partial \ln P_{12}}{\partial H_{12}} - \text{c.c.} = 0 . \tag{4.2.13}$$

Again, three differential equations can be extracted from this identity and the solutions are all Gaussians. If the zero of energy is chosen appropriately, the probability density $P(H)$ once more takes the form (4.2.9).

Let us turn finally to Hamiltonians with Kramers' degeneracy and with no geometric invariance. The smallest Hilbert space is now four dimensional. The 4×4 Hamiltonian is most conveniently written in quaternion notation,

$$H = \begin{pmatrix} h_{11} & h_{12} \\ h_{21} & h_{22} \end{pmatrix} , \qquad (4.2.14)$$

where each h_{ij} is a 2×2 block, representable as a superposition of unity and the triplet $\boldsymbol{\tau} = -i\boldsymbol{\sigma}$ [see (2.8.7)]. Due to (2.8.9,11), H is determined by six real amplitudes $h_{11}^{(0)}$, $h_{22}^{(0)}$, $h_{12}^{(\mu)}$, all of which are taken to be independent random numbers with a probability density of the form

$$P(H) = P_{11}\left(h_{11}^{(0)}\right) P_{22}\left(h_{22}^{(0)}\right) \prod_{\mu=0}^{3} P_{12}^{\mu}\left(h_{12}^{\mu}\right) ,$$
$$1 = \int dh_{11}^{(0)} dh_{22}^{(0)} \prod_{\mu=0}^{3} dh_{12}^{(\mu)} P(H) . \qquad (4.2.15)$$

To find the six respective probability densities, it suffices to require invariance of $P(H)$ when H is subjected to an arbitrary infinitesimal symplectic transformation. Such a change of basis is represented by

$$S = \begin{pmatrix} 1 - \boldsymbol{\xi} \cdot \boldsymbol{\tau} & \alpha \\ -\alpha & 1 + \boldsymbol{\xi} \cdot \boldsymbol{\tau} \end{pmatrix} \qquad (4.2.16)$$

with α an infinitesimal angle and $\boldsymbol{\xi}$ an infinitesimal real vector; of course, (4.2.16) is meant in quaternion notation, and the matrices $\boldsymbol{\tau} = -i\boldsymbol{\sigma}$ are as defined in (2.8.6). It is easy to verify that the 2×2 blocks of the Hamiltonian acquire the increments

$$dh_{11} = -dh_{22} = 2\alpha h_{12}^{(0)}$$
$$dh_{12} = \alpha\left(h_{22}^{(0)} - h_{11}^{(0)}\right) + \boldsymbol{\xi} \cdot \boldsymbol{h}_{12} - 2h_{12}^{(0)}\boldsymbol{\xi} \cdot \boldsymbol{\tau} . \qquad (4.2.17)$$

Now, we invoke $P(H + dH) = P(H)$ for arbitrary α and $\boldsymbol{\xi}$ and thus obtain

$$2h_{12}^{(0)}\left(\frac{d\ln P_{11}}{dh_{11}^{(0)}} - \frac{d\ln P_{22}}{dh_{22}^{(0)}}\right) + \left(h_{22}^{(0)} - h_{11}^{(0)}\right)\frac{d\ln P_{12}^{(0)}}{dh_{12}^{(0)}} = 0 ,$$
$$h_{12}^{(i)}\frac{d\ln P_{12}^{(0)}}{dh_{12}^{(0)}} - h_{12}^{(0)}\frac{d\ln P_{12}^{(i)}}{dh_{12}^{(i)}} = 0 \text{ for } i = 1, 2, 3 . \qquad (4.2.18)$$

The reader should note that the first of these identities has the same structure as (4.2.7) and (4.2.13). Since each of the six functions we are seeking has its own exclusive argument, the identities (4.2.18) imply six separate differential equations. As in the cases considered previously, the solutions are all Gaussians.

With a proper choice for the zero of energy, their product[1]

$$P(H) = C \, \exp\left\{ -2A\left[\left(h_{11}^{(0)}\right)^2 + \left(h_{22}^{(0)}\right)^2 + 2\sum_{\mu=0}^{3} \left(h_{12}^{(\mu)}\right)^2 \right] \right\} \qquad (4.2.19)$$

again gives the density $P(H)$ in the form (4.2.9).

To summarize, three different ensembles of random matrices follow from demanding (1) invariance of $P(H)$ under the three possible groups of canonical transformations and (2) complete statistical independence of all matrix elements. In view of the Gaussian form of $P(H)$ and the three groups of canonical transformations, these ensembles are called Gaussian orthogonal, Gaussian unitary, and Gaussian symplectic. Although $P(H)$ has been constructed here for the smallest possible dimensions, the result

$$P(H) = C \, e^{-A \, \mathrm{Tr} \, H^2} \, , \qquad (4.2.20)$$

common to all three ensembles, holds true independently of the dimensionality of H [73, 74]. Of course, in both $P(H)$ and in the integration measure, the appropriate number of independent real parameters must be accounted for; that number is $N(N+1)/2$ and N^2, respectively, for the real symmetric and complex Hermitian $N \times N$ matrices and $N(2N-1)$ for the quaternion real $2N \times 2N$ matrices.

Readers with mathematical needs will have realized that the invariance of the ensembles under the appropriate canonical transformations is not fully established before the invariance of the differential volume element in matrix space is shown. They are kindly asked to wait for satisfaction in Sect. 4.5.

4.3 Eigenvalue Distributions for the Gaussian Ensembles

The probability density (4.2.20) for the matrix elements of a random Hamiltonian H implies reduced probability densities for the eigenvalues of H. To perform the reduction, it is necessary to replace the H_{ij} as independent variables with another set of equal dimension which contains the eigenvalues as a subset; the remaining variables, over which to integrate, are parameters specifying the particular canonical transformation that diagonalizes H. The procedure will be illustrated again for the smallest possible number of dimensions.

The easiest case to deal with is that of real symmetric 2×2 Hamiltonians. The two eigenvalues read

$$E_\pm = \tfrac{1}{2}\left(H_{11} + H_{22}\right) \pm \tfrac{1}{2}\left[\left(H_{11} - H_{22}\right)^2 + 4H_{12}^2 \right]^{1/2} . \qquad (4.3.1)$$

---

[1] The reader's attention is drawn to the overall factor 2 in the exponent of (4.2.19); it stems from Kramers' degeneracy. When dealing with the GSE, some authors choose to redefine the trace operation as one half the usual trace, so as to account for only a single eigenvalue in each Kramers' doublet; correspondingly, these authors take the determinant as the square root of the usual one. Nowhere in this book will such tampering be permitted.

The simplest orthogonal transformation diagonalizing H,

$$O = \begin{pmatrix} \cos \Theta & -\sin \Theta \\ \sin \Theta & \cos \Theta \end{pmatrix} , \tag{4.3.2}$$

involves a single angle Θ. The reader should recall that the infinitesimal version of (4.3.2) was employed in Sect. 4.2. From $H = O \, \mathrm{diag}\,(E_+, E_-)\tilde{O}$, one can read off the relationships between the matrix elements H_{ij} and the quantities $E_\pm, \Theta$:

$$\begin{aligned} H_{11} &= E_+ \cos^2 \Theta + E_- \sin^2 \Theta \\ H_{22} &= E_+ \sin^2 \Theta + E_- \cos^2 \Theta \\ H_{12} &= (E_+ - E_-) \cos \Theta \sin \Theta . \end{aligned} \tag{4.3.3}$$

The Jacobian of this transformation is easily evaluated,

$$J = \det \frac{\partial (H_{11}, H_{22}, H_{12})}{\partial (E_+, E_-, \Theta)} = E_+ - E_- , \tag{4.3.4}$$

whereupon the reduced probability density of the eigenvalues for the Gaussian orthogonal ensemble of 2×2 Hamiltonians takes the form

$$P(E_+, E_-) = C \, |E_+ - E_-| \, \mathrm{e}^{-A(E_+^2 + E_-^2)} . \tag{4.3.5}$$

For complex Hermitian 2×2 Hamiltonians, the eigenvalues $E_\pm$ are also given by (4.3.1) but with $H_{12}^2 \to |H_{12}|^2$. Diagonalization may be achieved by the unitary transformation

$$U = \begin{pmatrix} \cos \Theta & -\mathrm{e}^{-\mathrm{i}\phi} \sin \Theta \\ \mathrm{e}^{\mathrm{i}\phi} \sin \Theta & \cos \Theta \end{pmatrix} . \tag{4.3.6}$$

The reader might note that this is not the most general unitary 2×2 matrix (see Problem 4.2). Together with the two eigenvalues $E_\pm$, the two angles Θ and ϕ suffice to represent the matrix elements of H. From $U^\dagger \, \mathrm{diag}\,(E_+, E_-) U = H$,

$$\begin{aligned} H_{11} &= E_+ \cos^2 \Theta + E_- \sin^2 \Theta \\ H_{22} &= E_+ \sin^2 \Theta + E_- \cos^2 \Theta \\ H_{12} &= H_{21}^* = (E_+ - E_-) \, \mathrm{e}^{\mathrm{i}\phi} \cos \Theta \sin \Theta \end{aligned} \tag{4.3.7}$$

and thus the Jacobian

$$\begin{aligned} J &= \det \frac{\partial (H_{11}, H_{22}, \mathrm{Re}\,\{H_{12}\}, \mathrm{Im}\,\{H_{12}\})}{\partial (E_+, E_-, \Theta, \phi)} \\ &= (E_+ - E_-)^2 \cos \Theta \sin \Theta . \end{aligned} \tag{4.3.8}$$

By integrating out the angles Θ and ϕ, one finally arrives at the eigenvalue distribution of the Gaussian unitary ensemble of random 2×2 Hamiltonians

$$P(E_+, E_-) = \mathrm{const}\,(E_+ - E_-)^2 \, \mathrm{e}^{-A(E_+^2 + E_-^2)} . \tag{4.3.9}$$

The procedure in the symplectic case is completely analogous: The 4×4 matrix H can be written in the quaternion form (4.2.14), and the pair of doubly degenerate eigenvalues is given by

$$E_\pm = \tfrac{1}{2} \left(h_{11}^{(0)} - h_{22}^{(0)} \right) \pm \tfrac{1}{2} \left[\left(h_{11}^{(0)} - h_{22}^{(0)} \right)^2 + 4 \sum_{\mu=0}^{3} \left(h_{12}^{(\mu)} \right)^2 \right]^{1/2} . \qquad (4.3.10)$$

Now, a symplectic matrix is needed for diagonalization. A convenient choice is the generalization of the infinitesimal transformation (4.2.16) to a finite one,

$$S = \begin{pmatrix} e^{-\boldsymbol{\xi} \cdot \boldsymbol{\tau}} \cos \alpha & \sin \alpha \\ -\sin \alpha & e^{\boldsymbol{\xi} \cdot \boldsymbol{\tau}} \cos \alpha \end{pmatrix} . \qquad (4.3.11)$$

Remarks similar to those above are appropriate at this point. The choice (4.3.11) is not the most general symplectic 4×4 matrix. However, it suits the present purposes because the four real parameters α, $\boldsymbol{\xi}$ provided by S, plus the two eigenvalues $E_\pm$, are equal in number to the independent matrix elements of H. From $S^\dagger \, \mathrm{diag}\,(E_+, E_-) S = H$, we obtain the relation

$$h_{11}^{(0)} = E_+ \cos^2 \alpha + E_- \sin^2 \alpha$$

$$h_{22}^{(0)} = E_+ \sin^2 \alpha + E_- \cos^2 \alpha$$

$$h_{12}^{(0)} = -(E_+ - E_-) \cos \alpha \sin \alpha \cos \xi \qquad (4.3.12)$$

$$\boldsymbol{h}_{12} = \hat{\boldsymbol{\xi}} \,(E_+ - E_-) \cos \alpha \sin \alpha \sin \xi \,,$$

where $\boldsymbol{\xi} = \xi \hat{\boldsymbol{\xi}}$, $\xi = |\boldsymbol{\xi}|$. The Jacobian of the transformation (4.3.11) from the six matrix elements $h_{mn}^{(\mu)}$ to the six parameters $E_\pm$, α, $\boldsymbol{\xi}$ is easily evaluated,

$$J = \det \frac{\partial(h_{11}^{(0)}, h_{22}^{(0)}, h_{12}^{(\mu)})}{\partial(E_+, E_-, \alpha, \boldsymbol{\xi})} \propto (E_+ - E_-)^4 . \qquad (4.3.13)$$

The coefficient unspecified in the latter proportionality is independent of the eigenvalues $E_\pm$ and thus irrelevant for the reduced distribution

$$P(E_+, E_-) = \mathrm{const}\,(E_+ - E_-)^4 e^{-A(E_+^2 + E_-^2)} . \qquad (4.3.14)$$

Comparing (4.3.5,9,14) one finds that the codimension n of a level crossing is related to the exponents of the powers in front of the Gaussians. The arguments of this section can be generalized to higher dimensions of the matrix H, and the general result for the joint distribution of eigenenergies E_μ is [73, 74]

$$P(E) = \mathrm{const} \prod_{\mu < \nu}^{1 \dots N} |E_\mu - E_\nu|^{n-1} \exp\left(-A \sum_{\mu=1}^{N} E_\mu^2 \right) \qquad (4.3.15)$$

with $n = 2$, 3, or 5. Clearly, the Gaussian factor in the foregoing density is nothing but the joint distribution (4.2.20) of the matrix elements expressed in the eigenrepresentation of H; the product of eigenvalue differences, on the other hand, is the energy dependent factor in the Jacobian for the transformation from matrix elements to eigenvalues and the "angles" in the diagonalizing matrix. The factorization of that Jacobian into two factors, one energy dependent and the other a function of the angles mentioned, signals statistical independence of eigenvalues and eigenvectors of the Gaussian ensembles under discussion.

4.4 Level Spacing Distributions

There is no problem in calculating the distributions of the level spacings for the three ensembles of random matrices, provided that the consideration is again restricted to the smallest number of dimensions, such that only a single pair of levels occurs. Taking the joint distribution (4.3.15) for $N = 2$,

$$
\begin{aligned}
P(S) \quad = \quad \text{const} \int_{-\infty}^{+\infty} dE_+ \int_{-\infty}^{+\infty} dE_- \delta\left(S - |E_+ - E_-|\right) \\
\times |E_+ - E_-|^{n-1} e^{-A(E_+^2 + E_-^2)} .
\end{aligned} \tag{4.4.1}
$$

With the unit of energy set such that the mean spacing is unity, the elementary integrals yield $\langle S \rangle = \int dS S P(S) = 1$, and with the normalization $\int_0^\infty dS \, P(S) = 1$,

$$
P(S) = \begin{cases}
(S\pi/2)e^{-S^2\pi/4} & \text{orthogonal} \\
(S^2 32/\pi^2)e^{-S^2 4/\pi} & \text{unitary} \\
(S^4 2^{18}/3^6\pi^3)e^{-S^2 64/9\pi} & \text{symplectic.}
\end{cases} \tag{4.4.2}
$$

These distributions show the expected degree of level repulsion in their behavior for $S \to 0$ and share a Gaussian fall-off at large S. They are plotted, together with their integrals $I(S) = \int_0^S dS' P(S')$, in Fig. 4.1. The $P(S)$ in (4.4.2) will be referred to as the Wigner surmises.

The spacing distributions can also be worked out for higher dimensions. The limit $N \to \infty$ which is of special interest will be dealt with in Sect. 4.11. Quite surprisingly, the asymptotic distributions differ only little from the Wigner distributions valid for $N = 2$, as is obvious from Fig. 4.2.

4.5 Invariance of the Differential Volume Element in Matrix Space

When introducing the three Gaussian ensembles of random Hermitian matrices in 4.2, I used the differential volume elements dH consisting of the products of

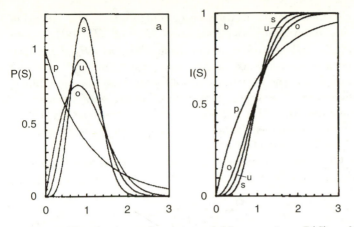

Fig. 4.1. Distributions of nearest neighbor spacings $P(S)$ and their integrals $I(S) = \int_0^S dS' P(S')$ for the Poissonian random process (p) the orthogonal (o), unitary (u), and symplectic (s) ensembles (Gaussian or circular) of random matrices

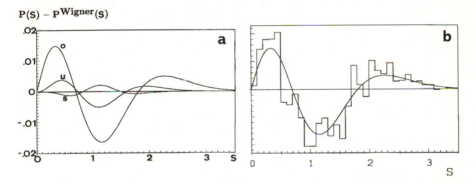

Fig. 4.2. (a) Deviations of the asymptotic ($N \to \infty$) level spacing distributions of the orthogonal (o), unitary (u), symplectic (s) (Gaussian or circular) ensembles from the respective Wigner surmises. **(b)** Numerical results for the difference $P(S) - P^{\mathrm{Wigner}}(S)$ for a kicked top pertaining to the orthogonal universality class. Histogram obtained from 101 Floquet matrices of the form (7.6.2) with dimension 1001, $\alpha = \pi/2$, and λ ranging from 8.50 to 9.50 in steps of 0.01. Smooth curve as in **(a)**. For the derivation of the asymptotic distributions, see Sect. 4.11

differentials of all independent matrix elements. These are

$$
dH = \begin{cases}
\displaystyle\prod_{i}^{1...N} dH_{ii} \prod_{i<j}^{1...N} dH_{ij} & \text{orthogonal} \\[2em]
\displaystyle\prod_{i}^{1...N} dH_{ii} \prod_{i<j}^{1...N} d^2 H_{ij} & \text{unitary} \\[2em]
\displaystyle\prod_{n}^{1...N} dh_{nn}^{(0)} \prod_{n<m}^{1...N} \prod_{\mu}^{0...3} dh_{nm}^{(\mu)} & \text{symplectic}
\end{cases}
\qquad (4.5.1)
$$

where for the complex elements in the unitary case, I introduced the convention $d^2 H_{ij} = d\mathrm{Re}\, H_{ij}\, d\mathrm{Im}\, H_{ij}$; equivalently, one may treat H_{ij} and H_{ij}^*, rather than $\mathrm{Re}H_{ij}$ and $\mathrm{Im}H_{ij}$, as independent and thus interpret $d^2 H_{ij}$ as $dH_{ij}^*\, dH_{ij}/(2\mathrm{i})$. It is about time to reveal the invariance of dH under the pertinent canonical transformation, thus completing the invariance proof for the Gaussian ensembles. To that end, I must show that a canonical transformation $H \to H'$ comes with a Jacobian equal to unity, $\det(\partial H'/\partial H) = 1$. For the sake of transparency, I shall write out the reasoning for the smallest nontrivial matrix dimension, $N = 2$; the validity for arbitrary N will become clear as well.

First, I turn to the GUE and once more employ the unitary matrix (4.3.6) to transform the Hermitian matrix H as $H' = UHU^\dagger$ or

$$
\begin{aligned}
H'_{11} &= c^2 H_{11} + s^2 H_{22} - c\,s\,(e^* H_{12} + e\, H_{12}^*) \\
H'_{22} &= s^2 H_{11} + c^2 H_{22} + c\,s\,(e^* H_{12} + e\, H_{12}^*) \\
H'_{12} &= c\,s\,e\,(H_{11} - H_{22}) + c^2 H_{12} - s^2 e^2 H_{12}^* \\
H'^{*}_{12} &= c\,s\,e\,(H_{11} - H_{22}) - s^2 e^{*2} H_{12} + c^2 H_{12}^*
\end{aligned}
\tag{4.5.2}
$$

where I have abbreviated $c = \cos\Theta$, $s = \sin\Theta$, and $e = \mathrm{e}^{\mathrm{i}\phi}$. The very appearance of (4.5.2) suggests reading the matrices H and H' as four-component vectors $\mathbf{h}$ and $\mathbf{h}'$ related by a 4×4 matrix $\mathbf{U}$ as $\mathbf{h}' = \mathbf{Uh}$. The latter matrix is unitary with respect to the scalar product $(\mathbf{a}, \mathbf{b}) = \mathrm{Tr}\, A^\dagger B$ of complex four-component vectors $\mathbf{a}$ and $\mathbf{b}$ related to 4×4 matrices A and B in the manner just introduced; the unitarity follows trivially from $(\mathbf{Ua}, \mathbf{Ub}) = \mathrm{Tr}\,(UAU^\dagger)^\dagger UBU^\dagger = \mathrm{Tr}A^\dagger B = (\mathbf{a}, \mathbf{b})$. The Jacobian of the transformation (4.5.2), simply the determinant $\det \mathbf{U}$, is thus unimodular; in fact, it is equal to unity since that is manifestly so for the infinitesimal version of (4.5.2), as is immediately checked with $c \to 1$, $se \to \Theta$ with infinitesimal Θ. With $\det \mathbf{U} = 1$, we have the desired invariance $dH' = dH$, at least for the GUE of complex Hermitian 2×2 matrices. But the argument is immediately extended to $N \times N$ matrices by letting $\mathbf{a}, \mathbf{b}, \mathbf{h}, \ldots$ be vectors with N^2 complex components and $\mathbf{U}$ an $N^2 \times N^2$ matrix.

The GOE is dealt with quite analogously. We have to consider the orthogonal transformation $H' = OH\tilde{O}$ between real symmetric matrices. Again securing nice looking formulae by choosing $N = 2$ and using the orthogonal matrix O from (4.3.2), we get the transformation

$$
\begin{aligned}
H'_{11} &= c^2 H_{11} + s^2 H_{22} - 2c\,s\,H_{12} \\
H'_{22} &= s^2 H_{11} + c^2 H_{22} + 2c\,s\,H_{12} \\
H'_{12} &= c\,s\,(H_{11} - H_{22}) + (c^2 - s^2)H_{12}
\end{aligned}
\tag{4.5.3}
$$

with $c = \cos\Theta$, $s = \sin\Theta$ as above. To reveal the Jacobian of this transformation as equal to unity, a little gymnastics is necessary since the 3×3 matrix in (4.5.3) is not orthogonal: Neither rows nor columns form pairwise orthogonal three-component vectors. The gymnastics will hardly produce sweat though since it amounts to no more than rewriting the 3×3 Jacobian as a 4×4 determinant,

$$
\begin{vmatrix} c^2 & s^2 & -2cs \\ s^2 & c^2 & 2cs \\ cs & -cs & c^2 - s^2 \end{vmatrix} = \begin{vmatrix} c^2 & s^2 & -cs & -cs \\ s^2 & c^2 & cs & cs \\ cs & -cs & c^2 & -s^2 \\ cs & -cs & -s^2 & c^2 \end{vmatrix} .
\tag{4.5.4}
$$

The equality is easily seen by first adding the fourth to the third column in the 4×4 determinant and then subtracting the third from the fourth row. It is not by chance that the foregoing 4×4 determinant results from the 4×4 Jacobian of the transformation (4.5.2) upon specializing as $e \to 1$; that specialization carries the transformation (4.5.2) into a slightly blown-up version of (4.5.3) which encompasses the latter for the at present real symmetric matrix H. But the 4×4 version of the transformation (4.5.3) has already been shown above to have a Jacobian equal to unity and due to the equality (4.5.4) the Jacobian for (4.5.3) equals unity as well. I trust the reader has refrained from indulging in blindly evaluating the determinants in (4.5.4). The faithless who may have arrived at $\det(\partial H'/\partial H) = 1$ in so deviant a manner must now work hard to generalize from $N = 2$ to arbitrary N. We others just proceed as in the unitary case and elevate the 3×3 Jacobian to an $N(N+1)/2 \times N(N+1)/2$ one which equals its blown-up $N^2 \times N^2$ version and thus unity.

The reader is invited to extend the invariance proof to the GSE as Problem 4.3.

4.6 Average Level Density

The joint distribution of energy levels (4.3.15) not only implies the distribution $P(S)$ of level spacings but also the ensemble-averaged level density

$$
\bar{\varrho}(E) = \int dE_2 \ldots dE_N P(E, E_2, \ldots, E_N) .
\tag{4.6.1}
$$

To secure a well-defined limit for the foregoing $(N-1)$-fold integral as $N \to \infty$, one sets the energy scale such that the second-order moments take the values for the GUE,

$$
\overline{H_{ij} H_{ji}} = \overline{H_{ii}^2} = \frac{1}{4N} .
\tag{4.6.2}
$$

Then, the density of levels has the semicircular form

$$
\bar{\varrho}(E) = \begin{cases} (2/\pi)\sqrt{1 - E^2} & \text{for } |E| < 1 \\ 0 & \text{for } |E| > 1 \end{cases}
\tag{4.6.3}
$$

known as *Wigner*'s semicircle law [76]. That law is also valid for the Gaussian orthogonal and symplectic ensembles, even with the same value unity of the radius, provided that the energy scale is appropriately chosen. On the other hand, if the width of the Gaussian distribution for the H_{ij} is chosen independently of N, the radius of the semicircle becomes proportional to $\sqrt{N}$, and the mean level spacing proportional to $1/\sqrt{N}$.

The most elegant and up-to-date method of deriving the semicircle law involves the representation of determinants by Gaussian integrals over anticommuting variables. Readers with indomitable curiosity about that derivation may jump to Sect. 4.12 below where Grassmann integration is introduced and then immediately on to Chap. 10. There we shall obtain the average level density from the average Green function

$$\overline{G}(E \pm i0^+) = \lim_{N \to \infty} \frac{1}{N} \mathrm{Tr} \overline{\frac{1}{E \pm i0^+ - H}} = \frac{E}{2\lambda^2} \mp \frac{i}{\lambda} \sqrt{1 - (E/2\lambda)^2} \qquad (4.6.4)$$

as $\bar{\varrho}(E) = \pm \frac{1}{\pi} \mathrm{Im} \overline{G}(E \mp i0^+)$, where the energy scale is set such that $\overline{|H_{ij}|^2} = \lambda^2/N$ instead of (4.6.2).

In contrast to the local fluctuations in the spectrum, the average level density (4.6.3) has no universal validity; rather it is a specific property of the Gaussian matrix ensembles. It differs, in particular, from the semiclassical average level density (Weyl's law, also known as the Thomas–Fermi distribution) for Hamiltonians of systems with f classical degrees of freedom,

$$\bar{\varrho}_{SC}(E) = \frac{1}{h^f} \int d^f p \, d^f q \, \delta \left(E - H(p, q) \right) , \qquad (4.6.5)$$

which estimates the number of energy levels in the interval $[E, E + \Delta E]$ as the number of Planck cells contained in the phase-space volume of that energy shell. Weyl's law will be derived in Sect. 5.7 for classically integrable dynamics and in Sect. 8.3.3 for the nonintegrable case.

4.7 Unfolding Spectra

It may be well to digress from random-matrix theory for a moment and to discuss the problem of how to extract an average level density from a given sequence of measured or calculated levels. The problem is analogous to that of finding a smoothed density of particles for a gas or a liquid. Easiest to cope with are spectra (or systems of particles) that appear homogeneous on energy (or length) scales exceeding a certain minimal range $\Delta E_{\min}$: the number ΔN of levels (degenerate ones counted with their degree of degeneracy) per energy interval ΔE yields the average density $\bar{\varrho} = \Delta N/(\Delta E N)$ provided $\Delta E \geq E_{\min}$.

More often, however, one encounters "macroscopically" inhomogeneous spectra (analogous to compressible fluids with density gradients) and then faces the problem of having to distinguish between local fluctuations in the level sequence and a systematic global energy dependence of the average density. When defining

$$\bar{\varrho}(E) = \frac{\Delta N}{\Delta E} \frac{1}{N} \qquad (4.7.1)$$

one must take an energy interval ΔE comprising many levels, $\Delta N \gg 1$, so that local fluctuations do not prevent the ratio $\Delta N/\Delta E$ from being smooth in E. On the other hand, ΔE must appear infinitesimal with respect to the scale on which $\bar{\varrho}(E)$ varies systematically. For the obvious conflict to be reasonably resolvable, a self-consistent separation of energy scales must be possible. By requiring that the average density $\bar{\varrho}(E)$ should vary little over an interval of the order of the local mean level spacing $1/\bar{\varrho}(E)N$, one obtains the self-consistency condition

$$|\bar{\varrho}'(E)| \ll N\bar{\varrho}(E)^2 . \tag{4.7.2}$$

In practice, the following procedure has proven viable. One considers the convolution of the density $\varrho(E) = (1/N)\sum_{i=1}^{N}\delta(E - E_i)$ with the Gaussian

$$g(E) = \frac{1}{\Delta\sqrt{\pi}}e^{-E^2/\Delta^2} , \tag{4.7.3}$$

$$\varrho_\Delta(E) = \frac{1}{N}\sum_{i=1}^{N}g(E - E_i) \tag{4.7.4}$$

and chooses the width Δ by minimizing the mean square deviation between the level staircase

$$\sigma(E) = \frac{1}{N}\sum_{i}\Theta(E - E_i) \tag{4.7.5}$$

and the integral of $\varrho_\Delta(E)$,

$$\sigma_\Delta(E) = \int_{-\infty}^{E}dE'\varrho_\Delta(E') . \tag{4.7.6}$$

If an optimal value of Δ can be found in this way, one may make the identification $\varrho_\Delta(E) = \bar{\varrho}(E)$ and identify $\sigma_\Delta(E)$ with the average level staircase $\bar{\sigma}(E)$. Fortunately, the mean square deviation between $\sigma(E)$ and $\sigma_\Delta(E)$ often depends weakly on Δ for Δ in the range of several typical level spacings; in such cases the self-consistency condition (4.7.2) is well obeyed.

The concept of level spacing fluctuations (and of other measures of local fluctuations in the spectrum) requires revision when the average density $\varrho(E)$ is energy-dependent. For instance, inasmuch as the terms "level clustering" and "level repulsion" are meant to describe local fluctuations, it would be quite inappropriate to describe a spectral region with high $\bar{\varrho}$ as one with less repulsion than a region with smaller $\bar{\varrho}$. Indeed, before it even begins to make sense to compare local fluctuations from two spectral regions with different average densities, one must secure uniformity of the average spacing throughout the spectrum by local changes of the unit of energy.

In stating, at the end of Sect. 4.4, that the Wigner distributions (4.4.2) hold for the three Gaussian ensembles of random matrices not only for the smallest dimensions possible but also, to a satisfactory degree of accuracy, in the limit $N \to \infty$, it was tacitly assumed that the levels rescale to a uniform average density with the help of Wigner's semicircle law.

A natural way of "unfolding" a spectrum to uniform density can be found as follows: One looks for a function $f(E)$ such that the rescaled levels

$$e_i = f(E_i) \tag{4.7.7}$$

have unit mean spacing, the mean evaluated with respect to energy intervals $[E - \Delta E/2, E + \Delta E/2]$. The corresponding rescaled interval goes from $f(E - \Delta E/2)$ to $f(E + \Delta E/2)$. The requirement

$$
\begin{aligned}
1 &= \tfrac{\Delta e}{\Delta N} = \tfrac{1}{\Delta N} \left[f(E + \Delta E/2) - f(E - \Delta E/2) \right] \\
&= \tfrac{\Delta E}{\Delta N} f'(E) = f'(E)/\bar{\varrho}(E)N
\end{aligned}
\tag{4.7.8}
$$

yields, upon integration, the function $f(E)$ as N times the average level staircase

$$\bar{\sigma}(E) = \int_{-\infty}^{E} dE' \, \bar{\varrho}(E') . \tag{4.7.9}$$

The rescaled levels are thus

$$e_i = N\bar{\sigma}(E_i) . \tag{4.7.10}$$

The unfolding (4.7.10) is certainly natural, but it is not the only one possible. For many applications, another unfolding,

$$e_i = E_i N \varrho(E_i) , \tag{4.7.11}$$

is convenient. The two rescalings (4.7.10, 4.7.11) are strictly equivalent only in the (uninteresting) case of homogeneous spectra $[\bar{\varrho}(E) = \text{const}]$; but with respect to spectral regions in which the energy dependence of $\varrho(E)$ is sufficiently weak to be negligible, there is practically no difference between (4.7.10) and (4.7.11).

4.8 Eigenvector Distributions

4.8.1 Single-Vector Density

For several applications, the distribution of the components of the eigenvectors of random matrices are of interest [77–80]. Considering first the Gaussian unitary ensemble of $N \times N$ matrices, one faces unit-norm eigenvectors with N in general complex components $c_n = c'_n + ic''_n$. For a given operator H, every eigenvector can be unitarily transformed into an arbitrary vector of unit norm. The only invariant characteristic of eigenvectors is therefore the norm itself, and the joint probability for the N complex components of an eigenvector must read

$$P_{\text{GUE}} (\{c_n\}) = \text{const} \, \delta \left(1 - \sum_{n=1}^{N} |c_n|^2 \right) , \tag{4.8.1}$$

where the constant is fixed by normalization. Evidently, $P_{\mathrm{GUE}}(\{c_n\})$ is non-zero only on the surface of a d-dimensional unit sphere with $d = 2N$. A convenient quantity to compare with numerical data is the reduced density of, say, $|c_1|^2$, i.e., of the probability of finding the first basis state populated in an eigenstate of H,

$$P_{\mathrm{GUE}}(y) = \int d^2c_1 \ldots d^2c_N \delta\left(y - |c_1|^2\right) P_{\mathrm{GUE}}(\{c\}) , \qquad (4.8.2)$$

which we shall calculate presently.

For the Gaussian orthogonal ensemble of $N \times N$ matrices, the eigenvectors can, without loss of generality, be assumed real. The joint probability density for the N components c_n is therefore concentrated on the surface of a d-dimensional unit sphere with $d = N$,

$$P_{\mathrm{GOE}}(\{c\}) = \mathrm{const}\, \delta\left(1 - \sum_{n=1}^{N} c_n^2\right) , \qquad (4.8.3)$$

and the reduced density of the probability c_1^2 that the first basis state is populated reads

$$P_{\mathrm{GOE}}(y) = \int dc_1 \ldots dc_N \delta\left(y - c_1^2\right) P_{\mathrm{GOE}}(\{c\}) . \qquad (4.8.4)$$

The Gaussian symplectic ensemble of random $N \times N$ Hamiltonians, finally, again has unit-norm eigenvectors with N complex components, and N is necessarily even. The joint density of the components c_n, $P_{\mathrm{GSE}}(\{c\})$, is given by (4.8.1); the reason is analogous to that given above for the unitary case: Every eigenvector of H can be symplectically transformed into an arbitrarily prescribed unit-norm vector. However, now the natural reduced density to compare to numerical data is slightly different from those above. Recalling that in the symplectic case there is no loss of generality in choosing a basis of the form,

$$|1\rangle , T|1\rangle , |2\rangle , T|2\rangle , \ldots |N/2\rangle , T|N/2\rangle , \qquad (4.8.5)$$

the two eigenvectors of H pertaining to one given eigenvalue can be written as

$$\begin{aligned}
|e_1\rangle &= c_1|1\rangle + \tilde{c}_1 T|1\rangle + c_2|2\rangle + \tilde{c}_2 T|2\rangle \ldots \\
T|e_1\rangle &= -\tilde{c}_1^*|1\rangle + c_1^* T|1\rangle - \tilde{c}_2^*|2\rangle + \tilde{c}_2^* T|2\rangle + \ldots .
\end{aligned} \qquad (4.8.6)$$

However, any linear combination of $|e_1\rangle$ and $T|e_1\rangle$ together with an orthogonal combination such as

$$\alpha|e_1\rangle + \beta T|e_1\rangle , \quad -\beta|e_1\rangle + \alpha T|e_1\rangle \qquad (4.8.7)$$

with $|\alpha|^2 + |\beta|^2 = 1$ is also a pair of unit-norm eigenvectors of H pertaining to the eigenvalue under consideration. A diagonalization code for H starting with the matrix H in the basis (4.8.5) may yield the two eigenvectors in the form (4.8.7) with any choice of α and β [rather than with $\alpha = 1$, $\beta = 0$ as was the case in

(4.8.6)]. What remains invariant, however, under the transition from (4.8.6) to (4.8.7) are the occupation probabilities for the subspaces spanned by $|i\rangle$ and $T|i\rangle$. Thus, one is led to compare the reduced density for, say, $|c_1|^2 + |\tilde{c}_1|^2$,

$$
\begin{aligned}
P_{\text{GSE}}(y) \;=\; & \int d^2 c_1 d^2 \tilde{c}_1 d^2 c_2 d^2 \tilde{c}_2 \ldots d^2 c_{N/2} d^2 \tilde{c}_{N/2} \\
& \times \delta\left(y - |c_1|^2 - |\tilde{c}_1|^2\right) P\left(\{c, \tilde{c}\}\right)
\end{aligned}
\tag{4.8.8}
$$

with the corresponding distribution obtained by diagonalizing H numerically. In the basis set of (4.8.5), the joint distribution of all components $\{c, \tilde{c}\}$ reads, of course,

$$
P_{\text{GSE}}\left(\{c, \tilde{c}\}\right) = \text{const } \delta\left(1 - \sum_{n=1}^{N/2}\left[|c_n|^2 + |\tilde{c}_n|^2\right]\right).
\tag{4.8.9}
$$

For all three Gaussian matrix ensembles, the joint distribution of all components of eigenvectors are uniformly concentrated on d-dimensional unit spheres, with d suitably related to the matrix dimension N ($d = N, 2N, 2N$ for the orthogonal, unitary, and symplectic cases, respectively). In a unified notation involving d real variables $x_1 \ldots x_d$, the properly normalized joint distribution reads

$$
P^{(d)}\left(x_1, \ldots, x_d\right) = \pi^{-d/2}\Gamma\left(\frac{d}{2}\right)\delta\left(1 - \sum_{n=1}^{d} x_n^2\right).
\tag{4.8.10}
$$

By integrating out $d - l$ of the variables x, one obtains the reduced density

$$
P^{(d,l)}\left(x_1, \ldots, x_l\right) = \pi^{-l/2}\frac{\Gamma(d/2)}{\Gamma((d-l)/2)}\left(1 - \sum_{n=1}^{l} x_n^2\right)^{(d-l-2)/2}.
\tag{4.8.11}
$$

The reduction from $P^{(d)}$ to $P^{(d,l)}$ is conveniently carried out by using a Fourier integral representation for the delta function in (4.8.10) so as to make the $(d-l)$-fold integral over $x_{l+1}, \ldots x_d$ the $(d-l)$th power of a single integral. Different choices for d and l now yield the reduced densities $P_{\text{GOE}}(y)$, $P_{\text{GUE}}(y)$, and $P_{\text{GSE}}(y)$ defined above.

For the orthogonal case [77, 78],

$$
\begin{aligned}
P_{\text{GOE}}(y) \;=\; & \int dx\, P^{(N,1)}(x)\delta(y - x^2) \\
=\; & \frac{1}{\sqrt{\pi}}\frac{\Gamma(N/2)}{\Gamma[(N-1)/2]}\frac{(1-y)^{(N-3)/2}}{\sqrt{y}}.
\end{aligned}
\tag{4.8.12}
$$

The analogous result for the unitary case is

$$
\begin{aligned}
P_{\text{GUE}}(y) \;=\; & \int dx_1 dx_2 P^{(2N,2)}(x_1, x_2)\delta\left(y - x_1^2 - x_2^2\right) \\
=\; & (N-1)(1-y)^{N-2},
\end{aligned}
\tag{4.8.13}
$$

and for the symplectic case

$$
\begin{aligned}
P_{\text{GSE}}(y) &= \int dx_1 dx_2 dx_3 dx_4 P^{(2N,4)}(x_1 \ldots x_4) \\
&\quad \times \delta\left(y - x_1^2 - x_2^2 - x_3^2 - x_4^2\right) \\
&= (N-1)(N-2)y(1-y)^{N-3} .
\end{aligned}
\tag{4.8.14}
$$

Considering that the mean $\langle y \rangle$ is of the order $1/N$, it is convenient to go over to the variable $\eta = yN$ and to perform the limit $N \to \infty$. After proper normalization, this yields the densities

$$
\begin{aligned}
P_{\text{GOE}}(\eta) &= \tfrac{1}{\sqrt{2\pi\eta}} e^{-\eta/2} , \\
P_{\text{GUE}}(\eta) &= e^{-\eta} , \\
P_{\text{GSE}}(\eta) &= \eta e^{-\eta} .
\end{aligned}
\tag{4.8.15}
$$

The first of these densities is known as the Porter–Thomas distribution [81, 82].

Interestingly, the densities $P(y)$ or $P(\eta)$ for the three ensembles are quite different in their y (or η) dependence, as illustrated in Fig. 4.3. This figure also compares these functions with data obtained by diagonalizing the Floquet operators of kicked tops with the appropriate symmetries and fully chaotic classical limits. The reliability of random-matrix theory is indeed impressive. Figure 4.3 also reveals, incidentally, that the eigenvector statistics yield a quantum mechanical chaos criterion.

Nowhere in this section has the Gaussian nature of the matrix ensembles been used. In fact, all results hold true unchanged for the eigenvectors of unitary matrices from Dyson's circular ensembles to which we shall turn in Sect. 4.10.

Quite different in status even though similar in content is Berry's conjecture for wave functions of billiards: Amplitudes sampled at independent points make

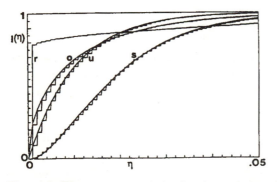

Fig. 4.3. Eigenvector statistics for the orthogonal (o), unitary (u), and symplectic (s) (Gaussian or circular) ensembles of random matrices (smooth curves), compared with numerically obtained statistics for kicked tops of the appropriate universality classes. Plotted are the integrated distributions, i.e., $I(\eta) = \int_0^\eta d\xi P(\xi)$. The curve labelled (r) refers to a kicked top under conditions of regular motion in the classical limit

up a Gaussian distribution; independent means further apart than a typical wavelength [83, 84]. That conjecture is well obeyed except for wavefunctions localized near periodic orbits. For a comprehensive discussion of these matters, including the so-called scars, the reader is referred to *Stöckmann's* book [85].

4.8.2 Joint Density of Eigenvectors

In concluding Sect. 4.3, we saw that eigenvectors and eigenvalues of matrices from the Gaussian ensembles are statistically independent. That observation allows us to write the joint densities of the eigenvectors. For a GUE matrix, the ith component of the μth eigenvector is a complex number to be denoted by $c_{i\mu}$. The joint density in question must express normalization and mutual orthogonality of all N eigenvectors while abstaining from any further prejudice,

$$
P_{GUE} = \left[\prod_{\mu=1}^{N} \delta\Big(\sum_{i=1}^{N} |c_{i\mu}|^2 - 1 \Big) \right] \left[\prod_{\mu<\nu}^{1...N} \delta\Big(\sum_{i=1}^{N} c_{i\mu}^* c_{i\nu} \Big) \right] . \tag{4.8.16}
$$

The product of delta functions is not in conflict with the angle-dependent part of the Jacobian discussed in Sect. 4.3. Those two densities differ only in the number and the character of the variables admitted: The Jacobian of Sect. 4.3 has normalization and orthogonality already worked in and parametrizes the eigenvectors in terms of the remaining variables, referred to as "angles" there.

The foregoing density applies to the circular unitary ensemble (CUE) to be met with in Sect. 4.10 as well, simply because all relevant properties are common to Hermitian and unitary matrices, i.e., normalizability and mutual orthogonality of eigenvectors as well as statistical independence of eigenvectors and eigenvalues. The modifications appropriate for the orthogonal and symplectic classes are obvious and will not be written here.

4.9 Ergodicity of the Level Density

When describing in Sect. 4.7 how a spectrum determined either experimentally or by computation can be unfolded to uniform mean density, it was necessary to define a mean level density by a local spectral average. Similar spectral averages, local or global, must be performed when extracting measures of fluctuations from a given spectrum. For instance, to construct the level spacing distribution, the distances between neighboring levels from the whole level sequence must be gathered.

Random-matrix theory, on the other hand, provides ensemble averages of quantities like the level density and the spacing distribution. From empirical and numerical evidence, one knows that for almost all individual Hamiltonians many observables related to the spectrum have spectral means that agree very well with their averages over the appropriate ensemble of random matrices, provided only that the number of levels is large ($N \gg 1$ or $N \to \infty$). Such (overwhelmingly likely) accurate agreement of spectral and ensemble averages is quite analogous to the equivalence of time and ensemble averages of certain observables in many-particle systems; in both cases one speaks of ergodic behavior of the observables in question. Proving ergodicity is often a hard task.

Now, I shall sketch the proof of ergodic behavior for the density of levels,

$$\varrho(E) = \frac{1}{N} \sum_{i=1}^{N} \delta(E - E_i) , \tag{4.9.1}$$

whose spectral average is

$$\langle \varrho(E) \rangle = \frac{1}{\Delta E} \int_{E-\Delta E/2}^{E+\Delta E/2} dE' \varrho(E') \tag{4.9.2}$$

and the ensemble mean reads

$$\bar{\varrho}(E) = \int d^N E \delta(E - E_1) P(E_1, E_2, \ldots, E_N) . \tag{4.9.3}$$

Ergodic behavior means that there is a limit, to be identified, in which the ensemble variance of $\langle \varrho(E) \rangle$ vanishes, and the ensemble mean of $\langle \varrho(E) \rangle$ is $\bar{\varrho}(E)$,

$$\overline{\langle \varrho(E) \rangle} \rightarrow \bar{\varrho}(E)$$
$$\mathrm{Var}\, \langle \varrho(E) \rangle = \overline{\langle \varrho(E) \rangle^2} - \left(\overline{\langle \varrho(E) \rangle} \right)^2 \rightarrow 0 . \tag{4.9.4}$$

Indeed, when the fluctuations of $\langle \varrho(E) \rangle$ within the ensemble vanish, one may identify $\langle \varrho(E) \rangle = \bar{\varrho}(E)$. Following the arguments of *Pandey* [86], I shall now show that the limit in question amounts to letting the number N of levels go to infinity, the energy interval ΔE used for the spectral average shrink to zero, and the mean number of levels ΔN contained in ΔE grow indefinitely,

$$N \rightarrow \infty , \qquad \Delta E \rightarrow 0 , \qquad \Delta N \rightarrow \infty . \tag{4.9.5}$$

This limit formally resolves the conflict discussed in Sect. 4.7: The interval ΔE must be sufficiently large that it comprises a large number of levels ΔN but small enough for $\langle \varrho(E) \rangle \approx \Delta N/(\Delta E\, N)$ and $\bar{\varrho}(E)$ to be constant within it. By writing $\Delta N = N \bar{\varrho}(E) \Delta E$, one sees that there is no contradiction between the three requirements (4.9.5), one may imagine

$$\Delta N \sim N^{\varepsilon} , \qquad \Delta E \sim N^{\varepsilon-1} , \qquad 0 < \varepsilon < 1 . \tag{4.9.6}$$

The first of the properties (4.9.4) becomes obvious when the ensemble-averaged density $\bar{\varrho}(E)$ is represented by a Taylor series,

$$\begin{aligned} \left\langle \overline{\varrho(E)} \right\rangle &= \frac{1}{\Delta E} \int_{E-\Delta E/2}^{E+\Delta E/2} dE' \bar{\varrho}(E') \\ &= \bar{\varrho}(E) + \frac{1}{24} \bar{\varrho}''(E)(\Delta E)^2 + \ldots \\ &\rightarrow \bar{\varrho}(E) \qquad \text{for } \Delta E \rightarrow 0 . \end{aligned} \tag{4.9.7}$$

A little more work is needed for the ensemble variance

$$\text{Var}\,\langle \varrho(E) \rangle \;=\; \left(\frac{1}{\Delta E}\right)^2 \int_{E-\Delta E/2}^{E+\Delta E/2} dE' dE'' \left[\overline{\varrho(E')\varrho(E'')}\right.$$
$$\left. -\bar{\varrho}(E')\bar{\varrho}(E'')\right]. \tag{4.9.8}$$

One encounters here the density–density correlation function for the matrix ensemble [73, 87–89],

$$S(E', E'') = \overline{\varrho(E')\varrho(E'')} - \bar{\varrho}(E')\bar{\varrho}(E'') . \tag{4.9.9}$$

By inserting (4.9.1),

$$S(E', E'') = \frac{1}{N^2} \sum_{ij} \overline{\delta(E' - E_i)\delta(E'' - E_j)} - \bar{\varrho}(E')\bar{\varrho}(E'') \tag{4.9.10}$$

and by separating diagonal from off-diagonal terms in the double sum over eigenvalues,

$$S(E', E'') \;=\; \frac{1}{N}\delta(E' - E'')\bar{\varrho}(E') - \bar{\varrho}(E')\bar{\varrho}(E'')$$
$$+\frac{1}{N^2} \sum_{i\neq j} \overline{\delta(E' - E_i)\delta(E'' - E_j)} . \tag{4.9.11}$$

The last term in (4.9.11) could be evaluated with the help of the reduced distribution function for two eigenvalues; it obviously has the meaning of the probability density for finding a level at E' and another one at E'', up to the normalization factor $(N - 1)/N$.

An interesting consequence of the inhomogeneity of the Gaussian ensembles with respect to energy is manifest in the correlation function $S(E', E'')$ at this point. Inasmuch as the ensemble-averaged density $\bar{\varrho}(E)$ is energy-dependent according to Wigner's semicircle law, the correlation function S depends on both E' and E'' separately, rather than on the difference between them. Before one can hope for stationarity in E, one must renormalize the correlation function

$$S(E', E'')/\bar{\varrho}(E')\bar{\varrho}(E'') = \delta(E' - E'')/N\bar{\varrho}(E') - 1$$
$$+ \sum_{i\neq j} \overline{\delta(E' - E_i)\delta(E'' - E_j)}/\bar{\varrho}(E')\bar{\varrho}(E'')N^2 \tag{4.9.12}$$

and rescale all energies by referring them to the local mean spacing $1/N\bar{\varrho}(E)$,

$$e = EN\bar{\varrho}(E) , \tag{4.9.13}$$

so as to have

$$S(E', E'')/\bar{\varrho}(E')\bar{\varrho}(E'') = \delta(e' - e'') - Y(e', e'') ,$$

$$Y(e', e'') = 1 - \sum_{i\neq j} \overline{\delta(e' - e_i)\delta(e'' - e_j)} , \tag{4.9.14}$$

where $Y(e', e'')$ is known as Dyson's two-level cluster function [73, 87–89]. Indeed, this function is stationary, i.e., it depends only on the difference $e' - e''$ for the three Gaussian ensembles and reads

$$\begin{aligned}
Y_{\text{GOE}}(e) &= s(e)^2 - J(e)D(e) \\
Y_{\text{GUE}}(e) &= s(e)^2 \\
Y_{\text{GSE}}(e) &= s(2e)^2 - I(2e)D(2e)
\end{aligned} \tag{4.9.15}$$

with

$$\begin{aligned}
s(e) &= \frac{\sin \pi e}{\pi e} \\
D(e) &= \frac{\partial s}{\partial e} \\
I(e) &= \int_0^e de'\, s(e') = \tfrac{1}{\pi} \operatorname{Si}(\pi e) \\
J(e) &= I(e) - \tfrac{1}{2}\operatorname{sgn}(e) \,.
\end{aligned} \tag{4.9.16}$$

A derivation of (4.9.15) will be presented in Sect. 4.13 and for the GUE in Sect. 10.

By inserting (4.9.15) in (4.9.8) and recalling that the averaging interval ΔE is chosen small enough to forbid any variation of $\bar{\varrho}(E)$ within, we can attack the double-energy integral of (4.9.8), starting with

$$\operatorname{Var}\langle \varrho(E) \rangle = \left[\frac{\bar{\varrho}(E)}{\Delta N} \right]^2 \int_0^{\Delta N} de''de' \left[\delta(e'' - e') - Y(e'' - e') \right]. \tag{4.9.17}$$

Neglecting relative corrections of order $1/\Delta N$,

$$\operatorname{Var}\langle \varrho(E) \rangle = \left[\frac{\bar{\varrho}(E)}{\Delta N} \right]^2 \times \begin{cases} \frac{2}{\pi^2}\left[\ln(2\pi \Delta N) + \gamma + 1 - \frac{\pi^2}{8} \right] & \text{GOE} \\ \frac{1}{\pi^2}\left[\ln(2\pi \Delta N) + \gamma + 1 \right] & \text{GUE} \\ \frac{1}{2\pi^2}\left[\ln(4\pi \Delta N) + \gamma + 1 + \frac{\pi^2}{8} \right] & \text{GSE} \end{cases} \tag{4.9.18}$$

where γ is Euler's constant. For all three cases, the ensemble variance of $\langle \varrho(E) \rangle$ vanishes as $(\ln \Delta N)/(\Delta N)^2$ as $\Delta N \to \infty$. The ergodicity of the level density, $\langle \varrho(E) \rangle \to \bar{\varrho}(E)$, is thus established.

A rougher argument leads to the same conclusion more rapidly. We just have to realize that the auxiliary functions constituting the cluster function are bounded everywhere and fall off with increasing e like $s(e) \propto D(e) \propto J(e) \propto 1/e$. It follows that the cluster functions themselves are bounded and enjoy the asymptotic large-e behavior $Y_{\text{GOE}}(e) \propto Y_{\text{GUE}}(e) \propto 1/e^2$ and $Y_{\text{GSE}}(e) \propto 1/e$. Like the delta function $\delta(e''-e')$ in the integrand of (4.9.17), the cluster function $Y(e''-e')$ can therefore contribute no worse than a term of the order ΔN to the double integral in the orthogonal and unitary cases and no more than a term $\propto \Delta N \ln \Delta N$ in the symplectic case. Such growth is obviously outdone by the common prefactor $\propto (\Delta N)^{-2}$. I shall come back to ergodicity in Sect. 4.14.

4.10 Dyson's Circular Ensembles

Without mentioning possible applications to the Floquet operators of periodically driven systems but certainly with scattering matrices in mind, Dyson introduced three so-called circular ensembles of random unitary matrices [73, 90]. In view of the respective canonical transformations of their member matrices, these ensembles carry the names orthogonal (COE), unitary (CUE), and symplectic (CSE). The joint densities of matrix elements are easily written; they are in fact identical in appearance to the joint densities of eigenvectors for Hermitian or unitary matrices of the appropriate class; in particular, for the CUE, the density in question is nothing but (4.8.16).

A somewhat nonconventional approach to the circular ensembles and their densities of matrix elements will be presented in Chap. 6. Here, we hurry to list some of their properties.

Most importantly, defining the unimodular eigenvalues of unitary matrices by their eigenphases ϕ_j,

$$F|\phi_j\rangle = \mathrm{e}^{-\mathrm{i}\phi_j}|\phi_j\rangle \,, \quad j = 1, 2, \dots N \,, \tag{4.10.1}$$

the joint distribution for the eigenphases is

$$
\begin{aligned}
P\left(\{\phi\}\right) &= (1/\mathcal{N}_\beta) \prod_{i<j} \left| \mathrm{e}^{-\mathrm{i}\phi_i} - \mathrm{e}^{-\mathrm{i}\phi_j} \right|^{\beta} \\
&= (1/\mathcal{N}_\beta) \prod_{i<j} \left| 2 \sin \frac{\phi_i - \phi_j}{2} \right|^{\beta}
\end{aligned}
\tag{4.10.2}
$$

where $\beta = n - 1$ denotes the degree of level repulsion which we have seen is smaller by 1 than the codimension n of a level crossing, i.e.,

$$
\beta =
\begin{cases}
1 & \text{COE} \\
2 & \text{CUE} \\
4 & \text{CSE}
\end{cases}
\tag{4.10.3}
$$

and $\mathcal{N}_\beta$ is a normalization constant. Since $P(\{\phi\})$ depends on the differences $\phi_i - \phi_j$ only, the circular ensembles, in contrast to the Gaussian ones, are homogeneous. Indeed, the reduced distribution of a single eigenphase ϕ is a constant,

$$\bar{\varrho}(\phi) = \int d^N \phi \, P\left(\{\phi\}\right) \delta(\phi - \phi_1) = \frac{1}{2\pi} \,. \tag{4.10.4}$$

Therefore, no unfolding is necessary in a spectrum of quasi-energies. The foregoing distribution is, of course, just the ensemble mean of the level density, the quasi-energy analogue of (4.6.1) and (4.9.1),

$$\varrho(\phi) = \frac{1}{N} \sum_{i=1}^{N} \delta(\phi - \phi_i). \tag{4.10.5}$$

Similarly, the density–density correlation function,

$$\overline{\varrho(\phi)\varrho(\phi')} = S(\phi - \phi') + \left(\frac{1}{2\pi}\right)^2, \tag{4.10.6}$$

the quasi-energy analogue of (4.9.10), is now a function of the difference $\phi - \phi'$ without any prior unfolding.

When dealing with the quantum spectra of classically integrable systems with at least two degrees of freedom in the next chapter, we shall meet with Poissonian fluctuations, i.e., that register no correlations between neighboring levels. Unitary matrices with such spectra are nicely modelled with the so-called Poissonian ensemble (PE) whose joint distribution of eigenphases is just the Nth power of the mean density,

$$P(\{\phi\}) = (\frac{1}{2\pi})^N \qquad \text{PE};$$

(4.10.7)

this fits into the scheme of (4.10.2,3) with a vanishing degree of level repulsion, $\beta = n - 1 = 0$; clustering as opposed to repulsion of levels is indeed an important quantum or wave signature of integrability.

4.11 Asymptotic Level Spacing Distributions

Sometimes the asymptotic ($N \to \infty$) level spacing distributions are needed for better accuracy than provided by the Wigner distributions (4.4.2) which are strictly valid for $N = 2$. Since it is known [73] that the Gaussian and circular ensembles have the same asymptotic level spacing distributions, it will be convenient here to take Dyson's circular ensembles as a starting point for a quantitative analysis of the limit $N \to \infty$. Our first goal will be to derive Taylor series for the spacing distributions $P(S)$ in powers of S; then, connecting these with asymptotic results for large S by suitable Padé approximants yields $P(S)$ for $0 \le S < \infty$ with any desired accuracy.

Easiest to treat is the case of the CUE which I propose to deal with in some detail now.

The product of eigenvalue differences in (4.10.2) can be related to a Vandermonde determinant [Ref. [73], Chap. 12],

$$\prod_{\substack{i<j}}^{1 \ldots N} (z_i - z_j) = \det \left(z_i^{k-1} \right) = \det \begin{pmatrix} 1 & z_1 & z_1^2 & \cdots & z_1^{N-1} \\ 1 & z_2 & z_2^2 & \cdots & z_2^{N-1} \\ \vdots & & & & \vdots \\ 1 & z_N & z_N^2 & \cdots & z_N^{N-1} \end{pmatrix},$$

(4.11.1)

where the abbreviation $z = e^{-i\phi}$ has been used. The reader may easily check the correctness of (4.11.1) by realizing that both the product and the determinant are polynomials in z_N of order $N-1$ with roots $z_1, z_2, \ldots, z_{N-1}$; a corresponding statement holds true for $z_{N-1}, z_{N-2}, \ldots$. Then, the squared modulus in (4.10.2) for $n - 1 = 2$ is also given by a determinant,

$$\prod_{\substack{i<j}}^{1 \ldots N} |z_i - z_j|^2 = \det \left(\sum_{l=1}^{N} z_l^{i-1} z_l^{*k-1} \right),$$

(4.11.2)

whereupon the eigenvalue distribution for the CUE takes the form

$$P(\{\phi\}) = \mathcal{N}_2^{-1}\det\left(\sum_{l=1}^{N} z_l^{i-1}z_l^{*k-1}\right). \tag{4.11.3}$$

For the discussion that follows, a little lemma about the mean of a symmetric function $f(\{\phi\})$ of all N phases ϕ with the weight (4.11.3) will be helpful,

$$\langle f(\{\phi\})\rangle = \mathcal{N}_2^{-1}N!\int_0^{2\pi} d^N\phi f(\{\phi\})\det\left(z_i^{i-1}z_i^{*k-1}\right). \tag{4.11.4}$$

To prove this identity,[2] we begin with

$$\langle f(\{\phi\})\rangle = \mathcal{N}_2^{-1}\int_0^{2\pi} d^N\phi f(\{\phi\})\det\left(\sum_{l=1}^{N} z_l^{i-1}z_l^{*k-1}\right)$$

and take out of the determinant the sums over l from all elements in its first row,

$$\langle f(\{\phi\})\rangle = \mathcal{N}_2^{-1}\int_0^{2\pi} d^N\phi f(\{\phi\})$$

$$\times \sum_{l=1}^{N}\det\begin{pmatrix} 1 & z_l^* & z_l^{*2} & \cdots \\ \sum_{j=1}^{N} z_j & \sum_{j=1}^{N} z_j z_j^* & \sum_{j=1}^{N} z_j z_j^{*2} & \cdots \\ \vdots & \vdots & \vdots & \vdots \end{pmatrix}.$$

Due to the symmetry of the integrand, the sum in question can be replaced by a factor N and the index l set to 1 in the first row. By subtracting from the lth row z_1^{l-1} times the first row,

$$\langle f(\{\phi\})\rangle = \mathcal{N}_2^{-1}N\int_0^{2\pi} d^N\phi f(\{\phi\})$$

$$\times\det\begin{pmatrix} 1 & z_1^* & z_1^{*2} & \cdots \\ \sum_{j=2}^{N} z_j & \sum_{j=2}^{N} z_j z_j^* & \sum_{j=2}^{N} z_j z_j^{*2} & \cdots \\ \vdots & \vdots & \vdots & \vdots \end{pmatrix}.$$

The foregoing argument is now repeated: The sum over j in the second row is replaced by a factor $(N-1)$ in front of the determinant and the index j set equal to 2 in the second row. One so proceeds row by row until (4.11.4) is established.

In a first application of (4.11.4), the normalization factor $\mathcal{N}_2$ in (4.11.3) can be determined as $\mathcal{N}_2 = N!(2\pi)^N$ since after setting $f(\{\phi\}) = 1$ in (4.11.4), the integral over the ith phase can be carried out in each element of the ith row:

$$1 = \mathcal{N}_2^{-1}N!\det\left(\int_0^{2\pi} d\phi_i z_i^{i-1}z_i^{*k-1}\right)$$

$$= \mathcal{N}_2^{-1}N!\det(2\pi\delta_{ik}) = \mathcal{N}_2^{-1}N!(2\pi)^N. \tag{4.11.5}$$

---

[2] In Sect. 8.8, an alternative method for dealing with problems of this kind will be presented.

In the next step toward constructing the level spacing distribution, we now consider the probability $E(S)$ that a phase interval of length S (in units of the mean spacing $2\pi/N$) is empty of levels,

$$E(S) = \left(\prod_{i=1}^{N} \int_{-\pi+\pi S/N}^{\pi-\pi S/N} d\phi_i \right) P(\{\phi\}) \,. \tag{4.11.6}$$

It should be noted that the homogeneity of the distribution $P(\{\phi\})$ justifies the particular location of the integration range in (4.11.6). The spacing distribution $P(S)$ equals the second derivative of the "gap probability" $E(S)$,

$$P(S) = \frac{\partial^2 E}{\partial S^2} \,. \tag{4.11.7}$$

Indeed, let $S = x + y$, and consider the difference $E(x+y) - E(x+y+\Delta x)$ which is the probability that $x + y$ is empty and the neighboring interval Δx not empty. For small Δx, the probability that Δx contains two or more levels is of second or higher order in Δx. Therefore, $-[\partial E(x+y)/\partial x]\Delta x$ is the probability that $x + y$ is empty and Δx contains one level. Reasoning analogously, $[\partial^2 E(x+y)/\partial x \partial y]\Delta x \Delta y$ is the probability that $x+y$ is empty and the surrounding intervals Δx and Δy each contain one level. The latter probability, on the other hand, may be expressed as the product of the "single-event" probability Δx and the conditional probability $P(S)\Delta y$ with $S = x + y$; note that in the units used, the mean density of levels is unity. The identity (4.11.7), together with (4.11.6), will find further applications in Chaps. 5 and 6.

Proceeding with the evaluation of $E(S)$ for the CUE, we observe that the integral in (4.11.6) affects all N phases symmetrically, whereupon the identity (4.11.4) is once more applicable. As in (4.11.5), the integral over the lth phase can be carried out by integrating all elements of the lth row in the determinant,

$$
\begin{aligned}
E(S) &= \int_{-\pi+\pi S/N}^{\pi-\pi S/N} \frac{d^N \phi}{(2\pi)^N} \det \left(z_l^{l-1} z_l^{*k-1} \right) \\
&= \det \left[\int_{-\pi+\pi S/N}^{\pi-\pi S/N} \frac{d\phi_l}{2\pi} e^{i\phi_l(l-k)} \right] \\
&= \det \left[\delta_{kl} - \frac{\sin\left[\pi(k-l)S/N\right]}{\pi(k-l)} \right] \,.
\end{aligned}
\tag{4.11.8}
$$

Now, this determinant and its second derivative $P(S)$ will be expanded as Taylor series in S,

$$E(S) = \sum_{l=0}^{\infty} E_l S^l \,, \quad P(S) = \sum_{l=0}^{\infty} (l+1)(l+2) E_{l+2} S^l \,. \tag{4.11.9}$$

As will become clear presently, a manageable expression for the expansion coefficient E_l can be constructed rather easily.

The starting point for the expansion in question is the usual representation of a determinant as a sum of products of N elements,

$$E(S) = \sum_{\mathcal{P}}(-1)^{\mathcal{P}} \prod_{k}^{1 \dots N} \left\{ \delta_{k,\mathcal{P}k} - \frac{\sin\left[\pi(k-\mathcal{P}k)S/N\right]}{\pi(k-\mathcal{P}k)} \right\} , \tag{4.11.10}$$

where $\mathcal{P}$ denotes a permutation of the set of indices $1, 2, 3, \dots, k, \dots N$ and $\mathcal{P}k$ the index associated with k by the permutation $\mathcal{P}$. Second, the Taylor series of $\sin x$ in powers of x must be employed for $x = \pi(k-\mathcal{P}k)S/N$,

$$\frac{\sin\left[\pi(k-\mathcal{P}k)S/N\right]}{\pi(k-\mathcal{P}k)} = \frac{1}{\pi} \sum_{l}^{0\dots\infty} (-1)^l \frac{(\pi S/N)^{2l+1}}{(2l+1)!} (k-\mathcal{P}k)^{2l} . \tag{4.11.11}$$

Finally, by inserting the binomial expansion of $(k-\mathcal{P}k)^{2l}$, we arrive at

$$\begin{aligned} E(S) &= \sum_{\mathcal{P}}(-1)^{\mathcal{P}} \prod_{k}^{1\dots N} \left[\delta_{k,\mathcal{P}k} + \frac{1}{\pi} \sum_{l}^{0\dots\infty} (-1)^{l+1} \frac{(\pi S/N)^{2l+1}}{(2l+1)!} \right. \\ &\left. \times \sum_{t}^{0\dots 2l} \binom{2l}{t} (-k)^{2l-t}(\mathcal{P}k)^{t} \right] . \end{aligned} \tag{4.11.12}$$

The zeroth-order Taylor coefficient may be read off as $E_0 = E(0) = 1$. That value is in fact immediately obvious from the definition of $E(S)$: An interval of zero length contains no eigenvalue with probability unity. To obtain the E_l with $l > 0$, consider the case where n out of the N factors in a summand in (4.11.12) are of nonvanishing order in S, and let these orders be $2l_1+1, 2l_2+1, \dots, 2l_n+1$. The remaining $(N-n)$ factors are of the type $\delta_{k,\mathcal{P}k}$. The contribution to $E(S)$ of such terms with fixed n and $l_1 \dots l_n$ reads[3]

$$\begin{aligned} &T\left(S, n, \{l_i\}\right) \\ &= \frac{1}{\pi^n n!} \sum_{k_1 \dots k_n}^{1\dots N} (-1)^{l_1+\dots+l_n+n} \frac{(\pi S/N)^{2l_1+1}}{(2l_1+1)!} \dots \frac{(\pi S/N)^{2l_n+1}}{(2l_n+1)!} \\ &\times \sum_{t_1}^{0\dots 2l_1} \dots \sum_{t_n}^{0\dots 2l_n} \binom{2l_1}{t_1} \dots \binom{2l_n}{t_n} (-k_1)^{2l_1-t_1} \dots (-k_n)^{2l_n-t_n} \\ &\times \sum_{\tilde{\mathcal{P}}} (-1)^{\tilde{\mathcal{P}}} \left(\tilde{\mathcal{P}}k_1\right)^{t_1} \dots \left(\tilde{\mathcal{P}}k_n\right)^{t_n} . \end{aligned} \tag{4.11.13}$$

The remaining $n!$ permutations $\tilde{\mathcal{P}}$ reshuffle the n indices on the n summation variables $k_1 \dots k_n$; they are the remnants of the $N!$ original permutations $\mathcal{P}$

---

[3] Actually, (4.11.13) requires $l_1 \neq l_2 \neq \dots \neq l_n$. When some of the l_i coincide such that k different values are assumed, $1 \leq k \leq n$, with multiplicities $n_1, n_2, \dots, n_k$ and $\sum n_i = n$, the factor $1/n!$ must be replaced by $1/(n_1!n_2!\dots n_k!)$. For the Taylor coefficients E_l, given in (4.11.18 or 4.11.21), the distinction in question is irrelevant, due to the appearance of another factor $n_1! \dots n_k!/n!$ accompanying the summation over the configurations $\{l_i\}$.

in (4.11.12) after requiring $k = \mathcal{P}k$ for $N - n$ of the original N factors in the product in (4.11.12). Since each $\tilde{\mathcal{P}}$ is nothing but a $\mathcal{P}$ with the $N - n$ constraints just mentioned, the respective parities are identical, $(-1)^{\mathcal{P}} = (-1)^{\tilde{\mathcal{P}}}$; the latter property of the $\tilde{\mathcal{P}}$ has already been used in writing (4.11.13). The sum over the $\tilde{\mathcal{P}}$ may be recognized as the $n \times n$ determinant $\det(k_i^{t_j})$, whereupon $T(S)$ takes the form

$$
T(S, n, \{l_i\})
$$
$$
= \frac{(-1)^{l_1 + l_2 + \cdots + l_n + n}}{\pi^n n!} \frac{(\pi S/N)^{2l_1 + 1}}{(2l_1 + 1)!} \cdots \frac{(\pi S/N)^{2l_n + 1}}{(2l_n + 1)!}
$$
$$
\times \sum_{t_1}^{0 \ldots 2l_1} \cdots \sum_{t_n}^{0 \ldots 2l_n} (-1)^{t_1 + t_2 + \cdots + t_n} \binom{2l_1}{t_1} \cdots \binom{2l_n}{t_n}
$$
$$
\times \sum_{\{k\}}^{1 \ldots N} \det\left(k_i^{2l_i - t_i + t_j}\right) . \tag{4.11.14}
$$

Clearly, $E(S) - 1$ results when T is summed first over all $l_1 \ldots l_n$ from 0 to ∞ and then over n from 1 to ∞.

At this point, it is well to realize that the n-fold sum over $k_1 \ldots k_n$ draws non-zero contributions only when all n summation variables k_i are different from one another in their values and, more importantly,

all t_i are different from one another, and

all $2l_i - t_i$ are different from one another. $\tag{4.11.15}$

Indeed, when two or more k_i or t_i coincide, the determinant has at least two identical rows and thus vanishes; when $(2l_i - t_i) = (2l_j - t_j)$, on the other hand, it is obvious from (4.11.3) that there is a pairwise cancelling of determinants in the sums over k_i and k_j. The constraint (4.11.15) yields an upper bound for the number n in $T(S, n, \{l_i\})$ for a fixed value of

$$
l = \sum_{i=1}^{n} (2l_i + 1) = 2 \sum_{i=1}^{n} l_i + n , \tag{4.11.16}
$$

the order of $T(S, n, \{l_i\})$ in S. The inequality in question,

$$
n^2 \leq l , \tag{4.11.17}
$$

will turn out to amount to a strong restriction of the number of terms $T(S, n, \{l_i\})$ contributing to the Taylor coefficient E_l. Postponing the proof of (4.11.17) until after its exploitation, we concentrate now on the Taylor coefficient[4]

$$
E_l = \left(\frac{\pi}{N}\right)^l \sum_{n=1,2,\ldots}^{n^2 \le l} \frac{(-1)^{(l+n)/2}}{\pi^n n!} \sum_{l_1 \ldots l_n}^{1,2,\ldots} \delta\left(\sum_{i=1}^{n} l_i, \frac{l-n}{2}\right)
$$

$$
\times \frac{1}{(2l_1+1)!} \cdots \frac{1}{(2l_n+1)!} \sum_{t_1}^{1 \ldots 2l_1} \sum_{t_n}^{1 \ldots 2l_n} (-1)^{t_1+\ldots+t_n} \binom{2l_1}{t_1} \cdots \binom{2l_n}{t_n}
$$

$$
\times \sum_{k_1 \ldots k_n}^{1 \ldots N} \det\left(k_i^{2l_i-t_i+t_j}\right) . \tag{4.11.18}
$$

The reader will appreciate the power of (4.11.17) in restricting the allowable values of n in (4.11.18) when observing that due to the definition (4.11.16) even (odd) values of l admit only even (odd) values of n. It follows that 16 (25) is the smallest even (odd) value of l for which two different values of n are allowed ($n = 2,4$, and $n = 3,5$, respectively). Note that $n = 1$ can contribute only to $l = 1$: For $n = 1$, no permutation $\mathcal{P}$ in (4.11.12) or $\tilde{\mathcal{P}}$ in (4.11.13) except the identity can contribute.

At this point, the asymptotic large-N behavior must be discussed. To that end, we observe that, to within corrections of relative order $1/N$, each of the n sums over a k_i may be replaced by an integral to yield

$$
\sum_{k=1}^{N} k^x = \frac{N^{x+1}}{x+1}\left[1 + \mathcal{O}\left(\frac{1}{N}\right)\right] \quad \text{for } x > 0 . \tag{4.11.19}
$$

Thus, the n-fold sum over the k_i in (4.11.18) gives

$$
\sum_{k_1 \ldots k_n}^{1 \ldots N} \det\left(k_i^{2l_i-t_i+t_j}\right)
$$

$$
= \det\left(\sum_k k^{2l_i-t_i+t_j}\right) = N^l \det\left(\frac{1}{2l_i-t_i+t_j+1}\right) \tag{4.11.20}
$$

whereupon the Taylor coefficient E_l takes the final form

$$
E_l^{(\mathrm{UE})} = \sum_{n=1,2,\ldots}^{n^2 \le l} \pi^{l-n} \frac{(-1)^{(l+n)/2}}{n!} \sum_{l_1 \ldots l_n}^{1,2,\ldots} \delta\left(\sum_{i=1}^{n} l_i, \frac{l-n}{2}\right)
$$

$$
\times \sum_{t_1}^{0 \ldots 2l_1} \cdots \sum_{t_n}^{0 \ldots 2l_n} \det\left(\frac{1}{2l_i-t_i+t_j+1}\right)
$$

$$
\times \prod_{k}^{1 \ldots n} \left[\frac{1}{(2l_k+1)!}\binom{2l_k}{t_k}(-1)^{t_k}\right] . \tag{4.11.21}
$$

---

[4] For the sake of typographical convenience, $\delta(m,n)$ here denotes the Kronecker symbol usually written as δ_{mn}.

The Taylor coefficients so determined indicate impressively rapid convergence of the Taylor series for $P(S)$ throughout the range of S from which the first few moments $\bar{S}, \overline{S^2}, \ldots$ draw important contributions, say $0 \leq S \lesssim 3$.

On the other hand, for $S \gtrsim 1.7$ *Dyson's* asymptotic result [92]

$$E_{\text{as}}^{(\text{UE})}(S) = \left(\frac{\pi}{2}\right)^{-1/4} e^{2B} S^{-1/4} e^{-(\pi^2/8)S^2}$$

$$B = \tfrac{1}{24}\ln 2 + \tfrac{3}{2}\zeta'(-1)\,, \tag{4.11.22}$$

holds with excellent accuracy. In particular, Dyson's asymptotics and the cheaply accessible Taylor series have an overlapping range of validity in the right wing of $P(S)$ such that Padé interpolations between the two approximations offer themselves for numerical evaluations of $P(S)$ with any desired accuracy [91].

Finally, we turn to the proof of the inequality (4.11.17) from the constraints (4.11.15) and the definition (4.11.16). For the sake of clarity, let us restate the task in question, reformulated slightly as a counting problem. With a fixed integer n, we wish to determine the configurations of two sets of integers, $t_1, t_2, \ldots, t_n$ and $l_1, l_2, \ldots, l_n$, that minimize the sum

$$l = \sum_{i=1}^{n}(2l_i + 1) = n + 2\left(\sum_{i=1}^{n} l_i\right) \stackrel{!}{=} \min \tag{4.11.23}$$

under the constraints

(a) $0 \leq t_i \leq 2l_i$,

(b) $l_i = 0, 1, 2, \ldots$,

(c) $t_1 \neq t_2 \neq \ldots \neq t_n$,

(d) $2l_1 - t_1 \neq 2l_2 - t_2 \neq \ldots \neq 2l_n - t_n$.

As a "warm-up" we consider the "primitive ladder" configurations $l_i = i - 1$ for $i = 1, 2, \ldots, n$. Then, the t_i are uniquely fixed by the above constraints: $t_1 = 0$ and $2l_1 - t_1 = 0$ by (a), $t_2 = 1$ and $2l_2 - t_2 = 1$ by constraints (c,d), and so forth, $t_i = 2l_i - t_i = i - 1$. The number l in this case takes the value $l = n^2$ which already corresponds to the minimum asserted.

The temptation may arise to search for lower values of l by cutting the primitive ladder at a lower height, say $l_i = i - 1$ for $i = 1, \ldots, n - \nu + 1$ and occupying one of its steps ν times. It is easy to see, however, that a violation of constraints (c,d) thus occurs. A conflict between (c,d) and a multiple occupancy of some integer by several l_i can be avoided only by leaving some lower integers unoccupied by any l_i. The extreme of such a configuration is that of "complete degeneracy", $l_i = L$ for $i = 1, 2, \ldots, n$. Due to constraint (a), this L must obey $2L + 1 \geq n$, which implies $l = n(2L + 1) \geq n^2$. For odd n, the smallest integer thus allowed is $L_{\min} = (n - 1)/2$ and in that case the totally degenerate configuration has the same l as the primitive ladder, $l = n^2$. Moreover, the $\{t_i\}$ and $\{2l_i - t_i\} = \{2L - t_i\}$ are equally densely packed in both configurations,

$\{t_i\} = \{2L - t_i\} = \{0, 1, 2, \ldots, n-1\}$. If n is even, $L_{\min} = n/2$, and therefore l is larger than n^2 for the totally degenerate configuration.

Now, it is clear how to show the impossibility that l will assume a value below n^2. Consider a general trial configuration

$$l_1 = l_2 = \ldots = l_{n_1} \equiv L_1$$

$$l_{n_1+1} = \ldots = l_{n_1+n_2} \equiv L_2$$

$$\vdots$$

$$l_{n_1+n_2+\ldots+n_{k-1}+1} = \ldots = l_{n_1+n_2+\ldots+n_k} = L_k$$

(4.11.24)

with

$$n_1 + n_2 + \ldots + n_k = n$$

and

$$L_1 < L_2 < \ldots < L_k \, .$$

Minimizing l means minimizing all of the L_i. By the foregoing argument for the fully degenerate configuration, L_1 is minimal when n_1 is chosen odd and then $L_{1,\min} = (n_1 - 1)/2$, $\{t_1, t_2, \ldots, t_{n_1}\} = \{2L_1 - t_1, \ldots, 2L_1 - t_{n_1}\} = \{0, 1, \ldots, n_1 - 1\}$. Having chosen n_1 as odd and $n_1 < n$ but otherwise arbitrary, we turn to the n_2-fold degenerate value L_2. The corresponding n_2 integers $t_{n_1+1}, \ldots, t_{n_1+n_2}$ must obey $t_i \geq n_1$ and $2L_2 - t_i \geq n_1$. Just as before, the minimum value of L_2 for fixed n_2 is found by assigning to the n_2 integers t_i in question the smallest values consistent with constraints (c,d), i.e.,

$$\begin{aligned}\{t_{n_1+1}, \ldots, t_{n_1+n_2}\} &= \{2L_2 - t_{n_1+1}, \ldots, 2L_2 - t_{n_1+n_2}\} \\ &= \{n_1, n_1 + 1, \ldots, n_1 + n_2 - 1\} \, .\end{aligned}$$

To accommodate these values of the t_i, the number L_2 must obey $2L_2 \geq 2n_1 + n_2 - 1$. By requiring n_2 to be odd, the minimal value of L_2 is $L_{2,\min} = n_1 + (n_2 - 1)/2$. Now, the argument can be continued to yield

$$L_{i,\min} = \sum_{j=1}^{i-1} n_j + \frac{1}{2}(n_i - 1) \, , \quad i = 1, 2, \ldots, k \, ,$$

(4.11.25)

provided that all n_i are chosen odd. The corresponding value of l is

$$l = \sum_{i=1}^{k} n_i (2L_{i,\min} + 1) = \left(\sum_{i=1}^{k} n_i\right)^2 = n^2 \, ,$$

(4.11.26)

independent of the n_i, provided only that the n_i are all odd. A larger value of l results when one or more n_i are chosen even, analogously to what was found for the totally degenerate configuration. The reader will realize that a special case of the configuration (4.11.24) with all n_i odd is the primitive ladder, for which $n_i = 1$ for $i = 1, 2, \ldots, n$.

Now, it is obvious that the lower bound $l_{min} = n^2$ cannot be beaten. It is realized in different configurations $\{l_i\}$ which all have in common the closest packing of the $\{t_i\}$ and $\{2l_i - t_i\}$ as

$$\{t_1, \ldots, t_n\} = \{2l_1 - t_1, \ldots, 2l_n - t_n\} = \{0, 1, \ldots, n-1\} . \qquad (4.11.27)$$

This completes our consideration of the spacing distribution for the CUE.

The circular orthogonal and circular symplectic ensembles can be treated analogously [91]. The corresponding probabilities $E(S)$ that an interval $2\pi S/N$ will be empty are represented as determinants [73] similar to (4.11.8), and these determinants are expanded in powers of S. In the limit $N \to \infty$, the following Taylor coefficients arise:

$$E_l^{(OE)} = \sum_{n=1,2,\ldots}^{l \geq 2n^2 - n} \left(\frac{\pi}{2}\right)^{l-n} \frac{(-1)^{(l+n)/2}}{n!} \sum_{l_1 \ldots l_n}^{1,2,\ldots} \delta\left(\sum_{i=1}^{n} l_i, \frac{l-n}{2}\right)$$

$$\times \sum_{t_1}^{0 \ldots l_1} \cdots \sum_{t_n}^{0 \ldots l_n} \det\left(\frac{1}{2l_i - 2t_i + 2t_j + 1}\right)$$

$$\times \prod_{k}^{1 \ldots n} \left[\binom{2l_k}{2t_k} \frac{1}{(2l_k + 1)!}\right] \qquad (4.11.28)$$

for the orthogonal ensemble, and

$$E_l^{(SE)} = \sum_{n=1,2,\ldots}^{l \geq 2n^2 - n} \pi^{l-n} \frac{(-1)^{(l+n)/2}}{n!} \sum_{l_1 \ldots l_n}^{1,2,\ldots} \delta\left(\sum_{i=1}^{n} l_i, \frac{l-n}{2}\right)$$

$$\times \sum_{t_1}^{0 \ldots l_1} \cdots \sum_{t_n}^{0 \ldots l_n} \det\left(\frac{1}{2l_i - 2t_i + 2t_j + 1}\right)$$

$$\times \prod_{k}^{1 \ldots n} \left[\binom{2l_k + 1}{2t_k} \frac{1}{(2l_k + 1)!}\left(1 - \frac{4t_k}{2l_k + 1}\right)\right] \qquad (4.11.29)$$

for the symplectic ensemble. Here, as in (4.11.21), the remaining determinants are $n \times n$. It may be worth noting that the restriction on n in (4.11.28, 29), $l \geq 2n^2 - n$, is even more stringent than was $l \geq n^2$ in (4.11.21): The range $1 \leq n \leq 4$ suffices to determine the E_l of the orthogonal and symplectic ensembles up to order $l = 44$; with the help of (4.11.9), the first 42 Taylor coefficients of the spacing distribution are thus accessible; they have been tabulated in Ref. [91]. The interested reader may also consult the reference for Padé interpolations between the small-S behavior inherent in the above Taylor expansions and the asymptotics at large S [92].

In concluding this section, it is appropriate to point out that, for most practical purposes, the Wigner distributions (4.4.2) are quite adequate. The rms deviations within a pair of competing distributions turn out to be

$$\left\{\int_0^\infty dS\,[P(S) - P_{\text{Wigner}}(S)]^2\right\}^{1/2} = \begin{cases} 1.6 \times 10^{-4} & \text{OE} \\ 0.4 \times 10^{-4} & \text{UE} \\ 1.5 \times 10^{-5} & \text{SE} . \end{cases} \tag{4.11.30}$$

The differences $P(S) - P_{\text{Wigner}}(S)$ for the three ensembles are plotted in Fig. 4.2. Spacing histograms drawn from experimentally or numerically determined (quasi) energy spectra are notoriously too rugged to allow for a distinction as small as that indicated in (4.11.30) or Fig. 4.2. However, the kicked top represented in Fig. 4.2b does provide evidence for the inaccuracy of the Wigner surmise.

Incidentally, the most urgent need for precise values of the spacing distribution arises in a context whose connection to quantum chaos has been a painfully open problem for quite a while. The nonreal zeros of Riemann's zeta function all have, according to Riemann's own hypothesis [93], one and the same real part 1/2. While the hypothesis remains unproven to this day [94], many millions of zeros have been computed numerically and do abide by the claim. The imaginary parts of the known zeros have a spacing distribution in fantastic agreement with that of the GUE [95, 96]. The GUE statistics of the Riemann zeros suggests that they might be related to the energy spectrum of some dynamic system with a fully chaotic classical limit. Search for a Hamiltonian is ongoing [97].

4.12 Determinants as Gaussian Integrals over Anticommuting Variables

Before proceeding with random matrices, we had better equip ourselves with a tool that will greatly facilitate certain types of averages. I aim at representing $N \times N$ determinants as Gaussian integrals over anticommuting variables [98–101],

$$\det A = \int \left(\prod_j^{1 \ldots N} d\eta_j^* d\eta_j\right) \exp\left(-\sum_{ik}^{1 \ldots N} \eta_i^* A_{ik} \eta_k\right) . \tag{4.12.1}$$

To explain this integral representation, the anticommuting (Grassmann) variables η_i and their conjugates η_i^* must be introduced through their algebra

$$\eta_i\,\eta_k^* + \eta_k^*\eta_i = 0$$

$$\eta_i\,\eta_k + \eta_k\eta_i = 0 \tag{4.12.2}$$

$$\eta_i^*\,\eta_k^* + \eta_k^*\eta_i^* = 0 .$$

It follows that $\eta_i^2 = \eta_i^{*2} = 0$. Thus, the most general function of a single such variable is linear, $F(\eta_1) = a + b\eta_1$ with a and b ordinary commuting numbers. Similarly, the most general function of two Grassmann variables is bilinear,

$$F(\eta_1, \eta_2) = a + b\eta_1 + c\eta_2 + \eta_1\eta_2 d . \tag{4.12.3}$$

Differentiation with respect to a variable η is defined analogously to differentiation with respect to usual commuting variables, $(\partial/\partial\eta_1)F(\eta_1) = b$, or

$$\frac{\partial}{\partial\eta_1}F(\eta_1,\eta_2) = b + \eta_2 d\ , \quad \frac{\partial}{\partial\eta_2}F(\eta_1,\eta_2) = c - \eta_1 d$$
$$\frac{\partial}{\partial\eta_1}\frac{\partial}{\partial\eta_2}F(\eta_1,\eta_2) = -\frac{\partial}{\partial\eta_2}\frac{\partial}{\partial\eta_1}F(\eta_1,\eta_2) = -d\ . \tag{4.12.4}$$

The sign of mixed derivatives is found by arranging the product of η's in F in the order inverse to that of the derivatives; obviously, $\partial/\partial\eta_i$ and $\partial/\partial\eta_j$ anticommute like η_i and η_j and $(\partial/\partial\eta_i)\eta_j = -\eta_j(\partial/\partial\eta_i)$ for $i \neq j$. It also follows that $\partial^2/\partial\eta_i^2 = 0$. When mixed derivatives with respect to η and η^* arise, all of the above rules hold analogously, i.e., $(\partial/\partial\eta_i)\partial/\partial\eta_j^* = -(\partial/\partial\eta_j^*)\partial/\partial\eta_i$.

Like differentiation, integration is defined as a linear operation. In addition, it is required that

$$\int d\eta_i\eta_i = \int d\eta_i^*\eta_i^* = 1\ , \quad \int d\eta_1 1 = \int d\eta_i^* 1 = 0 \tag{4.12.5}$$

and that the differentials $d\eta_i$ and $d\eta_i^*$ anticommute just like the η_i and η_i^* themselves. Coming back to $N = 2$ as in (4.12.3, 4), from the definition just given,

$$\int d\eta_1 F(\eta_1,\eta_2) = b + \eta_2 d$$
$$\int d\eta_2 F(\eta_1,\eta_2) = c - \eta_1 d \tag{4.12.6}$$
$$\int d\eta_2 d\eta_1 F(\eta_1,\eta_2) = -\int d\eta_1 d\eta_2 F(\eta_1,\eta_2) = d\ .$$

For arbitrary N, we may write the most general function as the series

$$F(\eta_1,\dots,\eta_N) = \sum_{\{m_i=0,1\}} f(m_1,\dots,m_N)\eta_1^{m_1}\eta_2^{m_2}\dots\eta_N^{m_N} \tag{4.12.7}$$

and then

$$\int d\eta_N \dots d\eta_1 F(\eta_1,\dots,\eta_N) = f(1,1,\dots,1)\ . \tag{4.12.8}$$

In short, *differentiation and integration are identical operations for Grassmann variables.* Grassmann integrals enjoy the privilege of not requiring boundaries to become definite.

Now, the integral representation (4.12.1) of a determinant is easily checked to be a consequence of the above definitions. Only the Nth order term of the Taylor expansion of the exponential on the r.h.s. in (4.12.1) can make a non-zero contribution to the $2N$-fold integral,

$$D \equiv \int \left(\prod_j d\eta_j^* d\eta_j\right) \exp\left(-\eta^* A\eta\right) = \int \left(\prod_j^{1\dots N} d\eta_j^* d\eta_j\right) \frac{(-1)^N}{N!}(\eta^* A\eta)^N$$

$$= \int \left(\prod_j^{1\dots N} d\eta_j d\eta_j^*\right) \frac{1}{N!}(\eta^* A\eta)^N\ . \tag{4.12.9}$$

For the integral to collect a non-zero contribution, the $2N$-fold sum in $(\eta^* A \eta)^N$ must be restricted so as to involve precisely all the N different η and their conjugates η^*. There are $(N!)^2$ such additive terms, according to the permutations of the N naturals $1, 2, \ldots, N$ as labels on the N factors η and N factors η^*,

$$
D = \int \left(\prod_{j}^{1 \ldots N} d\eta_j d\eta_j^* \right) \frac{1}{N!} \sum_{\{i\}} \sum_{\{j\}} \eta_{i_1}^* \eta_{j_1} \eta_{i_2}^* \eta_{j_2} \cdots \eta_{i_N}^* \eta_{j_N}
$$
$$
\times A_{i_1 j_1} A_{i_2 j_2} \cdots A_{i_N j_N} . \tag{4.12.10}
$$

Each of the $N!$ permutations $\{i\}$ yields one and the same contribution to D, once the $N!$ permutations $\{j\}$ are summed over. Thus, the factor $1/N!$ may be dropped and only one permutation $\{i\}$ admitted,

$$
D = \int \left(\prod_{j}^{1 \ldots N} d\eta_j d\eta_j^* \right) \sum_{\mathcal{P}} \eta_1^* \eta_{\mathcal{P}1} \eta_2^* \eta_{\mathcal{P}2} \cdots \eta_N^* \eta_{\mathcal{P}N} A_{1, \mathcal{P}1} \cdots A_{N, \mathcal{P}N} .
$$
$$
\tag{4.12.11}
$$

At this point, the $2N$-fold integral may be evaluated with the help of (4.12.5). The result is $+1$ or -1, depending on whether the permutation $\mathcal{P}1, \mathcal{P}2, \ldots \mathcal{P}N$ is even or odd with respect to the natural sequence $1\,2 \ldots N$; I shall denote such a sign factor by $(-1)^{\mathcal{P}}$. The final result,

$$
D = \sum_{\mathcal{P}} (-1)^{\mathcal{P}} A_{1, \mathcal{P}1} A_{2, \mathcal{P}2} \cdots A_{N, \mathcal{P}N} \equiv \det A , \tag{4.12.12}
$$

is in fact the familiar representation for the determinant of the matrix A.

We shall also encounter the so-called Pfaffian of real antisymmetric $2N \times 2N$ matrices $a = -\tilde{a}$ which is partially, i.e., up to the sign, defined by

$$
\mathrm{Pf}\,(a)^2 = \det a . \tag{4.12.13}
$$

A complete definition employs the Gaussian Grassmann integral

$$
\mathrm{Pf}\,(a) = \int d\eta_{2N} d\eta_{2N-1} \cdots d\eta_1 \exp \left(\frac{1}{2} \sum_{ik}^{1 \ldots 2N} \eta_i a_{ik} \eta_k \right) . \tag{4.12.14}
$$

The proof of this integral representation provides an opportunity to reveal the interesting behavior of Grassmann integrals under transformations of the integration variables.

It suffices to consider the homogeneous linear transformation

$$
\eta_i = \sum_{k} L_{ik} \xi_k \tag{4.12.15}
$$

and to invoke (4.12.8) to write the integral over all η's as

$$
\begin{aligned}
f(1, \ldots 1) &= \int d\xi_N \cdots \xi_1 J(\tfrac{\eta}{\xi}) F(\eta_1(\{\xi\}), \ldots, \eta_N(\{\xi\})) \\
&= J \int d^n \xi F(\{\eta(\{\xi\})\}) \\
&= J f(1, \ldots 1) \int d^N \xi \sum_{i_1 \ldots i_N}^{1 \ldots N} L_{1,i_1} L_{2,i_2} \cdots L_{N,i_N} \xi_{i_1} \xi_{i_2} \cdots \xi_{i_N} \\
&= J f(1, \ldots 1) \int d^N \xi \sum_{\mathcal{P}} L_{1,\mathcal{P}1} \cdots L_{N,\mathcal{P}N} \xi_{\mathcal{P}1} \cdots \xi_{\mathcal{P}N} \\
&= J f(1, \ldots 1) \det L \,.
\end{aligned}
\tag{4.12.16}
$$

The foregoing chain of equations should be mostly self-explanatory provided that the reader has cautiously gone through the derivation of the integral representation (4.12.1) of $\det A$ above. Nonetheless, a few remarks may be helpful. First, I have taken the Jacobian J in search out of the integral assuming that it is independent of the ξ's; that assumption will be borne out self-consistently and is of course due to the linearity of the transformation under consideration. Next, only those index configurations $i_1, i_2, \ldots i_N$ can contribute to the integral that are permutations of the naturals $1, 2, \ldots n$ since any Grassmann variable squares to zero. Finally, a sign factor $(-1)^{\mathcal{P}}$ arises from the identity $\xi_{\mathcal{P}1} \xi_{\mathcal{P}2} \cdots \xi_{\mathcal{P}N} = (-1)^{\mathcal{P}} \xi_1 \xi_2 \cdots \xi_N$, whereupon the definition (4.12.12) of a determinant is recognized. Comparing the first and last member of the chain, we arrive at the Jacobian alias Berezinian

$$
J(\tfrac{\eta}{\xi}) = \frac{1}{\det L} = \frac{1}{\det(\frac{\partial \eta}{\partial \xi})}
\tag{4.12.17}
$$

which is interesting indeed: While one is used to the Jacobian as the determinant of the derivatives of "old" with respect to "new" commuting variables, here we meet with the inverse of that determinant in the case of anticommuting variables.

Equipped with the Jacobian (4.12.17), now we can prove the integral representation (4.12.14) for a Pfaffian. To that end, we introduce a real orthogonal transformation from the $2N$ η's to $2N$ ξ's and choose the orthogonal $2N \times 2N$ matrix L such that (1) $\det L = 1$ and (2) so as to bring the antisymmetric matrix a to block diagonal form with 2×2 blocks

$$
\begin{pmatrix} 0 & a_i \\ -a_i & 0 \end{pmatrix}, \quad i = 1 \ldots N, \quad a_i = a_i^*
\tag{4.12.18}
$$

along the diagonal. Such an orthogonal transformation can be found (see Problem 4.10), and the $\pm i a_i$ are the (not necessarily different) eigenvalues. Now, the Grassmann integral in (4.12.14) becomes a product of twofold integrals,

$$
\prod_{i=1}^N \int d\xi_2 d\xi_1 \exp(a_i \xi_1 \xi_2) = \prod_{i=1}^N a_i = \sqrt{\det a} \,,
\tag{4.12.19}
$$

and we have reached our goal.

An equivalent definition of the Pfaffian is

$$
\text{Pf}\,(a_{ij}) = \frac{1}{(N)!}\sum_{\mathcal{P}}(-1)^{\mathcal{P}} a_{\mathcal{P}1,\mathcal{P}2} a_{\mathcal{P}3,\mathcal{P}4}\,\cdots\,a_{\mathcal{P}2N-1,\mathcal{P}2N} \tag{4.12.20}
$$

where the summation is over all permutations of $1\,2\,\ldots\,2N$ restricted by $\mathcal{P}1 < \mathcal{P}2, \mathcal{P}3 < \mathcal{P}4, \ldots$. It can be established by expanding the exponential in the original integral representation (4.12.14) and arguing analogously to our procedure from (4.12.9) to (4.12.10); readers with no prior familiarity with Grassmann algebra will want to exercise their new skills by solving Problem 4.11.

The reader is kindly invited to appreciate the difference between the Grassmann integrals presented above and the more familiar Gaussian integrals over commuting variables. The correspondent of (4.12.1) involves N pairs of complex conjugate variables $z_i, z_i^*, i = 1\ldots N$ and runs over all N complex planes; it reads

$$
\int\Big(\prod_{i=1}^{N}\frac{d^2z_i}{\pi}\Big)\exp\Big(-\sum_{ij}z_i^* A_{ij}z_j\Big) = \frac{1}{\det A} \tag{4.12.21}
$$

where the differential volume element is to be interpreted according to $d^2z = d\text{Re}\,z\,d\,\text{Im}\,z = dz^*dz/(2\text{i})$. The real part of the symmetric matrix A must be positive definite for the Gaussian integral to converge. Note that now an inverse determinant is given by a Gaussian integral. On the other hand, an inverse square root of a symmetric determinant can be written as the "real" Gaussian integral

$$
\int_{-\infty}^{+\infty}\Big(\prod_{i=1}^{N}\frac{dx_i}{\sqrt{2\pi}}\Big)\exp\Big(-\frac{1}{2}\sum_{ij}x_i a_{ij}x_j\Big) = \frac{1}{\sqrt{\det a}}\,, \tag{4.12.22}
$$

the counterpart of the Grassmann integral (4.12.14). Now, the matrix a is symmetric, may have even or odd dimension, and is restricted to have a positive definite real part to ensure convergence of the integral.[5] Note that no convergence precautions are necessary for the previous Gaussian Grassmann integrals. Simple proofs of the "bosonic" integral representations (4.12.21, 22) involve transformations of the integration variables diagonalizing the respective matrices A and a, just as in the "fermionic" cases.

The foregoing integral representations can be employed profitably when the matrices involved have random elements and when powers of the determinants with exponents ±1 or $\pm1/2$ are to be averaged. Several such cases will arise in the sequel.

The much in vogue art of writing ratios of determinants as mixed Gaussian integrals over both commuting ("bosonic") and anticommuting ("fermionic") variables, mostly referred to as "The Supersymmetry Method"[sic!], will be discussed in Chap. 10.

---

[5] Analytic continuation to imaginary a_{ij} yields the Fresnel integrals to be met in Chap. 8.

4.13 Two-Point Correlations of the Level Density

4.13.1 Two-Point Correlator and Form Factor

We first encountered the two-point correlation function of the level density in Sect. 4.10, when dealing with the ergodicity of the mean level density in Gaussian ensembles. Far more important is the use of the two-point function as a measure of spectral fluctuations, since in contrast to the semicircular mean density, these fluctuations can serve as models for the largely universal fluctuations we find in the spectra of dynamic systems. In view of such applications, the two-point function is worth special consideration.

The cluster functions (4.9.15) of the level density come out the same for the Gaussian and circular ensembles; they are, however, most easily calculated for the latter and that task will be taken up here.[6] We shall take advantage of a slightly rewritten version of the density of eigenphases: Representing the 2π-periodic delta function by a Fourier series,

$$\varrho(\phi) = \frac{1}{2\pi N} \sum_{i=1}^{N} \sum_{n=-\infty}^{\infty} e^{in(\phi-\phi_i)} = \frac{1}{2\pi N} \sum_{n=-\infty}^{\infty} t_n e^{in\phi}, \qquad (4.13.1)$$

where the nth Fourier coefficient is determined by

$$t_n = \sum_{i=1}^{N} e^{-in\phi_i} = \begin{cases} \operatorname{Tr} F^n & \text{for} \quad \beta = 0, 1, 2 \\ \frac{1}{2}\operatorname{Tr} F^n & \text{for} \quad \beta = 4 \end{cases}, \qquad (4.13.2)$$

i.e., the trace of the nth power of the unitary matrix F whose spectrum is at issue. For short, I shall refer to t_n as the nth trace. The factor $1/2$ in $t_n = \frac{1}{2}\operatorname{Tr} F^n$ for the symplectic case accounts for Kramers' degeneracy; it is indeed customary for that case to count every one of the doubly degenerate levels just once in the density $\varrho(\phi)$ and its Fourier coefficient t_n and to use the said factor $1/2$. Incidentally, inasmuch as unitary matrices from the circular ensembles model Floquet matrices, the exponent n may be associated with a dimensionless time counting the number of iterations of the "quantum map" F.

Now, the density–density correlation function (4.10.6) reads

$$\begin{aligned} \overline{\varrho(\phi)\varrho(\phi')} &= (\frac{1}{2\pi N})^2 \sum_n \overline{|t_n|^2} e^{in(\phi-\phi')} \\ &= (\frac{1}{2\pi N})^2 \sum_n \overline{|t_n|^2} \exp i\frac{2\pi}{N}n(e-e'); \end{aligned} \qquad (4.13.3)$$

here I have accounted for the homogeneity of the circular ensembles as $\overline{t_n t_{n'}} = \delta_{n,-n'}\overline{|t_n|^2}$ and rescaled the phase variable by referring to the mean spacing, $e = (N/2\pi)\phi$, analogously to (4.9.13). The mean squared-in-modulus traces $\overline{|t_n|^2}$, the so-called form factor, are now revealed as the Fourier coefficients of the density–density correlation function. Obviously, that form factor is even in n and obeys

---

[6] The two-point function of the Gaussian unitary ensemble will be treated in Chap. 10.

$\overline{|t_0|^2} = N^2$. Proceeding toward Dyson's two-point cluster function $Y(e)$, according to (4.9.14), we subtract the squared mean density $(1/2\pi)^2$ to get the "connected two-point correlator", rescale by dividing by the squared mean density, subtract the delta peak arising at zero spacing, change the overall sign, take a short breath, and then rejoice in

$$Y(e) = \delta(e) - \frac{2}{N^2} \sum_{n=1}^{\infty} \overline{|t_n|^2} \cos\left(\frac{2\pi n}{N} e\right). \tag{4.13.4}$$

It remains to calculate the form factor $\overline{|t_n|^2}$ for the various circular ensembles. The ensemble averages to be performed are easier for the characteristic function

$$\tilde{P}_{nN}^{\beta}(|k|) = \overline{\exp\left[-\frac{i}{2}\sum_i (ke^{-in\phi_i} + k^* e^{+in\phi_i})\right]}$$

$$= \overline{\prod_{i=1}^{N} \exp\left(-i|k|\cos n\phi_i\right)} \tag{4.13.5}$$

than for the form factor itself. The extra convenience gained is due to the multiplicative structure of the quantity to be averaged: Every phase ϕ_i enters through its exclusive exponential. As anticipated in writing out (4.13.5), the homogeneity of the joint density of eigenphases of the circular ensembles allows $\tilde{P}_{nN}^{\beta}$ to depend on the pair of complex conjugate variables k, k^* only through the modulus $|k|$. Thus, no error is incurred by assuming $k = k^* = |k|$ and simply writing $\tilde{P}_{nN}^{\beta}(k)$ for the characteristic function. The form factor is eventually obtained from the second derivative of $\tilde{P}_{nN}^{\beta}(k)$ at $k = 0$, i.e., the special case $m = 1$ of

$$\overline{|t_n|^{2m}} = (-4)^m \binom{2m}{m}^{-1} \frac{d^{2m}}{dk^{2m}} \tilde{P}_{nN}^{\beta}(k)\bigg|_{k=0}. \tag{4.13.6}$$

4.13.2 Form Factor for the Poissonian Ensemble

Upon invoking the Poissonian joint distribution (4.10.7), we immediately get the characteristic function in terms of the Bessel function $J_0(k)$,

$$\tilde{P}_{nN}^0(k) = J_0(k)^N \tag{4.13.7}$$

and thus the form factor, connected correlator, and cluster function as

$$\begin{aligned} \overline{|t_n|^2} &= N \quad \text{for} \quad n > 0, \quad \overline{|t_0|^2} = N^2 \\ S_{\text{PE}}(e) &= \left(\frac{1}{2\pi}\right)^2 \delta(e) \\ Y_{\text{PE}}(e) &= 0. \end{aligned} \tag{4.13.8}$$

Needless to say, these simple results signal the complete absence of correlations between levels; the trace $t_n = \sum_{i=1}^{N} e^{-i\phi_i}$ results from N steps of a random walk in a complex plane with each step of unit length but arbitrary direction.

4.13.3 Form Factor for the CUE

Now, I employ the joint eigenphase density in its determinantal form (4.11.3),

$$
P_{\mathrm{CUE}}(\{\phi\}) = \frac{1}{(2\pi)^N N!} \det\left(\sum_{i=1}^{N} e^{-\mathrm{i}\phi_i(l-m)}\right)_{l,m=1,\dots,N} \tag{4.13.9}
$$

together with the lemma (4.11.4) for the CUE mean of a symmetric function of all N eigenphases. Then, the desired average reads

$$
\tilde{P}_{nN}^2(k) = \int \frac{d^N\phi}{(2\pi)^N} e^{-\mathrm{i}k\sum_i \cos n\phi_i} \det\left(e^{-\mathrm{i}\phi_l(l-m)}\right). \tag{4.13.10}
$$

When the factor $\exp\left(-\mathrm{i}k\cos(n\phi_l)\right)$ is pulled into the lth row of the determinant, the lth phase appears only in that row. The average can thus be done row by row, whereupon the characteristic function takes the form of a Toeplitz determinant,

$$
\begin{aligned}
\tilde{P}_{nN}^2(k) &= \det\left[\int_0^{2\pi} \frac{d\phi}{2\pi} e^{-\mathrm{i}(l-m)\phi} e^{-\mathrm{i}k\cos(n\phi)}\right]\\
&= \det\left[\sum_{s=-\infty}^{s=+\infty} J_{|s|}(k)(-\mathrm{i})^{|s|}\delta_{l,m-ns}\right]. \tag{4.13.11}
\end{aligned}
$$

Readers wanting to see how the Bessel functions $J_{|s|}(k)$ with nonnegative integral orders $|s| = 1,2,\dots$ arise are invited to change the integration variable in the middle member of the foregoing chain of equations as $n\phi \to \phi$ and to split the resulting integral into n additive pieces each of which again runs between the limits 0 and 2π and equals any other one up to a phase factor. The sum of the n phase factors equals n if $(l-m)/n$ is an integer and vanishes otherwise.

Due to $J_{|s|}(0) = \delta_{s,0}$, the characteristic function reflects proper normalization, $\tilde{P}_n(0) = 1$. For $n \geq N$, the determinant in (4.13.11) becomes diagonal and yields

$$
\tilde{P}_{nN}^2(k) = J_0(k)^N \quad\Longrightarrow\quad \overline{|t_n|^2} = N \quad \text{for} \quad N \leq n. \tag{4.13.12}
$$

To find the form factor in the interval $1 \leq n \leq N$, we expand in powers of k,

$$
\begin{aligned}
\det M(k) &= \exp \mathrm{Tr}\, \ln M(k) \tag{4.13.13}\\
&= \exp \mathrm{Tr}\, \ln\left(1 + kM'(0) + \frac{1}{2}k^2 M''(0) + \mathcal{O}(k^3)\right)\\
&= \exp\left(k\mathrm{Tr}M'(0) + \frac{1}{2}k^2\mathrm{Tr}\left(M''(0) - M'(0)^2\right) + \mathcal{O}(k^3)\right).
\end{aligned}
$$

The derivatives of the Bessel functions at zero argument, $J'_{|s|}(0) = \frac{1}{2}\delta_{|s|,1}$ and $J''_{|s|}(0) = \frac{1}{4}\delta_{|s|,2} - \frac{1}{2}\delta_{s,0}$ yield $\mathrm{Tr}M'(0) = 0$, $\mathrm{Tr}M'(0)^2 = -\frac{1}{2}(N-n)$, and $\mathrm{Tr}M''(0) = -\frac{N}{2}$. The CUE form factor thus comes out as

$$
\overline{|t_n|^2} = \overline{|t_{-n}|^2} = \begin{cases} N^2\delta_{n,0} + n & \text{for } 0 \leq n \leq N\\ N & \text{for } N \leq n. \end{cases} \tag{4.13.14}
$$

Using this in the general expression (4.13.4) for the cluster function, we obtain

$$Y_{\text{CUE}}(e) = \frac{1}{N} + \frac{2}{N^2} \sum_{n=1}^{N} (N - n) \cos\left(\frac{2\pi n}{N} e\right). \tag{4.13.15}$$

Finally, on replacing the sum by an integral over $\tau = n/N$ in the limit $N \to \infty$, we get the result anticipated in (4.9.15),[7]

$$\frac{1}{N} \overline{|t_n|^2} \to K_{\text{CUE}}(\tau) = \begin{cases} \delta(\tau) + |\tau| & \text{for } |\tau| \leq 1 \\ 1 & \text{for } |\tau| > 1, \end{cases}$$

$$Y_{\text{CUE}}(e) = s(e)^2 = \left(\frac{\sin \pi e}{\pi e}\right)^2 = \frac{1 - \cos 2\pi e}{2\pi^2 e^2}. \tag{4.13.16}$$

Both the form factor $K_{\text{CUE}}(\tau)$ and the cluster function $Y_{\text{CUE}}(e)$ are depicted in Fig. 4.4, together with their variants for the orthogonal and symplectic classes.

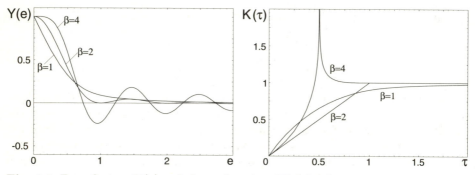

Fig. 4.4. Form factors $K(\tau)$ and cluster functions $Y(e)$ for the unitary, orthogonal, and symplectic ensembles according to (4.13.16, 29, 36)

The saturation of the form factor at $\overline{|t_n|^2} = N$ for $n \geq N$ is a peculiarity of the CUE. Easy to appreciate intuitively, basically as a consequence of the central limit theorem, is asymptotic saturation as $n \to \infty$, such as will also be encountered below for the COE and CSE. We must realize that the trace $t_n = \sum_{i=1}^{N} e^{-in\phi_i}$ sees the phases $n\phi_i$ only modulo 2π. Accidental close proximity now arises between pairs of phases whose counterparts in the original quasi-energy spectrum (i.e., for $n = 1$) are far apart and thus only weakly correlated, apart from a certain tendency to form a "lattice" with "sites" roughly spaced by $2\pi/N$. The phase factors $e^{-in\phi_i}$ will become ever more weakly correlated as n grows; their sum t_N must asymptotically become the end point of an N-step random walk in the complex plane starting at the origin, with each step of unit length but random direction. The central limit theorem would entail $t_n \to \sqrt{N}$ and irrespectively so of the provenance of the quasi-energies. In particular, this reasoning should (and does) also apply to quasi-energy spectra of dynamical systems. The rigidity of

---

[7] The ($N \to \infty$) limit $K(\tau)$ of the form factor is related to the Fourier transform $\tilde{Y}(\tau)$ of the cluster function $Y(e)$ by $K(\tau) = \delta(\tau) + 1 - \tilde{Y}(\tau)$.

spectra alluded to under the heading "lattice" does not stand in the way of the central limit theorem here but rather accentuates the onset of applicability only for $n > N$: The "lattice sites" are preferred with an uncertainty ϵ/N where ϵ is a number of order unity; that uncertainty is blown up to one of order 2π, i.e., complete uncertainty, for $n > N$.

The unit slope of the form factor at small "times" τ will be revealed as due to classical ergodicity in Chap. 9.7. It is not counterintuitive that the break between the linear small-time behavior and the large-time saturation takes place at the "Heisenberg" time $\tau_H = 1$ alias $n_H = N$, the time scale on which the discreteness of the quasi-energy spectrum becomes resolvable. After all, $n_H = N$ is the only time scale around in the random-matrix ensembles under study.[8]

In view of Chap. 8, a comment on the cluster function $Y_{\mathrm{CUE}}(e)$ is also indicated. The last member of (4.13.16) is split into a nonoscillatory and an oscillatory term both of which decay as $1/e^2$ as $e \to \infty$. The former is due to the corner of $|\tau|$ in the form factor at $\tau = 0$ and the latter to the corner at the Heisenberg time $\tau_H = 1$, as the reader is invited to check by solving Problem 4.13.

4.13.4 Form Factor for the COE

Modifying the foregoing calculation for the circular orthogonal ensemble provides an opportunity to explain the classic method of "integration over alternate variables" [73]. Again, we start from the characteristic function (4.13.5)

$$
\tilde{P}_{nN}^1(k) = (1/\mathcal{N}_1) \int d^N\phi \left[\left(\prod_{k<l} |e^{-i\phi_k} - e^{-i\phi_l}| \right) \prod_i e^{-ik\cos(n\phi_i)} \right]
$$

$$
= N! \, (1/\mathcal{N}_1) \int_{2\pi > \phi_1 > \phi_2 \ldots \phi_N > 0} d\phi_1 \ldots d\phi_N \{\ldots\} \qquad (4.13.17)
$$

where due to symmetry the hypercube of integration $\{0 < \phi_i < 2\pi\}$ could be traded against the hypertriangle $2\pi > \phi_1 > \phi_2 \ldots > \phi_N > 0$, making up by a factor $N!$. Within the hypertriangle, we write the product of moduli as

$$
\prod_{k<l} |e^{-i\phi_k} - e^{-i\phi_l}| = \prod_{k<l} 2\sin\left(\frac{\phi_k - \phi_l}{2}\right)
$$

$$
= i^{\frac{N(N-1)}{2}} \det\left(e^{im\phi_1}, \ldots, e^{im\phi_N} \right) \big|_{\underline{m} = (N-1)/2, \ldots, -(N-1)/2}
$$

$$
= i^{\frac{N(N-1)}{2}} \det\left(e^{im\phi_l} \right) . \qquad (4.13.18)
$$

Three remarks on the foregoing identity are in order. First, the modulus sign could be dropped since the phase ordering makes all sines positive; second, the underlined row index $\underline{m}$ in the determinants runs in unit steps from $(N-1)/2$

---

[8] By dynamic rather than random-matrix arguments, a case could be construed for a possibly competing scale, the "Ehrenfest" time $\tau_E = N^{-1}\ln N$ alias $n_E = \ln N$, i.e., the time scale on which a perturbation of the Floquet operator F becomes noticeable in expectation values of observables, given global chaos in the classical limit.

to $-(N-1)/2$ whereas indices not underlined run as usual through the naturals $1 \ldots N$; finally, the reader may check the last equality by patiently going backward from the product of sine functions to the product of eigenvalue differences and then on to the familiar Vandermonde determinant, without ever invoking a modulus operation to chase phase factors, as I sketch without further comment here:

$$
\begin{aligned}
\prod_{k<l} 2\sin(\frac{\phi_k - \phi_l}{2}) &= \mathrm{i}^{\frac{N(N-1)}{2}} \prod_{k<l}(\mathrm{e}^{-\mathrm{i}(\phi_k - \phi_l)/2} - \mathrm{e}^{\mathrm{i}(\phi_k - \phi_l)/2}) \\
&= \mathrm{i}^{\frac{N(N-1)}{2}} \prod_{k<l} \mathrm{e}^{\mathrm{i}(\phi_k + \phi_l)/2}(\mathrm{e}^{-\mathrm{i}\phi_k} - \mathrm{e}^{-\mathrm{i}\phi_l}) \\
&= \mathrm{i}^{\frac{N(N-1)}{2}} \prod_{i} \mathrm{e}^{\mathrm{i}(N-1)\phi_i/2} \det \mathrm{e}^{-\mathrm{i}(k-1)\phi_l} \\
&= \mathrm{i}^{\frac{N(N-1)}{2}} \det(\mathrm{e}^{\mathrm{i}\underline{m}\phi_1}, \ldots, \mathrm{e}^{\mathrm{i}\underline{m}\phi_N}). \quad (4.13.19)
\end{aligned}
$$

Importing (4.13.18) into the characteristic function (4.13.17) and assuming, for simplicity, even N, I fix the order of integrations to do first those over the phases with even indices and pull the $(2k)$th such integral into the $(2k)$th row of the determinant for $k = 1, \ldots, N/2$; whereas the $(2k)$th integral first appears over the interval $\phi_{2k+1} < \phi_{2k} < \phi_{2k-1}$, we may replace the lower limit with zero, simply by adding the Nth column of the determinant to the $(N-2)$th, the resulting $(N-2)$th to the $(N-4)$th, and so forth and thus obtain

$$
\tilde{P}_{nN}^1(k) = (1/\mathcal{N}_1)N! \int_{\phi_1 > \phi_3 > \ldots > \phi_{N-1}} d\phi_1 d\phi_3 \ldots d\phi_{N-1}\, \mathrm{i}^{N(N-1)/2}
$$

$$
\times \det \left(\mathrm{e}^{\mathrm{i}(\underline{m}\phi_1 - k\cos n\phi_1)}, \int_0^{\phi_1} d\phi\, \mathrm{e}^{\mathrm{i}(\underline{m}\phi - k\cos n\phi)}, \right.
$$

$$
\left. \mathrm{e}^{\mathrm{i}(\underline{m}\phi_3 - k\cos n\phi_3)}, \int_0^{\phi_3} d\phi\, \mathrm{e}^{\mathrm{i}(\underline{m}\phi - k\cos n\phi)}, \ldots \right). \quad (4.13.20)
$$

The remaining integral over $\phi_1, \phi_3, \ldots, \phi_{N-1}$ has a symmetric integrand so that we may extend the integration range to the $(N/2)$ dimensional hypercube of edge length 2π and make up by the factor $1/(N/2)!$.

At this point, it is convenient to represent the $N \times N$ determinant by a Gaussian Grassmann integral,

$$
\tilde{P}_{nN}^1(k) = (-1)^? \frac{N!}{(N/2)!\,\mathcal{N}_1} \int_0^{2\pi} d\phi_1 d\phi_3 \ldots d\phi_{N-1} \quad (4.13.21)
$$

$$
\times \left(\prod_l^{1\ldots N} d\eta_l^* \right) \left(\prod_{\underline{m}}^{\frac{(N-1)}{2} \cdots -\frac{(N-1)}{2}} d\eta_{\underline{m}} \right) \exp\left(-\sum_{l,\underline{m}} \eta_l^* M_{l,\underline{m}} \eta_{\underline{m}} \right),
$$

where M is the matrix whose determinant is under work. Note that I have lumped powers of i and -1 into an undetermined sign factor $(-1)^?$; this maneuver saves labor since the overall sign may change with every step of the following calculation and can be determined rather simply in the final result. As if retracing the proof of the Grassmann integral representation in Sect. 4.12, we replace the exponential by the Nth term of its Taylor expansion and realize that each of the N powers of the quadratic form $\eta^* M \eta$ must provide a different η^* and that there are $N!$ different ways of so drawing the product $\eta_1 \eta_2 \ldots \eta_N$ all of which contribute equally. Thus, the exponential in (4.13.21) is replaced by the N-fold sum over all the $\underline{m}$ of

$$\eta_1^* \eta_{\underline{m}_1} \eta_2^* \eta_{\underline{m}_2} \cdots \eta_N^* \eta_{\underline{m}_N}$$
$$\times e^{i(\underline{m}_1 \phi_1 - k \cos n\phi_1)} \int_0^{\phi_1} d\phi\, e^{i(\underline{m}_2 \phi - k \cos n\phi)}$$
$$\times e^{i(\underline{m}_3 \phi_3 - k \cos n\phi_3)} \int_0^{\phi_3} d\phi\, e^{i(\underline{m}_4 \phi - k \cos n\phi)}$$
$$\cdots\cdots\cdots\cdots \tag{4.13.22}$$
$$\times e^{i(\underline{m}_{N-1} \phi_{N-1} - k \cos n\phi_{N-1})} \int_0^{\phi_{N-1}} d\phi\, e^{i(\underline{m}_N \phi - k \cos n\phi)} .$$

When the N-fold integral over the η^*'s is now done, it leaves no trace outside the sign factor $(-1)^?$. To prepare for the remaining integral over the η's, we must take a closer look at the pairs $\eta_{\underline{m}_{2i-1}} \eta_{\underline{m}_{2i}}$ and their accompanying double phase integrals. The latter may be antisymmetrized in the indices $\underline{m}_{2i-1}, \underline{m}_{2i}$ since their symmetric parts do not contribute to the sum $\sum_{\underline{m}, \underline{m}'} \eta_{\underline{m}} \eta_{\underline{m}'} A_{\underline{m}, \underline{m}'}$ with

$$
\begin{aligned}
A_{\underline{m}, \underline{m}'}(k) &= \frac{1}{2\pi i} \int_0^{2\pi} d\phi \int_0^{\phi} d\phi' \left[e^{i(\underline{m}\phi - k \cos n\phi + \underline{m}'\phi' - k \cos n\phi')} \right]_{\text{antisym}} \\
&= \frac{1}{4\pi i} \int_0^{2\pi} d\phi \int_0^{2\pi} d\phi'\, \text{sign}(\phi - \phi') e^{i(\underline{m}\phi - k \cos n\phi + \underline{m}'\phi' - k \cos n\phi')} \\
&= \sum_{s, s'} J_{|s|}(k) J_{|s'|}(k) \frac{(-i)^{|s| + |s'|}}{\underline{m} + ns} \delta_{\underline{m}, -\underline{m}' - n(s+s')} \tag{4.13.23}
\end{aligned}
$$

manifestly antisymmetric. The Bessel functions in $A_{\underline{m}, \underline{m}'}$ arise when the phase integral is done after expanding the integrand in powers of k. Now, the characteristic function reads

$$\tilde{P}_{nN}^1(k) = (-1)^? \frac{N! (2\pi)^{\frac{N}{2}}}{(N/2)! \mathcal{N}_1} \int \left(\prod_{\underline{m}} d\eta_{\underline{m}} \right) \sum_{\{\underline{m}\}} \tag{4.13.24}$$
$$\times \left(\eta_{\underline{m}_1} \eta_{\underline{m}_2} A_{\underline{m}_1 \underline{m}_2} \right) \left(\eta_{\underline{m}_3} \eta_{\underline{m}_4} A_{\underline{m}_3 \underline{m}_4} \right) \cdots \left(\eta_{\underline{m}_{N-1}} \eta_{\underline{m}_N} A_{\underline{m}_{N-1} \underline{m}_N} \right) .$$

It is worth recalling the representation of the Pfaffian by a Gaussian Grassmann integral from Sect. 4.12: We are confronting such a beast since the integrand of the foregoing integral may be rewritten as $2^{\frac{N}{2}} \left(\frac{1}{2} \sum_{m, m'} \eta_m A_{mm'}(k) \eta_{m'} \right)^{\frac{N}{2}} = \left(\frac{N}{2} \right)! 2^{\frac{N}{2}} \exp \left(\frac{1}{2} \sum_{m, m'} \eta_m A_{mm'}(k) \eta_{m'} \right)$ such that

$$
\begin{aligned}
\tilde{P}_{nN}^1 &= \left[(-1)^? \frac{N!\,(4\pi)^{\frac{N}{2}}}{\mathcal{N}_1} \right] \int d^N \eta \exp\left[\frac{1}{2} \sum_{m,m'} \eta_m A_{mm'}(k)\eta_{m'} \right] \\
&= (") \operatorname{Pf}(A(k)) \\
&= (") \sqrt{\det A(k)} \\
&= (") \exp\left[\frac{1}{2}\operatorname{Tr}\ln A(k) \right].
\end{aligned}
\tag{4.13.25}
$$

Recalling that the form factor is given by the second derivative of the characteristic function at $k = 0$, according to (4.13.6), we may expand the matrix A in powers of k up to second order,

$$
\tilde{P}_{nN}^1 = \left[(") \sqrt{\det A(0)} \right] \exp \operatorname{Tr}
\tag{4.13.26}
$$

$$
\times \left\{ \frac{1}{2} k A^{-1}(0)A'(0) + \frac{1}{4}k^2 \left[A^{-1}(0)A''(0) - (A^{-1}(0)A'(0))^2 \right] \right\}.
$$

Here, the square bracket must equal unity; with the help of $J_{|s|}(0) = \delta_{s,0}$, we find $\sqrt{\det A(0)} = (-1)^? \, 2^{N/2}/(N-1)!!$; the final sign factor $(-1)^?$ thus turns out as $+1$ and the normalization factor $\mathcal{N}_1$ of the eigenphase density of the COE as

$$
\mathcal{N}_1 = (\frac{N}{2})!\,\pi^{\frac{N}{2}} 2^{2N}.
\tag{4.13.27}
$$

Had it not been for the secret desire to catch that normalization factor, I could of course have left an overall proportionality factor undetermined throughout the above calculation, as I have done with the overall sign factor. At any rate, the main goal of the present subsection is also reached quickly once we use the derivatives of the Bessel functions (given after (4.13.14)) to establish
(i) $\operatorname{Tr}A^{-1}(0)A'(0) = 0$ for $n \neq 0$,
(ii) $\operatorname{Tr}A^{-1}(0)A''(0) = -2N + n\sum_{m=1}^N [m + n - (N+1)/2]^{-1}$,
(iii) $\operatorname{Tr}[A^{-1}(0)A'(0)]^2 = 0$ for $n \geq N$, and
(iv) $\operatorname{Tr}\{ -A^{-1}(0)A''(0) + [A^{-1}(0)A'(0)]^2 \} = 2n - n\sum_{m=1}^n [m + (N-1)/2]^{-1}$ for $1 \leq n \leq N$.
The COE form factor thus comes out as

$$
\overline{|t_n|^2} = \overline{|t_{-n}|^2} = \begin{cases} N^2 & \text{for } n = 0 \\ 2n - n\sum_{m=1}^n \frac{1}{m+(N-1)/2} & \text{for } 1 \leq n \leq N \\ 2N - n\sum_{m=1}^N \frac{1}{m+n-(N+1)/2} & \text{for } N \leq n. \end{cases}
\tag{4.13.28}
$$

After approximating sums by integrals in the limit $N \to \infty$,

$$
\frac{1}{N}\overline{|t_n|^2} \to K_{\mathrm{COE}}(\tau) = \begin{cases} \delta(\tau) + 2|\tau| - |\tau|\ln(1 + 2|\tau|) & \text{for } |\tau| \leq 1 \\ 2 - |\tau|\ln\frac{2|\tau|+1}{2|\tau|-1} & \text{for } |\tau| > 1, \end{cases}
$$

$$
Y_{\mathrm{COE}}(e) = s(e)^2 - J(e)D(e)
\tag{4.13.29}
$$

with $D(e) = s'(e)$ and $J(e)$ an integral of $s(e)$, as defined in (4.9.16). Plots of $K_{\text{COE}}(\tau)$ and $Y_{\text{COE}}(\tau)$ are shown in Fig. 4.4. For future reference, I note the behavior of the cluster function at small and large e,

$$Y_{\text{COE}}(e) = \begin{cases} 1 - \frac{\pi^2}{6}|e| + \frac{\pi^4}{60}|e|^3 - \frac{\pi^4}{135}e^4 + \dots, & |e| \ll 1 \\ \frac{1}{\pi^2 e^2} - \frac{3 + \cos 2\pi e}{2\pi^4 e^4} + \dots, & |e| \gg 1. \end{cases} \tag{4.13.30}$$

The nonoscillatory part of the cluster function arises, as for the CUE, from the corner of $2|\tau|$ in the form factor while the jump of the third derivative of K_{COE} at the Heisenberg time yields the oscillatory part of $Y_{\text{COE}}(e)$; the respective leading terms are displayed in the large-e expansion; see Problem 4.13.

4.13.5 Form Factor for the CSE

Our starting point is the CSE analogue of (4.13.17) for the characteristic function,

$$\tilde{P}^4_{nN}(k) \quad \propto \quad \int_{2\pi > \phi_1 > \phi_2 \dots \phi_N > 0} d\phi_1 \dots d\phi_N \left(\prod_{k<l}^{1\dots N} \left| e^{-i\phi_k} - e^{-i\phi_l} \right|^4 \right)$$
$$\times \prod_i e^{-ik\cos(n\phi_i)}, \tag{4.13.31}$$

which differs from (4.13.17) only in the value of the level repulsion exponent. An overall factor was dropped and will be restored at the end by invoking normalization. (The interested reader may take up the extra labor of evaluating the normalization constant $\mathcal{N}_4$ of the joint density of eigenphases; see Problem (4.12).) Now, it is convenient to extend the N-fold integral to a $(2N)$-fold one,

$$\tilde{P}^4_{nN}(k) \quad \propto \quad \int_{2\pi > \phi_1 > \phi_2 \dots \phi_{2N} > 0} d\phi_1 \dots d\phi_{2N} \left(\prod_{k<l}^{1\dots 2N} \sin \frac{\phi_k - \phi_l}{2} \right)$$
$$\times \prod_{i=1}^{N} e^{-ik\cos(n\phi_i)} \frac{\partial}{\partial \phi_{2i-1}} \delta(\phi_{2i-1} - \phi_{2i} - \epsilon), \tag{4.13.32}$$

where ϵ is to be sent to zero from above. To check the equivalence of the two foregoing expressions, one may integrate by parts with respect to the phases with odd labels in the latter. Only those terms survive in the limit $\epsilon \to 0$ for which the N differentiations have turned precisely those N sines into cosines which are assigned vanishing arguments by the delta functions; the remaining sines then group to quadruples as, symbolically, $s_{13}s_{14}s_{23}s_{24} \to s_{24}^4$. Incidentally, the delta functions in (4.13.31) reflect Kramers' degeneracy. To do the $(2N)$-fold phase integral, we recall the identity (4.13.19) as extended to $2N$ ordered phases and integrate over every second phase to chase the delta functions. Renaming the remaining N integration variables as $\phi_{2k} \to \phi_k$ and letting $\epsilon \to 0$,

$$\tilde{P}^4_{nN} \quad \propto \quad \int_0^{2\pi} d^N\phi \, \det\left(\underline{m} \, e^{im\phi_1}, \, e^{im\phi_1}, \, \underline{m} \, e^{im\phi_2}, \, e^{im\phi_2}, \, \dots \right)$$
$$\times \prod_i e^{-ik\cos(n\phi_i)} \tag{4.13.33}$$

where the underlined row index $\underline{m}$ again goes in unit steps through half-integers, $\underline{m} \leq (2N - 1)/2$. Note that the integration range again is the hypercube.

As in (4.13.20), we encounter a determinant that contains every integration variable in a pair of columns. In fact, the further reasoning precisely parallels the above one for the COE, starting with the Grassmann integral representation of the determinant as in (4.13.21). The antisymmetric matrix A whose Pfaffian is incurred is now $2N \times 2N$ and takes the actually simpler form

$$A_{\underline{m},\underline{m}'}(k) = (\underline{m} - \underline{m}') \sum_{s=-\infty}^{\infty} J_{|s|}(k)(-i)^{|s|}\delta(\underline{m} + \underline{m}' + ns). \tag{4.13.34}$$

Again, we get the form factor from the k-expansion (4.13.26) as

$$\overline{|t_n^{\mathrm{CSE}}|^2} = \overline{|t_{-n}^{\mathrm{CSE}}|^2} = \begin{cases} N^2 & \text{for } n = 0 \\ \frac{1}{2}n + \frac{1}{4}n \sum_{m=1}^{n} \frac{1}{N+\frac{1}{2}-m} & \text{for } 1 \leq n \leq 2N \\ N & \text{for } 2N \leq n. \end{cases} \tag{4.13.35}$$

Finally, by employing $\tau = n/N$ and invoking the limit $N \to \infty$, we arrive at

$$\frac{1}{N}\overline{|t_n^{\mathrm{CSE}}|^2} \to K_{\mathrm{CSE}}(\tau) = \begin{cases} \delta(\tau) + \frac{|\tau|}{2} - \frac{|\tau|}{4}\ln|1 - |\tau|| & \text{for } |\tau| \leq 2 \\ 1 & \text{for } |\tau| > 2, \end{cases}$$

$$Y_{\mathrm{CSE}}(e) = s(2e)^2 - I(2e)D(2e) \tag{4.13.36}$$

with $I(e)$ the integral of $s(e) = (\pi e)^{-1}\sin(\pi e)$ given in (4.9.16). A peculiarity of the CSE is the logarithmic singularity of the form factor at $|\tau| = 1$. Saturation at the plateau unity sets in at $\tau_H = 2$ which time I choose to call the Heisenberg time here; the distinction of both τ_H and $\tau_H/2$ may be seen as due to Kramers' degeneracy. At small and large e, the cluster function reads

$$Y_{\mathrm{CSE}}(e) = \begin{cases} 1 - \frac{(2\pi e)^4}{135} + \dots, & |e| \ll 1 \\ -\frac{\pi}{2}\frac{\cos 2\pi e}{2\pi|e|} + \frac{1+(\pi/2)\sin 2\pi|e|}{(2\pi e)^2} - \frac{3+\cos 4\pi e}{2(2\pi e)^4} \dots, & e| \gg 1. \end{cases} \tag{4.13.37}$$

Plots of $K_{\mathrm{CSE}}(\tau)$ and $Y_{\mathrm{CSE}}(\tau)$ are shown in Fig. 4.4.

When comparing the form factor $K_{\mathrm{CSE}}(\tau)$ or the cluster function $Y_{\mathrm{CSE}}(e)$ with results for dynamic systems, one should not forget about Kramers' degeneracy, i.e., the factor $1/2$ in (4.13.2) for the symplectic case. One must also keep in mind that the time $n/N \to \tau$ is measured in units of half the Heisenberg time here, whereas the unit of time for the other ensembles is the Heisenberg time itself.

Again, the nonoscillatory terms in $Y_{\mathrm{CSE}}(e)$ are contributed by the corner of $\frac{1}{2}|\tau|$ in the form factor; the terms oscillating with frequencies 2π and 4π stem, respectively, from the logarithmic singularity at $\tau = 1$ and the jump of the third derivative of $K_{\mathrm{CS}}(\tau)$ at $\tau = 2$; see Problem 4.13.

4.14 Higher Correlations of the Level Density

4.14.1 Correlation and Cumulant Functions

I normalized the mean density of levels $\overline{\rho(\phi)}$ as a probability density, choosing $\rho(\phi) = N^{-1}\sum_{i=1}^{N}\delta(\phi - \phi_i)$. Common currency is also $R_1(\phi) = N\overline{\rho(\phi)}$, and this quantity actually has a better right to the name mean level density.

We likewise met with several variants of the two-point correlation function which it is well to put into systematic view here. First, there is the density–density correlator $\overline{\rho(\phi)\rho(\phi')}$ and its connected version $S(\phi, \phi') = \overline{\rho(\phi)\rho(\phi')} - \overline{\rho(\phi)}\,\overline{\rho(\phi')}$. By purging the density–density correlator of "self-correlation" terms (coinciding indices) and again changing normalization, one arrives at the second-order correlation function customarily defined as

$$R_2(\phi, \phi') = \sum_{i \neq j}^{1...N} \overline{\delta(\phi - \phi_i)\delta(\phi' - \phi_j)} \tag{4.14.1}$$
$$= N^2\overline{\rho(\phi)\rho(\phi')} - N\delta(\phi - \phi')\overline{\rho(\phi)}\,.$$

Finally, by subtracting the product of two mean densities from R_2, one gets the connected second-order correlation function or cumulant function or cluster function

$$C_2(\phi, \phi') = -R_2(\phi, \phi') + R_1(\phi)R_1(\phi')\,. \tag{4.14.2}$$

Inasmuch as the limit of infinite dimension N is of interest, one refers (quasi) energies to their mean spacing $1/R_1(\phi)$ and introduces the rescaled variable

$$e = R_1(\phi)\phi\,; \tag{4.14.3}$$

to secure meaningful limits for the two-point correlation functions, one proceeds to divide the product of two mean densities. Thus, Dyson's asymptotic two-point cluster function, for instance, is arrived at as

$$Y_2(e, e') = Y(e - e') = \lim_{N\to\infty} C_2(\phi, \phi')[R_1(\phi)R_1(\phi')]^{-1} \tag{4.14.4}$$

with the arguments of C_2, R_1 expressed in terms of the rescaled (quasi)energy e according to (4.14.3).

All of the above formulae hold for energy as well as quasi-energy spectra and the respective Gaussian and circular ensembles of random matrices. If the discussion is restricted to unitary matrices and their circular ensembles, as anticipated by the notation, of course, $R_1(\phi) = \frac{N}{2\pi}$, $e = \frac{N}{2\pi}\phi$ etc.; then, Dyson's asymptotic two-point cluster function results from the original second-order cumulant function as $C_2(e\frac{2\pi}{N}, e'\frac{2\pi}{N})(\frac{2\pi}{N})^2 \to Y(e - e')$.

Having sorted the various one- and two-point functions, we are prepared to face their higher order generalizations. Of importance will only be those not containing self-correlation terms. The nth order correlation function generalizing

R_2 may be obtained by integrating the joint probability density P_N of all N levels over $N - n$ variables,

$$R_n(\phi^1, \ldots, \phi^n) = \overset{\neq}{\sum_{i_1 \ldots i_n}} \overline{\delta(\phi^1 - \phi_{i_1}) \ldots \delta(\phi^n - \phi_{i_n})} \qquad (4.14.5)$$

$$= \frac{N!}{(N-n)!} \int_0^{2\pi} d\phi_{n+1} \ldots \int_0^{2\pi} d\phi_N \, P_N(\phi^1, \ldots, \phi^n, \phi_{n+1}, \ldots, \phi_N) \,.$$

I shall not write out the general expression relating correlation functions to their connected versions, the cumulant or cluster functions[73], since we shall need only those up to order four; these read[9]

$$\begin{aligned}
R_1(\phi) &= C_1(\phi), & (4.14.6) \\
R_2(\phi^1, \phi^2) &= -C_2(\phi^1, \phi^2) + C_1(\phi^1)C_1(\phi^2), \\
R_3(\phi^1, \phi^2, \phi^3) &= C_3(\phi^1, \phi^2, \phi^3) + C_1(\phi^1)C_1(\phi^2)C_1(\phi^3) \\
&\quad - \left(C_2(\phi^1, \phi^2)C_1(\phi^3) + \text{two permutations}\right) \\
R_4(\phi^1, \phi^2, \phi^3, \phi^4) &= -C_4(\phi^1, \phi^2, \phi^3, \phi^4) \\
&\quad + \left(C_1(\phi^1)C_3(\phi^2, \phi^3, \phi^4) + \text{three permutations}\right) \\
&\quad + \left(C_2(\phi^1, \phi^2)C_2(\phi^3, \phi^4) + \text{two permutations}\right) \\
&\quad - \left(C_2(\phi^1, \phi^2)C_1(\phi^3)C_1(\phi^4) + \text{five permutations}\right) \\
&\quad + C_1(\phi^1)C_1(\phi^2)C_1(\phi^3)C_1(\phi^4) \,.
\end{aligned}$$

Asymptotic cumulants are obtained by rescaling in the fashion of (4.14.3,4),

$$\begin{aligned}
Y_n(e^1, \ldots, e^n) &= \lim_{N \to \infty} \frac{C_n(\phi^1, \ldots, \phi^n)}{R_1(\phi^1) \ldots R_n(\phi^n)} & (4.14.7) \\
&= \lim_{N \to \infty} C_n(e^1 \frac{2\pi}{N}, \ldots, e^n \frac{2\pi}{N})(\frac{2\pi}{N})^n \,.
\end{aligned}$$

The first in the previous chain of equations holds for the Gaussian and circular ensembles, whereas the second is specialized to the circular case. At any rate, Y_n depends on the n variables e^i only through $n - 1$ independent differences.

The cluster functions $C_n(\phi^1, \ldots, \phi^n)$ are known [73] for the classic ensembles; needless to say, those with $n > 1$ vanish for the Poissonian ensembles thus signaling the absence of any correlations between levels. I shall not derive the C_n here but urge the reader to consult Chaps. 5 and 10 of Mehta's book [73] and enjoy the unsurpassable beauty of the orthogonal–polynomial method used there. For the CUE and the GUE, the asymptotic versions Y_n read

$$Y_n(e^1, \ldots, e^n) = \sum s(e_{12})s(e_{34}) \ldots s(e_{n1}) \qquad (4.14.8)$$

with $e_{ij} = |e^i - e^j|$; the sum runs over the $(n-1)!$ distinct cyclic permutations of the indices $(1, 2, \ldots, n)$, and $s(e)$ is the familiar $s(e) = (\pi e)^{-1} \sin(\pi e)$. In

---

[9] Like most authors in random-matrix theory, Mehta [73] included, I keep here to a sign tradition differing from the otherwise more widespread one by the factor $(-1)^{n-1}$ for the cumulant C_n; the Subsect. 4.1.4.3 will offer more about cumulants.

particular, $Y_2(e^1, e^2) = Y(e_{12}) = s(e_{12})s(e_{21}) = s(e_{12})^2$, and $Y_3(e^1, e^2, e^3) = s(e_{12})s(e_{23})s(e_{31}) + s(e_{13})s(e_{32})s(e_{21}) = 2s(e_{12})s(e_{23})s(e_{31})$.

A conclusion of relevance for what follows is that $(\frac{2\pi}{N})^n R_n(\frac{N}{2\pi}e^1, \ldots, \frac{N}{2\pi}e^n)$ remains finite in the limit $N \to \infty$ when the $\{e^i\}$ are fixed.

4.14.2 Four-Point Correlator and Ergodicity of the Two-Point Correlator

Taking up Pandey's thread of Sect. 4.9, I propose here to show in which sense every large random unitary matrix can be expected to display universal spectral fluctuations. For the sake of brevity, I shall confine myself to general unitary matrices drawn from the CUE and prove the ergodicity of the integral $\int_0^e de' Y(e')$ of the two-point cluster function $Y(e)$ which is the CUE average of the quantity

$$\langle T(e, \phi) \rangle = e - \sum_{i \neq j}^{1 \ldots N} (\frac{2\pi}{N})^2 \int_0^e de' \int_\phi^{\phi + \Delta\phi} \frac{d\phi'}{\Delta\phi} \delta(\phi' - \phi_i)\delta(\phi' + \frac{2\pi}{N}e' - \phi_j)$$

(4.14.9)

in the limit $N \to \infty$. The ϕ-integral amounts to a local spectral average of which the angular brackets around the quantity T are meant as a reminder. The CUE average of the product of the two delta functions in (4.14.9) is independent of ϕ, whereupon the ϕ-integral becomes trivial and $\lim_{N \to \infty} \langle T(e) \rangle = \int_0^e de' Y(e')$ results. The quantity $\langle T(e) \rangle$ is well suited for studying the ergodicity of the two-point cluster function since the two integrals over ϕ and e' smooth the product of delta functions such that one deals with a well-behaved object, even before any ensemble average. The integrated two-point cluster function in question is proven ergodic by showing that the ensemble variance of $\langle T \rangle$ vanishes as the matrix size N and subsequently the mean number of levels $\Delta N = \Delta\phi N/(2\pi)$ go to infinity. That variance reads

$$\text{Var}\langle T(\phi, e) \rangle = \sum_{i \neq j} \sum_{k \neq l} \left(\frac{2\pi}{N}\right)^4 \iint_0^e de' de'' \iint_\phi^{\phi + \Delta\phi} \frac{d\phi'}{\Delta\phi} \frac{d\phi''}{\Delta\phi}$$

(4.14.10)

$$\times \left\{ \delta(\phi' - \phi_i)\delta(\phi' + \frac{2\pi}{N}e' - \phi_j)\delta(\phi'' - \phi_k)\delta(\phi'' + \frac{2\pi}{N}e'' - \phi_l) \right.$$

$$\left. - \delta(\phi' - \phi_i)\delta(\phi' + \frac{2\pi}{N}e - \phi_j)\,\delta(\phi'' - \phi_k)\delta(\phi'' + \frac{2\pi}{N}e'' - \phi_l) \right\}.$$

The sum over double pairs $i \neq j$, $k \neq l$ in the foregoing variance has contributions from genuine quadruplets of levels, as well as from triplets and doublets, due to partial coincidences of indices,

$$\sum_{i \neq j} \sum_{k \neq l} = \sum_{ijkl}^{\neq} + \sum_{i=k,j,l}^{\neq} + \sum_{i=l,j,k}^{\neq} + \sum_{j=k,i,l}^{\neq} + \sum_{j=l,i,k}^{\neq} + \sum_{i=k,j=l}^{\neq} + \sum_{i=l,j=k}^{\neq},$$

(4.14.11)

where the summands are as in (4.14.10). As a first step, I shall show now that the genuine triplets and doublets sum to a contribution of order $1/\Delta N$ and can be dropped. The first of the two genuine-pair sums reads

$$
\sum_{i=k,j=l}^{\neq}(\ldots) = \sum_{i\neq j}\left(\frac{2\pi}{N}\right)^4 \iint_0^e de'de'' \iint_\phi^{\phi+\Delta\phi}\frac{d\phi'\,d\phi''}{\Delta\phi\,\Delta\phi} \tag{4.14.12}
$$

$$
\times\Big\{\delta(\phi'-\phi'')\frac{N}{2\pi}\delta(e'-e'')\overline{\delta(\phi'-\phi_i)\delta(\phi'+\frac{2\pi}{N}e'-\phi_j)}
$$

$$
-\overline{\delta(\phi'-\phi_i)\delta(\phi'+\frac{2\pi}{N}e'-\phi_j)\,\delta(\phi''-\phi_i)\delta(\phi''+\frac{2\pi}{N}e''-\phi_l)}\Big\}.
$$

Now, the definition of the second-order correlation function $R_2(\phi,\phi')$ comes in handy, as well as its property $R_2(\phi,\phi') = R_2(0,\phi-\phi')$ following from the homogeneity of the CUE. The sum in work simplifies to

$$
\sum_{i=k,j=l}^{\neq}(\ldots) = \left(\frac{2\pi}{N}\right)^4 \tag{4.14.13}
$$

$$
\times\left\{\int_0^e de'\,\frac{N}{2\pi\Delta\phi}R_2(0,\frac{2\pi}{N}e') - \frac{1}{N(N-1)}\left[\int_0^e de'R_2(0,\frac{2\pi}{N}e')\right]^2\right\}.
$$

Recalling that $(\frac{2\pi}{N})^2 R_2(\frac{2\pi}{N}e)$ has the finite limit $1-Y(e)$ as $N\to\infty$ and $\Delta N = \Delta\phi N/2\pi$, we immediately see that after the limit $N\to\infty$, only a term $\propto 1/\Delta N$ remains which vanishes as $\Delta N\to\infty$. The second genuine-pair sum in (4.14.11), similarly checked, vanishes in the same sense.

The first of the four genuine-triplet sums in (4.14.11) can be expressed in terms of the correlation functions R_2 and R_3 as

$$
\sum_{i=k,j,l}^{\neq}(\ldots) = \left(\frac{2\pi}{N}\right)^4 \tag{4.14.14}
$$

$$
\times\left\{\frac{1}{\Delta\phi}\iint_0^e de'de''R_3(0,\frac{2\pi}{N}e',\frac{2\pi}{N}e'') - \frac{N-2}{N(N-1)}\left[\int_0^e de'R_2(0,\frac{2\pi}{N}e')\right]^2\right\}.
$$

The finiteness of $(\frac{2\pi}{N})^n R_n(\frac{N}{2\pi}e^1,\ldots,\frac{N}{2\pi}e^n)$ again leaves an overall weight $1/\Delta N$ such that this, as well as the other three genuine-triple sums, vanishes in the ordered double limit announced.

The remaining sum over genuine quadruplets cannot be discarded so simply; it can be expressed in terms of R_4 and R_2 such that

$$
\mathrm{Var}\langle T(\phi,e)\rangle = \left(\frac{2\pi}{N}\right)^4 \iint_0^e de'de'' \iint_\phi^{\phi+\Delta\phi}\frac{d\phi'\,d\phi''}{\Delta\phi\,\Delta\phi} \tag{4.14.15}
$$

$$
\times\Big\{R_4(0,\frac{2\pi}{N}e',\phi''-\phi',\phi''-\phi'+\frac{2\pi}{N}e'')
$$

$$
-\frac{(N-2)(N-3)}{N(N-1)}R_2(0,\frac{2\pi}{N}e')R_2(0,\frac{2\pi}{N}e'') + \mathcal{O}(\frac{1}{N})\Big\}.
$$

Here the fraction $\frac{(N-2)(N-3)}{N(N-1)}$ may be replaced by unity, imagining the irrelevant $\mathcal{O}(\frac{1}{N})$ term suitably modified. Of the two phase integrals over ϕ' and ϕ'', one may be done right away since the integrand depends only on the difference $\phi' - \phi''$; introducing $\phi'' - \phi' = \frac{2\pi}{N}\epsilon$ and $\Delta N = \Delta\phi N/2\pi$,

$$
\text{Var}\langle T(\phi, e)\rangle = \left(\frac{2\pi}{N}\right)^4 \iint_0^e de' de'' \frac{1}{(\Delta N)^2} \int_0^{\Delta N} d\epsilon \, (\Delta N - \epsilon) \tag{4.14.16}
$$
$$
\times \left\{ R_4\left(0, \frac{2\pi}{N}e', \frac{2\pi}{N}\epsilon, \frac{2\pi}{N}\epsilon + \frac{2\pi}{N}e''\right) - R_2\left(0, \frac{2\pi}{N}e'\right)R_2\left(0, \frac{2\pi}{N}e''\right) + \mathcal{O}(\frac{1}{N}) \right.
$$
$$
\left. + \text{same with } \epsilon \to -\epsilon \right\}.
$$

At this point, the limit $N \to \infty$ with fixed ΔN can be taken, whereupon we must turn to studying the behavior of the remaining triple integral for large ΔN. Expressing all correlation functions in terms of asymptotic cumulant functions according to (4.14.6, 7), some further terms cancel, and we are left with

$$
\text{Var}\langle T(\phi, e)\rangle \xrightarrow{N \to \infty} \iint_0^e de' de'' \frac{1}{(\Delta N)^2} \int_0^{\Delta N} d\epsilon \, (\Delta N - \epsilon) \tag{4.14.17}
$$
$$
\times \left\{ -Y_4(0, e', \epsilon, \epsilon + e'') \right.
$$
$$
+ Y_3(e', \epsilon, \epsilon + e'') + Y_3(0, \epsilon, \epsilon + e'') + Y_3(0, e', \epsilon + e'') + Y_3(0, e', \epsilon)
$$
$$
+ Y_2(0, \epsilon)Y_2(0, \epsilon + e'' - e') + Y_2(0, \epsilon + e'')Y_2(0, \epsilon - e')
$$
$$
- Y_2(0, \epsilon) - Y_2(0, \epsilon + e'') - Y_2(0, \epsilon - e') - Y_2(0, \epsilon + e'' - e')
$$
$$
\left. + \text{same with } \epsilon \to -\epsilon \right\}.
$$

Upon inserting the CUE result (4.14.8) for the asymptotic cluster functions, we conclude from the symmetry of the integrand under exchange of e' and e'' and the homogeneity of the cluster functions that the terms labelled "same with $\epsilon \to -\epsilon$" equal the previous ones in the curly brackets. Moreover, by patiently checking term by term, we find at least two factors s with their arguments containing ϵ in each summand. It follows that the whole integrand falls off with growing ϵ as $(\Delta N - \epsilon)/\epsilon^2$ such that the integral over ϵ yields a sum of terms decaying asymptotically with ΔN like $\frac{1}{\Delta N}, \frac{\ln \Delta N}{(\Delta N)^2}$, or faster.

The announced ergodicity of the two-point cluster function with respect to the CUE is now established. It is well to keep in mind that the limit $N \to \infty$ was taken first and eventually also $\Delta N \to \infty$ while keeping the rescaled phase variable e finite. Only the limiting form $Y(e)$ is revealed as ergodic for finite arguments e; "finite e" means that given by the mean spacing $2\pi/N$ on the finest quasi-energy scale; on finite quasi-energy scales corresponding to infinite e, no ergodicity holds and no universality of the two-point cumulant can be expected.

Critical readers might have wondered why I worked with a local spectral average rather than a global one (which would have replaced $\Delta\phi$ by 2π and correspondingly ΔN by N) in the integrated form factor $\langle T(e)\rangle$; they are kindly invited to spot the point in the foregoing reasoning where an attempted merger of the two ordered limits into a single one would have to be regretted.

4.14.3 Traces and Secular Coefficients

Before digging further into more up-to-date issues in random-matrix theory, we had better recall some well-known facts from algebra. At the top of the list is the Hamilton–Cayley theorem which in a cavalier wording says that every matrix obeys its own secular equation. We need that theorem only for diagonalizable matrices like unitary and Hermitian ones, and for these the proof is trivial: Consider the secular polynomial of an $N \times N$ matrix F with eigenvalues λ_i,

$$P(\lambda) = \det(F - \lambda) = \prod_{i=1}^{N}(\lambda_i - \lambda) = \sum_{n=0}^{N}(-\lambda)^n a_{N-n} ; \qquad (4.14.18)$$

I shall refer to the coefficients a_n as the secular coefficients. By substituting the matrix F for the variable λ, a new $N \times N$ matrix $U^\dagger P(F)U$ results which is diagonalized by the same unitary matrix U as is F itself. In other words, if $U^\dagger F U = \mathrm{diag}(\lambda_1, \lambda_2, \ldots, \lambda_N)$, then $U^\dagger P(F)U$ is also diagonal with diagonal elements $P(\lambda_i)$, but these latter vanish by the definition of the secular polynomial, and thus the diagonal matrices $U^\dagger P(F)U$ and $P(F)$ vanish as well.

Upon taking the trace of $P(F)$, we get

$$\mathrm{Tr}\, P(F) = \sum_{n=0}^{N}(-1)^n t_n a_{N-n} = 0 , \qquad (4.14.19)$$

i.e., a linear relationship between the first N traces $t_n = \mathrm{Tr} F^n$ of the matrix F. All other traces can then be expressed in terms of the first $N-1$ ones, as becomes clear by multiplying $P(F)$ by arbitrary powers of F and then taking the trace.

Next, I turn to a little treasure left to posterity by Isaac Newton, a relationship between traces and secular coefficients.[10] A variation of a theme touched upon in the preceding subsection is to be played, the relationship between moments and cumulants of random variables. To briefly explain the latter, let us consider a function $M(x)$ of a single real variable x with $M(0) = 1$ and compare the Taylor expansions of that function and its logarithm,

$$M(x) = 1 + \sum_{n=1}^{\infty}\frac{1}{n!}M_n x^n ,$$

$$C(x) \equiv \ln M(x) = \sum_{n=1}^{\infty}\frac{1}{n!}C_n x^n . \qquad (4.14.20)$$

If $M(x)$ is the characteristic function of some random quantity, then it is said to generate the moments M_n and its logarithm $C(x)$ generates the cumulants C_n. By repeatedly differentiating the identity $M(x) = e^{C(x)}$ with respect to x and then setting $x = 0$, one immediately gets the linear implicit relationship between

---

[10] Newton was concerned with relationships between different symmetric functions of N variables [102].

moments and cumulants[11]

$$M_n = \sum_{m=1}^{n} \binom{n-1}{m-1} C_m M_{n-m} .$$ (4.14.21)

This may be solved either to express cumulants in terms of moments or vice versa. The first few moments are

$$
\begin{aligned}
M_1 &= C_1 \\
M_2 &= C_2 + (C_1)^2 \\
M_3 &= C_3 + 3C_2 C_1 + (C_1)^3 \\
M_4 &= C_4 + 4C_3 C_1 + 3(C_2)^2 + 6C_2 (C_1)^2 + (C_1)^4 .
\end{aligned}
$$ (4.14.22)

It will be instructive for the reader to ponder about "equivalence" and differences between the foregoing moment–cumulant relationships and (4.14.14, 15).

But on to Newton's formulae! They are just the above, read a little differently. Writing the secular polynomial (4.14.18) in the form

$$\det(F - \lambda) = \exp \operatorname{Tr} \ln(F - \lambda) = (-\lambda)^N \exp\left(-\sum_{n=1}^{\infty} \frac{1}{n} t_n \lambda^{-n} \right)$$ (4.14.23)

one recognizes that the secular coefficients a_n correspond to moments and the traces t_n to cumulants; the obvious differences in prefactors are easily accounted for and make the recursion relation (4.14.21) chameleon into the desired identities

$$n a_n = (-1)^{n-1} t_n + \sum_{m=1}^{n-1} (-1)^{m-1} t_m a_{n-m} .$$ (4.14.24)

A little recursive hocus-pocus solves Newton's relations for the secular coefficients in terms of the traces [103] ,

$$a_n = \frac{1}{n!} \begin{vmatrix} t_1 & 1 & 0 & \cdots & 0 & 0 \\ t_2 & t_1 & 2 & \cdots & 0 & 0 \\ \cdots & \cdots & \cdots & \cdots & \cdots & \cdots \\ t_{n-1} & t_{n-2} & t_{n-3} & \cdots & t_1 & n-1 \\ t_n & t_{n-1} & t_{n-2} & \cdots & t_2 & t_1 \end{vmatrix} ;$$ (4.14.25)

by expanding the determinant along the last row it is easy to check that the foregoing expression solves the recursion relationship. The solution plays an important role even outside random-matrix theory. In particular, if the traces t_n with $n = 1, \dots, N$ of a matrix F are known, Newton's formulae allow us to access the spectrum since they yield the full secular polynomial from the first N traces.

A few final remarks on the secular coefficients are in order. Their definition implies $a_0 = 1$ and $a_N = \det F$. If F is unitary, a further simple property,

---

[11] Note that in (4.14.20,21,22), in contrast to the proceeding subsection, I depart from Mehta's sign convention; readers not willing to put up with such wicked confusion may return to the random-matrix path of virtue by $C_n \to (-1)^{n-1} C_n$.

commonly called self-inversiveness [104–107], arises,

$$a_{N-n} = a_N a_n^*.$$ (4.14.26)

Self-inversiveness is checked as follows: $\det(F-\lambda) = (-\lambda)^N \det F \det(F^\dagger - \lambda^{-1}) = a_N \sum_{n=0}^{N}(-\lambda)^n a_n^*$. It reduces the number of independent real parameters in the secular polynomial to N, i.e., the number of eigenphases ϕ_i. The roots of a self-inversive polynomial are either unimodular or come in pairs such that $\lambda_i \lambda_{i'}^* = 1$, hence the name [104] (see Problem 4.14); the unitarity of F is, of course, a stronger property than self-inversiveness since it requires all eigenvalues to be unimodular. A certain variant of self-inversiveness, the so-called functional equation, plays an important role in the semiclassical quantization of autonomous systems [108, 109], as we shall see in Sect. 8.9.

One would sometimes like to check whether a given set of N traces t_n, $n = 1, \ldots, N/2$ comes from a unitary $N \times N$ matrix with even dimension N as $t_n = \text{tr} F^n$ and must therefore yield, through Newton's formulae, a secular polynomial with unimodular roots. A necessary but not sufficient condition is self-inversiveness. To formulate sufficient conditions [110], one may extend the given set of $N/2$ traces to all t_n with $-N + 1 < n < N - 1$ through $t_0 = N$, Newton's formulae, and self-inversiveness. Then, a Hermitian $N \times N$ "trace matrix"

$$t_{n,m} = t_{n-m} = t_{m-n}^*$$ (4.14.27)

can be introduced. Necessary and sufficient for unimodularity and noncoincidence of all roots of the secular polynomial defined by the t_n (and thus for the underlying matrix F to be unitary up to similarity transformations) [110] is strict positivity of the trace matrix $t_{n,m}$ as secured by the inequalities

$$\begin{vmatrix} t_0 & t_1 \\ t_1^* & t_0 \end{vmatrix} > 0, \quad \begin{vmatrix} t_0 & t_1 & t_2 \\ t_1^* & t_0 & t_1 \\ t_2^* & t_1^* & t_0 \end{vmatrix} > 0, \ldots \quad \begin{vmatrix} t_0 & \cdots & t_{N-1} \\ \vdots & & \vdots \\ t_{N-1}^* & \cdots & t_0 \end{vmatrix} > 0.$$ (4.14.28)

It is indeed easy to check that the trace matrix must be positive *semi*definite since it could not otherwise come from N unimodular numbers $e^{i\phi_i}$ with $t_m = \sum_i e^{im\phi_i}$ for all m. Strict positivity is required to secure noncoinciding eigenphases.

To prove the foregoing claim, we may look for conditions under which (for even N; the case of odd N may be treated similarly) the $N/2$ equations

$$t_n = \sum_{i=1}^{N} e^{in\phi_i} \quad \text{with} \quad n = 1, \ldots, N/2$$ (4.14.29)

uniquely determine the $\phi_1, \phi_2, \ldots, \phi_N$ as functions of the traces t_n with $1 \leq n \leq N/2$. Since any permutation for one such set is also a solution, we may restrict ourselves to the manifold

$$0 \leq \phi_1 < \phi_2 < \ldots < \phi_N < 2\pi \text{ "="} 0,$$ (4.14.30)

where 2π "=" 0 indicates closure of the 2π interval. The boundary of the manifold appears when two neighboring eigenphases coincide and thus is given by

$$0 = \Big| \prod_{j<k} (e^{i\phi_j} - e^{i\phi_k}) \Big|^2 = \det(t_{n-m}). \tag{4.14.31}$$

The trace matrix enters the game through the by now familiar identification (4.11.1,2) of the product of eigenvalue differences with a Vandermonde determinant. Our goal will be achieved once we have shown that the transformation (4.14.29) between phases and traces is one-to-one and maps the region (4.14.30) onto the region (4.14.28). It will in fact suffice to show that the boundary of (4.14.30) is mapped onto that of (4.14.30) since by continuity the interiors must then map onto one another as well.

Let us look at the self-inversive polynomial of degree N constructed from the traces t_n with $1 \le n \le N/2$ with the help of Newton's formulae (4.14.25) and self-inversiveness (4.14.26). The roots λ must then lie on the unit circle or come in pairs $(\lambda, \frac{1}{\lambda^*})$, whereupon the full $N \times N$ trace matrix can be written as $t_{n,m}^{\mathrm{SI}} = \sum_{j=1}^{N} e^{i(n-m)(\phi_j + i\epsilon_j)}$ with either $\epsilon = 0$ or pairwise $(\epsilon_j = -\epsilon_k, \phi_j = \phi_k)$; complex "phases" thus come in complex conjugate pairs; note that the Hermiticity of $t_{n,m}$ is preserved when such pairs are admitted. The determinant of the trace matrix with self-similarity worked in generalizes the one defined in (4.14.31) as

$$\begin{aligned}
\det(t_{n-m}^{\mathrm{SI}}) &= \prod_{j<k} \Big\{ \big[e^{i(\phi_j + i\epsilon_j)} + e^{i(\phi_k + i\epsilon_k)} \big] \big[e^{-i(\phi_j + i\epsilon_j)} + e^{-i(\phi_k + i\epsilon_k)} \big] \Big\} \\
&= \prod_{j<k} \Big\{ 2\sin[(\phi_j - \phi_k + i\epsilon_j - i\epsilon_k)/2] \Big\}^2.
\end{aligned} \tag{4.14.32}$$

Now, imagine that we start with real nondegenerate phases (i.e., all $\epsilon_j = 0$) and then let two phases collapse into degeneracy, say $\phi_1 = \phi + \eta$, $\phi_2 = \phi - \eta$ with $\eta \to 0$, and thereafter proceed to a complex conjugate pair $(\eta \to i\epsilon)$. It is easy to see that the resulting $t_{n,m}^{\mathrm{SI}}$ immediately loses positivity: Its trace matrix becomes

$$t_{n-m}^{\mathrm{SI}} = e^{i(n-m)\phi} 2\cos[\eta(n-m)] + \sum_{j=3}^{N} e^{i(n-m)\phi_j} \tag{4.14.33}$$

and its determinant (4.14.32) can be written as

$$\begin{aligned}
\det(t_{n-m}^{\mathrm{SI}}) &= 4\sin^2 \eta \prod_{k=3}^{N} \Big\{ 4\big[\sin^2(\tfrac{\phi - \phi_k}{2}) - \sin^2(\tfrac{\eta}{2}) \big] \Big\}^2 \\
&\quad \times \prod_{l<m}^{3\ldots N} 4\sin^2(\tfrac{\phi_l - \phi_m}{2}).
\end{aligned} \tag{4.14.34}$$

As η goes imaginary, $\eta \to i\epsilon$, the latter determinant becomes $\propto -\sinh^2 \epsilon < 0$ and never changes sign again even when ϵ grows to infinity. The lowest eigenvalue of the matrix $t_{n,m}$ thus changes sign at $\eta = 0$ and is negative for $\epsilon \ne 0$. Equations (4.14.33,34) together imply that lowest eigenvalue is $\propto -\epsilon^2$ for small ϵ. It is also easy to see from the trace matrix (4.14.33) that the 2×2 determinant $t_0^2 - |t_1|^2$ becomes negative at least for sufficiently large ϵ. From this fact and the nonvanishing of the determinant (4.14.34), we may already conclude that the lowest eigenvalue is negative.

The foregoing reasoning can be generalized to a further coalescence of two phases into complexity. In such a coalescence, the determinant $\det (t_{n,m}^{\mathrm{SI}})$ changes sign once more and may thus contain two or no negative eigenvalues. Again, one sees that the determinant preserves its sign as the second pair moves away from the real axis of the complex "phase" plane. The preserved sign together with the eventual negativity of $t_0^2 - |t_1|^2$ implies that there is at least one negative eigenvalue. The same argument works for any number of separated coalescences and invariably shows negativity of the lowest eigenvalue as long as separated pairs of phases have become complex. All other cases, such as coalescences of pairs in the complex plane can be included as limits by continuous changes of the parameters ϕ_j, ϵ_j and therefore again the lowest eigenvalue is nonpositive. At any rate, strict positivity of the trace matrix is established as necessary and sufficient for all roots of the secular polynomial to be unimodular and nondegenerate.

4.14.4 Cross-Correlations Between Traces and Ergodicity of the Form Factor

The title of this subsection may sound like too much of a mouthful. Indeed, (if N is even) the sequence of $N/2$ traces $t_n = \mathrm{Tr} F^n$ of a unitary $N \times N$ matrix F with $n = 1, \ldots, N/2$ uniquely determines the spectrum. By following the n dependence of the form factor $|t_n|^2$ of a single matrix F, one thus sets the eye on "system specific" rather than only universal properties. There is no reason to expect much similarity between the form factor $|t_n|^2$ of a single matrix and its ensemble average $\overline{|t_n|^2}$. However, just as a local spectral average is involved in establishing the ergodicity of, say, the two-point cluster function of the level density we may subject the form factor of a single matrix F to a local "time average," i.e., a smoothing of the n-dependence over some interval Δn. Moreover, so as not to confuse fluctuations and systematic n dependence, it is reasonable to consider the normalized form factor $|t_n|^2/\overline{|t_n|^2}$ and its local time average

$$\langle |t_n|^2 / \overline{|t_n|^2} \rangle = \frac{1}{\Delta n} \sum_{n'=n-\Delta n/2}^{n+\Delta n/2} |t_{n'}|^2 / \overline{|t_{n'}|^2} \,. \tag{4.14.35}$$

If it can be shown that this smoothed normalized version of the form factor has a small ensemble variance within the appropriate circular ensemble, we can indeed expect some degree of universality in that quantity for a single Floquet matrix F, provided, of course, that the pertinent classical dynamics is globally chaotic. We shall in fact establish such ergodicity [111] by showing that the ensemble variance

$$\mathrm{Var}\langle |t_n|^2 / \overline{|t_n|^2} \rangle = \left(\frac{1}{\Delta n} \right)^2 \sum_{n',n''=n-\Delta n/2}^{n+\Delta n/2} \left(\frac{\overline{|t_{n'}|^2 |t_{n''}|^2}}{\overline{|t_{n'}|^2}\,\overline{|t_{n''}|^2}} - 1 \right) \tag{4.14.36}$$

vanishes in the limit $N \to \infty$, $\Delta n \to \infty$, again restricting ourselves to the CUE. The yet simpler Poissonian case is recommended to the reader as Problem 4.15.

Obviously, we need to calculate the fourth moments $\overline{|t_n|^4}$ and $\overline{|t_n|^2 |t_{n'}|^2}$. The interested reader who has labored through the calculation of the mean squared traces will have no difficulties in extending the Taylor expansion of the characteristic function (4.13.5) to the next nonvanishing term ($\propto |k|^4$) which yields

$$
\overline{|t_n|^4} = \overline{|t_{-n}|^4} = \begin{cases} N^4 & \text{for } n = 0 \\ 2n^2 & \text{for } 1 \le n \le N/2 \\ N + 2n(n-1) & \text{for } N/2 \le n \le N \\ 2N(N - \tfrac{1}{2}) & \text{for } N \le n \end{cases}
$$

$$
\overset{n \ge 0}{=} 2\left(\overline{|t_n|^2}\right)^2 + \overline{|t_{2n}|^2} - 2\overline{|t_n|^2} . \tag{4.14.37}
$$

This is consistent, in the limit of large dimension N and for $n > 0$, with a Gaussian distribution of the t_n with vanishing mean and variance $\overline{|t_n|^2}$ given by (4.13.14), a first hint at a more general result to be established below. More precisely speaking, the moments $t_n^\mu (t_n^*)^\nu$ up to order $\mu + \nu = 4$ display strictly Gaussian relationships for $0 < n \le N/2$ while for larger orders, such behavior prevails to within corrections of relative order $1/N$.

At any rate, the ensuing standard deviation $(\overline{|t_n|^4} - \overline{|t_n|^2}^2)^{1/2}$ has no tendency to become small as the dimension N increases and defines an n dependent "band" around the variance $\overline{|t_n|^2}$ within which Gaussian statistics would suggest that the $|t_n|^2$ of an individual matrix lies with probability $\mathrm{e}^{-1/2} - \mathrm{e}^{-3/2} \approx 0.38$. This quantifies the above remarks about the nonuniversality of the n dependence of the form factor.

To establish the cross-correlation $\overline{|t_m|^2 |t_n|^2}$, we may extend the definition (4.13.5) of the characteristic function to two different traces

$$
\tilde{P}(k_m, k_n) = \overline{\exp\left\{ -\frac{\mathrm{i}}{2} \sum_i \left[(k_m \mathrm{e}^{-\mathrm{i}m\phi_i} + k_n \mathrm{e}^{-\mathrm{i}n\phi_i}) + \text{c.c.} \right] \right\}} \tag{4.14.38}
$$

and analogously to (4.13.11) arrive at a Toeplitz determinant,

$$
\tilde{P}(k_m, k_n) = \det M = \exp \operatorname{Tr} \ln M , \tag{4.14.39}
$$

$$
M_{\mu\nu} = \int_0^{2\pi} \frac{d\phi}{2\pi} \mathrm{e}^{\mathrm{i}\phi(\mu-\nu)} \exp\left\{ -\frac{\mathrm{i}}{2}\left[(k_m \mathrm{e}^{-\mathrm{i}m\phi} + k_n \mathrm{e}^{-\mathrm{i}n\phi}) + \text{c.c.} \right] \right\}
$$

$$
\text{with} \quad \mu, \nu = 1, \dots, N .
$$

I shall have to say something about such Toeplitz determinants in the following subsection. For now we can be content with a single term in its Taylor expansion, $\partial^4 \tilde{P} / \partial k_m \partial k_m^* \partial k_n \partial k_n^* |_{k_m = k_n = 0} = (-\mathrm{i}/2)^4 \overline{|t_m|^2 |t_n|^2}$. With a little patience and the guidance of Sec. 4.13.2, the reader will check for $m, n > 0$ that

$$\overline{|t_m|^2 |t_n|^2} = \overline{|t_m|^2 |t_n|^2} + \overline{|t_{m+n}|^2} + \overline{|t_{m-n}|^2} - 2\overline{|t_{\max(m,n)}|^2}$$
$$= N^2 + (N-m)(N-n)\Theta(N-m)\Theta(N-n)$$
$$- N(N-m)\Theta(N-m) - N(N-n)\Theta(N-n)$$
$$- (N-m-n)\Theta(N-m-n) - (N-|m-n|)\Theta(N-|m-n|)$$
$$+ 2[N - \max(m,n)]\Theta[N - \max(m,n)] . \tag{4.14.40}$$

Thus, we are led to the following measure of the relative cross-correlation between two traces:

$$\frac{\overline{|t_m|^2 |t_n|^2}}{\overline{|t_m|^2}\ \overline{|t_n|^2}} - 1 = \begin{cases} \frac{N-m-n}{mn} & \text{for} \quad N-n \leq m \leq N \\ -\frac{N-m+n}{N\min(N,n)} & \text{for} \quad N < m \leq N+n \\ 0 & \text{otherwise} \end{cases} \tag{4.14.41}$$

where without loss of generality I have assumed that $m > n > 0$. Statistical independence of the two traces would entail vanishing relative cross-correlation, and that is in fact the behavior prevailing in most ranges of m, n. Otherwise, where the cross-correlation does not vanish, it at least turns out small, i.e., of order $1/N$; more precisely, $-1/N \leq \overline{|t_m|^2 |t_n|^2}/(\overline{|t_m|^2}\ \overline{|t_n|^2}) - 1 \leq 0$.

We may conclude from the above investigations that the traces t_n of CUE matrices behave, at least as far as moments of orders 1 to 4 are concerned, as if they were independent Gaussian random variables to within corrections of relative weight $1/N$. That same statement holds true for matrices from the COE and the CSE; in the latter case the precision is only $\mathcal{O}(1/\ln N)$ when n is near half the Heisenberg time $n_H^{\mathrm{CSE}} = N$; the interested reader is referred to Ref. [111] for these cases and to Ref. [110] for higher order moments in the case of the CUE.

Equipped with $\overline{|t_m|^2 |t_n|^2}$ and $\overline{|t_m|^4}$, we can estimate the CUE variance (4.14.36) of the time-averaged form factor. Since the cross-correlation (4.14.41) of two traces is nonpositive for $m \neq n$, we get an upper limit for $\mathrm{Var}\langle |t_n|^2 / \overline{|t_n|^2}\rangle$ by dropping the "off-diagonal" terms with $n' \neq n''$ in (4.14.36). For the diagonal terms with $n > 0$, we get an estimate from the second line of (4.14.37), $\overline{|t_n|^4} - (\overline{|t_n|^2})^2 \leq (\overline{|t_n|^2})^2$ since $2\overline{|t_n|^2} \geq \overline{|t_{2n}|^2}$. It follows that

$$\mathrm{Var}\langle |t_n|^2 / \overline{|t_n|^2}\rangle \leq \frac{1}{\Delta n} \stackrel{\Delta n \to \infty}{\longrightarrow} 0 . \tag{4.14.42}$$

Thus, the (time-averaged) form factor turns out to be every bit as ergodic as is, e.g., the (locally spectrally averaged) two-point cluster function, its unruly fluctuations from one n to the next notwithstanding.

4.14.5 Joint Density of Traces of Large CUE Matrices

A powerful theorem about Toeplitz determinants, due originally to Szegö and Kac and extended by Hartwig and Fischer [112], helps to find the marginal and joint distributions of the traces t_n of CUE matrices with finite "times" n in the limit, as the dimension N goes to infinity [107]. In that limit, the finite-time traces will turn out to be statistically independent and to have Gaussian distributions with the means and variances already determined above, $\overline{t_n} = 0$, $\overline{|t_n|^2} = n$.

Once more I invoke the ubiquitous theorem (4.11.4) for the CUE mean of a symmetric function of all eigenphases, again for that function a product. The integral over the phase ϕ_i can then be pulled into the ith row of the determinant in (4.11.4), whereupon the average becomes the Toeplitz determinant

$$\overline{\prod_{m=1}^{N} f(\phi_m)} = \det(f_{l-m}) \equiv T(\{f\}),$$

$$l, m = 1, 2, \ldots, N,$$

(4.14.43)

the elements of which are the Fourier coefficients

$$f_m = \int_0^{2\pi} \frac{d\phi}{2\pi} e^{im\phi} f(\phi)$$

(4.14.44)

of the function $f(\phi)$. The Hartwig–Fischer theorem assigns to the determinant the asymptotic large-N form

$$T(\{f\}) = \exp(N l_0 + \sum_{n=0}^{\infty} n l_n l_{-n})$$

(4.14.45)

with the l_n the Fourier coefficients of $\ln f(\phi)$, i.e., $\ln f(\phi) = \sum_{n=-\infty}^{\infty} l_n e^{im\phi}$. The function $f(\phi)$ must meet the following four conditions for the theorem to hold: (1) $f(\phi) \neq 0$ for $0 \leq \phi < 2\pi$, (2) $\arg f(2\pi) = \arg f(0)$, (3) $\sum_{n=-\infty}^{\infty} |n| |f_n|^2 < \infty$, and (4) $\sum_{n=-\infty}^{\infty} |f_n| < \infty$.

For a first application, I consider the marginal distribution of the nth trace t_n whose Fourier transform is the characteristic function (4.13.11) and involves

$$f(\phi) = \exp\left[-\frac{i}{2}(k e^{-in\phi} + k^* e^{in\phi})\right].$$

(4.14.46)

Provided that n remains finite as $N \to \infty$, that function fulfills all conditions of the theorem (see below for an explicit check), and the only nonvanishing Fourier coefficients of its logarithm are $l_n = l_{-n}^* = -ik/2$. The asymptotic form of the Toeplitz determinant thus reads $e^{-nk^2/4}$ and yields the density of the nth trace as its Fourier transform as the Gaussian already anticipated several times,

$$P(t_n) = \frac{1}{\pi n} e^{-|t_n|^2/n}.$$

(4.14.47)

The foregoing reasoning is easily extended to the joint density of the first n traces $P(t_1, \ldots, t_n) = \overline{\prod_{m=1}^{n} \delta^2(t_m - \sum_{l=1}^{N} e^{im\phi_l})}$ whose Fourier transform $\tilde{P}(k_1, \ldots, k_n)$ is once more of the form (4.14.43) with the function $f(\phi)$ from (4.14.46) generalized to the sum

$$f(\phi) = \exp\left[-\frac{i}{2} \sum_{m=1}^{n} (k_m e^{-im\phi} + k_m^* e^{im\phi})\right].$$

(4.14.48)

The nonvanishing Fourier coefficients of $\ln f(\phi)$ are now $l_m = l^*_{-m} = -\mathrm{i}k_m/2$ with $m = 1, \dots, n$ and entail a limiting form of the characteristic function which is just the product of the marginal ones met above. The joint density then comes out as the announced product of marginal distributions

$$P(t_1, \dots, t_n) = \frac{1}{n!\pi^n} \exp\left(-\sum_{m=1}^n |t_m|^2/m\right). \tag{4.14.49}$$

The result obviously generalizes to the joint density of an arbitrary set of finite-time traces. The previously resulting hints of statistical independence and Gaussian behavior of all finite-time traces in the limit $N \to \infty$ are thus substantiated. To appreciate the importance of the restriction to finite n and to avoid a bad mathematical conscience, it is well to verify the aforementioned conditions of the Hartwig–Fischer theorem on the function $f(\phi)$ in (4.14.48). The first two of them are clearly fulfilled since $\mathrm{i}\ln f(\phi)$ is real, continuous, and (2π)-periodic. The third and fourth conditions are met since the derivative $f'(\phi)$ is square integrable (provided n remains finite!),

$$\int_0^{2\pi} d\phi |f'(\phi)|^2 = \sum_{m=-\infty}^\infty m^2|f_m|^2 = \pi \sum_{m=1}^n m^2|k_m|^2 < \infty. \tag{4.14.50}$$

Since $|m| \leq |m|^2$ in the foregoing Parseval inequality, we check condition (3), $\sum_{m=-\infty}^\infty |m||f_m|^2 < \infty$. Finally, Cauchy's inequality yields $\sum_{m=-\infty}^\infty |f_m| = |f_0| + \sum_{m\neq 0} |\frac{1}{m}||mf_m| \leq |f_0| + (\sum_{m\neq 0} |\frac{1}{m}|^2)^{1/2}(\sum_{m\neq 0} |mf_m|^2)^{1/2}$; due to Parsefal's inequality and the convergence of $\sum_{m\neq 0} 1/m^2$, we find condition (4) met as well.

For finite dimension N, the independence as well as the Gaussian character of the traces are only approximate; both properties tend to get lost as the sum of the orders (times) of the traces involved in a set increases. This is quite intuitive a finding since already all traces for a single unitary matrix are uniquely determined by the N eigenphases such that only $N/2$ traces are linearly independent.

4.15 Correlations of Secular Coefficients

I propose here to study the autocorrelation of the secular polynomial of random matrices from the circular ensembles [106, 107, 113]. These correlations provide a useful reference for the secular polynomials of Floquet or scattering matrices of dynamic systems and their semiclassical treatment. In particular, we shall be led to an interesting quantum distinction between regular and chaotic motion.

A starting point is the secular polynomial of an $N \times N$ unitary matrix F,

$$\det(F - \lambda) = \sum_{n=0}^N (-\lambda)^n a_{N-n} = \prod_{i=1}^N (\mathrm{e}^{-\mathrm{i}\phi_i} - \lambda), \tag{4.15.1}$$

which enjoys the self-inversiveness discussed in the preceding subsection.

An autocorrelation function of the secular polynomial for any of the circular ensembles may be defined as

$$P_N^\beta(\psi, \chi) = \overline{\prod_{i=1...N} (e^{-i\phi_i} - e^{-i\psi})(e^{i\phi_i} - e^{i\chi})}, \tag{4.15.2}$$

with the overbar indicating an average over the circular ensemble characterized by the level repulsion exponent β. For the orthogonal and unitary circular ensembles, this correlation function may be identified with $\overline{\det(F - e^{-i\psi})(F^\dagger - e^{i\chi})}$. In the symplectic case, however, the definition (4.15.2) accounts only for one level per Kramers' doublet such that we have $\det(F - \lambda) = \left[\sum_{n=0}^N (-\lambda)^n a_{N-n}\right]^2 = \prod_{i=1}^N (e^{-i\phi_i} - \lambda)^2$ and thus $P_N^4(\psi, \chi) = \overline{[\det(F - e^{-i\psi})(F^\dagger - e^{i\chi})]^{\frac{1}{2}}}$.

Due to the homogeneity of the circular ensembles, the correlation function in question depends on the two phases ψ and χ only through the single unimodular variable $x = e^{i(\chi - \psi)}$; thus it may be written in terms of the auxiliary function

$$f(\phi, x) = (e^{-i\phi} - x)(e^{i\phi} - 1) \tag{4.15.3}$$

as

$$P_N^\beta(x) = \overline{\prod_i f(\phi_i, x)} = \sum_{n=0}^N x^n \overline{|a_n|^2}. \tag{4.15.4}$$

In writing the last member of the foregoing equation, I have once more invoked the homogeneity of the circular ensembles to conclude that

$$\overline{a_m a_n^*} = \delta_{mn}. \tag{4.15.5}$$

Obviously, our correlation function may serve as a generating function for the ensemble variances of the secular coefficients, $\overline{|a_n|^2} = \frac{1}{n!} d^n P_N^\beta(x)/dx^n|_{x=0}$.

The correlation function $P_N^\beta(x)$ and the variances $\overline{|a_n|^2}$ are most easily calculated in the Poissonian case $\beta = 0$,

$$P_N^0(x) = \left[\int_0^{2\pi} \frac{d\phi}{2\pi} f(\phi, x)\right]^N = (1 + x)^N,$$

$$\overline{|a_n|^2} = \binom{N}{n}. \tag{4.15.6}$$

The treatment of the CUE, COE, and CSE proceeds in parallel to the calculation of the characteristic function of the traces in the preceding section. As in (4.13.5), we must average a product, now with factors $f(\phi_i, x)$ rather than $\exp(-i\cos n\phi_i)$. For the CUE, that replacement changes (4.13.11) into

$$P_N^2(x) = \det\left[\int_0^{2\pi} \frac{d\phi}{2\pi} e^{i(m-n)\phi} f(\phi, x)\right]$$

$$= \det\left[(1 + x)\delta(m - n) - \delta(m - n + 1) - x\delta(m - n - 1)\right]$$

$$= \sum_{n=0}^N x^n. \tag{4.15.7}$$

The CUE variance of the secular coefficient thus comes out independent of n,

$$\overline{|a_n|^2} = 1\,, \tag{4.15.8}$$

and that independence is in marked contrast to the binomial behavior (4.15.6) found for the Poissonian case.

Similarly, for the orthogonal and symplectic ensembles, we are led to Pfaffians of antisymmetric matrices A^β, $P_N^\beta(x) = \mathrm{Pf}(A^\beta(x))/\mathrm{Pf}(A^\beta(0))$. In the orthogonal case where for simplicity I assume even N, one finds (4.13.23) replaced with

$$
\begin{aligned}
A^1_{\underline{m}\underline{m}'}(x) &= \frac{1}{4\pi\mathrm{i}} \int_0^{2\pi} d\phi\, d\phi'\, \mathrm{sign}(\phi - \phi') f(\phi, x) f(\phi', x) \mathrm{e}^{\mathrm{i}(\underline{m}\phi + \underline{m}'\phi')} \\
&= \left[\frac{(1+x)^2}{\underline{m}} + x\Big(\frac{1}{\underline{m}+1} - \frac{1}{\underline{m}'+1}\Big)\right]\delta_{\underline{m},-\underline{m}'} \\
&\quad - (1+x)\Big(\frac{1}{\underline{m}} - \frac{1}{\underline{m}'}\Big)\Big(\delta_{\underline{m},-\underline{m}'+1} + x\,\delta_{\underline{m},-\underline{m}'-1}\Big) \\
&\quad + \frac{1}{\underline{m}-1}\,\delta_{\underline{m},-\underline{m}'+2} + \frac{x^2}{\underline{m}+1}\,\delta_{\underline{m},-\underline{m}'-2}
\end{aligned}
\tag{4.15.9}
$$

where again the underlined index runs in unit steps through the half-integers in $0 < |\underline{m}| \le (N-1)/2$. The analogue of (4.13.33) in the symplectic case reads

$$
\begin{aligned}
A^4_{\underline{m}\underline{m}'}(x) &= (\underline{m} - \underline{m}') \int_0^{2\pi} \frac{d\phi}{2\pi}\, \mathrm{e}^{\mathrm{i}(\underline{m}+\underline{m}')\phi} f(\phi, x) \\
&= (\underline{m} - \underline{m}')\Big((1+x)\delta(\underline{m} + \underline{m}') - \delta_{\underline{m},-\underline{m}'-1} - x\delta_{\underline{m},-\underline{m}'+1}\Big)
\end{aligned}
\tag{4.15.10}
$$

with $0 < |\underline{m}| \le (2N-1)/2$.

The evaluation of the Pfaffian is a lot easier in the latter case than in the former since A^1 is[12] "pentadiagonal" but A^4 is only "tridiagonal." Moreover, the $2N \times 2N$ matrix A^4 falls into four $N \times N$ blocks, and the two diagonal ones vanish such that the Pfaffian $\mathrm{Pf}(A^4(x))$ equals, up to the sign, the determinant of either off-diagonal $N \times N$ block. One reads off the following recursion relation for the correlation function $P_N^4(x) = \mathrm{Pf}(A^4(x))/\mathrm{Pf}(A^4(0))$ in search,

$$P_N^4(x) = (1+x)P_{N-1}^4(x) - x\frac{(2N-2)^2}{(2N-1)(2N-3)}P_{N-2}^4(x)\,; \tag{4.15.11}$$

this in turn is easily converted into a recursion relation for the variances,

$$\overline{|a_n^{(N)}|^2} = \overline{|a_n^{(N-1)}|^2} + \frac{1}{n}\overline{|a_{n-1}^{(N-1)}|^2} - \frac{(2N-2)^2}{n(2N-1)(2N-3)}\overline{|a_{n-1}^{(N-2)}|^2}\,. \tag{4.15.12}$$

With $a_0^{(N)} = 1$ for all naturals N, one finds the solution (with $a_n^{(N)}$ typographically stripped back down to a_n)

$$\overline{|a_n|^2} = \binom{N}{n}^2 \binom{2N}{2n}^{-1} \overset{N \gg 1}{\approx} \sqrt{\frac{N}{\pi n(N-n)}}\,. \tag{4.15.13}$$

---

[12] To remove the following quotation marks, think of the rows of the determinant swapped pairwise, the first with the last, the second with the last but one, etc.

Finally mustering a good measure of courage, I turn to the orthogonal case, i.e., the Pfaffian of the "pentadiagonal" antisymmetric matrix $A^1_{\underline{m},\underline{m}'}(x)$ given in (4.15.10). A little convenience is gained by actually removing those quotation marks in the way indicated in the footnote, i.e., by looking at the matrix $A_{\underline{m},\underline{m}'} \equiv A^1_{\underline{m},-\underline{m}'}$; recall that the underlined indices take the half-integer values $-\underline{M}, -\underline{M}+1, \ldots, \underline{M}$ with $\underline{M} = (N-1)/2$; the determinants of A and A^1 can differ at most in their sign which is irrelevant for our correlation function $P^1_N(x) = \sqrt{\det A^1(x)/\det A^1(0)} = \sqrt{\det A(x)/\det A(0)}$. The key to the latter square root is to observe that the $N \times N$ matrix $A_{\underline{m},\underline{m}'}(x)$ fails to be the product BC of the two tridiagonal matrices

$$
\begin{aligned}
B_{\underline{m}\,\underline{m}'} &= -\frac{1+x}{\underline{m}}\delta_{\underline{m},\underline{m}'} + \frac{x}{\underline{m}+1}\delta_{\underline{m},\underline{m}'-1} + \frac{1}{\underline{m}-1}\delta_{\underline{m},\underline{m}'+1} \\
C_{\underline{m}\,\underline{m}'} &= -(1+x)\delta_{\underline{m},\underline{m}'} + x\delta_{\underline{m},\underline{m}'-1} + \delta_{\underline{m},\underline{m}'+1}
\end{aligned}
\tag{4.15.14}
$$

only in the two elements in the upper left and the lower right corner. As a remedy, I momentarily extend the matrix dimension to $N+2$, bordering the matrices A, B, C by frames of single rows and columns as

$$
\boxed{B} = \left(
\begin{array}{c|ccccc|c}
1 & 1 & 1 & \cdots & 1 & 1 & 1 \\
\hline
(-\underline{M}-1)^{-1} & & & & & & 0 \\
0 & & & & & & 0 \\
\vdots & & & B & & & \vdots \\
0 & & & & & & 0 \\
0 & & & & & & x(\underline{M}+1)^{-1} \\
\hline
0 & 0 & 0 & \cdots & 0 & 0 & 1
\end{array}
\right),
\tag{4.15.15}
$$

$$
\boxed{C} = \left(
\begin{array}{c|ccccc|c}
-\underline{M}-1 & x & 0 & \cdots & 0 & 0 & -\underline{M}-1 \\
\hline
0 & & & & & & -\underline{M} \\
0 & & & & & & -\underline{M}+1 \\
\vdots & & & C & & & \vdots \\
0 & & & & & & \underline{M}-1 \\
0 & & & & & & \underline{M} \\
\hline
0 & 0 & 0 & \cdots & 0 & 1 & \underline{M}+1
\end{array}
\right),
\tag{4.15.16}
$$

$$
\boxed{A} = \left(
\begin{array}{c|ccccc|c}
-\underline{M}-1 & 0 & 0 & \cdots & 0 & 0 & 0 \\
\hline
1 & & & & & & 0 \\
0 & & & & & & 0 \\
\vdots & & & A & & & \vdots \\
0 & & & & & & 0 \\
0 & & & & & & 0 \\
\hline
0 & 0 & 0 & \cdots & 0 & 1 & \underline{M}+1
\end{array}
\right).
\tag{4.15.17}
$$

The bordered matrix $\boxed{A}$ now enjoys the product structure in full, $\boxed{A} = \boxed{B}\boxed{C}$. The various determinants obviously obey $\det\boxed{C} = \left(\prod_{\underline{m}=-\underline{M}-1}^{\underline{M}+1} \underline{m}\right)\det\boxed{B}$ and

$\det \boxed{A} = -(\underline{M}+1)^2 \det A$ such that upon drawing the square root of $\det \boxed{A}$, we get

$$P_N^1(x) = \sqrt{\frac{\det A(x)}{\det A(0)}} = (-1)^?(\frac{2}{N+1})^2 \det \boxed{C}. \tag{4.15.18}$$

It remains to evaluate the $(N+2) \times (N+2)$ determinant $\det \boxed{C}$. Since the first column contains just a single non-zero element, we reduce the dimension to $N+1$,

$$\frac{\det \boxed{C}}{-(\underline{M}+1)} = \left(\begin{array}{ccccc|c} -(1+x) & x & \ldots & 0 & & -\underline{M} \\ 1 & -(1+x) & \ldots & 0 & & -\underline{M}+1 \\ 0 & 1 & \ldots & 0 & & -\underline{M}+2 \\ \vdots & \vdots & \ldots & \vdots & & \vdots \\ 0 & 0 & \ldots & x & & \underline{M}-1 \\ \hline 0 & 0 & \ldots & -(1+x) & & \underline{M} \\ 0 & 0 & \ldots & 1 & & \underline{M}+1 \end{array}\right).$$

Here the original $N \times N$ matrix C appears in the upper left corner. Starting with the uppermost one, we now add to each row the sum of all of its lower neighbors such that in the elements of the new last column the sums $f(\underline{n}) = \sum_{\underline{m}=-\underline{M}-1}^{\underline{n}} \underline{m}$ appear and the determinant reads

$$\frac{\det \boxed{C}}{\underline{M}+1} = \left(\begin{array}{cccc|c} -x & 0 & \ldots & 0 & f(\underline{M}) \\ 1 & -x & \ldots & 0 & f(\underline{M}-1) \\ 0 & 1 & \ldots & 0 & f(\underline{M}-2) \\ \vdots & \vdots & \ldots & \vdots & \vdots \\ 0 & 0 & \ldots & 0 & f(-\underline{M}+1) \\ 0 & 0 & \ldots & -x & f(-\underline{M}) \\ \hline 0 & 0 & \ldots & 1 & f(-\underline{M}-1) \end{array}\right).$$

We finally annul the secondary diagonal with unity as elements: Beginning at the top, we add to each row the $\frac{1}{x}$-fold of its immediate upper neighbor. The arising determinant equals the product of its diagonal elements, $\det \boxed{C} = (\underline{M}+1)(-x)^{2\underline{M}+1} \sum_{n=0}^{2\underline{M}+1}(\frac{1}{x})^n f(-\underline{M}-1+n)$. Recalling $\underline{M} = (N-1)/2$ with N even, we get from (4.15.18) the correlation function in search as

$$P_N^1(x) = \sum_{n=0}^{N} x^n \left[1 + \frac{n(N-n)}{N+1}\right] \tag{4.15.19}$$

and read off the mean squared secular coefficients for the COE,

$$\overline{|a_n|^2} = 1 + \frac{n(N-n)}{N+1}. \tag{4.15.20}$$

Let us summarize and appreciate the results for the four circular ensembles. The mean squared secular coefficients (4.15.6, 8, 13, 20) are special cases of

$$\overline{|a_n^\beta|^2} = \binom{N}{n} \frac{\Gamma(n+2/\beta)\Gamma(N-n+2/\beta)}{\Gamma(2/\beta)\Gamma(N+2/\beta)} \tag{4.15.21}$$

for the appropriate values of the repulsion exponent β. Actually, the latter general formula has been shown [107] to be valid for arbitrary real β, provided the joint density of eigenvalues (4.10.2) is so extended.

With the family relationship between the $\overline{|a_n^\beta|^2}$ pointed out, the differences should be commented on as well. For fixed $n \neq 0, N$, we face a growth of that function with decreasing β. That trend is not surprising; an increase of the degree of level repulsion implies a tendency toward equidistant levels. For $\beta \to \infty$, one must expect a perfectly rigid spectrum[13] according to the secular equation $a^N + e^\Phi = 0$, i.e., $a_n = 0$ except for $n = 0, N$. Conversely, the Poissonian statistics arising for $\beta \to 0$ entails the weakest spectral stiffness and thus the largest mean squared secular coefficients. For a large dimension N, the Poissonian limit $\overline{|a_n^0|^2} = \binom{N}{n}$ is exponentially larger than the three other distinguished $\overline{|a_n^\beta|^2}$ with $\beta = 1, 2, 4$. The difference in question may in fact be seen as one of the quantum criteria to distinguish regular ($\beta = 0$) from chaotic dynamics ($\beta = 1, 2, 4$), as will become clear in the following section where a detailed comparison of random matrices and Floquet matrices of dynamical systems will be presented; see in particular Fig. 4.9.

The correlation function $P_N^\beta(x)$ itself is due some attention. To give it a nice appearance, we return to (4.15.4) and realize that due to the symmetry of the $\overline{|a_n|^2}$ under $n \leftrightarrow N - n$, we may switch from $P_N^\beta(x)$ to a function [113]

$$C^\beta(\eta) = \frac{P_N^\beta(e^{-i2\pi\eta/N})e^{i\pi\eta}}{P_N^\beta(1)} = \frac{\sum_{n=0}^{N} |a_n|^2 e^{-i2\pi\eta(n/N-1/2)}}{\sum_{n=0}^{N} |a_n|^2} \tag{4.15.22}$$

which is real for real η; note the normalization to $C_N^\beta(0) = 1$. Using our results for the variances $\overline{|a_n|^2}$ and doing sums as integrals,

$$C^\beta(\eta) = \begin{cases} \left(\cos\frac{2\pi\eta}{N}\right)^N \approx \exp\left(-\frac{2\pi^2\eta^2}{N}\right) & \text{for} \quad \beta = 0 \\ \frac{3}{2}\left(1 + \frac{1}{\pi^2}\frac{\partial^2}{\partial\eta^2}\right)\frac{\sin\pi\eta}{\pi\eta} & \text{for} \quad \beta = 1 \\ \frac{\sin\pi\eta}{\pi\eta} & \text{for} \quad \beta = 2 \\ J_0(\pi\eta) & \text{for} \quad \beta = 4 \end{cases} \tag{4.15.23}$$

Checking for the small-η behavior, we find that the decay of the correlation function proceeds faster, the larger the level-repulsion exponent β, and actually much more slowly in the Poissonian case than in any of the three cases corresponding to classical chaos.

---

[13] Indeed, generalizing the joint distribution of eigenvalues (4.3.15) to arbitrary real $\beta = n - 1$, one is led to a Wigner distribution $P_\beta(S) = AS^\beta e^{-BS^2}$ with A and B fixed by normalization and $\overline{S} = 1$ which for $\beta \to \infty$ approaches the delta function $\delta(S - 1)$.

4.16 Fidelity of Kicked Tops to Random-Matrix Theory

Periodically kicked tops are unsurpassed in their faithfulness to random-matrix theory, and that fact will be put into view here. As already mentioned several times in this book, tops can be designed so as to be members of any universality class.

The Floquet operators employed for the unitary class read

$$
F = \exp\left(-i\frac{\tau_z J_z^2}{2j+1} - i\alpha_z J_z\right) \exp\left(-i\frac{\tau_y J_y^2}{2j+1} - i\alpha_y J_y\right)
$$
$$
\times \exp\left(-i\frac{\tau_x J_x^2}{2j+1} - i\alpha_x J_x\right). \tag{4.16.1}
$$

Indeed, if all torsion constants τ_i and rotational angles α_i are non-zero and of order unity, no time-reversal invariance (nor any geometric symmetry) reigns even approximately and the classical dynamics is globally chaotic. To secure an antiunitary symmetry as time-reversal invariance while keeping global classical chaos, we just have to erase torsion and rotation with respect to one axis; the resulting Floquet operator then pertains to the orthogonal class, as discussed in Sect. 2.12. Finally, simple symplectic tops are attained by choosing a representation of half-integer j for

$$
F = \exp\left[-i\frac{\tau_1 J_z^2}{2j+1} - i\frac{\tau_2(J_x J_z + J_z J_x)}{2(2j+1)} - i\frac{\tau_3(J_x J_y + J_y J_x)}{2(2j+1)}\right]
$$
$$
\times \exp\left(-i\frac{\tau_4 J_z^2}{2j+1}\right), \tag{4.16.2}
$$

since no geometric symmetries are left while the appearance of only two unitary factors secures an antiunitary symmetry.[14]

Fig. 4.5 shows the spacing distributions $P(S)$ and their integrals $I(S) = \int_0^S dS' P(S')$ for tops with the Floquet operators just listed. The graphs pertaining to the unitary ($\tau_z = 10, \tau_y = 0, \tau_x = 4, \alpha_z = \alpha_y = 1, \alpha_x = 1.1$), orthogonal ($\tau_z = 10, \alpha_z = 1, \alpha_y = 1$, all others zero), and Poissonian ($\tau_z = 10, \alpha_z = 1$, all others zero) cases were obtained for $j = 1000$. The symplectic case is a little more obstinate for large matrix dimensions for which reason the pertinent graph was constructed by averaging over the distributions obtained from all half-integers j between 49.5 and 99.5, keeping the control parameters fixed to $\tau_1 = 10, \tau_2 = 1, \tau_3 = 4, \tau_4 = 2.1$. In all cases, agreement with the predictions of random-matrix theory is good. Incidentally, for the databases involved, the Wigner surmises (4.4.2) suffice as representatives of random-matrix theory.

When trying to build the two-point cluster function from a set of $2j+1$ quasi-energies, one must first worry about how to generate a smooth function from the

---

[14] The reader might (and should!) wonder whether the "orthogonal" version of (4.16.1) goes "symplectic" for half-integer j. It does not. The reason is $T^2 = +1$ for *all* j in that case since two components of an angular momentum can simultaneously be given real representations.

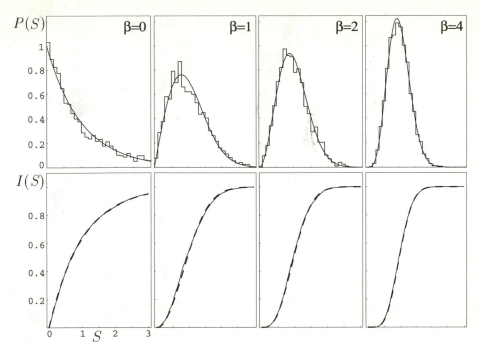

Fig. 4.5. Spacing distributions $P(S)$ and their integrals $I(S)$ for kicked tops with the Floquet operators (4.16.1) and (4.16.2), with rotational angles and torsion constants as given in the text. Hilbert space dimension $2j + 1 = 2001$ for the Poissonian ($\beta = 0$), orthogonal ($\beta = 1$), and unitary ($\beta = 2$) cases; in the symplectic case ($\beta = 4$), distributions for all half-integer j between 49.5 and 99.5 were superimposed while keeping the coupling constants τ_i and thus the classical dynamics unchanged

sum over products of two delta functions provided by the naive definition

$$y(\phi, e) = 1 - \left(\frac{2\pi}{N}\right)^2 \sum_{i \neq j} \delta(\phi - \phi_i)\delta(\phi + \frac{2\pi e}{N} - \phi_j). \qquad (4.16.3)$$

We encountered a similar problem in proving the ergodicity of the two-point cluster function for the CUE in Sect. 4.14.1. The delta functions there were smoothed by subjecting $y(\phi, e)$ to both a local spectral average and an integral over e (see (4.14.9)). I resort here to the same strategy with one slight modification: Inasmuch as the spectrum to be examined has a density of levels fluctuating about the *uniform* mean $\frac{N}{2\pi}$, we make optimal use of the data by employing a global rather than only local spectral average[15] and consider the integral

$$I(e) = \int_0^e de' \int_0^{2\pi} \frac{d\phi}{2\pi} y(\phi, e') = e - \frac{1}{N} \sum_{i \neq j} \Theta\left(\frac{e2\pi}{N} + \phi_i - \phi_j\right). \qquad (4.16.4)$$

---

[15] While the technicalities of the proof of ergodicity demanded two independent, large parameters, N and ΔN, no such needs arise here.

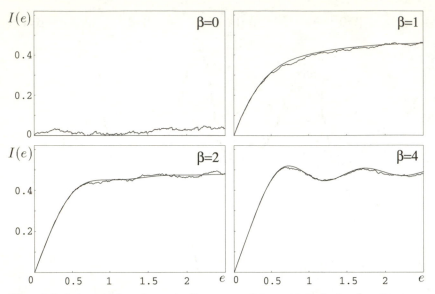

Fig. 4.6. Integrated and spectrally averaged two-point cluster function according to (4.16.4) for kicked tops (*rugged curves*) with Hilbert space dimensions and coupling constants as for (4.5). The *full curves* represent the predictions of the pertinent circular ensembles

For this "self-averaging" quantity, Fig. 4.6 indeed reveals fine agreement between tops and the random-matrix predictions for the Poissonian, unitary, orthogonal, and symplectic universality classes.

A glance at Fig. 4.7 suggests more delicacies in the contrast between the form factor $|t_n|^2$ of a single top and the mean over the pertinent circular ensemble, here the CUE. As already argued in Sect. 4.14.11, the form factor derived from a single spectrum performs a rather unruly dance about the ensemble mean in its n dependence. No disobedience to random-matrix theory is to be lamented, however, since that theory itself predicts such fluctuations [111]. The shortdashed lines in the graphs of part (a) of Fig. 4.7 define a band of one standard deviation

$$\Delta(n,\beta) = \left(\overline{|t_n|^2}\right)^{-1} \sqrt{\overline{|t_n|^4} - \overline{|t_n|^2}^2} \tag{4.16.5}$$

around the CUE mean, the latter shown as the full curve. Inasmuch as the underlying matrix dimension is large, the asymptotic Gaussian behavior found in Sect. 4.14.4 can be expected to prevail. Thus, we may expect the fraction $e^{-\frac{1}{2}} - e^{-\frac{3}{2}} \approx 0.3834$ of all points $|t_n|^2$ to lie within that band and actually find the fraction 0.3966 for the top. Similarly, the abscissa and the longdashed curve surround a band of two standard deviations which should contain 86.47% of all points and does 87.0%. To further corroborate the obedience of the top to random-matrix theory, we display, in part (b) of Fig. 4.7, a histogram for the distribution of $\left(|t_n|^2 / \overline{|t_n|^2_{\mathrm{CUE}}}\right)^{1/2}$ which indeed closely follows the full curve provided by the Gaussian distribution of the traces.

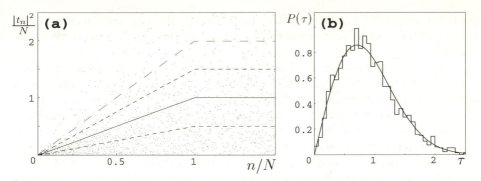

Fig. 4.7. (a) Time dependence of form factor of unitary kicked top (*dots*) with Hilbert space dimension and control parameters as for (4.5). The full curve depicts the CUE form factor (4.13.16). In good agreement with random-matrix theory, about 40% (87%) of all dots lie within a band of one (two) standard deviation(s) (4.16.5) around the CUE mean, the respective bands are delimited by the *shortdashed* (*longdashed*) lines (b) Histogram for the normalized form factor $\tau = |t_n|^2/\overline{|t_n|^2_{\mathrm{CUE}}}$ for unitary kicked top with $j = 500$ and coupling constants as for (4.5). The *full curve* depicts the prediction of random-matrix theory based on the Gaussian distribution (4.14.46) of the traces

Fig. 4.8 presents a check on the ergodicity of the form factor discussed in Sect. 4.14.11. Part (a) of the figure pertains to the "unitary" top specified above (here with $j = 500$) and shows a local time average $\langle |t_n|^2/\overline{|t_n|^2} \rangle$ taken over a time window $\Delta n = 20 \approx N/50$. The fluctuations from one n to the next are much smoothed compared to the nonaveraged form factor in Fig. 4.7; they define a band whose width should be compared with the CUE variance $\mathrm{Var}\langle |t_n|^2/\overline{|t_n|^2} \rangle$ estimated in (4.14.42). In fact, the CUE bound $1/\Delta n$ of (4.14.42) is respected by

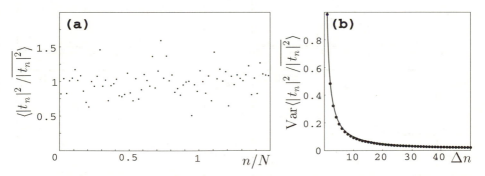

Fig. 4.8. (a) Time dependence of local time average of normalized form factor $\langle |t_n|^2/\overline{|t_n|^2_{\mathrm{CUE}}} \rangle$ according to (4.14.35) for unitary kicked top with $j = 500$ and coupling constants as for (4.5); the time window is $\Delta n = 20 \approx (2j+1)/50$ (b) The *full curve* gives the upper bound $1/\Delta n$ of CUE variance of $\langle |t_n|^2/\overline{|t_n|^2_{\mathrm{CUE}}} \rangle$ according to (4.14.42) vs. the time window Δn; the *dots* depict the standard deviation from the mean of $\langle |t_n|^2/\overline{|t_n|^2_{\mathrm{CUE}}} \rangle$ for the same unitary top as in (a), determined from clouds as in (a).

the top: Part (b) of Fig. 4.8 shows that bound as the full curve and the numerically found width of the fluctuations of the local time average for various values of the time window Δn as dots.

The statistics of secular coefficients of Floquet matrices remain to be scrutinized. A simple indicator is the n dependence of the coefficients a_n whose mean squared moduli were evaluated in Sect. 4.16. Remarks similar to those above on the n dependence of the traces t_n apply here: A single Floquet matrix produces a sequence of secular coefficients fluctuating so wildly that the underlying systematic n dependence becomes visible only after some careful smoothing. As done for the traces above, one could average the $|a_n|^2$ over some interval Δn or over different dimensions N and plot versus n/N, while keeping rotational angles and torsion constants fixed. Instead, Fig. 4.9 is based on averages over small volume elements of control parameter space, while the dimension $N \propto 1/\hbar$ remains fixed. All four universality classes are seen to yield n dependences of $|a_n|^2$ not dissimilar to the averages over the pertinent circular ensemble. In particular, the integrable case has its $|a_n|^2$ so overwhelmingly larger than any of the other three classes that we may indeed see that difference as one of the quantum indicators of the alternative regular/chaotic. The term "not dissimilar" needs to be qualified: The differences in the circular-ensemble averages are appreciably larger for the secular coefficients than for any of the other quantities shown above. Worse yet, the differences change when the underlying volume element in control parameter space is shifted. The reason for this strong nonuniversality of the secular coeffi-

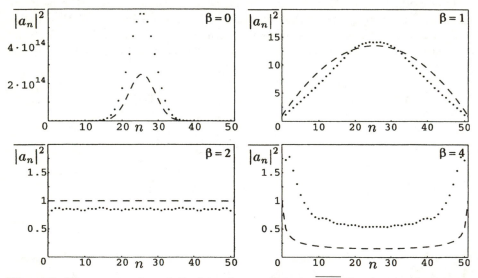

Fig. 4.9. Mean squared moduli of secular coefficients $\overline{|a_n|^2}$ for the circular ensembles (*dashes*) and for kicked tops (*dots*) vs n. The averages for the tops are based on 20,000 unprejudiced random points in control parameter space from the intervals $\tau_z \in [10, 15], \alpha_z \in [0.6, 1.2], \alpha_y \in [0.6, 1.2]$ for the orthogonal top; for the unitary top, $\alpha_x = 1.1, \alpha_y = \alpha_z = 1, \tau_y = 0,$ and $\tau_x, \tau_z \in [10, 25]$; and for the symplectic top, $\tau_1 \in [15, 20], \tau_4 \in [10, 15], \tau_2 = 1, \tau_3 = 4$. All cases involve spectra of 51 eigenvalues

cients can be understood on the basis of the semiclassical periodic-orbit theory to be discussed in Chap. 8. Periodic orbits of period n will turn out to determine the nth trace t_n and through Newton's formulae (4.14.25) the secular coefficients a_m with all $m \geq n$; system specific properties show up predominantly in short periodic orbits with periods not much larger than unity which do not affect the large-m traces t_m but can and indeed do affect the large-m coefficients a_m.

To summarize, no greater faithfulness to random-matrix theory could be expected than that met in tops. The importance of these systems as standard models of both classical and quantum chaotic behavior will be further underscored in Chap. 7 by showing that they comprise, in a certain limit, the prototypical system with localization, the kicked rotator. A rough characterization of that limit at which we shall arrive there is that a top will begin to show quantum localization as soon as an appropriate localization length is steered to values smaller than the dimension $2j + 1$.

4.17 Problems

4.1. Show that S in (4.2.16) is indeed a symplectic matrix.

4.2. Show that the most general unitary 2×2 matrix can be specified with the help of four real parameters α, ε as $U = \exp(i\alpha + i\varepsilon \cdot \sigma)$, and characterize the specialization to (4.3.6). In which geometric sense does (4.3.6) generalize (4.2.11) to a finite (rather than infinitesimal) change of basis?

4.3. Show the invariance of the differential volume element in matrix space for the GSE. Proceed analogously to the case of the GUE and the GOE, write the GSE matrix H for $N = 2$ as a 6×6 matrix, and calculate the Jacobian for symplectic transformations in a way allowing generalization to arbitrary N.

4.4. Present an argument demonstrating that Wigner's semicircle law holds for all three of the Gaussian ensembles.

4.5. Calculate the local mean level density for the spin Hamiltonian $H = -kj/2 + pJ_z + kJ_z^2/2j$ with $-j \leq J_z \leq +j$ and $j \gg 1$, where k and p are arbitrary coupling constants. Compare with Wigner's semicircle law.

4.6. Calculate the mean value and the variance of the squared moduli of the components y of the eigenvectors of random matrices for all three ensembles.

4.7. Calculate the ensemble mean for the density of levels defined in (4.9.3) by using the eigenvalue distribution of the CUE.

4.8. Calculate the ensemble mean of $(1/N) \sum_i^{1 \ldots N} \prod_{j(\neq i)}^{1 \ldots N} \Theta(|\phi_j - \phi_i| - s)$ using (4.11.4). What meaning does the resulting probability density have?

4.9. Generalize (4.11.7) to $P(S) = (\partial^2 E/\partial S^2)/\varrho^2$ for a constant but nonunit mean density of levels ϱ.

4.10. Show that a real antisymmetric matrix a of even dimension can be brought to block diagonal form with the blocks (4.12.18) along the diagonal by a real orthogonal transformation L, $L\tilde{L} = 1$, $L = L^*$. Hint: Why are the eigenvalues imaginary and why must they come in pairs $\pm ia_i$? What can be said about the two eigenvectors pertaining to a pair $\pm ia_i$ of eigenvalues? Reshuffle the diagonalizing unitary matrix by suitably combining columns so as to make it orthogonal. What can finally be done to enforce $\det L = 1$? Why is $\det a$ a perfect square and positive? What changes if the dimension is odd?

4.11. Prove the representation (4.12.20) for a Pfaffian.

4.12. Evaluate the normalization constant $\mathcal{N}_4$ as defined in (4.10.2) of the joint density of eigenphases of the CSE. Proceed as done for the COE in Sect. 4.13.4 by keeping track of all proportionality factors, powers of i apart. The result is $\mathcal{N}_4 = \pi^N N! (2N-1)!!$

4.13. Leisure permitting, the reader will want to check the Fourier transforms from the form factor to the cluster function in (4.13.16, 30, 36); only the orthogonal case requires ambition. A more worthwhile exercise, *quasi-mandatory* for those intending to acquire later a semiclassical understanding of random-matrix type behavior of generic dynamic systems, is to prove the following: (1) If the $(2n)$th derivative $K^{(2n)}(\tau)$ of an even real function $K(\tau)$ has a delta-function singularity, $K^{(2n)}(\tau) = a[\delta(\tau - \tau_0) + \delta(\tau + \tau_0)] + \ldots$, that singularity contributes to the Fourier transform as $\int_{-\infty}^{\infty} d\tau K(\tau) \cos 2\pi e\tau = (-1)^n 2a(2\pi e)^{-2n} \cos 2\pi e\tau_0 + \ldots$ (2) A logarithmic singularity in $K^{(2n)}(\tau) = a(\ln|\tau - \tau_0| + \ln|\tau + \tau_0|) + \ldots$ goes hand in hand with $\int_{-\infty}^{\infty} d\tau K(\tau) \cos 2\pi e\tau = (-1)^n a(2\pi)^{-2n} |e|^{-(2n+1)} \cos 2\pi e\tau_0 + \ldots$.

Use this to explain the leading nonoscillatory and oscillatory terms in the cluster functions, checking, in particular, correctness of the coefficients. Why are there corrections to the leading terms for the COE and the CSE but not for the CUE?

4.14. Show that the roots of a self-inversive polynomial are either unimodular or come in pairs such that $\lambda_i \lambda_{i'}^* = 1$. This is an easy task for $N = 2$, and you may be satisfied with that.

4.15. Evaluate the variance of the time-averaged form factor (4.14.36) for the Poissonian ensemble, and thus show that the inequality (4.14.42) holds here as well.

5. Level Clustering

5.1 Preliminaries

Here I propose to consider classical autonomous systems with f degrees of freedom and $2f$ pairs of canonical variables p_i, q_i. We shall meet with invariant tori, caustics, and Maslov indices and proceed to the semiclassical torus quantization à la *Einstein, Brillouin* and *Keller* (EBK) and its modern variant, a periodic-orbit theory. The latter will allow us to understand why there is no repulsion but rather clustering of levels for generic integrable systems with two or more degrees of freedom; the density of level spacings therefore usually takes the form of a single exponential, $P(S) = e^{-S}$. Single-freedom systems, if autonomous, are always integrable and do not respect any general rule for their spacing statistics; we shall postpone a discussion of their behavior to the subsequent chapter.

Readers with more curiosity about EBK than can be satisfied here are referred to *Percival*'s review [114] or the recent book by *Brack* and *Bhaduri* [115]. A semiclassical treatment of classically chaotic dynamics will follow in Chap. 8.

5.2 Invariant Tori of Classically Integrable Systems

The dynamics in question have f independent constants of the motion $C_i(p, q)$ with vanishing Poisson brackets:

$$\{C_i, C_j\} = \frac{\partial C_i}{\partial p} \frac{\partial C_j}{\partial q} - \frac{\partial C_i}{\partial q} \frac{\partial C_j}{\partial p} = 0 . \tag{5.2.1}$$

Each phase-space trajectory $p(t)$, $q(t)$ thus lies on an f-dimensional surface embedded in the $2f$-dimensional phase space. Usually, this surface has the topology of an f-dimensional torus.

A nontrivial example is provided by a point mass moving in a plane under the influence of an attractive central potential $V(r)$. The energy E and the angular momentum L are two independent constants of the motion and define a two-dimensional torus in the four-dimensional phase space.

A phase-space trajectory starting on such a torus will keep winding around it forever. In general, trajectories will not be closed; only for the harmonic $[V(r) \sim r^2]$ and the Kepler $[V(r) \sim 1/r]$ potentials are all trajectories closed. For other potentials, periodic orbits are possible but constitute, as will become clear presently, a subset of measure zero in the set of all trajectories.

When projected onto the two-dimensional configuration space, every trajectory of the present model remains between two circles, which are concentric with respect to the center of force. These circles, called caustics, are the singularities of the projection of the 2-torus onto the configuration plane. When touching a caustic the trajectory has a momentarily vanishing radial momentum, i.e., a turning point of its radial libration.

For a circular "billiard table," every configuration-space orbit also remains in between two concentric circles of which only the inner one is a caustic, however; the outer one is a boundary at which the particle bounces off rather than coming to rest momentarily.

Note that no caustic is ever encountered in a rotation for which an angle increases indefinitely, where all other coordinates, if any, remain constant.

Classical motion on an f-torus embedded in a $(2f)$-dimensional phase space is conveniently described in terms of f pairs of action and angle variables I_i, Θ_i. The actions characterize the f-torus via

$$I_i = \frac{1}{2\pi} \oint_{\Gamma_i} p \, dq > 0 \,, \tag{5.2.2}$$

where Γ_i is one of the f independent irreducible loops around the torus. Being properties of invariant tori, the I_i are constants of the motion. By a suitable canonical transformation, the actions I_i become new momenta with angles Θ_i as their conjugate coordinates. The Hamiltonian, a constant of the motion itself, can be expressed as a function of the f actions and does not depend on the angles. Hamilton's equations thus give

$$\dot{\Theta}_i = \frac{\partial H}{\partial I_i} \equiv \omega_i \,,$$
$$\Theta_i(t) = \omega_i t + \Theta_i(0) \,. \tag{5.2.3}$$

The frequencies ω_i are the angular velocities with which the phase space point travels around the torus. The angle Θ_i changes by 2π as the loop Γ_i orbits once.

When the frequencies ω_i are incommensurate, the trajectory starting from $\Theta(0)$ will tend to fill the torus densely as time elapses. We might even speak of ergodic motion in that case since time averages (of suitable quantities) will tend to ensemble averages with uniform probability density on the torus. Periodic orbits result, on the other hand, when the ω_i are related rationally, i.e., when all ω_i are integral multiples of a certain fundamental frequency ω_0,

$$\boldsymbol{\omega} = \boldsymbol{M}\omega_0 \,, \quad M_i \text{ integer} \,. \tag{5.2.4}$$

Such an orbit closes in on itself after after the fundamental period $2\pi/\omega_0$, having completed M_1 trips around the loop Γ_1, M_2 around Γ_2 etc. The corresponding torus is called rational. Every rational torus accommodates an f-parameter continuum of closed orbits related to one another by shifts of the initial angles $\boldsymbol{\Theta}(0)$. Rational tori are as exceptional among the tori in phase space as the rational numbers among the reals.

5.3 Einstein–Brillouin–Keller Approximation

To solve the Schrödinger equation of a classically integrable system in the semi-classical limit, one may use the ansatz [114–120]

$$\psi(q) = a(q)e^{iS(q)/\hbar} \ . \tag{5.3.1}$$

By demanding that $\psi(q)$ be single valued in the q, one obtains a quantum condition for each of the f actions,

$$I_i = \left(m_i + \tfrac{1}{4}\alpha_i\right)\hbar \ , \tag{5.3.2}$$

with nonnegative integer quantum numbers m_i and the so-called Maslov indices α_i. The ith Maslov index α_i equals the number of caustics encountered along the ith irreducible loop Γ_i around the classical torus with respect to which the action I_i was defined. The quantization prescription (5.3.2) is often referred to as semiclassical or torus quantization.

The quantum condition on the action originates from the phase change in the semiclassical wave function along the loop Γ_i,

$$\frac{\Delta S}{\hbar} = \oint_{\Gamma_i} \frac{p\,dq}{\hbar} \ . \tag{5.3.3}$$

Another contribution to the total change of phase may come from the amplitude $a(q)$, which can be shown to diverge at caustics. Such a singularity means, of course, that the ansatz (5.3.1) breaks down near a caustic. It was a brilliant idea of Maslov to switch to momentum space when q approaches a caustic and to replace (5.3.1) with a similar ansatz for $\tilde{\psi}(p)$. Since a caustic in q-space is not a caustic in p-space, no singularity threatens $\tilde{\psi}(p)$ where the semiclassical form (5.3.1) of $\psi(q)$ runs into trouble. Replacing the latter by the Fourier transform of $\tilde{\psi}(p)$ near the q-space caustic (see Chap. 8.4.2), one finds the total phase increment picked up along Γ_i as

$$\frac{\Delta S}{\hbar} = \frac{2\pi\,I_i}{\hbar} - \frac{\pi}{2}\alpha_i \ . \tag{5.3.4}$$

That change must be an integral multiple of 2π for the wave to close in on itself uniquely, i.e., not to destroy itself by destructive interference.

By expressing the Hamiltonian $H(\boldsymbol{I})$ in terms of the quantized actions (5.3.2), one obtains the semiclassical approximation for the energy levels

$$E_{\boldsymbol{m}} = H\left(\boldsymbol{I_m}\right) = H\left[\left(\boldsymbol{m} + \tfrac{1}{4}\boldsymbol{\alpha}\right)\hbar\right] \ . \tag{5.3.5}$$

In integrable systems, these semiclassical levels usually give an excellent approximation for sufficiently high degrees of excitation. Their accuracy is often surprisingly good for low energies, too, and in some exceptional cases even the whole spectrum is reproduced rigorously.

The f-dimensional harmonic oscillator, for instance, has the Hamiltonian

$$H = \sum_{i=1}^{f} \left(\frac{1}{2m} p_1^2 + \frac{1}{2} m\omega_i^2 q_i^2 \right) = \sum_i \omega_i I_i \ . \tag{5.3.6}$$

Since a libration encounters two turning points in every period, one must set $\alpha_i = 2$ and thus obtain

$$E_{\boldsymbol{m}} = \sum_i \hbar\omega_i \left(m_i + \tfrac{1}{2} \right) \ , \quad m_i = 0, 1, 2, \ldots \tag{5.3.7}$$

which is indeed the exact result. Quite amusingly, the contribution of the Maslov indices gives the correct zero-point energy.

As a second example, the reader is invited to check that the Hamiltonian of the hydrogen atom can be expressed in terms of the radial, polar, and azimuthal actions as

$$H = -\frac{me^4}{2(I_r + I_\Theta + I_\phi)^2} \ . \tag{5.3.8}$$

The radial and polar motions are librations and the azimuthal motion is a rotation, so one has $\alpha_r = \alpha_\Theta = 2$ and $\alpha_\phi = 0$ and thus the semiclassical energies

$$E_{\boldsymbol{m}} = -\frac{me^4}{2\hbar^2 n^2} \ , \quad n = m_r + m_\Theta + m_\phi + 1 \tag{5.3.9}$$

which is again exact.

Finally, a particle in an f-dimensional box is described by

$$H = \frac{1}{2m} \sum_{i=1}^{f} p_i^2 \ , \quad 0 \le q_i \le a_i \ . \tag{5.3.10}$$

The squared momenta p_i^2 are f independent constants of the motion and the actions read

$$I_i = \frac{1}{2\pi} \left[\int_0^{a_i} dq |p_i| + \int_{a_i}^0 dq \left(-|p_i| \right) \right] = a_i \frac{|p_i|}{\pi} \ . \tag{5.3.11}$$

Reexpressed in the actions, the Hamiltonian

$$H = \frac{\pi^2}{2m} \sum_i \frac{I_i^2}{a_i^2} \tag{5.3.12}$$

yields the semiclassical levels

$$E_{\boldsymbol{m}} = \frac{\pi^2 \hbar^2}{2m} \sum_i \frac{m_i^2}{a_i^2}$$

which again coincide with the exact ones, except that the quantum numbers m_1 start from 1 instead of 0 [115].

For further examples and a more systematic treatment, the reader is referred to Ref. [115].

5.4 Level Crossings for Integrable Systems

It is easy to see that the levels obtained by torus quantization generally are not inhibited from crossing when $f \geq 2$; the codimension of a crossing is $n = 1$. To this end, one imagines that the Hamiltonian and thus the asymptotic eigenvalues

$$E_{\boldsymbol{m}}(k) = H\left[\hbar\left(\boldsymbol{m} + \tfrac{1}{4}\boldsymbol{\alpha}\right), k\right] , \qquad (5.4.1)$$

depend on a single parameter k. In an f-dimensional space with axes $m_1, m_2, \ldots,$ m_f, the quantum numbers in (5.4.1) are represented by the points of a primitive cubic lattice where all components of the lattice vectors are nonnegative integers. In this space, the function H of f continuous variables $\boldsymbol{m}$ defines a continuous family of energy surfaces. An allowed energy eigenvalue arises for every such energy surface intersecting one of the discrete lattice points, say $\boldsymbol{m}^*$. In general, an energy surface intersecting $\boldsymbol{m}^*$ will not intersect any other lattice point. There is a whole bundle of energy surfaces through $\boldsymbol{m}^*$, labelled by the parameter k, however, and by properly choosing k, a special surface can, in general, be picked which does intersect a second lattice point. The restriction "in general" of the foregoing statement excludes funny exceptions like the harmonic oscillator for which the energy surface is a hyperplane. Barring such exceptional surfaces, one concludes that a level crossing can be enforced by controlling a single parameter in the Hamiltonian. It follows from the arguments of Sect. 3.4 that the spacing density typically approaches a non-zero constant at zero spacing,

$$P(S) \to P(0) \neq 0 \text{ for } S \to 0 . \qquad (5.4.2)$$

This is in contrast to the power-law behavior typical of nonintegrable systems.

The foregoing reasoning does not apply to a single ladder or "multiplet" of levels arising if we let one of the f quantum numbers run while keeping all others fixed or, equivalently, if we have $f = 1$ to begin with. The parametric changes possible for such single multiplets will be discussed in Sect. 6.5.

5.5 Poissonian Level Sequences

Before proceeding to show that the generic integrable system has an exponential distribution of its level spacings, it will be useful to introduce some probabilistic concepts. In particular, I intend to demonstrate here that exponentially distributed level spacings allow an interpretation of the sequence of levels as a random sequence without any correlation between the individual "events."

Let us assume that the spectrum is already unfolded to unit mean spacing of neighboring levels everywhere, $\bar{S} = 1$. Then, we employ, as an auxiliary concept, the conditional probability $g(S)dS$ of finding a level in the interval $[E + S, E + S + dS]$ given one at E. By virtue of the assumed homogeneity of the level distribution, $g(S)$ does not depend on E. The probability density $P(S)$ of finding the nearest neighbor of the level at E in dE at $E + S$ is evidently the product of

$g(S)dS$ with the probability that there is no other level between E and $E + S$,

$$P(S) = g(S) \int_S^\infty dS' P(S') . \tag{5.5.1}$$

Differentiating with respect to S,

$$\frac{\partial P}{\partial S} = \frac{\partial g}{\partial S} \int_S^\infty dS' P(S') - gP = \left(g^{-1}\frac{\partial g}{\partial S} - g\right) P . \tag{5.5.2}$$

This differential equation is solved by

$$P(S) = \text{const } g(S) \mathrm{e}^{-\int_0^S dS' g(S')} . \tag{5.5.3}$$

For a Poissonian level sequence, now, there are no correlations at all between the individual levels. Thus, the conditional probability $g(S)$ is a constant, and $P(S) = C \exp(-AS)$. Normalizing and scaling so that $\bar{S} = 1$, one obtains the exponential distribution

$$P(S) = \mathrm{e}^{-S} . \tag{5.5.4}$$

As a by-product of the above reasoning, the result is that the conditional probability density $g(S)$ behaves like a power

$$g(S) \to S^\beta \tag{5.5.5}$$

with $\beta = 0, 1, 2, 4$ for, respectively, integrable systems and nonintegrable systems with orthogonal, unitary, and symplectic canonical transformations.

5.6 Superposition of Independent Spectra

In this section I shall present an important limit theorem which will help us to understand level spacings of integrable systems: L independent spectra, each with its own homogeneous density and spacing distribution, superpose to a spectrum with exponentially distributed spacings in the limit $L \to \infty$. The arguments presented below go back in essence to *Rosenzweig* and *Porter* [121] and *Berry* and *Robnik* [122].

For the present purpose, it is again convenient to employ the probability $E(S)$ that an interval of length S is empty of levels. Assuming that the mean distance is unity, the spacing distribution $P(S)$ can be obtained from (4.11.7) as the second derivative of the gap probability $E(S)$,

$$P(S) = \frac{\partial^2 E}{\partial S^2} . \tag{5.6.1}$$

Now consider L independent ladders of levels, the μth of which has the constant mean spacing $1/\varrho_\mu$ and the spacing distribution $P_\mu(S)$ normalized as

$$\int_0^\infty dS\, P_\mu(S) = 1$$
$$\int_0^\infty dS\, S\, P_\mu(S) = \frac{1}{\varrho_\mu} . \tag{5.6.2}$$

The probability of finding no level of the μth ladder in an interval of length S is, in accordance with (5.6.1),

$$
\begin{aligned}
E_\mu(S) &= \varrho_\mu \int_S^\infty dx \int_x^\infty dy\, P_\mu(y) \\
&= \varrho_\mu \int_S^\infty dx (x - S) P_\mu(x) \,.
\end{aligned}
\tag{5.6.3}
$$

It follows that $E_\mu(S)$ falls off monotonically from $E_\mu(0) = 1$ to $E_\mu(\infty) = 0$, and the "initial" slope is

$$
\begin{aligned}
E'_\mu(0) &= -\varrho_\mu \,, \\
E_\mu(S) &= 1 - \varrho_\mu S + \dots \,.
\end{aligned}
\tag{5.6.4}
$$

Due to the assumed independence of the ladders, the joint probability that there is no level from any ladder in an interval of length S is just the product

$$
E(S) = \prod_{\mu=1}^{L} E_\mu(S) = \exp \sum_\mu \ln E_\mu(S) \,.
\tag{5.6.5}
$$

Now, we assume that $L \gg 1$ and that the ϱ_μ all have roughly equal magnitudes, $\varrho_\mu \sim 1/L$, normalizing for convenience such that the total density is unity,

$$
\sum_{\mu=1}^{L} \varrho_\mu = 1 \,.
\tag{5.6.6}
$$

The joint probability $E(S)$ falls off more rapidly by a factor of the order L than the typical single-ladder probability $E_\mu(S)$. Therefore, it is legitimate to replace the $E_\mu(S)$ on the right-hand side of (5.6.5) by the first two terms of their Taylor series (5.6.4). The exponential form for $E(S)$ results in the limit $L \to \infty$,

$$
E(S) \to \mathrm{e}^{-S} \,.
\tag{5.6.7}
$$

It is hard to resist speculating, by naive appeal to the limit theorem just established, that the semiclassical spectrum (5.3.5) must have, for $f \geq 2$ and generic functions $H(\boldsymbol{I})$, an exponential spacing distribution. After all, the semiclassical levels tend to follow one another quite erratically with respect to the direction of the vector $\boldsymbol{m}$, as is obvious from Fig. 5.1. There is, no doubt, an element of truth in such naive reasoning. However, the only reasonably sound derivation of the exponential spacing distribution for integrable systems known at present is lengthy and technical and does not use the limit theorem (5.6.7).

5.7 Periodic Orbits and the Semiclassical Density of Levels

The starting point for deriving the spacing distribution is the density of semi-classical energy levels

$$
\varrho(E) = \sum_{\{m_i > 0\}} \delta \left\{ E - H \left[\hbar \left(\boldsymbol{m} + \tfrac{1}{4} \boldsymbol{\alpha} \right) \right] \right\}
\tag{5.7.1}
$$

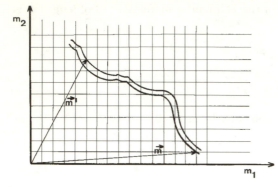

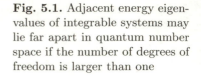

Fig. 5.1. Adjacent energy eigenvalues of integrable systems may lie far apart in quantum number space if the number of degrees of freedom is larger than one

which will be shown to be representable as a sum of contributions from all classical periodic orbits.

The presentation will closely follow *Berry* and *Tabor*'s original one [123], [124]. Note that the density $\varrho(E)$ is normalized here such that its integral over an energy interval ΔE gives the number of levels contained in the interval (rather than the fraction of the total number of atoms). To pursue the goal indicated it is convenient to rewrite (5.7.1) as

$$\varrho(E) = \int_{\boldsymbol{I}>0} d^f I \delta \left[E - H(\boldsymbol{I})\right] \sum_{\boldsymbol{m}} \delta^f \left[\boldsymbol{I} - \hbar \left(\boldsymbol{m} + \frac{\alpha}{4}\right)\right] . \tag{5.7.2}$$

The previous restriction on the f quantum numbers $\boldsymbol{m} > 0$ is dropped here in favor of the restriction $\boldsymbol{I} > 0$ on the f-fold integral. Next, we employ Poisson's summation formula,

$$\sum_{M=-\infty}^{+\infty} e^{i2\pi Mx} = \sum_{m=-\infty}^{+\infty} \delta(x - m) , \tag{5.7.3}$$

to replace the f-dimensional train of delta functions by an f-fold sum over exponentials,

$$\begin{aligned}\varrho(E) &= \frac{1}{\hbar^f} \sum_{\boldsymbol{M}=-\infty}^{+\infty} e^{-i\pi\boldsymbol{\alpha}\cdot\boldsymbol{M}/2} \\ &\times \int_{\boldsymbol{I}>0} d^f I \delta \left[E - H(\boldsymbol{I})\right] e^{i2\pi\boldsymbol{M}\cdot\boldsymbol{I}/\hbar} ,\end{aligned} \tag{5.7.4}$$

and each of the f summation variables M_i runs through all integers.

The $\boldsymbol{M} = 0$ term in the series representation (5.7.4),

$$\bar{\varrho}(E) \equiv \frac{1}{\hbar^f} \int_{\boldsymbol{I}>0} d^f I \delta \left[E - H(\boldsymbol{I})\right] , \tag{5.7.5}$$

is the venerable Thomas–Fermi result already alluded to in Sect. 4.6 which is also known as Weyl's law. [To see the equivalence of (4.6.4) and (5.7.5) note that $\hbar^{-f} = h^{-f}(2\pi)^f$ and that $H(\boldsymbol{I})$ is independent of the f angle variables $\boldsymbol{\Theta}$ conjugate to the actions $\boldsymbol{I}$.] As already indicated in the symbol $\bar{\varrho}(E)$, the Thomas–Fermi level density (5.7.5) may be considered a local spectral average of $\varrho(E)$, that does not reflect local fluctuations in the level sequence such as local clustering or repulsion. An important and intuitive interpretation of (5.7.5) becomes accessible through its integral, the average level staircase

$$\bar{\sigma}(E) = \frac{1}{\hbar^f} \int_{\boldsymbol{I}>0} d^f I \Theta \left[E - H(\boldsymbol{I})\right] , \qquad (5.7.6)$$

which counts the number of levels below E as the phase-space volume "below" the corresponding classical energy surface divided by $\hbar^f$; each quantum state is thus assigned the phase-space volume $\hbar^f$.

The $\boldsymbol{M} \neq 0$ terms in (5.7.4) describe local fluctuations in the spectrum with scales ever finer as $|\boldsymbol{M}|$ increases. Defining

$$\Delta\varrho(E) = \varrho(E) - \bar{\varrho}(E) = \sum_{\boldsymbol{M}\neq0} e^{-i\pi\boldsymbol{\alpha}\cdot\boldsymbol{M}/2}\varrho_{\boldsymbol{M}} \qquad (5.7.7)$$

one has the $\boldsymbol{M}$th density fluctuation

$$\varrho_{\boldsymbol{M}} = \frac{1}{\hbar^f} \int_{\boldsymbol{I}>0} d^f I \delta \left[E - H(\boldsymbol{I})\right] e^{i2\pi\boldsymbol{M}\cdot\boldsymbol{I}/\hbar} . \qquad (5.7.8)$$

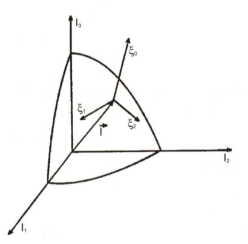

Fig. 5.2. Energy surface in action space

To evaluate the I integral, we introduce a new set of orthogonal coordinates as integration variables, $\xi_0, \xi_1, \dots, \xi_{f-1}$, the "zeroth" one of which measures the perpendicular distance from the energy surface, and the remaining $f-1$ variables parametrize the energy surface (Fig. 5.2). The integral then takes the form

$$\varrho_{\boldsymbol{M}} = \int d^{f-1}\xi e^{i2\pi\boldsymbol{M}\cdot\boldsymbol{I}(\boldsymbol{\xi})/\hbar} \int d\xi_0 \frac{\delta(\xi_0)}{|\partial H/\partial\xi_0|} \qquad (5.7.9)$$

where it is understood that $d^{f-1}\xi$ contains the Jacobian of the transformation of the integration variables and the vector $\boldsymbol{\xi}$ lies in the energy surface so as to have the $f-1$ components $\xi_1, \ldots, \xi_{f-1}$. By noting that

$$\left(\frac{\partial H}{\partial \xi_0}\right)_{\xi_0 = 0} = |\nabla_{\boldsymbol{I}} H(\boldsymbol{I})|_{H=E} = |\boldsymbol{\omega}(\boldsymbol{I}(\boldsymbol{\xi}))| \tag{5.7.10}$$

is the length of the f-dimensional frequency vector, one obtains

$$\varrho_{\boldsymbol{M}} = \int d^{f-1}\xi \frac{e^{i2\pi \boldsymbol{M} \cdot \boldsymbol{I}(\boldsymbol{\xi})/\hbar}}{|\boldsymbol{\omega}(\boldsymbol{I}(\boldsymbol{\xi}))|} . \tag{5.7.11}$$

Without sacrificing consistency with the semiclassical approximation for the energy levels, the remaining integral over the energy surface can be evaluated in the limit $\hbar \to 0$ in which the phase $2\pi \boldsymbol{M} \cdot \boldsymbol{I}(\boldsymbol{\xi})/\hbar$ is a rapidly oscillating function of $\boldsymbol{\xi}$. The integral will thus be negligibly small unless there is a point $\boldsymbol{\xi}^M$ on the energy surface for which the phase in question is stationary,

$$\boldsymbol{M} \cdot \frac{\partial \boldsymbol{I}}{\partial \xi_i} = 0 \text{ for } \boldsymbol{\xi} = \boldsymbol{\xi}^M , \quad i = 1, 2, \ldots, f-1 . \tag{5.7.12}$$

Since the $f-1$ vectors $\partial \boldsymbol{I}/\partial \xi_i$ are all tangential to the energy surface, the stationary-phase condition (5.7.12) requires the lattice vector $\boldsymbol{M}$ to be perpendicular to the energy surface. An equally significant consequence of the vector $\boldsymbol{\xi}$ lying in the energy surface is

$$\frac{\partial H}{\partial \xi_i} = \nabla_{\boldsymbol{I}} H \cdot \frac{\partial \boldsymbol{I}}{\partial \xi_i} = \boldsymbol{\omega} \cdot \frac{\partial \boldsymbol{I}}{\partial \xi_i} = 0 , \tag{5.7.13}$$

i.e., the orthogonality of the frequency vector to the energy surface. It follows from (5.7.12) and (5.7.13) that the lattice vector $\boldsymbol{M}$ and the frequency vector $\boldsymbol{\omega}$ must be parallel at the points of stationarity of the phase $2\pi \boldsymbol{I} \cdot \boldsymbol{M}/\hbar$,

$$\omega_1 : \omega : \ldots : \omega_f = M_1 : M_2 : \ldots M_f . \tag{5.7.14}$$

Because the M_i are integers, one concludes that the ω_i must be commensurate and the corresponding tori, defined by $\boldsymbol{I}^M = \boldsymbol{I}(\boldsymbol{\xi}^M)$, are rational.

Obviously, a nonnegligible Mth density fluctuation $\varrho_{\boldsymbol{M}}$ corresponds to periodic orbits on the rational torus $\boldsymbol{I}^M$. The lattice vector $\boldsymbol{M}$ defines the topology of the closed orbits inasmuch as an orbit obeying (5.7.14) closes in on itself after M_1 periods 2π of Θ_1, M_2 periods of Θ_2, etc. Figure 5.3 illustrates closed orbits of simple topologies (M_r, M_ϕ) for a two-dimensional potential well, M_r counting the number of librations of the radial coordinate and M_ϕ the number of revolutions around the center before the orbit closes.

When a lattice vector $\boldsymbol{M}$ induces a nonnegligible density fluctuation $\varrho_{\boldsymbol{M}}$ so do all its integer multiples since these also fulfill the stationary-phase condition (5.7.12). Therefore, it is useful to introduce the primitive version $\boldsymbol{\mu}$ of $\boldsymbol{M}$ such that the components μ_i are relatively prime, together with its multiples $\boldsymbol{M} = q\boldsymbol{\mu}$,

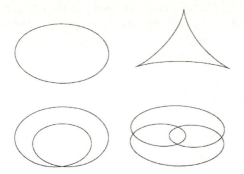

Fig. 5.3. Closed orbits of simple topologies in a two-dimensional configuration space

where q is a positive integer. The density fluctuation $\varrho_{q\mu}$ corresponds to a closed orbit which is just the primitive orbit $\boldsymbol{\mu}$ traversed q times. The total action along a closed orbit $\boldsymbol{M} = q\boldsymbol{\mu}$,

$$S(\boldsymbol{M}) = 2\pi\boldsymbol{M} \cdot \boldsymbol{I}^M \ , \tag{5.7.15}$$

is obviously q times the action along the primitive orbit $\boldsymbol{\mu}$,

$$S(\boldsymbol{M}) = qS(\boldsymbol{\mu}) \ .$$

It remains to write the stationary-phase approximation to ϱ_M. The general formula for the stationary-phase approximation of a multiple integral reads

$$\int d^n x e^{\mathrm{i}f(\boldsymbol{x})} \varPhi(\boldsymbol{x})$$

$$= \sum_{\boldsymbol{x}^s} \sqrt{\frac{(2\pi)^n}{|\det(\partial^2 f(\boldsymbol{x})/\partial x_i \partial x_j)|_{\boldsymbol{x}=\boldsymbol{x}^s}}} \varPhi(\boldsymbol{x}^s) e^{\mathrm{i}f(\boldsymbol{x}^s)+\mathrm{i}\beta\pi/4} \tag{5.7.16}$$

where $\boldsymbol{x}^s$ are the points of stationary phase, defined by $\partial f/\partial x_i^s = 0$, and $\beta(\boldsymbol{x}^s)$ is the difference in the number of positive and negative eigenvalues of the $n \times n$ matrix $\partial^2 f/\partial x_i \partial x_j$.

The determinant in (5.7.16) deserves special comment. According to (5.7.11), in this case it is $(f-1) \times (f-1)$ and reads

$$\det\left(\frac{2\pi}{\hbar}\boldsymbol{M} \cdot \frac{\partial^2 \boldsymbol{I}}{\partial \xi_i \partial \xi_j}\right) = \left(\frac{2\pi q}{\hbar}\right)^{f-1} \det\left(\boldsymbol{\mu} \cdot \frac{\partial^2 \boldsymbol{I}}{\partial \xi_i \partial \xi_j}\right)$$

$$\equiv \left(\frac{2\pi q|\boldsymbol{\mu}|}{\hbar}\right)^{f-1} K\left(\boldsymbol{I}^\mu\right) \ , \tag{5.7.17}$$

where

$$K\left(\boldsymbol{I}^\mu\right) = \det\left(\hat{\boldsymbol{\mu}} \cdot \frac{\partial^2 \boldsymbol{I}}{\partial \xi_i \partial \xi_j}\right) \tag{5.7.18}$$

is the scalar curvature of the energy surface at the point $\boldsymbol{I}^M$. To appreciate this interpretation of K, let us consider $f = 2$; the vector $d\boldsymbol{I}/d\xi_1$ is tangential to the energy surface and $d(d\boldsymbol{I}/d\xi_1) = (d^2\boldsymbol{I}/d\xi_1^2)d\xi_1$ is the increment of that tangent vector along $d\xi_1$; the larger the value of $\hat{\mu} \cdot d^2I/d\xi_1^2$, the stronger the curvature of the energy surface (Fig. 5.4). For $f = 3$, it is easy to see that K is the Gaussian curvature of the then two-dimensional energy surface.

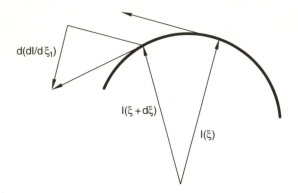

d(dI/d ξ₁)

I(ξ + dξ)

I(ξ)

Fig. 5.4. Curvature of the energy surface

The final result for the $\boldsymbol{M}$th density fluctuation now reads

$$\varrho_{\boldsymbol{M}} = \left(\frac{\hbar}{q|\boldsymbol{\mu}|}\right)^{(f-1)/2} \frac{\exp\{i[qS(\boldsymbol{\mu})/\hbar + \pi\beta(\boldsymbol{\mu})/4]\}}{|\boldsymbol{\omega}(\boldsymbol{I}^{\boldsymbol{\mu}})| \cdot |K(\boldsymbol{I}^{\boldsymbol{\mu}})|^{1/2}} \ . \tag{5.7.19}$$

When summing up according to (5.7.7), one must realize that (5.7.14) implies that the $M_i = q\mu_i$ all have the same sign because all ω_i are nonnegative. Moreover, from (5.7.11)

$$\varrho_{\boldsymbol{M}} = \varrho^*_{-\boldsymbol{M}} \ , \tag{5.7.20}$$

and thus the total density fluctuation takes the form

$$\Delta\varrho(E) \;\; = \;\; 2\hbar^{-(f+1)/2} \sum_{\boldsymbol{\mu}>0} |\boldsymbol{\mu}|^{(f-1)/2} \left|\omega\left(\boldsymbol{I}^{\boldsymbol{\mu}}\right) K\left(\boldsymbol{I}^{\boldsymbol{\mu}}\right)^{1/2}\right|^{-1}$$

$$\times \sum_{q=1}^{\infty} q^{(f-1)/2} \cos\left[\frac{qS(\boldsymbol{\mu})}{\hbar} - \frac{q\boldsymbol{\mu} \cdot \boldsymbol{\alpha}\pi}{2} + \frac{\beta(\boldsymbol{\mu})\pi}{4}\right] \ . \tag{5.7.21}$$

This remarkable formula expresses the semiclassical level density fluctuations in terms of purely classical quantities, the latter related to periodic orbits. *Berry* and *Tabor* [123] followed the synthesis of the level density for the Morse potential in two dimensions according to (5.7.21) by accounting for more and more closed orbits. Starting from the smooth Thomas–Fermi background (5.7.5), ever finer variations of $\varrho(E)$ become visible as more closed orbits are included. In the limit, because all such orbits are allowed to contribute, $\varrho(E)$ develops a delta-function peak at each energy level E_i.

It is quite remarkable that the periodic-orbit result (5.7.21) breaks down for the harmonic oscillator since the energy surfaces for this "pathological" system are planes with zero curvature. In fact, the harmonic oscillator was already excluded in Sect. 5.4 for the same reason.

5.8 Level Density Fluctuations for Integrable Systems

It is convenient to start by rescaling the semiclassical energy levels

$$E_m = H\left(I = m\hbar\right) \tag{5.8.1}$$

to uniform mean density (Note that I have dropped the Maslov indices in (5.8.1); they may be imagined eliminated by a shift of the origin in m space; in the semiclassical limit, $m_i \gg \alpha_i$, they are quite unimportant anyway). The unfolding functions (4.7.10) and (4.7.11) are less suitable for the present purpose than one designed by *Berry* and *Tabor* [124] which endows the rescaled levels e_m with the homogeneity property

$$e_{\beta m} = \beta^f e_m \,, \quad \beta > 0 \,. \tag{5.8.2}$$

The construction of the e_m proceeds in two steps. First, one replaces Planck's constant $\hbar$ with a continuous variable h and follows the energy levels $E_m(h) = H(mh)$ as h varies. Their intersections with a fixed reference energy E define a sequence of discrete values h_m. The mapping of the original energy levels $E_m(\hbar)$ onto the numbers h_m will in general be nonlinear (Fig. 5.5) and may even reshuffle the ordering. A second nonlinear mapping gives the rescaled levels as

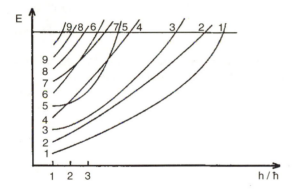

Fig. 5.5. Reshuffling of energy levels in the Berry–Tabor rescaling

$$e_m = \left(\frac{1}{h_m}\right)^f \int_{I>0} d^f I\, \Theta\left[E - H(I)\right] = \left(\frac{\hbar}{h_m}\right)^f \bar{\sigma}(E) \tag{5.8.3}$$

where $\bar{\sigma}(E)$ is the Thomas–Fermi or Weyl form of the average level staircase, normalized such that $\bar{\sigma}(E)/N \to 1$ for $E \to \infty$. The homogeneity (5.8.2) follows trivially from $E = H(mh_m) = H(\beta m h_m/\beta)$. It also follows that

$$h_m \geq \hbar \text{ and } e_m \leq \bar{\sigma}(E) \text{ for } E_m \leq E. \tag{5.8.4}$$

The rescaled levels e_m have the density

$$\varrho(e) = \sum_{m>0} \delta(e - e_m) \tag{5.8.5}$$

the local spectral average of which is obtained when the summation over $\boldsymbol{m}$ is replaced with an f-fold integration,

$$\bar{\varrho}(e) = \int_{\boldsymbol{m}>0} d^f m \delta(e - e_{\boldsymbol{m}}) \,, \tag{5.8.6}$$

as was explained in Sect. 5.7. [The reader may recall the derivation of (5.7.5) from (5.7.2) with the help of Poisson's summation formula.] By virtue of the homogeneity property (5.8.2), the average density is a constant. Indeed, by changing the integration variables in (5.8.6) as $\boldsymbol{m} = \alpha^{1/f}\boldsymbol{x}$ with arbitrary positive α,

$$\begin{aligned} \bar{\varrho}(e) &= \alpha \int_{\boldsymbol{x}>0} d^f x \delta(e - e_{\alpha^{1/f}\boldsymbol{x}}) = \alpha \int_{\boldsymbol{x}>0} d^f x \delta(e - \alpha e_{\boldsymbol{x}}) \\ &= \int_{\boldsymbol{x}>0} d^f x \delta\left(\frac{e}{\alpha} - e_{\boldsymbol{x}}\right) = \bar{\varrho}\left(\frac{e}{\alpha}\right) \,. \end{aligned} \tag{5.8.7}$$

The constant $\bar{\varrho}$, however, must have the value unity since the rescaled average staircase

$$\int_0^e de' \bar{\varrho}(e') = \bar{\varrho}e \tag{5.8.8}$$

grows from 0 to $\bar{\sigma}(E)$ as e grows from 0 to $\bar{\sigma}(E)$, so that $\bar{\sigma}(E) = \bar{\varrho}\bar{\sigma}(E)$ gives

$$\bar{\varrho} = 1 \,. \tag{5.8.9}$$

The remainder of the argument parallels that given in Sect. 5.7. Poisson's summation formula is invoked again to write the density fluctuations as

$$\Delta\varrho(e) = \varrho(e) - 1 = \sum_{\boldsymbol{M}\neq 0} \varrho_{\boldsymbol{M}} \,, \tag{5.8.10}$$

$$\varrho_{\boldsymbol{M}} = \int d^f m \delta(1 - e_{\boldsymbol{m}}) e^{i2\pi \boldsymbol{M}\cdot\boldsymbol{m}e^{1/f}} \,. \tag{5.8.11}$$

In the $\boldsymbol{M}$th density fluctuation, the "standard" energy surface $e_{\boldsymbol{m}} = 1$ occurs in the delta function, and the current energy e is elevated, with the help of the homogeneity (5.8.2), into the phase factor. Again introducing orthogonal coordinates $\boldsymbol{\xi} = (\xi, \dots, \xi_{f-1})$ on the energy surface and using the stationary-phase approximation,

$$\varrho_{\boldsymbol{M}} = \frac{\exp[i2\pi \boldsymbol{M}\cdot\boldsymbol{m}e^{1/f} - i\pi(f-1)/4]}{|\nabla_{\boldsymbol{m}}e_{\boldsymbol{m}}| \cdot |\det(\boldsymbol{M}\cdot(\partial^2 \boldsymbol{m}/\partial\xi_i\partial\xi_j)|^{1/2}e^{(f-1)/2f}}\Bigg|_{\boldsymbol{m}=\boldsymbol{m}^{\boldsymbol{M}}} \,, \tag{5.8.12}$$

provided there is a point $\boldsymbol{m}^{\boldsymbol{M}}$ on the energy surface at which the phase is stationary

$$\boldsymbol{M}\cdot\frac{\partial\boldsymbol{m}}{\partial\xi_i} = 0 \,; \tag{5.8.13}$$

otherwise, if there is no solution m^M to (5.8.13), the Mth density fluctuation is negligible.

In a spectral region near some energy e_0, the density fluctuation ϱ_M displays oscillations of the form

$$\varrho_M(e) = A_M(e_0) \exp\left\{\mathrm{i}\left[K_M(e_0)(e - e_0) + \Phi_M(e_0)\right]\right\} \tag{5.8.14}$$

with an amplitude

$$A_M(e_0) = \left[|\nabla_m e_m|^{-1}\left|\det\left(M\frac{\partial^2 m}{\partial\xi_i\partial\xi_j}\right)\right|^{-\frac{1}{2}} e_0^{-(f-1)/2f}\right]_{m=m^M}, \tag{5.8.15}$$

a "wave number"

$$K_M(e_0) = \frac{2\pi M \cdot m^M}{f e_0^{1-1/f}}, \tag{5.8.16}$$

and a phase

$$\Phi_M(e_0) = 2\pi M \cdot m^M e_0^{1/f} - \frac{\pi}{4}(f - 1). \tag{}$$

For a fixed large energy e_0, the phase Φ_M varies more rapidly (by a factor e_0) than the wave number K_M when one of the components of the vector M changes by unity. Consecutive phase jumps, taken modulo 2π, will tend to fill the interval $[0, 2\pi]$ randomly. Therefore, the total density fluctuation $\Delta\varrho = \varrho(e) - 1$ may be expected to depend quite erratically on $e - e_0$; correspondingly, the density–density correlation function

$$S(e_0 + \tau, e_0) = \frac{1}{\Delta\sigma}\int_{-\Delta\sigma/2}^{+\Delta\sigma/2} d\sigma\,\Delta\varrho(e_0 + \tau + \sigma)\Delta\varrho(e_0 + \sigma) \tag{5.8.17}$$

should decay rapidly with increasing τ.

The reader may recall that $S(e', e'')$ was already studied in Sect. 4.9, with a slightly different normalization and with an ensemble average instead of the local spectral average in (5.8.17). The averaging interval $\Delta\sigma$ must be large enough to contain many levels but small enough to allow approximation (5.8.14).

To check on the expected behavior of S, we insert (5.8.10, 5.8.14) into (5.7.17). Only the "diagonal" terms in the double sum over two vectors M, M' survive. For these,

$$M' = -M, \quad K_{-M} = K_M, \quad \Phi_{-M} = -\Phi_M. \tag{5.8.18}$$

It follows that

$$S(e_0 + \tau, e_0) = \sum_M A_M^2 e^{\mathrm{i}K_M\tau}. \tag{5.8.19}$$

For convenience, now let us consider the Fourier transform of S with respect to the variable τ,

$$
\begin{aligned}
\tilde{S}(e_0, K) &= \frac{1}{2\pi} \int_{-\infty}^{+\infty} d\tau e^{-iK\tau} S(e_0 + \tau, e_0) \\
&= \sum_M A_M^2 \delta(K - K_M) .
\end{aligned}
\tag{5.8.20}
$$

In the limit of large energies, $e_0 \to \infty$, both A_M and K_M are, as already mentioned, smooth functions of $\boldsymbol{M}$. Thus no appreciable error results when the sum over $\boldsymbol{M}$ is replaced by an integral,

$$
\tilde{S}(e_0, K) = \int d^f M A_M^2 \delta(K - K_M) .
\tag{5.8.21}
$$

As a first step toward evaluating the $\boldsymbol{M}$-integral, the length of the frequency vector in the amplitude A_M must be reexpressed as

$$
|\nabla_{\boldsymbol{m}} e_{\boldsymbol{m}}|_{\boldsymbol{m} = \boldsymbol{m}^M} = f / \left(\boldsymbol{m}^M \cdot \hat{\boldsymbol{M}}\right)
\tag{5.8.22}
$$

where $\hat{\boldsymbol{M}} = \boldsymbol{M}/|\boldsymbol{M}|$. Indeed, the energy surface relevant for the gradient in question is $e = 1$, and a nearby energy surface is characterized by the point $\boldsymbol{m}^M(1 + \varepsilon)$, with a small real number ε, at which

$$
1 + \Delta e = e_{(1+\varepsilon)\boldsymbol{m}^M} = (1 + \varepsilon)^f e_{\boldsymbol{m}^M} = (1 + \varepsilon)^f \approx 1 + f\varepsilon .
\tag{5.8.23}
$$

On the other hand, the distance between the two surfaces is $\hat{\boldsymbol{M}} \cdot \varepsilon \boldsymbol{m}$ such that the quotient $\Delta e / \varepsilon \hat{\boldsymbol{M}} \cdot \boldsymbol{m}$ gives (5.8.22). Combining (5.8.15), (5.8.22), and (5.8.21) one gets the Fourier transformed density–density correlation function as

$$
\tilde{S}(e_0, K) =
$$
$$
\int d^f M \frac{(\boldsymbol{m}^M \cdot \hat{\boldsymbol{M}})^2 \delta(K - 2\pi \boldsymbol{M} \cdot \boldsymbol{m}^M) / f e_0^{1-1/f}}{f^2 e_0^{(f-1)/f} |\boldsymbol{M}|^{f-1} |\det(\partial^2 \boldsymbol{m} \cdot \hat{\boldsymbol{M}} / \partial \xi_i \partial \xi_j)|_{\boldsymbol{m} = \boldsymbol{m}^M}} .
\tag{5.8.24}
$$

A most remarkable simplification results now if one changes the integration variables: $\boldsymbol{M} \to K e_0^{1-1/f} \boldsymbol{M}$; the integral is independent of both e_0 and K,

$$
\tilde{S}(e_0, K) = \text{const} .
\tag{5.8.25}
$$

The fact that $\tilde{S}$ is independent of the energy e_0 is not really a big surprise since the rescaling of the energy (5.8.3) was designed so as to make the spectrum statistically homogeneous. The lack of dependence of $\tilde{S}$ on K, however, implies merely the expected rapid falloff of $S(\tau)$ with increasing $|\tau|$:

$$
S(\tau) = 2\pi \tilde{S}(0) \delta(\tau) .
\tag{5.8.26}
$$

A little more geometry reveals that the prefactor $2\pi \tilde{S}(0)$ of the delta function is unity.

The f integration variables M in (5.8.24) may be chosen as $|M|$ and $\hat{M}$ with the unit vector determined in terms of $f-1$ angles; the delta function allows one to perform the $|M|$ integral,

$$\tilde{S} = \frac{1}{2\pi f} \int d^{f-1}\hat{M} \frac{\hat{M} \cdot m^M}{|\det(\partial^2 m \cdot \hat{M}/\partial \xi_i \partial \xi_j)|_{m=m^M}} \ . \tag{5.8.27}$$

Now every direction $\hat{M}$ of M determines a direction $\hat{m}$ of $m = |m|\hat{m}$. Switching to the $f-1$ integration variables $\hat{m}$,

$$\tilde{S} = \frac{1}{2\pi f} \int d^{f-1}\hat{m} \frac{|\partial \hat{M}/\partial \hat{m}| \, |m\hat{m} \cdot \hat{M}|}{|\det(\hat{M} \cdot (\partial^2 m\hat{m}/\partial \xi_i \partial \xi_j)|_{m=m^M}} \ . \tag{5.8.28}$$

As already mentioned in Sect. 5.7, the determinant in the denominator of (5.8.28) is related to the curvature of the energy surface. To establish an interpretation more useful in the present context, recall that the $f-1$ vectors $t_i = \partial m/\partial \xi_i$ are tangential whereas $\hat{M}$ is orthogonal to the energy surface,

$$\hat{M} \cdot \frac{\partial m}{\partial \xi_i} = \hat{M} \cdot t_i = 0 \ . \tag{5.8.29}$$

This identity holds independently of ξ so that the increments of M and t_i along an infinitesimal $d\xi$ are constrained to obey

$$\hat{M} dt_i + t_i d\hat{M} = 0 \ . \tag{5.8.30}$$

A previous stipulation was that the $f-1$ coordinates ξ_i be orthogonal to one another such that an arbitrary infinitesimal vector dm on the energy surface obeys

$$dm = \sum_i t_i d\xi_i \ ,$$
$$(dm)^2 = \sum_i t_i^2 d\xi_i^2 \tag{5.8.31}$$

Now, it is convenient to require that the ξ_i are also locally Cartesian,

$$dm^2 = \sum_i d\xi_i^2 \Leftrightarrow t_i^2 = 1 \ , \tag{5.8.32}$$

i.e., to make the t_i unit vectors. Then, the identity (5.8.30) gives

$$\hat{M} \frac{\partial^2 m}{\partial \xi_i \partial \xi_j} = \hat{M} \frac{\partial \hat{t}_i}{\partial \xi_j} = -\hat{t}_i \frac{\partial \hat{M}}{\partial \xi_j} = -\frac{\partial \hat{M}_i}{\partial \xi_j} \tag{5.8.33}$$

where $d\hat{M}_i$ denotes the ith Cartesian component of the vector $d\hat{M}$. The determinant in question,

$$\det \hat{M} \frac{\partial^2 m}{\partial \xi_i \partial \xi_j} = \det \frac{\partial \hat{M}_i}{\partial \xi_j} = \left| \frac{\partial \hat{M}}{\partial \xi} \right| \ , \tag{5.8.34}$$

is thus revealed as the Jacobian for a transformation from the $f-1$ angular coordinates $\hat{\boldsymbol{M}}$ to the $f-1$ Cartesian coordinates $\boldsymbol{\xi}$.

The integral (5.8.28) now takes the form

$$
\begin{aligned}
\tilde{S} &= \frac{1}{2\pi f} \int_{\hat{m}>0} d^{f-1}\hat{\boldsymbol{m}} \left(\hat{\boldsymbol{m}} \cdot \hat{\boldsymbol{M}}\right) m \frac{|\partial \hat{\boldsymbol{M}}/\partial \hat{\boldsymbol{m}}|}{|\partial \hat{\boldsymbol{M}}/\partial \boldsymbol{\xi}|} \\
&= \frac{1}{2\pi f} \int d^{f-1}\hat{\boldsymbol{m}} \left(\hat{\boldsymbol{m}} \cdot \hat{\boldsymbol{M}}\right) m \left|\frac{\partial \boldsymbol{\xi}}{\partial \hat{\boldsymbol{m}}}\right| ,
\end{aligned}
\tag{5.8.35}
$$

which allows for the following geometric interpretation. The infinitesimal $d^{f-1}\hat{\boldsymbol{m}}$ is a surface element on the f-dimensional unit sphere and $d^{f-1}\hat{\boldsymbol{m}}|\partial \boldsymbol{\xi}/\partial \boldsymbol{m}|$ is a Cartesian surface element on the energy surface $e=1$ with the vectorial representation $d^{f-1}\hat{\boldsymbol{m}}|\partial \boldsymbol{\xi}/\partial \hat{\boldsymbol{m}}|\hat{\boldsymbol{M}}$. This vector has the component $d^{f-1}\hat{\boldsymbol{m}}|\partial \boldsymbol{\xi}/\partial \hat{\boldsymbol{m}}|\hat{\boldsymbol{M}} \cdot \hat{\boldsymbol{m}}$ along $\hat{\boldsymbol{m}}$, which is simply the differential area on the energy surface "above" the element $d^{f-1}\hat{\boldsymbol{m}}$ of the unit sphere. Written with the help of the polar representation of the energy surface, that latter element is $d^{f-1}\hat{\boldsymbol{m}}[m(\hat{\boldsymbol{m}})]^{f-1}$, and the integral in (5.8.35) takes the form

$$
\tilde{S} = \frac{1}{2\pi f} \int_{\hat{m}>0} d^{f-1}\hat{\boldsymbol{m}} \, [m(\hat{\boldsymbol{m}})]^f .
\tag{5.8.36}
$$

The physical meaning of (5.8.36) becomes apparent when the $(f-1)$-fold integral is blown up to the f-fold integral:

$$
\tilde{S} = \frac{1}{2\pi} \int_{m>0} d^f m \, \Theta \left[m - m(\hat{\boldsymbol{m}})\right] .
\tag{5.8.37}
$$

One recognizes the average level staircase $\bar{\sigma}(e)$ evaluated at $e=1$. Since $\bar{\sigma}(e) = e$, the final result reads

$$
\tilde{S} = \frac{1}{2\pi} .
\tag{5.8.38}
$$

In view of (5.8.26), the density–density correlation function becomes

$$
S(\tau) = \delta(\tau) .
\tag{5.8.39}
$$

The local fluctuations of the level density at different energies bear no correlations, just as if the levels followed one another as the independent events of a Poisson process.

5.9 Exponential Spacing Distribution for Integrable Systems

Now, I propose to show that the correlation function $S(\tau)$ is related to the conditional probability density $g(\tau)$ of finding a level in the interval $[e+\tau, e+\tau+d\tau]$, given one at e. It was shown in Sect. 5.5 that this conditional probability density determines the spacing distribution $P(\sigma)$ according to

$$P(\sigma) = \text{const } g(\sigma) e^{-\int_0^\sigma d\tau g(\tau)} . \tag{5.9.1}$$

Denoting the local spectral average by the brackets $\langle \dots \rangle$, let us consider

$$1 - \delta(\tau) + S(\tau)$$

$$= 1 - \delta(\tau) + \left\langle \left\{ \sum_m \delta(e_0 + \tau - e_m) - 1 \right\} \left\{ \sum_{m'} \delta(e_0 - e_{m'}) - 1 \right\} \right\rangle$$

$$= -\delta(\tau) + \sum_{m,m'} \langle \delta(e_0 + \tau - e_m) \delta(e_0 - e_{m'}) \rangle$$

$$= \sum_{m \neq m'} \langle \delta(e_0 + \tau - e_m) \delta(e_0 - e_{m'}) \rangle . \tag{5.9.2}$$

The little calculation in (5.9.2) uses $\langle \varrho \rangle = 1$. By writing out the spectral average explicitly, one obtains

$$1 - \delta(\tau) + S(\tau)$$

$$= \left\{ \frac{1}{\Delta\sigma} \int_{-\frac{\Delta\sigma}{2}}^{+\frac{\Delta\sigma}{2}} d\sigma \sum_m \delta(e_0 + \tau + \sigma - e_m) \right\} \sum_{m'(\neq m)} \delta(\tau - e_m + e_{m'}) . \tag{5.9.3}$$

The sum over m' gives the probability density of having a level at a distance τ away from e_m and the curly bracket averages over all levels e_m in the interval $\Delta\sigma$ around e_0. The function considered in (5.9.3) is therefore the probability density for finding a level in $[e + \tau, e + \tau + d\tau]$ given one at e, where e may lie anywhere in the spectral range $\Delta\sigma$ around e_0,

$$g(\tau) = 1 - \delta(\tau) + S(\tau) . \tag{5.9.4}$$

Together with $S(\tau) = \delta(\tau)$, which is typical for integrable systems, one has

$$g(\tau) = 1 \tag{5.9.5}$$

and thus the exponential distribution of level spacings

$$P(\tau) = e^{-\tau} . \tag{5.9.6}$$

5.10 Equivalence of Different Unfoldings

The unfolding employed in Sect. 5.8 makes the rescaled levels homogeneous with degree f in the f quantum numbers $\boldsymbol{m}$. That homogeneity was of critical importance at several stages of the argument. For one thing, all geometric considerations were referred to the single standard energy surface $e = 1$; moreover, the oscillatory dependence of the Mth density fluctuation ϱ_M on the energy could be made manifest. The question arises whether the exponential spacing distribution is really a general property of integrable systems with $f \geq 2$ and not simply an artefact due to a peculiar unfolding.

It was argued in Sect. 4.7 that the most natural unfolding involves the average level staircase $\bar{\sigma}(E)$ as

$$e' = \bar{\sigma}(E) . \tag{5.10.1}$$

In practical applications, the average density is often used to rescale

$$e'' = E\bar{\varrho}(E) . \tag{5.10.2}$$

This unfolding is only locally equivalent to (5.10.1), i.e., with respect to spectral regions within which the density $\bar{\varrho}$ is practically constant.

The unfoldings (5.10.1, 5.10.2) do not in general provide homogeneity of degree f to the rescaled semiclassical levels e'_m and e''_m. An exception arises only for integrable Hamiltonians which themselves are homogeneous in the f actions $\boldsymbol{I}$ (see Problem 5.6). For all other integrable systems, it is not immediately obvious that (5.10.1, 5.10.2) also yield exponentially distributed spacings.

An important property of (5.10.1) is the monotonicity of $e'(E)$ since this ensures that $e'_i > e'_j$ provided $E_i > E_j$. The density-based unfolding (5.10.2) need not strictly have that monotonicity but does in almost all cases of practical relevance. Not so for Berry and Tabor's unfolding used in Sect. 5.8! It was in fact pointed out that the original levels E_m may be reshuffled in their order by the rescaling (5.8.3) rather than just squeezed together or stretched apart to secure uniform mean spacing. Could the e_m owe their exponential spacing distribution to such reshuffling?

I cannot formally prove that (5.10.1) and (5.10.2) yield exponentially distributed spacings whenever (5.8.3) does so. Nonetheless, a reasonably convincing argument may be drawn from the following example. Consider a particle with $f = 2$ subjected to a harmonic binding in one coordinate and free in a finite interval with respect to the second coordinate. Then, the Hamiltonian takes the form

$$H = \alpha I_1 + \beta I_2^2 \tag{5.10.3}$$

and the semiclassical levels read (again dropping the Maslov index for the oscillator part!)

$$E_m = \alpha\hbar m_1 + \beta\hbar^2 m_2^2 . \tag{5.10.4}$$

Clearly, intersections of levels become possible once Planck's constant is replaced by a continuous variable h. Therefore, the Berry–Tabor levels e_m will not have precisely the same ordering as the E_m. The possible reshuffling, however, is not due to any externally imposed randomness; rather, one is facing the fact that a single free parameter in H suffices to enforce a level crossing, i.e., a generic property of integrable systems. All effective randomness in the sequence of the e_m thus appears entirely due to the assumed integrability of the Hamiltonian H.

5.11 Problems

5.1. Give $H(\boldsymbol{I})$ for the Kepler problem.

5.2. Show that the hydrogen spectrum is as pathological (nongeneric) as that of the harmonic oscillator. Give other examples of nongeneric spectra of integrable systems.

5.3. Show that the probability density for finding the kth neighbor of a level in the distance increment [S, S+dS], for a stationary Poissonian "process" is

$$P_k(S) = \frac{S^{k-1}}{(k-1)!} e^{-S} .$$

For $k = 2$ the so-called semi-Poissonian distribution results which is usually written as $P(S) = 4Se^{-2S}$, so as to secure $\langle S \rangle = 1$; see Ref. [125].

5.4. Show that the conditional probability $g(S)$ defined in Sect. 5.5 is related to Dyson's two-level cluster function (Sect. 4.9) $Y(S) = 1 - g(S)$, provided that the spectrum is homogeneous.

5.5. Prove rigorously that L independent spectra with $E_\mu(S) = (1/L^2)\exp(-LS)$ superpose to yield a spectrum with exponentially spaced levels.

5.6. Show that the Thomas–Fermi level staircase is homogeneous of degree f for integrable systems whose Hamiltonian is a homogeneous function of the f actions $\boldsymbol{I}$. Give the explicit form of $\bar{\sigma}(E)$.

5.7. In which sense may the frequencies ω_i defined by (5.2.2, 5.2.3) be taken to be positive? (see the remark on (5.7.14) following (5.7.19)).

6. Level Dynamics

6.1 Preliminaries

As *Pechukas* discovered [126], the fate of the eigenvalues and eigenvectors of a Hamiltonian $H = H_0 + \lambda V$ upon variation of λ can be described by a set of ordinary first-order differential equations. That set can be interpreted as Hamilton's equations for fictitious classical particles moving in one dimension as the "time" λ elapses [127, 128]. The number N of these fictitious particles equals the number of levels of H. However, the phase space of the fictitious many-body systems has a dimension larger than $2N$ due to the fact that the coupling strengths for particle pairs become dynamic variables themselves. The nontrivial interactions notwithstanding, the fictitious-particle dynamics is integrable; this integrability follows from the equivalence between the dynamics in question and the quantum mechanical problem of diagonalizing the finite-dimensional matrix $H = H_0 + \lambda V$.

The same strategy works for Floquet operators $F = \exp(-i\lambda V) \exp(-iH_0)$ of periodically kicked quantum systems [129, 130] on which this chapter largely concentrates. Here, as the fictitious time λ elapses, the phase-space trajectory of the fictitious classical system winds around an N torus and tends to cover that torus ergodically [131].

The reformulation of level dynamics as the classical Hamiltonian flow of a fictitious one-dimensional gas is particularly useful for dynamic systems whose classical limit is globally chaotic. If there are two or more classical degrees of freedom and if $H_0 + \lambda V$ is classically integrable, the quantum energy levels do not generically repel, at least not in any way reminiscent of repulsive interactions between particles. As we have seen in the previous chapter and shall check again here from a different perspective, levels from different multiplets cross freely. Intra-multiplet encounters of levels, it shall be revealed here, typically take the form of just barely avoided crossings with the closest approach spacing often difficult to resolve, at least in the limit $N \to \infty$.

For chaotic dynamics, on the other hand, the limit $N \to \infty$ allows applying statistical mechanics to the fictitious particles and thus indirectly to the original quantum problem represented by the Hamiltonian H or the Floquet operator F. We shall in fact see that random-matrix theory for the original quantum problem arises as equilibrium statistical mechanics for the fictitious particles: Spectral characteristics like the distribution of nearest-neighbor spacings or the two-point cluster function of the level density calculated with a uniform probability coverage of the torus come out as predicted by random-matrix theory to within corrections of relative weight $\frac{1}{N}$.

Thus, an important step toward understanding the universality of spectral fluctuations according to the Bohigas–Giannoni–Schmit conjecture [132] can be made. Ergodic coverage of an N-torus means equality of time (here, "time" is the control parameter λ) and ensemble averages of spectral characteristics like the distribution of nearest neighbor spacings. Since the "time" interval needed for the average is of the order of a collision time of the fictitious gas and as such of the order $\frac{1}{N^\nu} \propto \hbar^{f\nu}$ with some positive exponent ν, all averaged dynamical systems have the same classical limit.

Concepts and methods of nonequilibrium statistical mechanics also come into play. For instance, the transition from level clustering to level repulsion (which accompanies the classical transition from predominantly regular motion to global chaos) appears as a relaxation into equilibrium [133]. Another example is the transition from one universality class of level repulsion to another, when H or F changes its antiunitary symmetries as λ grows. Such relaxation processes will be illustrated by calculating the level spacing distribution $P(S, \lambda)$ for some simple cases. In perhaps the most interesting application of irreversible statistical mechanics, I shall show that Dyson's Brownian-motion model [134] is a rigorous consequence of level dynamics for autonomous systems.

Throughout this chapter, Planck's constant will be set equal to unity, $\hbar = 1$, except where the disregard of Planck might hurt the feelings of even the toughest theorists.

6.2 Fictitious Particles

We shall deal with time-dependent Hamiltonians of the form

$$H(t) = H_0 + \lambda V \sum_{n=-\infty}^{+\infty} \delta(t - n) \tag{6.2.1}$$

which entail the single-period Floquet operators

$$F = \mathrm{e}^{-\mathrm{i}\lambda V} \mathrm{e}^{-\mathrm{i}H_0} . \tag{6.2.2}$$

The dimension of the Hilbert space is assumed finite and will be denoted by N. When the kick strength λ is varied, the eigenvalues and eigenvectors of F, defined by

$$F|m\rangle = \mathrm{e}^{-\mathrm{i}\phi_m}|m\rangle , \quad m = 1, 2, \ldots N , \tag{6.2.3}$$

will change, and a description of that change is to be established. Our goal is similar to that of perturbation theory, but we demand more than just the first few terms of a power series in λ.

Differentiating the identity $F_{mm} = \mathrm{e}^{-\mathrm{i}\phi_m}$ with respect to the kick strength,

$$\dot{\phi}_m = V_{mm} + \mathrm{i}\left\{\langle \dot{m}|m\rangle + \langle m|\dot{m}\rangle\right\} = V_{mm} , \tag{6.2.4}$$

assuming that the eigenstates $|m\rangle$ are normalized to unity for all λ (a dot above a bra or ket denotes differentiation of that bra or ket). To find the variation of V_{mm} with λ, we differentiate again:

$$
\begin{aligned}
\dot{V}_{mm} &= \langle\dot{m}|V|m\rangle + \langle m|V|\dot{m}\rangle \\
&= \sum_n \left\{ \langle\dot{m}|n\rangle V_{nm} + V_{mn}\langle n|\dot{m}\rangle \right\} .
\end{aligned}
\tag{6.2.5}
$$

The quantities $\langle\dot{m}|n\rangle$ and $\langle n|\dot{m}\rangle$ are obtained by differentiating the eigenvalue equation (6.2.3) and taking the scalar product with $|n\rangle (\neq |m\rangle)$,

$$
\langle n|\dot{m}\rangle = \langle\dot{m}|n\rangle^* = \frac{-\mathrm{i}V_{nm}}{1 - \mathrm{e}^{-\mathrm{i}\phi_{nm}}} ,
\tag{6.2.6}
$$

so that the rate of change of the diagonal element V_{mm} reads

$$
\dot{V}_{mm} = \mathrm{i} \sum_{n(\neq m)} V_{mn} V_{nm} \left(\frac{1}{1 - \mathrm{e}^{-\mathrm{i}\phi_{mn}}} - \frac{1}{1 - \mathrm{e}^{+\mathrm{i}\phi_{mn}}} \right) .
\tag{6.2.7}
$$

Note the shorthand

$$
\phi_{mn} = \phi_m - \phi_n
\tag{6.2.8}
$$

for the difference between two quasi-energies. The off-diagonal elements V_{mn} are treated similarly,

$$
\dot{V}_{mn} = \langle\dot{m}|V|n\rangle + \langle m|V|\dot{n}\rangle = \sum_l \left\{ \langle\dot{m}|l\rangle V_{ln} + V_{ml}\langle l|\dot{n}\rangle \right\} .
\tag{6.2.9}
$$

Separating $l = m$, $l = n$ from $l \neq m, n$ and considering that $\langle n|\dot{n}\rangle = \langle m|\dot{m}\rangle = 0$ can be realized by properly choosing a phase factor for each eigenvector (see Problem 6.5), we finally obtain

$$
\begin{aligned}
\dot{V}_{mn} =\ & -\mathrm{i}\frac{V_{mn}(V_{mm} - V_{nn})}{1 - \mathrm{e}^{-\mathrm{i}\phi_{mn}}} \\
& + \mathrm{i} \sum_{l(\neq m,n)} V_{ml} V_{ln} \left(\frac{1}{1 - \mathrm{e}^{-\mathrm{i}\phi_{ml}}} - \frac{1}{1 - \mathrm{e}^{-\mathrm{i}\phi_{ln}}} \right) ,\quad m \neq n .
\end{aligned}
\tag{6.2.10}
$$

The differential equations (6.2.4, 7, 10) form a complete set. We shall simply refer to that set as level dynamics. Together with "initial" conditions set at, say, $\lambda = 0$, they suffice to determine the quasi-energies ϕ_m and all matrix elements of the perturbation V for arbitrary values of λ. Perturbation theory could be extracted by solving in terms of a power series in λ.

The flow so defined in a space spanned by the ϕ_m, V_{mm}, V_{mn} cannot, however, be a classical Hamiltonian flow since it is not free of sources,

$$
\sum_m \left(\frac{\partial\dot{\phi}_m}{\partial\phi_m} + \frac{\partial\dot{V}_{mm}}{\partial V_{mm}} \right) + \sum_{m\neq n} \frac{\partial\dot{V}_{mn}}{\partial V_{mn}} = -\mathrm{i} \sum_{m\neq n} \frac{V_{mm} - V_{nn}}{1 - \mathrm{e}^{-\mathrm{i}\phi_{mn}}} \neq 0 .
\tag{6.2.11}
$$

The net divergence of the flow originates from the first term in (6.2.10). Thus, an intuitive next step is to replace the off-diagonal elements V_{mn} by

$$l_{mn} = V_{mn} f_{mn} \tag{6.2.12}$$

and to try to determine the factors f_{mn} so as to cancel the first term in (6.2.10) in the "equation of motion" of l_{mn}. Via the relationships $\dot{l} = \dot{V} f + V \dot{f}$ and (6.2.4), this requirement yields

$$\frac{\dot{f}_{mn}}{f_{mn}} = \frac{i\dot{\phi}_{mn}}{1 - e^{-i\phi_{mn}}} = \frac{i\dot{\phi}_{mn}}{2} + \left[\ln \sin \left(\frac{\phi_{mn}}{2} \right) \right]^{\cdot} \tag{6.2.13}$$

which is easily integrated to

$$f_{mn} = \mathrm{const}\, e^{i\phi_{mn}/2} \sin \left(\frac{\phi_{mn}}{2} \right) . \tag{6.2.14}$$

For convenience, the integration constant is chosen as -2 and thus

$$l_{mn} = -2 V_{mn} e^{i\phi_{mn}/2} \sin \left(\frac{\phi_{mn}}{2} \right) = -l_{nm}^{*} . \tag{6.2.15}$$

After a little algebra, the now source-free flow can be written as

$$\dot{\phi}_m = p_m$$

$$\dot{p}_m = - \sum_{n(\neq m)} \frac{1}{4} l_{mn} l_{nm} \frac{\cos (\phi_{mn}/2)}{\sin^3 (\phi_{mn}/2)} \tag{6.2.16}$$

$$\dot{l}_{mn} = \sum_{l(\neq m,n)} \frac{1}{4} l_{ml} l_{ln} \left[\sin^{-2} \left(\frac{\phi_{ml}}{2} \right) - \sin^{-2} \left(\frac{\phi_{ln}}{2} \right) \right]$$

where $p_m = V_{mm}$ has been introduced as an obvious shorthand.

The first two equations in the set (6.2.16) can clearly be interpreted as classical Hamiltonian equations

$$\dot{\phi}_m = \frac{\partial \mathcal{H}}{\partial p_m} = \{ \mathcal{H}, \phi_m \}$$
$$\dot{p}_m = -\frac{\partial \mathcal{H}}{\partial \phi_m} = \{ \mathcal{H}, p_m \} \tag{6.2.17}$$

with λ as the "time" and the Hamiltonian function

$$\mathcal{H} = \frac{1}{2} \sum_{m=1}^{N} p_m^2 + \sum_{m \neq n} \frac{1}{8} \frac{|l_{mn}|^2}{\sin^2(\phi_{mn}/2)} . \tag{6.2.18}$$

Thus, one is confronting N fictitious classical particles of unit mass located on the unit circle, $0 \leq \phi_m < 2\pi$, and interacting pairwise via a repulsive potential $\sim 1/\sin^2 (\phi_{mn}/2)$. The coupling strengths $|l_{mn}|^2$ are not constants, however, but dynamic variables obeying the third equation in the set (6.2.16). The l_{mn} cannot be looked upon as single-particle properties; each refers to a pair of particles. Nonetheless, they have come to be called "angular momenta." Their equations of motion are revealed as having Hamiltonian structure by assigning Poisson brackets to the l_{mn}.

To distinguish between Poisson brackets for the angular momenta and those of the canonical pairs p_m, ϕ_m,

$$\{p_m, \phi_n\} = \delta_{mn} , \tag{6.2.19}$$

the former will be denoted by round brackets $(\dots , \dots)$. The two Poisson brackets in question are independent in the sense that

$$(l_{ij}, p_m) = (l_{ij}, \phi_m) = 0$$
$$\{p_m, l_{ij}\} = \{\phi_m, l_{ij}\} = 0 . \tag{6.2.20}$$

The Leibniz product rule is assumed to hold for the angular-momentum brackets as it does for the p and ϕ:

$$(f, gh) = (f, g)h + g(f, h) . \tag{6.2.21}$$

The appropriate Poisson brackets look slightly different for the three universality classes (orthogonal, unitary, and symplectic) of Floquet operators. When F enjoys $O(N)$ as its group of canonical transformations,[1]

$$l_{mn} = -l_{nm} , \tag{6.2.22}$$

and

$$(l_{mn}, l_{ij}) = \tfrac{1}{2} \left(\delta_{mj} l_{ni} + \delta_{ni} l_{mj} - \delta_{nj} l_{mi} - \delta_{mi} l_{nj} \right) , \tag{6.2.23}$$

whereas $U(N)$ entails

$$l_{mn} = -l_{nm}^* \tag{6.2.24}$$

and

$$(l_{mn}, l_{ij}) = \delta_{in} l_{mj} - \delta_{mj} l_{in} . \tag{6.2.25}$$

I shall not pause to construct the brackets (6.2.23, 25) and that pertaining to the symplectic group $Sp(N)$ here [135]; Problem 6.3 and Chap. 9.10 offer different

---

[1] The Poisson brackets (6.2.23, 25) may be read as those of generators of infinitesimal rotations in, respectively, real or complex N-dimensional vector spaces; hence the name "angular momenta"; see Problem 6.3. Note that the antisymmetry (6.2.22) in the orthogonal case is not manifest in the definition (6.2.15); however, suitable constant phase factors in the "initial" Floquet eigenstates $|m, \lambda = 0\rangle$ make the initial matrix $l_{mn}(\lambda = 0)$ real antisymmetric and the equations of motion then preserve that property for $\lambda > 0$.

methods for the derivation. For now it is more important and, in fact, easy to verify that the above angular momentum Poisson brackets yield the equation of motion for the l_{mn} in (6.2.16) as

$$\dot{l}_{mn} = (\mathcal{H}, l_{mn}) \ . \tag{6.2.26}$$

The reader is invited in Problem 6.4 to establish the fictitious-particle dynamics corresponding to time-independent Hamiltonians $H = H_0 + \lambda V$,

$$\dot{x}_m = \frac{\partial \mathcal{H}}{\partial p_m}$$
$$\dot{p}_m = -\frac{\partial \mathcal{H}}{\partial x_m}$$
$$\dot{l}_{mn} = (\mathcal{H}, l_{mn}) \tag{6.2.27}$$
$$\mathcal{H} = \tfrac{1}{2} \sum_m p_m^2 + \tfrac{1}{2} \sum_{m \neq n} \frac{|l_{mn}|^2}{(x_m - x_n)^2} \ ,$$

where the x_m denote the eigenenergies of H. The classical Hamiltonian flow (6.2.27), first found by *Pechukas* [126] and *Yukawa* [127, 128], can formally be obtained from (6.2.16) by linearizing via $\sin x \to x$, $\cos x \to 1$; thus the periodic potential proportional to $\sin^{-2}(\phi_{mn}/2)$ is replaced by an inverse-square potential proportional to $1/x_{mn}^2$.

Prior to the discovery of their relevance to the diagonalization of quantum Hamiltonians and Floquet operators, the classical flows (6.2.27) and (6.2.16) had been known in the mathematical literature. The special cases of constant l_{mn} are the Calogero–Moser and Sutherland–Moser dynamics [136–138]. Even the case of dynamic l_{mn} had been investigated by *Wojciechowski* [139] as a "marriage of the Euler equations with the Calogero–Moser system." From the classical point of view, an interesting property of these systems is their integrability. In the present context, the integrability of (6.2.27) is not surprising since fictitious-particle dynamics is equivalent to the diagonalization of an $N \times N$ matrix.

A more elegant and powerful method of establishing level dynamics will be presented in Chap. 9.10; in particular, the derivation of the Poisson brackets for the variables x, p, l will become easily accessible there. The curious reader will be excused for momentarily visiting that exposition before digging deeper into this chapter.

6.3 Conservation Laws

To find conserved quantities for classical flow (6.2.16), it is convenient to rederive that flow slightly more abstractly. Let us imagine the matrices $F = \mathrm{e}^{-\mathrm{i}\lambda V} F_0$, V and F_0 specified in some fixed representation independent of λ. In this section, lower case Latin letters will be reserved for matrices expressed in the eigenrepresentation of the Floquet operator F, defined by

$$\mathrm{e}^{-\mathrm{i}\phi} = W^{\dagger} F W \ ,$$
$$\phi = \mathrm{diag}\,(\phi_1, \phi_2, \ldots, \phi_N) \ . \tag{6.3.1}$$

The diagonalizing unitary transformation W depends, of course, on λ. The matrix

$$v = W^\dagger V W \tag{6.3.2}$$

has the elements

$$v_{mn} = \delta_{mn} p_m + (1 - \delta_{mn}) V_{mn} = \delta_{mn} p_m + \frac{(1 - \delta_{mn}) l_{mn}}{\mathrm{i}(e^{\mathrm{i}\phi_{mn}} - 1)} , \tag{6.3.3}$$

and the matrix

$$l = \mathrm{i} \left(e^{\mathrm{i}\phi} v e^{-\mathrm{i}\phi} - v \right) \tag{6.3.4}$$

has as its elements the angular momenta l_{mn} given in (6.2.15). We will also employ the matrix

$$a = \dot{W}^\dagger W , \tag{6.3.5}$$

where the dot again means differentiation with respect to λ. It is easily seen that a is anti-Hermitian,

$$a^\dagger = W^\dagger \dot{W} = -W^\dagger W \dot{W}^\dagger W = -a ; \tag{6.3.6}$$

thus its diagonal elements are imaginary; in fact, they can be set equal to zero by properly choosing phase factors for the eigenvectors of F (Problem 6.5).

By differentiating $e^{-\mathrm{i}\phi}$, v, and l with respect to λ, we obtain

$$\dot{\phi} = \mathrm{i}a + v - \mathrm{i}e^{-\mathrm{i}\phi} a e^{\mathrm{i}\phi} ,$$
$$\dot{v} = [a, v] , \tag{6.3.7}$$
$$\dot{l} = [a, l] .$$

The first of these equations yields $\dot{\phi}_m = v_{mm} = p_m$ and, upon taking off-diagonal elements, one obtains the a_{mn} in terms of the V_{mn} and the quasi-energies ϕ_m as

$$a_{mn} = \frac{\mathrm{i}v_{mn} e^{\mathrm{i}\phi_{mn}}}{e^{\mathrm{i}\phi_{mn}} - 1} . \tag{6.3.8}$$

The equation of motion for v in (6.3.7) comprises (6.2.9) and (6.2.10) and thus especially the canonical equation for the momenta p_m from (6.2.16), while the equation of motion of l is the canonical equation of the angular momenta (6.2.26). According to the last two equations in (6.3.7), the matrix a may be looked upon as a generator of infinitesimal "time" translations. Most importantly, the commutator structure in these equations, commonly called the Lax form [140] of the canonical equations, immediately yields an infinity of constants of the motion,

$$C_{\mu_1 \nu_1 \mu_2 \nu_2 \ldots} = \mathrm{Tr}\, \{ l^{\mu_1} v^{\nu_1} l^{\mu_2} v^{\nu_2} \ldots \}$$
$$\mu_1, \nu_1, \mu_2, \nu_2, \ldots = 0, 1, 2, 3, \ldots . \tag{6.3.9}$$

Of course, not all of these constants are functionally independent. It is not known at present whether the $C_{\mu_1\nu_1\mu_2\nu_2,...}$ contain all functionally independent constants of the flow (6.2.16) or (6.3.7).

As we shall see in Sect. 6.7 below, the flow (6.2.16) is generically ergodic on an N-torus. Then, it will be natural to apply statistical mechanics to the N fictitious particles. Clearly, the appropriate statistical ensemble to be used is a generalized microcanonical ensemble represented by a product of delta functions, one factor for each of the constants of the motion defining the N-torus. But before embarking onto statistical analyses, we need to clarify some more basic issues.

6.4 Intermultiplet Crossings

Even nonintegrable classical Hamiltonian flows or maps can have constants of the motion C, possibly symmetry based. For convenience, I formulate the following for a Hamiltonian flow and assume a single conserved quantity, but everything is easily transcribed to maps and any number of conservation laws. Such a C has a vanishing Poisson bracket with the Hamiltonian function H, $\{H, C\} = 0$. The two corresponding quantum operators, to be distinguished by a hat in the present section, then commute, $[\hat{H}, \hat{C}] = 0$. The quantum energy spectrum consists of multiplets, each of which is labelled by an eigenvalue of $\hat{C}$.

Now let us consider a Hamiltonian of the structure $\hat{H}(\lambda) = \hat{H}_0 + \lambda\hat{V}$. If the single conservation law we are allowing for holds for all values of λ, the conserved quantity will in general also depend on λ and be denoted by $\hat{C}(\lambda)$. Of course, $\hat{C}$ remains independent of λ if commuting with both $\hat{H}_0$ and $\hat{V}$, and this is the natural situation when one and the same symmetry reigns for $\hat{H}_0$ and $\hat{V}$. I shall first discuss that simple case but still insist on $[\hat{H}_0, \hat{V}] \neq 0$.

Levels from different multiplets have no reason not to cross when the control parameter λ is varied. In fact, since $\hat{H}_0$ and $\hat{C}$ have common eigenvectors, $\hat{C}|m, i; \lambda\rangle = C_i|m, i; \lambda\rangle$, $\hat{H}|m, i; \lambda\rangle = E_{mi}(\lambda)|m, i; \lambda\rangle$, off-diagonal matrix elements of the perturbation $\hat{V}$ pertaining to different C_i, i.e., to different multiplets, vanish, $\langle m, i; \lambda|\hat{V}|n, j; \lambda\rangle = 0$. Therefore, the level dynamics (6.2.16) has vanishing intermultiplet couplings and splits into independent dynamics for the separate multiplets. Then, an intermultiplet crossing of levels has codimension 1 which means uninhibited crossings as λ varies. Loosely speaking, every multiplet has its own Pechukas–Yukawa gas.

Intermultiplet crossings are still unhindered if the conserved quantity in question depends on λ, provided only a pair of common eigenfunctions $|i, \lambda\rangle$ of $\hat{H}(\lambda)$ and $\hat{C}(\lambda)$ with close by energies, $E_1(\lambda) \approx E_2(\lambda)$, do not have the eigenvalues $C_1(\lambda)$ and $C_2(\lambda)$ in simultaneous close approach for, say $\lambda \approx \lambda_0$. In that case, we may start from the the eigenfunctions $|i, \lambda_0\rangle$ of $\hat{H}(\lambda_0)$ and $\hat{C}(\lambda_0)$ and calculate their successors $|i, \lambda_0 + \delta\lambda\rangle$ perturbatively, treating $\delta\lambda\hat{C}'(\lambda_0)$ as a perturbation of $\hat{C}(\lambda_0)$. Since by assumption the eigenvalues $C_1(\lambda_0)$ and $C_2(\lambda_0)$ are not close to one another, the denominator $C_1(\lambda_0) - C_2(\lambda_0)$ in the perturbation expansion is not small such that we can even calculate the shifted energies employing the zero-order eigenfunctions as $E_i(\lambda) = E_i(\lambda_0) + \delta\lambda\langle i, \lambda_0|\hat{V}|i, \lambda_0\rangle \equiv E_i(\lambda_0) + \delta\lambda\hat{V}_{ii}$. Thus, we expect the nearest crossing for

$$\delta\lambda = -\frac{E_1(\lambda_0) - E_2(\lambda_0)}{V_{11} - V_{22}}. \tag{6.4.1}$$

It is worth underscoring the responsibility of the conservation $\hat{C}(\lambda)$ for close encounters of levels from different multiplets to be associated with true crossings rather than anticrossings. Moreover, conserved quantities dependent on the strength of a perturbation is no more outlandish a phenomenon than hydrogen in an electric field [140].

6.5 Level Dynamics for Classically Integrable Dynamics

An integrable classical Hamiltonian flow with f freedoms has (at least) f conserved quantities which may be chosen as the actions of f action angle pairs of phase-space coordinates. If the f conservation laws remain rigorously intact quantum mechanically, the quantum energy spectrum consists of multiplets within each of which only the quantum number associated with a single conserved observable labels levels. What we have just learned about intermultiplet crossings remains true here as well. It remains to clarify whether intramultiplet crossings can generically occur or are avoided. That question can be addressed for $f = 1$, without loss of generality, and a satisfactory answer can be found with the help of a few examples.

There are single-freedom systems without level crossings like the harmonic oscillator with $H_0 = \frac{1}{2m}p^2 + \frac{1}{2}m\omega^2 x^2$ and $V = -\lambda x$ with λ a constant external force. The well-known levels $E_n(\lambda) = (n + \frac{1}{2})\omega - \lambda^2/2m\omega^2$ retain their spacing ω as λ varies. Another instructive example is the Hamiltonian

$$H = \omega J_z + \lambda J_x = \sqrt{\omega^2 + \lambda^2}\left(\frac{\omega}{\sqrt{\omega^2 + \lambda^2}}J_z + \frac{\lambda}{\sqrt{\omega^2 + \lambda^2}}J_x\right) \tag{6.5.1}$$

which generates uniform rotation rotation with angular velocity $\sqrt{\omega^2 + \lambda^2}$ about the axis defined by the unit vector $\hat{e} = (\frac{\omega}{\sqrt{\omega^2 + \lambda^2}}, 0, \frac{\lambda^2}{\sqrt{\omega^2 + \lambda^2}})$. In a Hilbert space with total angular momentum quantum number j, the levels of that Hamiltonian are $E_m(\lambda) = m\sqrt{\omega^2 + \lambda^2}$ with $m = -j, -j+1, \ldots, j$. These never cross as λ grows. We could even speak of a single avoided crossing for each pair of nearest neighbor levels at $\lambda = 0$ since the spacing $\sqrt{\omega^2 + \lambda^2}$ does have a minimum there. However, the spectrum has no fluctuations at all inasmuch as we can find an explicit (λ-dependent) rescaling of the energy to get a rigid ladder of equidistant levels. Such rescalings are achieved by $H' = (\omega J_z + \lambda J_x)/\sqrt{\omega^2 + \lambda^2}$ or, equivalently, $H'' = J_z \cos\lambda + J_x \sin\lambda$.

But let us not overhastily take complete rigidity of single multiplets as generic! Some Hamiltonians composed of angular momentum operators immediately tell us to beware of such a misconception. For instance,

$$H(\lambda) = J_z \cos\lambda + \left[\tfrac{4}{2j+1}J_z^2 - \tfrac{1}{2}J_z - (j + \tfrac{1}{2})\right]\sin\lambda \tag{6.5.2}$$

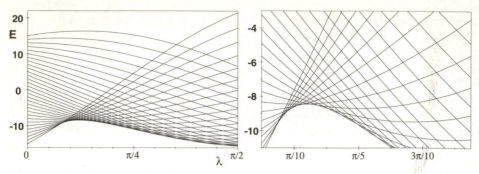

Fig. 6.1. Level crossings for the single-freedom Hamiltonian $H(\lambda) = J_z \cos \lambda + (\frac{1}{2}J_z + \frac{4}{2j+1}J_z^2 - j) \sin \lambda$ with $j = 15$. The right part is a blowup of the left

has the $2j + 1$ levels

$$E_m(\lambda) = m \cos \lambda + \left[\frac{4}{2j+1}m^2 - \frac{1}{2}m - (j + \frac{1}{2})\right] \sin \lambda \qquad (6.5.3)$$

with $m = -j, -j + 1, \ldots, j$, if the quantum number j is kept fixed. Figure 6.1 displays the level dynamics for the latter Hamiltonian as $H(\lambda)$ is converted from $H_0 = J_z$ into $V = \frac{4}{2j+1}J_z^2 - \frac{1}{2}J_z - (j + \frac{1}{2})$. It is easy to see that the number of crossings in the interval $0 \leq \lambda \leq \pi/2$ is $\propto N^2$. These crossings arise since E_m is a quadratic function of the quantum number m. Momentarily letting m range continuously, we encounter a minimum of E_m at $m_{\min}(\lambda) = (\frac{1}{8} - \frac{1}{4}\cot \lambda)(j + \frac{1}{2})$. As long as that $m_{\min}(\lambda)$ lies outside the interval $[-j, j]$, the levels E_m form a monotonic sequence and cannot cross as λ varies; but for λ exceeding the critical value λ_c determined by $\tan \lambda = \frac{2}{9}$, the minimum of E_m lies within the interval $-j \leq m \leq j$ and then a further increase of λ leads to crossings.

An even simpler example of a single-freedom Hamiltonian with level crossings is $H(\lambda) = J_z \cos \lambda - \frac{1}{j^2}J_z^3 \sin \lambda$. I leave to the reader to show that within each multiplet of fixed j, each level crosses all others as λ grows from 0 to $\pi/2$.

It is well to illustrate the absence of level repulsion for the quasi-energies of integrable quantum maps. Quasi-energies are defined modulo 2π, and taking the modulus is very effective in bringing about level crossings. The eigenphases of $F = \exp[-i(J_z + \frac{\lambda}{2j+1}J_z^2)]$ display a never ending sequence of crossings as do those of the even simpler Floquet operator $F = \exp(-i\lambda J_z)$. See Fig. 6.2.

The common feature of all of these examples is the absence of repeated avoided crossings between all neighboring levels, with closest approach spacings not much smaller than the mean spacing. If we still want to hold on to an analogy with a gas that PYG would have to be imagined as a rather ideal one. Frequent strongly avoided crossings seem to be the exclusive privilege of nonintegrable dynamics. For instance, if we destroy integrability by allowing for a kick such that the Floquet operator of the linear rotation, $F_0 = \exp(-i\lambda J_z)$, is accompanied by the torsion $F_1(\tau) = \exp(-i\frac{\tau}{2j+1}J_y^2)$, one gets two multiplets, each of which displays avoided crossings but no crossings; needless to say, as long as the torsion constant τ is sufficiently small, the avoided crossings are narrow and may be hard to

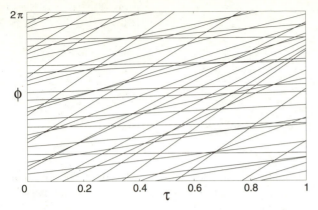

Fig. 6.2. Level crossings for the integrable Floquet operator $F(\lambda) = \exp\left(J_z + \frac{\lambda}{2j+1}J_z^2\right)$ with $j = 15$

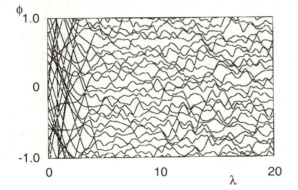

Fig. 6.3. Level dynamics for the kicked top with the Floquet operator $F = \exp(-i\frac{\lambda}{2j+1}J_z^2)\exp(-ipJ_y)$, $j = 25$, and $p = 1.7$. Only eigenvalues pertaining to the positive eigenvalue of $\exp(i\pi J_y)$ are displayed. The classical transition from regular motion ($\lambda = 0$) to global chaos upward of $\lambda \approx 5$ looks like an equilibration process whereas equilibrium appears to reign for $\lambda \gtrsim 5$

resolve, but non-zero they definitely are; as soon as $\tau = \mathcal{O}(1)$, we confront the predominance of chaos and universal level repulsion; Fig. 6.3 reveals the transition toward universal level repulsion for the Floquet operator $F = F_0F_1(\tau)$ under discussion.

We ought to dig deeper into the difference between integrable and noninte-grable dynamics manifest in avoided intramultiplet crossings. As becomes clear from Fig. 6.4, the single-freedom and thus integrable Hamiltonian

$$H(\lambda) = \left(J_z + \epsilon J_x\right)\cos\lambda + \left[\frac{4}{2j+1}J_z^2 - \frac{1}{2}J_z - (j + \frac{1}{2})\right]\sin\lambda \qquad (6.5.4)$$

differs from (6.5.3) by the perturbation $\epsilon J_x \cos\lambda$ which does not commute with the remainder; for Fig. 6.4 that perturbation was chosen small, $\epsilon = 0.4$, so as

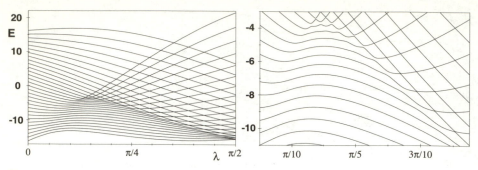

Fig. 6.4. Some crossings are strongly avoided for the single-freedom Hamiltonian $H(\lambda) = \left(J_z + \frac{2}{5} J_x \right) \cos \lambda + \left(\frac{1}{2} J_z + \frac{4}{2j+1} J_z^2 - j \right) \sin \lambda$ with $j = 15$. The right part is a blowup of the left.

to retain some similarity to the case $\epsilon = 0$ displayed in Fig. 6.2. Now, we encounter avoided crossings of a less trivial kind than those for (6.5.1). It will be noted, though, that the apparent crossings (which are not true crossings but unresolved avoided ones) outnumber the visibly avoided ones, and here lies the clue to the distinction in search. A good tool for the intended digging is the Einstein–Brillouin–Keller or WKB quantization presented in the previous chapter. For a comprehensive exposition of the techniques and ideas behind the more qualitative reasoning to be followed here, the reader is referred to *P.A. Braun's* review [142].

For a semiclassical treatment of the Hamiltonian (6.5.4), we must first degrade it to a classical Hamiltonian function. To that end, we introduce a pair of canonically conjugate phase-space coordinates p, φ through

$$J_z/j \to p, \quad J_x/j \to \sqrt{1 - p^2} \cos \varphi, \quad J_y/j \to \sqrt{1 - p^2} \sin \varphi; \qquad (6.5.5)$$

this amounts to replacing the angular momentum $\boldsymbol{J}$ by the classical unit vector $\boldsymbol{X} = (X, Y, Z) = \lim_{j \to \infty} \boldsymbol{J}/j$ and representing the latter by a polar angle θ and an azimuthal angle φ with $p = \cos \theta$. The canonical conjugacy of p and φ can be checked as follows: We start from the commutator $ij[J_x/j, J_y/j] = -J_z/j$. The usual replacement of commutators by Poisson brackets yields the Poisson brackets $\{X, Y\} = -Z$ of a classical angular momentum;[2] the latter in turn can be obtained from $\{p, \varphi\} = 1$ and $X = \sqrt{1 - p^2} \cos \varphi, Y = \sqrt{1 - p^2} \sin \varphi, Z = p$ through the identity $\{f(p), g(\varphi)\} = f'(p) g'(\varphi)$. The phase space thus turns out to be the rectangle $-1 \le p \le 1, 0 \le \varphi < 2\pi$ in the plane spanned by p, φ; the rectangle may of course be seen as a projection of the unit sphere onto the said plane; therefore, the phase space is sometimes said to be spherical. It is well to keep the foregoing reasoning in mind in view of the ubiquity of angular-momentum dynamics (the kicked top, for instance) in this book.

---

[2] Note that we have set Planck's constant equal to unity here such that its role is taken over by $1/j$; in more conventional notation the transition from commutators of quantum observables $\hat{A}, \hat{B}, \ldots$ to Poisson brackets of the associated classical observables $A, B, \ldots$ reads $\frac{1}{\hbar}[\hat{A}, \hat{B}] \to \{A, B\} = \frac{\partial A}{\partial p} \frac{\partial B}{\partial q} - \frac{\partial B}{\partial p} \frac{\partial A}{\partial q}$.

The classical Hamiltonian function $H/(j + \frac{1}{2}) \to h(p, \varphi)$ reads

$$
\begin{aligned}
h(p, \varphi, \lambda) &= \left(p + \epsilon\sqrt{1 - p^2}\cos\varphi\right)\cos\lambda + \left(2p^2 - \tfrac{1}{2}p - 1\right)\sin\lambda \\
&\equiv h_0(p, \varphi)\cos\lambda + h_1(p)\sin\lambda.
\end{aligned}
\tag{6.5.6}
$$

The ensuing classical dynamics is open to the qualitative investigation of single-freedom systems familiar from elementary textbooks, even though the Hamiltonian is not the sum of a kinetic and a potential term. Since $|\cos\varphi| \le 1$, it is helpful to look into the plane spanned by the momentum p and the energy E and to draw the graphs of the two functions

$$
U_+(p, \lambda) = h(p, 0, \lambda), \quad U_-(p, \lambda) = h(p, \pi, \lambda).
\tag{6.5.7}
$$

Figure 6.5 reveals that these two curves join smoothly with vertical slope at $p = \pm 1$. All points in the enclosed area correspond to classically allowed states. The shape of that area changes from a pancake to a banana as we increase the control parameter λ from zero toward $\pi/2$; at $\lambda = \pi/2$, the banana has shrunk in width to the parabolic segment $h_1(p)$ in $-1 \le p \le 1$. The lower curve, $U_-(p)$, has a single minimum irrespective of λ; the upper one, $U_+(p)$, has a single maximum for λ below a critical value λ_c but two maxima and an intervening minimum if $\lambda > \lambda_c$. It follows that p oscillates back and forth within only a single allowed

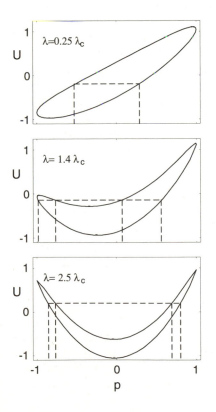

Fig. 6.5. The classically accessible region in the energy-momentum plane for the Hamiltonian (6.5.6) is surrounded by the curves $U_\pm(p)$ given in (6.5.7). The top plot refers to a subcritical value of λ for which only a single interval of oscillation exists while two separate such intervals arise for supercritical λ (middle and bottom); in the bottom case, the coalescence of the curves $U_\pm(p)$ into the single line $E_m|_{m \to p}$ according to (6.5.3) is imminent.

interval if $\lambda < \lambda_c$ whatever value the energy takes, whereas for $\lambda > \lambda_c$, there is an energy range with two separate accessible intervals for p.

The classically forbidden region in between separate intervals of oscillation becomes penetrable by quantum tunneling. Consequently, energy levels within the energy range allowing for classically separate motions but well away from the "top of the barrier" (the minimum of $U_+(p)$ in the present example) can come in close pairs with the small splitting proportional to the factor $\exp\{-|\int_{\text{barrier}} d\varphi p|/\hbar\}$; in other words, tunneling through a high and thick barrier is exponentially suppressed and thus very slow.

The three parts of Fig. 6.5 refer to three different values of λ, one of which is subcritical so as to allow for only one classically accessible p-interval; quantum mechanically. no level spacing is made small by tunneling if λ varies in the subcritical range, as is obvious from the level dynamics in Fig. 6.4. The middle part of Fig. 6.5 pertains to a slightly supercritical value in the neighborhood of which near crossings appear in the level dynamics which persist onto the third part which has its λ closer to $\pi/2$.

We should clarify why some and which fraction of the avoided crossings are strongly avoided. These distinguished encounters happen near a line $E(\lambda)$, not drawn into the level dynamics of Fig. 6.4, which gives the energy of the minimum of the upper boundary $U_+(p, \lambda)$ of the banana of Fig. 6.5, corresponding to the "top of the potential barrier." Clearly, pairs of levels near that "top" cannot come very close to one another since here tunneling processes have but a small and thin barrier to penetrate; thus the factor $\exp\{-|\int_{\text{barrier}} d\varphi p|/\hbar\}$ determining the splitting is not small. A substantial fraction (finite even when $N \to \infty$) of all levels will get close to the curve $E(\lambda)$ such that the number of strongly avoided crossings is proportional to N, quite small a minority indeed compared to the unresolved crossings whose number, we have seen, is of the order N^2. It may be well to note that for a given value of λ the energy $E(\lambda)$ defines a contour line in phase space which is the separatrix between periodic orbits encircling either one or the other of the two stable fixed points (the latter corresponding to the two minima of $U_+(p)$) and those encircling both fixed stable points; the separatrix itself of course intersects itself at the hyperbolic fixed point.

One final example of an integrable Hamiltonian is needed to convince us that strongly avoided crossings are exceptional events in the semiclassical limit $N \to \infty$. In search of a Hamiltonian $H(\lambda)$ whose N levels display roughly N^2 strongly avoided crossings and whose minimal spacings are a sizable fraction of the mean spacing, we might try a large-order polynomial in angular momentum components like a Chebyshev polynomial of the first kind, $T_n(J_z/j)$ with $n \gg 1$, in

$$H = \left[T_n(J_z/j) + 2J_x/j\right]\cos\lambda + (J_z/j)\sin\lambda. \tag{6.5.8}$$

It is easily checked that the ensuing level dynamics has the majority of crossings strongly avoided, if the dimension $2j+1$ of the Hilbert space is comparable to the order n; but as soon as $j \gg n$, the majority of crossings become barely resolvable as avoided. If for the sake of illustration we fix $n = 10$ and count the percentage of closest approach spacings whose ratio with the mean spacing is in excess of e^{-4}, we find that percentage as 98, 44, and 31, respectively, for $j = 10, 20, 100$. To conclude, integrable dynamics have level dynamics corresponding to near ideal PYG's. The weak nonideality lies in the rare strongly avoided crossings and the exponential smallness of most closest-approach spacings.

6.6 Two-Body Collisions

As a final prelude to the promised statistical analyses, it is useful to consider what can be learned from the close encounter of two particles of the Pechukas–Yukawa gas corresponding to a single multiplet in a system with global classical chaos. We shall recover the insight already obtained by almost degenerate perturbation theory in Sect. 3.4.

When two of the N particles, to be labeled 1 and 2, approach each other so closely that their mutual force by far exceeds the force exerted on either by any other particle, the collision effectively decouples from the rest of the dynamics. Restricting the consideration to the Floquet case, from (6.2.16),

$$\ddot{\phi}_1 = -\ddot{\phi}_2 = \tfrac{1}{4}|l_{12}|^2 \frac{\cos(\phi_{12}/2)}{\sin^3(\phi_{12}/2)}$$
$$\dot{l}_{12} = 0 ,$$

(6.6.1)

the latter equation is due to the absence of small denominators $\sin^2(\phi_{12}/2)$ in $\dot{l}_{12}$ given by the third of the equations (6.2.16). Obviously, the center of mass $(\phi_1 + \phi_2)/2$ moves uniformly, while the relative coordinate $\phi = \phi_{12}/2$ obeys (setting $l_{12} = l$ for short)

$$\ddot{\phi} = \frac{1}{4}|l|^2 \frac{\cos \phi}{\sin^3 \phi} = -\frac{\partial V}{\partial \phi}$$

(6.6.2)

where the potential energy

$$V(\phi) = \frac{1}{8} \frac{|l|^2}{\sin^2 \phi}$$

(6.6.3)

is π-periodic in ϕ and confines the coordinate ϕ to $[0, \pi]$ if ϕ is in that interval initially. By using energy conservation, one easily constructs the solution

$$\phi(\lambda) = \arccos \left[\cos \hat{\phi} \cos \left(\sqrt{2E}(\lambda - \hat{\lambda}) \right) \right] .$$

(6.6.4)

Here $\hat{\lambda}$ denotes one of the times of closest approach of the two particles, $2\hat{\phi}$ is the corresponding angular distance, and

$$E = \frac{|l|^2}{8 \sin^2 \hat{\phi}}$$

(6.6.5)

is the energy of the nonlinear oscillation.

Consistent with the neglect of all other particles, we may assume that $\hat{\phi} \ll 1$ and study the close encounter at times $\lambda \approx \hat{\lambda}$ by expanding all cosines as $\cos x = 1 - x^2/2$. The result is

$$\phi = \pm \left[\hat{\phi}^2 + |l|^2 \frac{(\lambda - \hat{\lambda})^2}{4\hat{\phi}^2} \right]^{1/2}. \tag{6.6.6}$$

The discriminant in (6.6.6) is a sum of two or three nonnegative terms depending on whether the angular momentum l is real or complex. Thus, The codimension of a level crossing is found once more as $n = 2$ and $n = 3$ for systems with, respectively, $O(N)$ and $U(N)$ as their canonical groups. As could have been expected on intuitive grounds, isolating a pair of colliding particles from the remaining $N - 2$ particles is equivalent to the nearly degenerate perturbation theory applied to an avoided level crossing in Sect. 3.2.

The case in which $Sp(N)$ is the canonical group can be treated similarly, even though Kramers' degeneracy looks slightly strange in the fictitious-particle picture: the particles have pairwise identical positions ϕ and an avoided crossing corresponds to a collision of two pairs of particles with $\phi_1 = \phi_{\bar{1}}$, $\phi_2 = \phi_{\bar{2}}$. Now, the equations of motion read (see Problem 6.6)

$$\ddot{\phi}_1 = -\ddot{\phi}_2 = \frac{1}{4} \left(|l_{12}|^2 + |l_{1\bar{2}}|^2 \right) \frac{\cos(\phi_{12}/2)}{\sin^3(\phi_{12}/2)} \tag{6.6.7}$$

with the coupling strength $|l_{12}|^2 + |l_{1\bar{2}}|^2$ again effectively constant. The solution $\phi(\lambda)$ given in (6.6.6) for the nondegenerate cases holds here, too, but with the replacement $|l_{12}|^2 \rightarrow |l_{12}|^2 + |l_{1\bar{2}}|^2$, which immediately yields the correct codimension of a level crossing, $n = 5$.

6.7 Ergodicity of Level Dynamics and Universality of Spectral Fluctuations

6.7.1 Ergodicity

Now, we can take up one of the most intriguing questions of the field. Why do dynamical systems with global classical chaos display universal spectral fluctuations in fidelity to random-matrix theory, and for what reasons do exceptions occur? Our discussion will focus primarily on periodically driven systems and their quasi-energy spectra. A motivating glance at Fig. 6.3 is recommended; there we see the quasi-energy spectrum of a kicked top ($F = \exp(-i\frac{\lambda}{2j+1} J_z^2) \exp(-ipJ_y)$) with the torsion strength λ varied from the integrable case $\lambda = 0$ to the range for which the classical motion is globally chaotic, $\lambda \gtrsim 5$. Clearly, equilibrium reigns in the chaotic range while for λ up to, say, 5 we encounter an equilibration process.

An elementary argument due to *Kuś* [131] shows that level dynamics, i.e., the Hamiltonian flow (6.2.16), is generically ergodic on an N-torus. The λ dependence of the eigenvalues and eigenvectors of a Floquet operator is uniquely determined by the λ dependence of the matrix representative of $F(\lambda)$ in any representation. It is convenient to employ momentarily the eigenrepresentation of V where

$$V|\mu\rangle = \omega_\mu|\mu\rangle, \qquad F(\lambda)\,\mathrm{e}^{\mathrm{i}H_0}|\mu\rangle = \mathrm{e}^{-\mathrm{i}\lambda V}|\mu\rangle = \mathrm{e}^{-\mathrm{i}\lambda\omega_\mu}|\mu\rangle. \tag{6.7.1}$$

Then, the matrix $F(\lambda)$ depends on λ only through the N exponentials $\mathrm{e}^{-\mathrm{i}\lambda\omega_\mu}$ which we assume all differ from one another. We conclude that $F(\lambda)$ moves through the space of unitary matrices along a "trajectory" lying on an N-torus. If we reparametrize matrix space by a change of representation, say, to the eigenrepresentation of $F(\lambda)$, we get a transformed trajectory described by the Hamiltonian flow of level dynamics. However, the topology of the manifold to which the trajectory is confined cannot have changed and still is that of an N-torus. Moreover, if the eigenvalues ω_μ are incommensurate, the motion on the N-torus will indeed be ergodic.

Ergodicity means that we can equate "time" averages, i.e., λ averages of certain observables with ensemble averages. The appropriate ensemble to use is the generalized microcanonical ensemble nailing down the (deformed) N-torus on which level dynamics takes place. In particular, it is *not* legitimate to work with the primitive microcanonical ensemble $\mathcal{Z}^{-1}\delta(\mathcal{H} - \mathcal{E})$ which fixes only the energy of the Pechukas–Yukawa gas, simply because the trajectory of level dynamics explores only an N-torus rather than the $[N(N+1) - 1]$-dimensional energy shell. As to the observables for which ensemble average equals "time" average (and ultimately even typical instantaneous values), the same thoughts as for other many-body systems apply. Quantities like spectrally averaged products of the level density or the spacing distribution can be expected to be self-averaging and thus to qualify for statistical description. Others, like the not spectrally averaged products of the level density, the (neither spectrally nor temporally averaged) form factor, or the localization length we shall meet in Chapters 7, 8, and 10 allow for weaker statements only: Averages over the "time" λ still equal ensemble averages but there are strong ensemble fluctuations. Here, we are concerned mostly with self-averaging quantities.

Time averages for the Pechukas–Yukawa gas are control parameter averages over a family of original dynamical systems: A λ average over, say, the spacing distribution for $F(\lambda) = \mathrm{e}^{-\mathrm{i}(\frac{\lambda}{2j+1}J_z^2 + \alpha J_z)}\,\mathrm{e}^{-\mathrm{i}\beta J_x}$ involves not a single kicked top but a whole one-parameter family of such. How can we draw conclusions for the actually observed universality of spectral fluctuations of individual dynamic systems like a single kicked top? The clue to the answer lies in the further question over how long an interval $\Delta\lambda$ must we calculate time averages before the ensemble means are approached.

One often stipulates that the interval $\Delta\lambda$ is large in the sense $\Delta\lambda \to \infty$. Such an extreme request allows us to accommodate and render weightless equilibration processes starting from initial conditions far from equilibrium. For level dynamics, such equilibration would arise if $F(\lambda)|_{\lambda=0} = \mathrm{e}^{-\mathrm{i}H_0}$ were the quantized version of an integrable classical dynamics and the switch-on of $\mathrm{e}^{-\mathrm{i}\lambda V}$ paralleled the expansion of chaos to (almost) global coverage of phase space. On the other hand, if we bar such initial situations and assume that level dynamics displays avoided crossings, right away the interval need not be much larger than the mean control parameter distance between subsequent avoided crossings of a pair of neighboring levels. For the Pechukas–Yukawa gas, we could call that interval a collision time, i.e., the mean temporal distance between two subsequent collisions of a particle with its neighbors.

6.7.2 Collision Time

The collision time just alluded to should scale with the number of particles (i.e., the number of levels N) like a power [143],

$$\lambda_{\text{coll}} \propto N^{-\nu} \text{ with } \nu > 0. \tag{6.7.2}$$

It is well to devote a little consideration to that power-law behavior and to indicate that the exponent ν cannot be expected to be universal, save for its positivity under conditions of chaos.

First, let us consider a Hamiltonian of the structure $H(\lambda) = H_0 + \lambda V$ and take H_0 and V as independent random $N \times N$ matrices from, say, the GUE. Both should have zero mean and the same variance,

$$\overline{H_{0ij}} = \overline{V_{ij}} = 0, \quad \overline{|H_{0ij}|^2} = \overline{|V_{ij}|^2} = 1/N. \tag{6.7.3}$$

According to Wigner's semicircle law, see (4.6.2, 3), the mean level spacing is $\Delta \equiv \overline{E_{i+1} - E_i} \sim \frac{1}{N}$. The level velocity vanishes in the ensemble mean and has the root-mean-square

$$\overline{p_i^2} = \overline{\langle \psi_i | V | \psi_i \rangle^2} \sim \frac{1}{N}. \tag{6.7.4}$$

Thus, a typical level velocity is

$$p \sim N^{-\frac{1}{2}} \tag{6.7.5}$$

and the simple estimate $\lambda_{\text{coll}} \approx \Delta/p \sim N^{-1/2}$ yields the exponent $\nu = 1/2$.

The same value of the exponent results if we consider a Floquet operator $F(\lambda) = e^{-iH_0} e^{-i\lambda V}$ and again take H_0 and V random as before.

For the Floquet operator of the kicked top with $V = J_z^2/(2j+1)$, the following argument will reveal $\nu = \frac{3}{2}$. Fixing the number of levels as $N = 2j+1$ we get the mean level velocity $\overline{v} \equiv \frac{1}{N}\sum_i^N V_{ii} = \frac{1}{N}\text{Tr}V = \frac{N^2}{12}(1 + \mathcal{O}(\frac{1}{N}))$; but this describes a drift common to all levels and is irrelevant for collisions. A typical relative velocity p results from the variance $p^2 = \frac{1}{N}\sum_{i=1}^N (V_{ii} - \overline{v})^2 \equiv \frac{1}{N}\sum_{i=1}^N \tilde{V}_{ii}^2$, where again the matrix elements are meant in the eigenrepresentation of the Floquet operator $F(\lambda)$; but in that representation the perturbation $\tilde{V}$ looks like a full matrix with $\sum_i \tilde{V}_{ii} \approx 0$ and $\sum_i \tilde{V}_{ii}^2 \approx \frac{1}{N}\sum_{ij} \tilde{V}_{ij}^2 = \frac{1}{N}\text{Tr}\tilde{V}^2 = \frac{N^2}{180}(1 + \mathcal{O}(\frac{1}{N}))$; we conclude $p \propto N^{\frac{1}{2}}$. The mean level spacing $\frac{2\pi}{N}$ yields the collision time $\propto N^{-\frac{3}{2}}$ and the exponent $\nu = \frac{3}{2}$; the latter value was confirmed by numerically following level dynamics for $10 < j < 160$ [143].

At any rate, the asymptotic disappearance of the collision time under conditions of chaos is in sharp contrast to the parametric change of levels in integrable systems. No collision time could at all sensibly be established for the latter, as is clear from our discussion of Sect. 6.5.

6.7.3 Universality

But inasmuch as Weyl's law implies $N \propto (2\pi\hbar)^{-f}$, the minimal control parameter window under discussion can be specified as $\Delta\lambda \sim \hbar^{f\nu}$ and thus vanishingly small from a classical perspective. *Thus, all of the different quantum systems involved in the control parameter alias "time" average can be said to have the same classical limit.* Such classically vanishing control parameter intervals have recently been suggested by *Zirnbauer* [144] in an attempt at demonstrating the universality of spectral fluctuations by the superanalytic method to be expounded in Chap. 10.

The assumption of a close-to-equilibrium initial condition for the Pechukas–Yukawa gas means roughly, as already indicated above, that the original dynamical system should have global chaos to begin with. That assumption is crucial since an equilibration of the gas corresponding to the classical transition from regular to dominantly chaotic behavior takes a time of classical character independent of $N \propto \hbar^{-f}$ and thus much larger than λ_{coll}.

Back to the assumption of incommensurate eigenvalues ω_μ of V which entails ergodicity on the N-torus. We can and should partially relax that assumption; we should since we often encounter commensurate eigenvalues, most notably perhaps for the kicked top with $V \propto J_z^2$. The unitary operator $e^{-i\frac{\lambda}{2j+1}J_z^2}$ is periodic in λ with period $2\pi(2j+1)$ if $2j+1$ is odd and $8\pi(2j+1)$ if $2j+1$ is even. We may be permissive since, inasmuch as that period is $\mathcal{O}(N)$ while the collision time is $\mathcal{O}(N^{-1})$, the distinction of strict and approximate ergodicity on the torus is quite irrelevant. In such cases we can still equate time averages with ensemble averages for the Pechukas–Yukawa gas, choosing $\Delta\lambda$ of the order of λ_{coll} and the generalized microcanonical ensemble to describe equilibrium on the torus.

To understand finally why spectral fluctuations for systems with global classical chaos are universal and faithful to random-matrix theory, we now have to take the most difficult step and show that equilibrium statistical mechanics for the Pechukas–Yukawa gas entails random-matrix type behavior for the relevant spectral characteristics. This will be the object of the next section.

Before precipitating ourselves into that rather serious adventure, it is well to mention exceptional classes of dynamical systems that, although globally chaotic in their classical behavior, do not display level repulsion and therefore do not fall into the universality classes we are principally dealing with here. Prototypical for one exceptional class is the kicked rotor which we shall treat in some detail in Chap. 7. Classical chaos notwithstanding, the eigenfunctions of the Floquet operator are exponentially localized in the H_0 basis. It follows that matrix elements of V with respect to different quasi-energy eigenstates are exponentially small if the states involved do not overlap; should the corresponding levels engage in an avoided crossing, the distance of closest approach would be exponentially small compared to the mean spacing. But since almost all pairs of states have that property in a Hilbert space whose dimension N is large compared to the (dimensionless) localization length l, almost all avoided crossings become visible only after a blow-up of the eigenphase scale by a factor of the order $e^l \gg 1$; on the scale of a mean spacing such narrow anticrossings look like crossings and this is why quantum localization comes with Poissonian spectral fluctuations.

The qualitative reasoning just presented about localization may be formalized in the fashion of our above treatment of two-body collisions. By appeal to conservation of the energy $\frac{1}{2}\dot{\varphi}^2 + U(\varphi)$ according to (6.6.2, 3), we can express the distance of closest approach $2\hat{\varphi}$ in terms of an initial distance $2\varphi_0$ taken to equal the mean distance $2\pi/N$ and an initial velocity $-\dot{\varphi}_0 = v_0$ as $\left(\frac{\sin \varphi_0}{\sin \hat{\varphi}}\right)^2 = 1 + \frac{v_0^2}{|V_{12}|^2}$ where the matrix element V_{12} refers to the two Floquet eigenstates of the colliding levels at the initial "time"; that matrix element is exponentially small for nonoverlapping states and makes the avoided crossing look like a crossing.

A second class of exceptions is constituted by some billiards on surfaces of constant negative curvature that display what is sometimes called arithmetic chaos. These have certain symmetries of the so-called Hecke type that produce effectively independent multiplets of quantum energy levels; though their classical effect is not to provide integrals of the motion which would make for integrability; they rather provide degenerate periodic orbits whose degeneracy increases exponentially with the period [145].

Quantum symmetries without classical counterparts have also been found for cat maps [146] and shown to cause spectral statistics not following the usual association of classical symmetries with quantum universality classes.

6.8 Equilibrium Statistics

In the limit of a large number of levels N, the fictitious N-particle system calls for a statistical description [147, 148]. As already mentioned, the phase-space trajectory ergodically fills an N-torus and the appropriate equilibrium phase-space density is therefore the generalized microcanonical one

$$\varrho(\phi, p, l) \sim \prod_{\mu} \delta\left(C_{\mu}(\phi, p, l) - \bar{C}_{\mu}\right) . \tag{6.8.1}$$

All the constants of the motion $C_{\mu}(\phi, p, l)$, which define the N-torus through their initial values $\overline{C}_{\mu}$, should be accounted for in the distribution function (6.8.1). We shall have to make do with the $C(\phi, p, l) = \mathrm{Tr}\left\{l^{\mu}v^{\nu}l^{\sigma}v^{\tau}\ldots\right\}$ since these are the only ones known at present. (In view of their overabundance, they might allow one to specify the N-torus – but that remains to be proven.)

The phase-space distribution (6.8.1) should be regarded as a competitor of Dyson's circular ensembles of random unitary matrices. Of foremost interest is the reduced distribution of the N phases ϕ_m obtained by integrating out all N momenta and all independent angular momenta l_{mn},

$$P(\phi) = \int dp\, dl \varrho(\phi, p, l) . \tag{6.8.2}$$

The ϕ dependence of $P(\phi)$ arises through the off-diagonal elements of the matrix v given in (6.3.3). As is obvious from that definition, for a fixed finite value of the angular momentum l_{mn}, the associated v_{mn} tends to infinity when the two coordinates ϕ_m and ϕ_n become equal; the phase-space functions $C_\mu(\phi, p, l)$ involving v diverge then, too, and the coordinate distribution thus vanishes. To find out how $P(\phi)$ approaches zero as two particles suffer a close encounter, we may change the integration variables in (6.8.2) according to

$$l_{mn} \to v_{mn} = \frac{l_{mn}}{i(e^{i\phi_{mn}} - 1)} \,, \quad m \neq n \,. \tag{6.8.3}$$

The Jacobian of this transformation,

$$J(\phi) = \left| \det\left(\frac{\partial l}{\partial v}\right) \right| = \prod_{(m,n)} \left| e^{-i\phi_m} - e^{-i\phi_n} \right| \,, \tag{6.8.4}$$

is a function of the coordinates ϕ deserving special attention. The product in (6.8.4) is over all distinct pairs of particles, each pair counted with a multiplicity characteristic of the group of canonical transformations of the underlying Floquet operator. In the orthogonal case, the $l_{mn} = -l_{nm}$ are real, and $N(N-1)/2$ in number and the multiplicity in question is one. When $U(N)$ applies, however, the multiplicity is two since the $l_{mn} = -l_{nm}^*$ are complex and the original l integral is over $N(N-1)/2$ complex planes. The symplectic case, finally, produces the multiplicity four, as can be seen from the following symmetry argument.

The time reversal covariance $TFT^{-1} = F^\dagger$ yields, on differentiation w.r.t. λ,

$$TVT^{-1} = F^\dagger V F \,. \tag{6.8.5}$$

We assume that the eigenbasis of F is organized such that $|m\rangle$ and $T|m\rangle \equiv |\bar{m}\rangle$ are the two eigenvectors pertaining to the quasi-energy ϕ_m. Taking matrix elements in (6.8.5) between the states pertaining to a pair of quasi-energies, one obtains

$$V_{mn} = e^{-i\phi_{mn}} V_{\bar{n}\bar{m}} \,, \qquad V_{m\bar{n}} = -e^{-i\phi_{mn}} V_{n\bar{m}} \,. \tag{6.8.6}$$

Now, it is obvious that, of the four matrix elements of V associated with the pair of levels ϕ_m, ϕ_n with $m < n$, $V_{mn}, V_{\bar{m}n}, V_{m\bar{n}}, V_{\bar{m}\bar{n}}$, only two are independent. Thus, the integral in (6.8.2) is over the complex plane for both l_{mn} and $l_{m\bar{n}}$, and the Jacobian (6.8.4) acquires four factors $|e^{-i\phi_m} - e^{-i\phi_n}|$ for each pair $m < n$.

To summarize, the Jacobian in question can be written as

$$J(\phi) = \prod_{i<j} \left| e^{-i\phi_i} - e^{-i\phi_j} \right|^{n-1} = \prod_{i<j} \left| 2\sin\left(\frac{\phi_{ij}}{2}\right) \right|^{n-1} \tag{6.8.7}$$

where n is the codimension of a level crossing,

$$n = \begin{cases} 2 & O(N) \\ 3 & U(N) \\ 5 & Sp(N) \,. \end{cases} \tag{6.8.8}$$

It is quite noteworthy that $J(\phi)$ is, to within a normalization constant, the joint probability density (4.10.2) for the eigenphases of unitary random matrices from

Dyson's circular ensembles. Moreover, the reader will appreciate that the factor $J(\phi)$ arose as a Jacobian for Dyson's ensembles as well, in that case from the transformation from matrix elements to eigenvalues and the parameters in the diagonalizing transformation. (Actually, in Chap. 4 the argument was given only for the Gaussian ensembles, but it is easy to carry over to the circular ensembles.)

When the substitution (6.8.3) is made in (6.8.2), the generalized microcanonical distribution ϱ depends on the coordinates ϕ only through the $l_{mn} = v_{mn} \mathrm{i} (\exp(\mathrm{i}\phi_{mn}) - 1)$ in the phase-space functions $C = \mathrm{Tr}\{l^\mu v^\nu \ldots\}$. Now, a crossing $\phi_{mn} \to 0$ (at fixed v_{mn}) no longer implies a divergence of any of these C. When the reduced distribution $P(\phi)$ is written as

$$P(\phi) = J(\phi)\tilde{P}(\phi) , \qquad \tilde{P}(\phi) \sim \int dp\, dv\, \varrho , \tag{6.8.9}$$

the integral $\tilde{P}(\phi)$ does not in general vanish at a crossing. Therefore, the behavior of $P(\phi)$ near crossings is dominated by the Jacobian $J(\phi)$, i.e., it is the same as that postulated by random-matrix theory. As an immediate consequence, one recovers a level spacing distribution obeying the repulsion law

$$P(S) \sim S^{n-1} \sim S^\beta \text{ for } S \to 0 . \tag{6.8.10}$$

To fully establish random-matrix theory as an implication of equilibrium statistical mechanics of the fictitious N-particle system, one would have to show that $\tilde{P}(\phi)$ is effectively a constant, at least for those particle configurations to which $J(\phi)$ assigns appreciable weight. It is amusing to note that one would strictly obtain $P(\phi)/J(\phi) = \mathrm{const}$ if one restricted the generalized microcanonical ensemble (6.8.1) so as to admit only constants of the flow (6.3.7) of the form $\mathrm{Tr}\{v^\mu\}$; evidently, the ϕ dependence of $P(\phi)$ in that case would be exclusively due to the Jacobian $J(\phi)$. Even more amusingly, the Hamiltonian function $\mathcal{H}$ is of precisely that form:

$$\mathcal{H} = \frac{1}{2}\mathrm{Tr}\{v^2\} = \frac{1}{2}\sum_m p_m^2 + \frac{1}{8}\sum_{m \neq n} \frac{|l_{mn}|^2}{\sin^2(\phi_{mn}/2)} . \tag{6.8.11}$$

Thus, the usual microcanonical ensemble is among the ensembles equivalent to random-matrix theory with respect to the statistics of eigenvalues, as first remarked by *Yukawa* in the analogous case of autonomous quantum systems [127]. Needless to say, this observation does not constitute a proof that $P(\phi)/J(\phi) \approx \mathrm{const}$ for the ergodic motion on the N-torus. Rather, the functionally independent constants of the flow determining the N-torus can definitely not all be of the form $\mathrm{Tr}\{v^m\}$, since, of the latter quantities, only N are linearly independent. Arguments supporting the effective constancy of $P(\phi)/J(\phi)$ will be given in the next section. We shall see that "effective" means that any ϕ dependence of $P(\phi)/J(\phi)$ affects the spacing distribution and low-order cluster functions of the level density by no more than corrections of relative order $1/N$. With that result, established our reasoning of Sect. 6.7 to support the universality of spectral fluctuations will be completed.

6.9 Random-Matrix Theory as Equilibrium Statistical Mechanics

6.9.1 General Strategy

This section is devoted to showing that equilibrium statistical mechanics for the fictitious-particle dynamics of Sect. 6.2 becomes equivalent to random-matrix theory as the number of levels N goes to infinity. For technical convenience, we consider Floquet operators $F(\lambda) = e^{-i\lambda V} e^{-iH_0}$ without antiunitary symmetries and shall aim to calculate their distribution of nearest neighbor spacings; for a more complete treatment, the reader is referred to [147, 148]. Since the microcanonical ensemble (6.8.1) is hard to work with, we employ a generalized canonical ensemble,

$$\varrho(\phi, p, l) = Z^{-1} \exp\left[-\sum \alpha C(\phi, p, l)\right] , \tag{6.9.1}$$

which fixes the constants of the motion C not sharply but only in the ensemble mean. The coefficients α in (6.9.1) are Lagrange multipliers determined by the mean values of the corresponding C,

$$\langle C \rangle = -\frac{\partial}{\partial \alpha} \ln Z . \tag{6.9.2}$$

As already indicated after (6.8.1), the argument must be confined to the constants of the motion constructed in Sect. 6.3,

$$C(\phi, p, l) = \operatorname{Tr}\left\{l^\mu v^\nu l^\sigma v^\tau \ldots\right\} . \tag{6.9.3}$$

Again, the spacing distribution $P(S) = E''(S)$ will be characterized by its second integral, the gap probability $E(S)$. To demonstrate that the generalized canonical ensemble yields the same $E(S)$ as Dyson's circular unitary ensemble (CUE),

$$
\begin{aligned}
E(S) &= \int_{-\pi+\pi S/N}^{\pi-\pi S/N} d\phi \int dl \int dp\, \varrho(\phi, p, l) \\
&= E_{\text{CUE}}(S)\left[1 + \mathcal{O}\left(\frac{1}{N}\right)\right] ,
\end{aligned}
\tag{6.9.4}
$$

after some further preliminaries, we shall proceed in close analogy with the calculation of $E_{\text{CUE}}(S)$ in Sect. 4.11. The latter calculation amounted to performing the N-fold integral in

$$E_{\text{CUE}}(S) = \int_{-\pi+\pi S/N}^{\pi-\pi S/N} d^N\phi\, P_{\text{CUE}}(\phi) \tag{6.9.5}$$

over the joint distribution of all eigenvalues

$$P_{\text{CUE}}(\phi) = \frac{1}{N!(2\pi)^N} \prod_{i<j}^{1 \ldots N} \left|e^{-i\phi_i} - e^{-i\phi_j}\right|^2 . \tag{6.9.6}$$

As a warm-up for treating the more numerous integrals in (6.9.4), a look back at Sect. 4.11 may be advisable. If afraid of technicalities, the reader might even prefer to jump ahead to the concluding remarks in Subsect. 6.9.5.

As a first step toward evaluating $E(S)$, we change integration variables as in (6.8.3), thereby incurring the Jacobian $J(\phi) \sim P_{\mathrm{CUE}}(\phi)$,

$$E(S) = \int_{-\pi+\pi S/N}^{\pi-\pi S/N} d^N \phi P_{\mathrm{CUE}}(\phi) \tilde{P}(\phi) , \tag{6.9.7}$$

$$\tilde{P}(\phi) = \mathcal{N}^{-1} \int d^{(N^2-N)} v d^N p e^{-\sum \alpha C} . \tag{6.9.8}$$

Needless to say, for the N^2 real integrals in (6.9.8) to exist, the constants of the motion C must be selected from the set (6.9.3) such that the highest powers of $|v_{ij}|$ and p_j in $\sum \alpha C$ have even exponents and positive coefficients. A further precaution will be necessary: The coordinate integrals in (6.9.7) become manageable only after formally extracting all ϕ-dependent terms in $\exp(-\sum \alpha C)$ from the exponent by a Taylor expansion in powers of (some of) the Lagrange multipliers α. It must be ensured that this series converges uniformly, otherwise term-by-term integration over the v, p, and ϕ would not be permissible. Therefore, a suitable part of $\sum \alpha C$, to be identified presently, must remain unexpanded.

None of the N independent constants of the motion of the type

$$C_n = \mathrm{Tr}\{v^n\} , \quad n = 1, 2, \dots, N \tag{6.9.9}$$

depends on ϕ and hence these may be kept in the exponent. The N^2-N remaining C_i, $i = N + 1, N + 2, \dots, N^2 - N$, do depend on the coordinates through $l_{mn} = iv_{mn}(e^{i\phi_{mn}} - 1)$. Let the highest power of v_{mn} in any of the C be $|v_{mn}|^M$ with M even and necessarily $M \geq N$. Now, consider the coefficient of $|v_{mn}|^M$, and let its maximum value, attained for some coordinate configuration(s), be denoted by

$$T = \sum_{i=N+1}^{N^2} \alpha_i T_i , \tag{6.9.10}$$

where $\alpha_i T_i$ is the contribution from C_i; of course $T_i = 0$ for all C_i not containing $|v_{mn}|^M$. Then, by separating as

$$\sum_{i=N+1}^{N^2} \alpha_i C_i = T \mathrm{Tr}\{v^M\} + \sum_{i=N+1}^{N^2} \alpha_i \left(C_i - T_i \mathrm{Tr}\{v^M\}\right) \tag{6.9.11}$$

and expanding $\exp(-\alpha_i C_i)$ in powers of α_i, a uniformly (for all values of v and ϕ) converging series is obtained.

The following remark is an aside, meant to explain the separation (6.9.11) in terms of an elementary example in which only a single integral over a real variable v arises with an integrand depending on a parameter ϕ,

$$I = \int_{-\infty}^{+\infty} dv\, e^{-T(\phi)v^2}\,. \tag{6.9.12}$$

For this "toy" integral to exist, the function $T(\phi)$ must be positive for all admissible values of ϕ. Let T be the maximum value attained by $T(\phi)$, such that $0 < T(\phi) \leq T$. The latter bounds for $T(\phi)$ imply

$$\frac{|T(\phi) - T|}{T} < 1 \tag{6.9.13}$$

and thus allow an expansion of I in powers of $T(\phi) - T$ with subsequent term-by-term integration over v. Due to (6.9.13), the resulting series

$$I = \sum_l \frac{1}{l!} \left(\frac{T - T(\phi)}{T}\right)^l \sqrt{\frac{\pi}{T}}(2l - 1)!!\,, \tag{6.9.14}$$

converges to the correct value $I = \sqrt{\pi/T(\phi)}$.

On the basis of the foregoing remarks, now we rewrite the gap probability $E(S)$ as

$$E(S) = \int_{-\pi+\pi S/N}^{\pi-\pi S/N} d^N\phi\, P_{\text{CUE}}(\phi) \int d^{N^2}v\, \mathcal{N}^{-1} \tag{6.9.15}$$

$$\times \exp\left(-\sum_{i=1}^{N} \alpha_i C_i - T\,\text{Tr}\,\{v^M\}\right) \prod_{j=N+1}^{N^2} \sum_{l=0}^{\infty} \frac{\alpha_j^l}{l!} \left(T_j\,\text{Tr}\,\{v^M\} - C_j\right)^l$$

where the shorthand $d^{N^2}v$ indicates that the integral over the off-diagonal elements v_{ij} and the momenta $p_i = v_{ii}$ are nothing but the integral over all elements[3] of the matrix v. The product of the Taylor series in (6.9.15) may be rearranged in a unique way as a sum of two terms,

$$\prod_{j=N+1}^{N^2} \sum_l \frac{1}{l!}\alpha_j^l \left(T_j\,\text{Tr}\,\{v^M\} - C_j\right)^l = s(v) + \delta s(\phi, v) \tag{6.9.16}$$

where $s(v)$ is independent of the coordinates ϕ and $\delta s(\phi, v)$ has an oscillatory ϕ dependence, $\delta s(\phi + 2\pi, v) = \delta s(\phi, v)$. Moreover, due to the symmetry of both the integration range in (6.9.15) and the density $P_{\text{CUE}}(\phi)$, the term $\delta s(\phi, v)$ may be imagined totally symmetrized in all N coordinates ϕ. Before specifying s and δs explicitly, it is useful to remark on the two additive contributions to the gap probability $E(S)$: As is clear from (6.9.5, 15), the part originating from $s(v)$ differs from $E_{\text{CUE}}(S)$ only by a constant factor; surprisingly, the same statement will turn out to hold true for the contribution from $\delta s(\phi, v)$, after taking the limit $N \gg 1$. By eventually normalizing, $E(S) = E_{\text{CUE}}(S)$ will result.

---

[3] More precisely, over all independent real parameters in the Hermitian matrix v.

The ϕ dependence of the C_i enters through the angular momentum matrix l. A power $(l^\mu)_{ij}$ gives rise to a ϕ-dependent factor of the type

$$\sin\frac{\phi_{i2}}{2}\sin\frac{\phi_{23}}{2}\ldots\sin\frac{\phi_{\mu j}}{2}\,e^{i\phi_{ij}/2}\,, \tag{6.9.17}$$

where, as before, $\phi_{mn}=\phi_m-\phi_n$ and where the intermediate indices, denoted simply by numbers, must eventually be turned into summation variables. Products like (6.9.17) can be rewritten as sums of sine and cosine functions,

$$\left(\sin\frac{\phi_{12}}{2}\sin\frac{\phi_{23}}{2}\ldots\sin\frac{\phi_{n1}}{2}\right)2^{n-1}i^n \tag{6.9.18}$$

$$=\begin{cases} 1+\displaystyle\sum_{l=1}^{n/2}\;\sum_{i_1<j_1\,\ldots\,<i_l<j_l}^{1\ldots n}(-1)^{\sum_{t=1}^{l}(i_t-j_t)}\cos\left(\sum_{t=1}^{l}\phi_{i_tj_t}\right) & \text{for } n \text{ even}\\[2em] -i\displaystyle\sum_{l=1}^{(n-1)/2}\;\sum_{i_1<j_1\,\ldots\,<j_l}^{1\ldots n}(-1)^{\sum_{t=1}^{l}(i_t-j_t)}\sin\left(\sum_{t=1}^{l}\phi_{i_tj_t}\right) & \text{for } n \text{ odd}\end{cases}$$

and

$$\left(\sin\frac{\phi_{12}}{2}\sin\frac{\phi_{23}}{2}\ldots\sin\frac{\phi_{n-1,n}}{2}\cos\frac{\phi_{n1}}{2}\right)2^{n-1}i^{n-1}$$

$$=\begin{cases} 1+\sum_{l=1}^{(n-1)/2}(\text{sum of cosines as above}) & \text{for } n \text{ odd}\\ -i\sum_{l=1}^{n/2}(\text{sum of sines as above}) & \text{for } n \text{ even}.\end{cases}$$

When several powers like $l^\mu,l^\sigma\ldots$ appear in a given constant of the motion C_i or in higher orders of the α expansion of (6.9.15), in a product of constants of the motion, one must deal with a product of several terms of the type (6.9.18). Again, such agglomerates allow decompositions of the type (6.9.18). Eventually, the multiple Taylor series in (6.9.15) takes the schematic form

$$s(v)+\delta s(\phi,v)=\sum\left[F(v)\left(1+\sum\cos\text{ines}\right)+G(v)\left(\sum\sin\text{es}\right)\right] \tag{6.9.19}$$

where the sines and cosines in the brackets have coefficients ± 1, as in (6.9.18); the symbols $F(v)$ and $G(v)$ denote v-dependent coefficients. With the coordinate integrals in (6.9.15) done, the sine terms in (6.9.19) vanish due to the symmetry of the integration range and of $P_{\mathrm{CUE}}(\phi)$ under $\phi\to-\phi$. Thus, one arrives at $s(v)=\sum F(v)$ and $\delta s(\phi,v)=\sum F(v)\sum\cos\text{ines}$. To verify that $\delta s(\phi,v)$ has no effect on $E(S)$, for $N\to\infty$, we shall have to establish three statements: (1) For each fixed coefficient $F(v)$, the ϕ integral over $P_{\mathrm{CUE}}(\phi)(1+\sum\cos\text{ines})$ draws a vanishingly small contribution from each individual cosine term. (2) Some of the cosine terms arise with a multiplicity so large that they survive in the limit $N\to\infty$; these, however, turn out to be proportional to $E_{\mathrm{CUE}}(S)$ in their S dependence. (3) Those cosine terms that would destroy the independence from S of $E(S)/E_{\mathrm{CUE}}(S)$ have multiplicities so small that they disappear, as $N\to\infty$.

The general cosine-type contribution to $E(S)$ arising for fixed $F(v)$ reads

$$
I_\mu\left(S,\{r\}\right) = \int_{-\pi+\pi S/N}^{\pi-\pi S/N} d\phi^N P_{\mathrm{CUE}}(\phi) \cos\left(\sum_{k=1}^{\mu} r_k \phi_{i_k}\right) \tag{6.9.20}
$$

$$
= \frac{(N-\mu)!}{N!} \sum_{i_1 \neq i_2 \neq \ldots \neq i_\mu} \int_{-\pi+\pi S/N}^{\pi-\pi S/N} d\phi^N P_{\mathrm{CUE}}(\phi) \cos\left(\sum_{k=1}^{\mu} r_k \phi_{i_k}\right)
$$

with μ positive or negative integers $r_1 \ldots r_\mu$ and $\mu = 1, 2, \ldots N$. We may extend (6.9.20) to $\mu = 0$ by letting $\cos(\ )\to 1$ in which case $I_0(S) = E_{\mathrm{CUE}}(S)$.

6.9.2 A Typical Coordinate Integral

The discussion of I_μ may be approached by considering the special case for which $\mu = 2m$ is even, m of the r_k equal to $+1$, and the remaining r_k equal to -1,

$$
I_{2m}\left(S,\{\pm 1\}\right) = \frac{(N-2m)!}{N!} \sum_{i_1 \neq \ldots \neq i_{2m}}^{1\ldots N} \int_{-\pi+\pi S/N}^{\pi-\pi S/N} d\phi^N P_{\mathrm{CUE}}(\phi)
$$
$$
\times \cos\left(\sum_{k=1}^{m} \phi_{i_{2k-1} i_{2k}}\right). \tag{6.9.21}
$$

The N-fold integral in (6.9.21) is easily evaluated after the fashion of (4.11.4, 8). The result is an $N \times N$ determinant differing in $2m$ lines from that in (4.11.8),

$$
I_{2m}\left(S,\{\pm 1\}\right) = \frac{(N-2m)!}{N!} \sum_{i_1 \neq \ldots \neq i_{2m}}^{1\ldots N} \tag{6.9.22}
$$

$$
\times \det \left\{ \begin{array}{ll} \delta(k,l) - \frac{\sin[\pi(k-l)S/N]}{\pi(k-l)} & \text{for } k \notin \{i_1 \ldots i_{2m}\} \\[2mm] \delta(k+1,l) - \frac{\sin[\pi(k+1-l)S/N]}{\pi(k+1-l)} & \text{for } k \in \{i_1, i_3, \ldots, i_{2m-1}\} \\[2mm] \delta(k-1,l) - \frac{\sin[\pi(k-1-l)S/N]}{\pi(k-1-l)} & \text{for } k \in \{i_2, i_4, \ldots, i_{2m}\} \end{array} \right\}.
$$

The remainder of the discussion of $I_{2m}(S,\{\pm 1\})$ parallels that of Sect. 4.11. To find the expansion of I_{2m} in powers of S [see (4.11.9)], the determinant is written as a sum over products of N elements [see (4.11.10)]; the sines are represented by their Taylor series [see (4.11.11)]; and finally the arguments of the sines are expressed as a binomial expansion [see (4.11.12)]. The zeroth-order Taylor coefficient is most easily obtained by setting $S = 0$ in (6.9.22),

$$
I_{2m,0} = \frac{(N-2m)!}{N!} \sum_{i_1 \neq \ldots \neq i_{2m}}^{1\ldots N} \sum_{\mathcal{P}} (-1)^{\mathcal{P}}
$$

$$
\times \prod_{k=1}^{m} \delta\left(i_{2k-1}, i_{\mathcal{P}(2k-1)} - 1\right) \delta\left(i_{2k}, i_{\mathcal{P}(2k)} + 1\right). \tag{6.9.23}
$$

The sum over all permutations of $2m$ indices represents the determinant of a $2m \times 2m$ matrix whose elements are either zero or one. It is easy to see that only $m!$

out of the total $(2m)!$ permutations $\mathcal{P}$ yield non-zero contributions to $I_{2m}(0)$: To avoid a zero for any of the $2m$ Kronecker factors, $\mathcal{P}$ must consist of m index swaps, each of which interchanges an odd with an even index. There are indeed $m!$ such permutations, all of the sign $(-1)^{\mathcal{P}} = (-1)^m$, and all contributing identically. It follows that

$$I_{2m,0} = (-1)^m \frac{m!(N-2m)!}{N!} \sum_{i_1 \neq \ldots \neq i_{2m}}^{1 \ldots N} \prod_{k=1}^{m} \delta\left(i_{2k-1}, i_{2k} - 1\right) . \tag{6.9.24}$$

The remaining sum is easily evaluated by the following series of self-explanatory steps:

$$\sum_{i_1 \neq \ldots \neq i_{2m}}^{1 \ldots N} \prod_{k=1}^{m} \delta\left(i_{2k-1}, i_{2k} - 1\right)$$

$$= m! \sum_{\substack{i_1 < i_3 < \ldots < i_{2m-1} \\ (i_1 \neq i_2 \neq i_3 \neq \ldots \neq i_{2m})}} \prod_{k=1}^{m} \delta(\ldots)$$

$$= m! \sum_{i_1=1}^{N-2m+1} \sum_{i_3=i_1+2}^{N-2m+3} \sum_{i_5=i_3+2}^{N-2m+5} \ldots \sum_{i_{2m-1}=i_{2m-3}+2}^{N-1} 1$$

$$= m! \sum_{\substack{i_1 < i_3 < \ldots < i_{2m-1}}}^{1 \ldots (N-m)} 1 = \sum_{\substack{i_1 \neq i_3 \neq \ldots \neq i_{2m-1}}}^{1 \ldots (N-m)} 1 = \frac{(N-m)!}{(N-2m)!} . \tag{6.9.25}$$

Therefore, the zeroth-order Taylor coefficient of $I_{2m}(S)$ emerges as

$$I_{2m,0} = (-1)^m / \binom{N}{m} , \tag{6.9.26}$$

i.e., smaller in modulus than the zeroth-order Taylor coefficient of $E_{\text{CUE}}(S)$, $E_{\text{CUE},0} = 1$ [see (4.11.12)]. The reason that $I_{2m,0}/E_{\text{CUE},0}$ vanishes, at least as N^{-m} for $N \to \infty$, is obvious: The index shifts of ± 1 in $2m$ rows of the determinant in (6.9.22) cancel m of the $2m$ summations over $i_1 \ldots i_{2m}$.

To obtain the higher Taylor coefficients $I_{2m,l}$, we consider the expansion of $I_{2m}(S)$ analogous to (4.11.12), which was already commented on after (6.9.22). As in (4.11.13), we first focus on those additive terms of the expansion for which n out of the total N elements of the determinant enter with nonvanishing orders in S and let these orders be $(2l_1+1), (2l_2+1), \ldots (2l_n+1)$. Then, the remaining $N - n$ elements enter with their zeroth-order (in S) parts and are thus of the type [see (6.9.22)] $\delta(i_k, i_{\mathcal{P}k})$ or $\delta(i_k \pm 1, i_{\mathcal{P}k})$. Of course, each element entering with $\delta(i_k, i_{\mathcal{P}k})$ reduces by one the number of indices still subject to reshuffling by the permutations $\mathcal{P}$. Now, the nontrivial permutations in contrast to both (6.9.23) and (4.11.13), may mix indices in zeroth-order factors $\delta(i_k \pm 1, i_{\mathcal{P}k})$ and in nonzeroth-order ones. However, similarly, as in (6.9.23), not all of these permutations need be included; nor must all possibilities of assigning one of the three types of Kroneckers just mentioned to each of the $N - n$ elements entering in zeroth order be kept, to leading order in N. Similarly distinguished as above for $I_{2m,0}$ are those permutations that are index swaps within pairs of Kroneckers

$$\delta\left(i_{2k-1}+1, i_{\mathcal{P}(2k-1)}\right)\delta\left(i_{2k}-1, i_{\mathcal{P}(2k)}\right) = \delta\left(i_{2k-1}+1, i_{2k}\right)$$
$$\text{for } \mathcal{P}(2k) = 2k - 1 \text{ and } \mathcal{P}(2k-1) = 2k ,$$

(6.9.27)

for all Kroneckers of the type $\delta(i_k \pm 1, i_{\mathcal{P}k})$ when their total number is even, and for all but one when their number is odd. The distinction is constituted by the fact that the two Kroneckers in a pair of the type (6.9.27) eliminate only a single one of the $2m$ summations over the $i_1 \ldots i_{2m}$ in (6.9.22); if unmatched in such pairs, two Kroneckers cancel two of these summations and thus, to leading order in N, do not contribute to $I_{2m,l}$.

Now, consider the case where the number of such pairs is $m - \nu$ with $\nu \leq m$ and where no elements of the determinant contribute an unmatched[4] $\delta(i_k \pm 1, i_{\mathcal{P}k})$. There are $\binom{m}{\nu}^2$ possibilities to select $m - \nu$ indices from $\{i_1 i_3 \ldots i_{2m-1}\}$ and $m - \nu$ others from $\{i_2 i_4 \ldots i_{2m}\}$ to make up $m - \nu$ pairs, all of which yield the same contribution to the sum over all $i_1 i_2 \ldots i_{2m}$ in $I_{2m,l}$. Of the $2m$ rows of the determinant in (6.9.22) with indices $i_1 \ldots i_{2m}$, there are thus 2ν rows that contribute an element in nonzeroth order in S. For the sake of notational convenience, we let these rows carry the indices $i_1 i_2 \ldots i_{2\nu}$; moreover, we let the other $n - 2\nu$ rows providing an element in nonzeroth order have the indices $k_1, k_2, \ldots, k_{n-2\nu}$. With all these remarks in mind the reader will be able to appreciate the analogue of (4.11.13):[5]

$$
\begin{aligned}
T_m\left(S, n, \{l_i\}\right) = & \sum_{\nu=0}^{m} \frac{1}{(n-2\nu)!\pi^n}\binom{m}{\nu}^2\frac{(N-2m)!}{N!} \\
& \times \left[\prod_{k=1}^{n}(-1)^{l_k+1}\frac{(\pi S/N)^{2l_k+1}}{(2l_k+1)!}\right]\sum_{i_1 \neq \ldots i_{2m} \neq k_1 \neq \ldots k_{n-2\nu}} \\
& \times (-1)^{m-\nu}(m-\nu)!\left[\prod_{p=\nu+1}^{m}\delta\left(i_{2p-1}+1, i_{2p}\right)\right]\sum_{\mathcal{Q}}(-1)^{\mathcal{Q}} \\
& \times \sum_{t_1=0}^{2l_1}\cdots\sum_{t_n=0}^{2l_n}\binom{2l_1}{t_1}\cdots\binom{2l_n}{t_n}(-1)^{t_1+\ldots+t_n}
\end{aligned}
$$

---

[4] Terms with $m - \nu$ pairs plus an unmatched $\delta(i_k \pm 1, i_{\mathcal{P}k})$ are smaller by at least a factor $1/N$; for the sake of brevity, I forego the treatment of that case.

[5] To simplify the appearance of $T_m(S, n, \{l_i\})$ I have assumed $l_1 \neq l_2 \neq \ldots \neq l_n$. See the footnote to (4.11.13).

$$\times \left[\prod_{q=1}^{n-2\nu} (k_q)^{2l_q - t_q} \left(\mathcal{Q} k_q \right)^{t_q} \right]$$

$$\times \left[\prod_{r=1}^{\nu} (i_{2r-1} + 1)^{2l_{n-2\nu+r} - t_{n-2\nu+r}} \left(\mathcal{Q} i_{2r-1} \right)^{t_{n-2\nu+r}} \right.$$

$$\left. \times (i_{2r} - 1)^{2l_{n-\nu+r} - t_{n-\nu+r}} \left(\mathcal{Q} i_{2r} \right)^{t_{n-\nu+r}} \right]. \tag{6.9.28}$$

Upon summing over all $l_1 \ldots l_n$ from 0 to ∞ and then over n from 1 to ∞, the Taylor expansion of $I_{2m}(S) - 1$ results, with all Taylor coefficients stripped of corrections of relative weight $1/N$ or smaller. The $n!$ permutations $\mathcal{Q}$ in (6.9.28) reshuffle the summation variables $k_1 \ldots k_{n-2\nu}$, $i_1 \ldots i_{2\nu}$; their sum yields an $n \times n$ determinant whereupon T_m takes a form analogous to (4.11.14),

$$T_m (S, n, \{l_i\}) = \sum_{\nu=0}^{m} \frac{1}{(n-2\nu)! \pi^n} \binom{m}{\nu}^2 (-1)^{m-\nu} (m-\nu)! \frac{(N-2m)!}{N!}$$

$$\times \left[\prod_{k=1}^{n} (-1)^{l_k+1} \frac{(\pi S/N)^{2l_k+1}}{(2l_k+1)!} \right]$$

$$\times \sum_{t_1=0}^{2l_1} \cdots \sum_{t_n=0}^{2l_n} \binom{2l_1}{t_1} \cdots \binom{2l_n}{t_n} (-1)^{t_1+ \ldots + t_n}$$

$$\times \sum_{i_1 \neq \ldots \neq i_{2\nu} \neq k_1 \neq, \ldots \neq k_{n-2\nu}}^{1 \ldots N}$$

$$\times \det \begin{pmatrix} (k_\sigma)^{2l_\sigma - t_\sigma + t_\tau} & \sigma = 1, \ldots, n-2\nu \\ (i_{2\sigma-1} + 1)^{2l_{n-2\nu+\sigma} - t_{n-2\nu+\sigma}} (i_{2\sigma-1})^{t_\tau} & \sigma = 1, \ldots, \nu \\ (i_{2\sigma} - 1)^{2l_{n-\nu+\sigma} - t_{n-\nu+\sigma}} (i_{2\sigma})^{t_\tau} & \sigma = 1, \ldots, \nu \end{pmatrix}$$

$$\times \sum_{\substack{i_{2\nu+1} \neq \ldots \neq i_{2m} \\ (\neq i_1 \ldots \neq i_{2\nu} \neq k_1 \neq \ldots \neq k_{n-2\nu})}}^{1 \ldots N} \left[\prod_{p=\nu+1}^{m} \delta (i_{2p-1} + 1, i_{2p}) \right], \tag{6.9.29}$$

where the column index τ of the determinant ranges from 1 to n, while the row labelling is displayed explicitly.

The summation over $i_{2\nu+1} \ldots i_{2m}$ in (6.9.29) is analogous to (6.9.25), except that the number of effectively independent summation variables is $m - \nu$ [instead of m as in (6.9.25)] and that the additional n restrictions allow at most $N - n$ different values for each summation variable [instead of N in (6.9.25)]. Unfortunately, the present sum cannot be evaluated explicitly for a general configuration of the $i_1 \ldots i_{2\nu}$, $k_1 \ldots k_{n-2\nu}$. Only an upper limit is provided by (6.9.25) with $m \to m - \nu$ and $N \to N - n$,

$$\sum_{\substack{i_{2\nu+1} \neq \ldots \neq i_{2m} \\ \neq i_1 \neq \ldots \neq i_{2\nu} \neq k_1 \neq \ldots k_{n-2\nu}}}^{1 \ldots N} \prod_{p=\nu+1}^{m} \delta (i_{2p-1} + 1, i_{2p}) \leq \frac{(N-n-m+\nu)!}{(N-n-2m+2\nu)!}. \tag{6.9.30}$$

That upper bound is attained by special configurations of the $\{i_1 \dots k_{n-2\nu}\}$ such as $\{1, 2, \dots, n\}$ or $\{N-n+1, N-n+2, \dots, N\}$. At any rate, the bound (6.9.30) will suffice to show that the Taylor coefficient $I_{2m,l}$ vanishes as $N \to \infty$. The modulus of T_m is bounded from above when the sum in question is replaced by the r.h.s. of (6.9.30). With that simplification, the n-fold sum of determinants can be carried out to within corrections of relative weight $1/N$, considering that in the limit $N \to \infty$ the restrictions $i_1 \neq i_2 \neq \dots$ become irrelevant. All but 2ν of these sums are immediately given by (4.11.19) while for the remaining 2ν sums the analogous result

$$\sum_{i=1}^{N} (i \pm 1)^r i^s = \frac{N^{r+s+1}}{r+s+1} \left[1 + \mathcal{O}\left(\frac{1}{N}\right) \right] \tag{6.9.31}$$

is easily obtained with $r, s \ll N$. Then, the sum of determinants comes out precisely as in (4.11.20) whereupon $T_m(S, n, \{l_i\})$ turns out to be bounded as

$$|T_m(S, n, \{l_i\})| \leq \left| \left\{ \sum_{\nu=0}^{m} \frac{n!}{(n-2\nu)!} \binom{m}{\nu}^2 (-1)^{m-\nu}(m-\nu)! \right. \right.$$
$$\left. \left. \times \frac{(N-2m)!}{N!} \frac{(N-n-m+\nu)!}{(N-n-2m+2\nu)!} \right\} \right| \times |T(S, n, \{l_i\})| \tag{6.9.32}$$

with $T(S, n, \{l_i\})$ the corresponding term in the Taylor expansion of $E_{\text{CUE}}(S)$ given in (4.11.14, 4.11.15). Since, in analogy with Sect. 4.11

$$I_{2m,l} S^l = \sum_{n=1,2,\dots}^{n^2 \leq l} \sum_{l_1 \dots l_n}^{1,2,\dots} \delta\left(\sum_{i=1}^{n} l_i, \frac{l-n}{2} \right) T_m(S, n, \{l_i\}) , \tag{6.9.33}$$

the negligibility of $I_{2m}(S)$ will become obvious once the factor $\{\dots\}$ in (6.9.32) is revealed as vanishing for $N \to \infty$.

The latter goal is most easily achieved for $m \ll N$. In that case, the highest power in N arises for $\nu = 0$; Stirling's formula implies $|\{\dots\}| \to m! N^{-m}$. Note that the integer n may be restricted to values $n \ll N$ since the Taylor series of $I_{2m}(S)$, like the one of $E_{\text{CUE}}(S)$ in Sect. 4.11, may be truncated at a finite order $l_{\max}$ and $n^2 \leq l_{\max}$. On the other hand, if m is near its maximum value $N/2$ for even N or $(N-1)/2$ for odd N, $|\{\dots\}|$ is smaller still, in fact, exponentially small. For the sake of illustration, let N and n be even; then $\nu = n/2$ gives the dominant contribution, and Stirling's formula yields $|\{\dots\}| \to 2^{-N}\sqrt{\pi N/2}$.

Besides demonstrating that $I_{2m}(S) \to 0$ with $N \to \infty$, the foregoing considerations also yield a more specific result for the special case $m \ll N$ which will shortly turn out to be important. Consider the Taylor coefficients $I_{2m,l}$ with $l \ll N$; these should suffice to determine the function $I_{2m}(S)$ within the interesting interval $0 \leq S \lesssim \mathcal{O}(1)$. For $l \ll N$, the number n in $T_m(S, n, \{l_i\})$ is similarly restricted since $n^2 \leq l$. To within corrections of order l/N, m/N, the sum in (6.9.30) is given by $N^{-(m-\nu)}$, and instead of the inequality (6.9.33) one can write an asymptotic equality, $T_m = \{\dots\} T$, with the curly bracket $\{\dots\}$ as in (6.9.32) except for the replacement $(N-n-m+\nu)!/(N-n-2m+2\nu)! \rightarrow N^{-(m-\nu)}$. Moreover, for $m \ll N$, the leading term in the curly bracket in question arises for $\nu = 0$ such that $T_m(S, n, \{l_i\}) = (-1)^m m! N^{-m} T(S, n, \{l_i\})$. Since the ratio T_m/T turns out to be independent of n and of the l_i, the Taylor coefficients of $I_{2m}(S)$ and of $E_{\text{CUE}}(S)$ have a ratio independent of the order, $I_{2m,l}/E_{\text{CUE},l} = (-1)^m m! N^{-m}$, as long as $l \ll N$. Inasmuch as higher order Taylor coefficients are not needed for $0 \leq S \lesssim \mathcal{O}(1)$, we may conclude

$$I_{2m}(S, \{\pm 1\}) = (-1)^m m! N_{\text{CUE}}^{-m}(S)$$
$$\text{for } m \ll N \text{ and } 0 \leq S \lesssim \mathcal{O}(1). \tag{6.9.34}$$

Whereas $I_{2m}(s) \rightarrow 0$ with $N \rightarrow \infty$ and $m \geq 1$ might be considered intuitive, the fact that $I_{2m}(S, \{\pm 1\})/E_{\text{CUE}}(S)$ is independent of S for $0 \leq S \lesssim \mathcal{O}(1)$ is quite a nontrivial and perhaps surprising result.

6.9.3 Influence of a Typical Constant of the Motion

At this point, we are more than halfway toward demonstrating the asymptotic equivalence of equilibrium statistical mechanics and random-matrix theory with respect to the level spacing distribution. Having ascertained the asymptotic negligibility of the coordinate integral in a typical correction term, it remains to take up a remark made above relating to the structure (6.9.19): Can the number of such correction terms grow so large as to upset the smallness of the individual correction term? To answer this question, I propose to focus on a particular constant of the motion,

$$C_{\mu\nu} = \text{Tr}\left(l^\mu v^\nu\right) \text{ with } \mu \text{ odd}, \tag{6.9.35}$$

and to consider its contribution to $E(S)$ in first order in the α expansion of (6.9.15). According to (6.9.17) and the first of the multiple Fourier decompositions (6.9.18), one is led to a sum of coordinate integrals of the type $I_{2m}(S, \{\pm 1\})$,

$$I(S) \equiv 2^{-\mu} i^{-\mu-1}$$
$$\times \left[E_{\text{CUE}}(S) + \sum_{m=1}^{(\mu+1)/2} (-1)^m \binom{(\mu+1)/2}{m} I_{2m}(S, \{\pm 1\}) \right] \tag{6.9.36}$$

after using the elementary sum formula (for $\mu + 1$ even)

$$\sum_{i_1 < i_2 < \ldots < i_{2m}}^{1 \ldots (\mu+1)/2} (-1)^{i_1 + i_2 + \ldots + i_{2m}} = (-1)^m \binom{(\mu+1)/2}{m}. \tag{6.9.37}$$

The easiest case to discuss is that of μ remaining finite as $N \rightarrow \infty$. With the help of (6.9.34), we immediately obtain

$$I(S) = E_{\text{CUE}}(S) 2^{-\mu} i^{-\mu-1} \left[1 + \sum_{m=1}^{(\mu+1)/2} \binom{(\mu+1)/2}{m} m! N^{-m} \right]$$
$$= E_{\text{CUE}}(S) 2^{-\mu} i^{-\mu-1} \left[1 + \mathcal{O}\left(\frac{1}{N}\right) \right]. \tag{6.9.38}$$

Obviously, the number of correction terms, given by (6.9.37), itself remains finite for $N \to \infty$ and $I(S)$ remains faithful to $E_{\text{CUE}}(S)$ in its S dependence.

A more intricate situation is encountered for μ of the order N, say $\mu + 1 = N$. Now, the smallness of $I_{2m}(S)$ with $m \ll N$ can be compensated for by the binomial coefficient $\binom{N/2}{m}$ in (6.9.36). Indeed, according to (6.9.34), the terms in question,

$$(-1)^m \binom{N/2}{m} I_{2m}\left(S, \{\pm 1\}\right) / E_{\text{CUE}}(S)$$
$$= \frac{(N/2)!(N-m)!}{(N/2-m)!N!} = \left(\frac{1}{2}\right)^m \quad \text{for } m \ll N, \tag{6.9.39}$$

are of order unity as long as m is. Fortunately, however, no alteration of the S dependence of $I(S)$ away from $E_{\text{CUE}}(S)$ is brought about by these terms, since the ratio (6.9.39) is independent of S. As the summation variable m in (6.9.36) grows to become comparable to $N/2$, on the other hand, the ratios $I_{2m}(S)/E_{\text{CUE}}(S)$ cease to be independent of S; it is only by virtue of their exponential smallness that they do not noticeably spoil the constancy of $I(S)/E_{\text{CUE}}(S)$ with respect to S. To illustrate the negligibility of the large-m contributions on the right-hand side in (6.9.36), let $m = \alpha N$ with $1/N \ll \alpha < 1/2$. Upon using (6.9.32) and Stirling's formula, one easily finds

$$\binom{N/2}{m} |I_{2m}(S)/E_{\text{CUE}}(S)| = \mathcal{O}\left[2^{-N\alpha} \left(\frac{1-\alpha}{\sqrt{1-2\alpha}}\right)^N \left(\frac{1-2\alpha}{1-\alpha}\right)^{N\alpha} \right],$$
$$\tag{6.9.40}$$

i.e., an exponentially small value for all α in the permissible range. The final conclusion is that in the limit, as $N \to \infty$, $I(S)/E_{\text{CUE}}(S)$ is a constant independent of S in the range $0 \le S \lesssim \mathcal{O}(1)$.

6.9.4 The General Coordinate Integral

Completing the argument of this section, we proceed to show that the general coordinate integral $I_\mu(S, \{r\})$ defined in (6.9.20) vanishes with $N \to \infty$ at least as fast as the special integral $I_{2m}(S, \{\pm 1\})$. The necessary modification of Sect. 6.9.2 will merely require a few remarks. For the sake of brevity, we confine the discussion to even values of μ, $\mu = 2m$.

In analogy with (6.9.22), one encounters $N \times N$ determinants

$$I_{2m}\left(S, \{r\}\right) = \frac{(N-2m)!}{N!} \sum_{i_1 \neq i_2 \, \cdots \, \neq i_{2m}}^{1 \ldots N} \tag{6.9.41}$$

$$\times \det \begin{pmatrix} \delta(k,l) - \frac{\sin(\pi(k-l)S/N)}{\pi(k-l)} & \text{for } k \notin \{i_1 i_2 \ldots i_{2m}\} \\ \delta(k+r_k, l) - \frac{\sin\left(\pi(k+r_k-l)S/N\right)}{\pi(k+r_k-l)} & \text{for } k \in \{i_1 i_2 \ldots i_{2m}\} \end{pmatrix}.$$

While $2m$ rows in the determinant in (6.9.22) had index shifts $k \to k \pm 1$, now we have to deal with $2m$ shifts $k \to k + r_k$. The Taylor coefficient of order zero takes the form

$$I_{2m,0}\left(\{r\}\right)=\frac{(N-2m)!}{N!}\sum_{i_1\neq\,\dots\,\neq i_{2m}}^{1\,\dots\,N}\sum_{\mathcal{P}}(-1)^{\mathcal{P}}\prod_{k=1}^{2m}\delta\left(i_k+r_k,i_{\mathcal{P}k}\right),\quad(6.9.42)$$

in obvious generalization of (6.9.24). Now, we will argue that for fixed m, the $I_{2m,0}(\{r\})$ that have the highest order in N are those for which the $2m$ integers r_k form m pairs $\pm r_k$ with $k=1,2,\dots m$. The reason is simply that, for a general configuration $\{r_k, k=1,2,\dots 2m\}$, the $2m$ Kroneckers in (6.9.42) are all independent such that all of the $2m$ summations over $i_1,\dots i_{2m}$ are cancelled. For the special configurations $\{\pm r_k, k=1,2,\dots m\}$, however, the $2m$ Kroneckers coincide in pairs and thus leave m independent summation variables.

To prove the statement just made and at the same time to identify the particular permutations $\mathcal{P}$ contributing in leading order in N, one can resort to the reasoning already familiar from Sect. 6.9.2. Imagine a particular index k in (6.9.42) fixed, say $k=1$, and consider its associate $\mathcal{P}1$ by the permutation $\mathcal{P}$. The two summation variables so labelled are subjected to a Kronecker condition, $i_{\mathcal{P}1}=i_1+r_1$. Then look at $\mathcal{P}\mathcal{P}1$; there are the obvious possibilities $\mathcal{P}\mathcal{P}1=1$ or $\mathcal{P}\mathcal{P}1\neq 1$ for this index. The second case may be discarded since it brings in one more independent Kronecker than the first and thus contributes to one order lower in N. Indeed, the Kronecker condition $i_{\mathcal{P}\mathcal{P}1}=i_{\mathcal{P}1}-r_{\mathcal{P}1}=i_1-r_1-r_{\mathcal{P}1}$ reduces to the identity $i_1=i_1$ if $\mathcal{P}\mathcal{P}1=1$ and $r_1+r_{\mathcal{P}1}=0$; only one of the two Kronecker conditions for $i_{\mathcal{P}1}$ and $i_{\mathcal{P}\mathcal{P}1}$ remains effective then. The argument may be continued through all $2m$ indices and yields the asserted predominance of shift configurations with pairs of opposite integers $\pm r_k$. The m integers r_k may all be different from one another, in which case only one permutation in (6.9.42) contributes, namely, the one consisting of an index swap within each pair pertaining to a given r_k; for convenience, these pairs may be labelled such that $\mathcal{P}(2k-1)=2k$, $\mathcal{P}(2k)=2k-1$, and $r_{2k-1}\to +r_k$, $r_{2k}\to -r_k$. This m-fold index swap has the sign $(-1)^m$, but coincidences of some r_k may occur with multiplicities $\mu_1\dots\mu_m$, $\sum_{i=1}^{m}\mu_i=m$, $\mu_i\in(0,1,\dots m)$, then the number of permutations with nonvanishing summands in (6.9.42) is $\prod_{i=1}^{m}\mu_i!$, and all of these summands are equal. Thus, the Taylor coefficient largest in magnitude (for fixed m) (6.9.42) takes the form

$$I_{2m,0}\left(\{\pm r\}\right)\;=\;(-1)^m\underbrace{\left(\prod_{i=1}^{m}\mu_i!\right)}_{(\Sigma_i\,\mu_i=m)}\frac{(N-2m)!}{N!}$$

$$\times\sum_{i_1\neq\,\dots\,\neq i_{2m}}^{1\,\dots\,N}\prod_{k=1}^{m}\delta\left(i_{2k-1}-r_k,i_{2k}\right).\qquad(6.9.43)$$

The remaining sum in (6.9.43) is obviously bounded by N^m. Moreover, since the product of multiplicity factorials is bounded by $m!$, the coefficient in question has an asymptotic weight no larger than

$$I_{2m,0}\left(\{\pm r\}\right)=(-1)^m m!N^m\frac{(N-2m)!}{N!},\qquad(6.9.44)$$

just as in the special case $r_k=1$, [see (6.9.26)].

The construction of the higher Taylor coefficients $I_{2m,l}(\{\pm r\})$ proceeds analogously. As long as all integers r_k obey $r_k \ll N$, the estimates of Sect. 6.9.2 apply once more, and the replacement $\pm 1 \rightarrow \pm r_k$ amounts to no more than notational changes. If, however, one or more of the r_k are of the order N or larger, one nontrivial modification of the reasoning does become necessary: Consider a row of the determinant in (6.9.41) involving a large r_k with $|r_k| \gg N$,

$$\delta(k+r_k,l) - \frac{\sin[\pi(k+r_k-l)S/N]}{\pi(k+r_k-l)} = -\frac{1}{\pi r_k} \sin\left[\pi(k+r_k-l)\frac{S}{N}\right].$$

$$(6.9.45)$$

Note that $\delta(k+r_k,l)$ is necessarily equal to zero since $0 \leq k,l \leq N$ and $|r_k| \gg N$. Moreover, the sine function in such a row should not be expanded in powers of S due to its comparatively slow convergence for $0 \leq S \lesssim \mathcal{O}(1)$. However, the prefactor $1/r_k \ll 1/N$ may be pulled out of the row and immediately reveals the asymptotic negligibility of the corresponding $I_{2m}(S,\{\pm r_k\})$.

6.9.5 Concluding Remarks

To summarize, all of the coordinate integrals in (6.9.4) capable of an S dependence differing from that of $E_{\text{CUE}}(S)$ by more than an overall scale factor are negligible in the limit $N \rightarrow \infty$. Thus, the generalized canonical ensemble (6.9.1) for our fictitious gas yields the same spacing distribution $E''(S)$ as Dyson's circular unitary ensemble of random matrices.

I should add that the discussion of this section can be extended in several ways [148]. Most notably, Floquet operators with antiunitary symmetries allow for similar derivations of the spacing distributions of the orthogonal and symplectic ensembles. Statistical properties of the circular ensembles of random matrices other than the level spacing distributions, like low-order cluster functions of the level density, can be derived from equilibrium statistical mechanics as well. It may be especially noteworthy that Pandey's proof of the ergodicity of the classical ensembles of random matrices can be extended to the generalized canonical ensembles treated here.

Unfortunately, some technical difficulties are incurred in generalizing the present arguments to the energy levels of autonomous dynamical systems. One obvious obstacle lies in the unstable nature of the Pechukas–Yukawa gas described by (6.2.27): Due to the repulsive interparticle interaction, the gas tends to explode as the "time" λ grows; thus any application of equilibrium statistical mechanics appears to be a questionable enterprise. Actually, this first difficulty will be overcome in Sect. 6.10 below: A suitable rescaling of the energy levels will amount to confining the gas in an external harmonic potential [151]. Even then, the energy levels range in the infinite interval $-\infty < x_i < +\infty$; as a consequence, an α expansion like (6.9.15), while convergent for "particles" confined to the unit circle, would necessarily be divergent. Only a much more severely restricted canonical ensemble lends itself to easy analysis, namely, the one involving only constants of the motion of the type $\text{Tr}\{v^\mu\}$, $\mu = 1,2,\ldots$. As was the case for unitary operators in Sect. 6.8, such an ensemble turns out to rigorously imply the joint distribution of N coordinates known from the Gaussian ensembles of Hermitian random matrices.

6.10 Dynamics of Rescaled Energy Levels

The levels $E_m(\lambda)$ of a quantum Hamiltonian $H_0 + \lambda V$ fly apart as $E_m \sim \lambda$ for $\lambda \to \infty$. Correspondingly, the fictitious gas introduced by Pechukas keeps expanding indefinitely with increasing λ, due to the repulsive two-body interaction in the Hamiltonian function $\mathcal{H}(E, p, l)$ in (6.2.27). This slightly nasty behavior of the level dynamics (6.2.27) was the main reason why I preferred to treat the Floquet spectrum of periodically kicked systems in most of the previous sections of this chapter.

In fact, the fictitious classical gas associated with the Floquet spectrum in Sect. 6.2 finds its most natural analogue for autonomous quantum systems not in the original Pechukas–Yukawa gas but in a slightly modified gas to be introduced presently. The modification will amount to confinement of the gas [151].

When treating λ dependent matrix ensembles in Sect. 6.9, it proved advantageous to rescale the Hamiltonian,

$$H = g(H_0 + \lambda V) , \tag{6.10.1}$$

so as to secure a mean density of levels independent of λ. Now, I shall show that there is a particular rescaling, given by a pair of functions $g(\tau)$ and $\lambda(\tau)$ of a fictitious time τ, such that the τ dependence of the eigenvalues and eigenvectors of $H(\tau)$ is equivalent again to the classical Hamiltonian dynamics of a fictitious many-particle system. This dynamics will turn out to differ from the original dynamics (6.2.27) only in the appearance of a harmonic confining potential in the Hamiltonian function $\mathcal{H}(E, p, l)$. Interestingly, the confining potential is the same as that occurring in Dyson's stochastic Coulomb gas model, cf. (6.9.19).

To establish the promised classical dynamics, we will consider the eigenvalue problem of $H(\tau)$,

$$H(\tau)|m, \tau\rangle = E_m(\tau)|m, \tau\rangle \tag{6.10.2}$$

and differentiate with respect to τ. Proceeding as in Sect. 6.2 but using $\dot{H} = (\dot{g}/g)H + g\dot{\lambda}V$, we obtain the set of differential equations

$$
\begin{aligned}
\dot{E}_m &= \frac{\dot{g}}{g}E_m + g\dot{\lambda}V_{mm} , \\
\dot{V}_{mm} &= 2g\dot{\lambda}\sum_{n(\neq m)}\frac{V_{mn}V_{nm}}{E_m - E_n} , \\
\dot{V}_{mn} &= -g\dot{\lambda}V_{mn}\frac{V_{mm} - V_{nn}}{E_m - E_n} \\
&\quad -g\dot{\lambda}\sum_{l(\neq m,n)}V_{ml}V_{ln}\left(\frac{1}{E_l - E_m} + \frac{1}{E_l - E_n}\right) \quad \text{for } m \neq n .
\end{aligned}
\tag{6.10.3}
$$

These correspond to (6.2.4, 7, 10). In pursuit of a Hamiltonian version of the flow (6.10.3), we associate "momenta" with the "coordinates" E_m via

$$p_m = \dot{E}_m \ . \tag{6.10.4}$$

The rate of change of these "momenta" is determined by the flow (6.10.3) according to

$$\dot{p}_m = \left[\left(\frac{\dot{g}}{g} \right)^{\cdot} - \frac{\dot{g}}{g} \frac{(g\dot{\lambda})^{\cdot}}{g\dot{\lambda}} \right] E_m + \left[\frac{\dot{g}}{g} + \frac{(g\dot{\lambda})^{\cdot}}{g\dot{\lambda}} \right] p_m + g\lambda \dot{V}_{mm} \ . \tag{6.10.5}$$

Next, we must eliminate the sources $\partial \dot{p}_m / \partial p_m$ and $\partial \dot{V}_{mn} / \partial V_{mn}$. This goal is partly attained by changing from the off-diagonal elements V_{mn} to angular momenta l_{mn} as in of (6.2.12, 13):

$$l_{mn} = (E_m - E_n) \frac{V_{mn}}{g} \ . \tag{6.10.6}$$

Thus, the flow (6.10.3) takes the form [with $(\ )^{\cdot}$ denoting $\frac{\partial}{\partial t}(\ )$]

$$\begin{aligned}
\dot{E}_m &= p_m \ , \\
\dot{p}_m &= \left[\left(\frac{\dot{g}}{g} \right)^{\cdot} - \frac{\dot{g}}{g} \frac{(g\dot{\lambda})^{\cdot}}{g\dot{\lambda}} \right] E_m + \left[\frac{\dot{g}}{g} + \frac{(g\dot{\lambda})^{\cdot}}{g\dot{\lambda}} \right] p_m \\
&\quad - 2g^3 \dot{\lambda}^2 \sum_{n(\neq m)} l_{mn} l_{nm} / (E_m - E_n)^3 \ , \\
\dot{l}_{mn} &= g^2 \dot{\lambda} \sum_{l(\neq m,n)} l_{ml} l_{ln} \left[(E_l - E_m)^{-2} - (E_l - E_n)^{-2} \right] \ .
\end{aligned} \tag{6.10.7}$$

The remaining source $\partial \dot{p} / \partial p$ is annulled by requiring

$$\frac{\dot{g}}{g} + \frac{(g\dot{\lambda})^{\cdot}}{g\dot{\lambda}} = 0 \Leftrightarrow g^2 \dot{\lambda} = \text{const} \tag{6.10.8}$$

whereupon the coefficient of E_m in the force takes the simple form $\ddot{g}/g$.

At this point, the Hamiltonian character of the flow (6.10.7) is obvious. The Hamiltonian function differs from Pechukas', i.e., that given in (6.2.27), by the harmonic addition $-(\ddot{g}/2g) \sum_m E_m^2$. The Poisson brackets can be taken from Sect. 6.2.

We still have some freedom in choosing the function $g(\tau)$, despite the constraints

$$\begin{aligned}
g &= 1 \text{ for } \lambda = 0 \ , \\
g\lambda &\to 1 \text{ for } \lambda \to \infty \ ,
\end{aligned} \tag{6.10.9}$$

which any reasonable rescaling in (6.10.1) must obey. As in the previous section, we require that the average density of levels of H does not depend on τ, and the average here is to be understood as a local spectral average. However, that requirement is even harder to implement in the present context than it was in Sect. 6.9. Thus, we take the liberty of choosing that the fictitious many-particle system is autonomous, i.e., the Hamiltonian function $\mathcal{H}$ becomes independent of τ,

$$\frac{\ddot{g}}{g} = \text{const}.\tag{6.10.10}$$

Our next choice concerns the sign of the constant ratio $\ddot{g}/g$. The much more intuitive situation is that of a negative sign since the harmonic part of the Hamiltonian function $\mathcal{H}$ is binding in this case. By adopting a suitable unit of τ, $\ddot{g}/g = -1$ and this, together with the boundary condition (6.10.9), uniquely yields

$$\left.\begin{array}{l} g(\tau) = \cos\tau \\ \lambda(\tau) = \tan\tau \end{array}\right\} \quad \text{for } 0 \leq \tau \leq \tau_0 \equiv \frac{\pi}{2}.\tag{6.10.11}$$

Interestingly, the monotonic decrease of the weight $g(\tau)$ of H_0 from 1 to 0, as τ grows from 0 to $\tau_0 = \pi/2$, is the mirror image about $\tau = \tau_0/2$ of the monotonic increase in the weight $g(\tau)\lambda(\tau) = \sin\tau$ of the perturbation V. Moreover, by eliminating τ between g and λ, the relationship

$$\lambda = \frac{\sqrt{1-g^2}}{g} \Leftrightarrow g = \frac{1}{\sqrt{1+\lambda^2}}\tag{6.10.12}$$

results, which we shall meet again in a rather different context in Sect. 6.15. Thus, the Hamiltonian (6.10.1) and its Pechukas–Yukawa partner become

$$\begin{aligned} H &= H_0\cos\tau + V\sin\tau, \\ \mathcal{H} &= \sum_m \frac{1}{2}\left(p_m^2 + E_m^2\right) + \frac{1}{2}\sum_{m\neq n}\frac{|l_{mn}|^2}{(E_m - E_n)^2}. \end{aligned}\tag{6.10.13}$$

As already mentioned above, the new fictitious-gas Hamiltonian function $\mathcal{H}$ differs from that in (6.2.27) only in the presence of the confining harmonic potential.

The positive sign of the constant in (6.10.10) is also worth looking at. As is easy to verify, the unique solution of $\ddot{g}/g = +1$ obeying the boundary condition (6.10.10) is

$$\left.\begin{array}{l} g(\tau) = \sinh(\tau_0 - \tau) \\ g(\tau)\lambda(\tau) = \sinh\tau \end{array}\right\} \quad \text{for } 0 \leq \tau \leq \tau_0 = \ln\left(1+\sqrt{2}\right).\tag{6.10.14}$$

Here again, $g(\tau)$ and $g(\tau)\lambda(\tau)$ are mirror symmetric to one another about $\tau = \tau_0/2$. While $g(\tau)$ falls monotonically from 1 to 0 as τ evolves from 0 to τ_0, $\lambda(\tau)$ grows monotonically from 0 to ∞. The Hamiltonian function

$$\mathcal{H} = \sum_m \frac{1}{2}\left(p_m^2 - E_m^2\right) + \frac{1}{2}\sum_{m\neq n}\frac{|l_{mn}|^2}{(E_m - E_n)^2}\tag{6.10.15}$$

has a nonbinding harmonic component in this case. An important difference from the case of the binding harmonic potential lies in the relationship $g = (1+2\sqrt{2}\lambda+\lambda^2)^{-1/2}$ which follows from (6.10.14).

Remarkably enough, the classical many-particle dynamics generated by either Hamiltonian function $\mathcal{H}$, (6.10.13) or (6.10.15), is integrable; the integrability follows from the equivalence to the problem of diagonalizing the quantum Hamiltonian $H = g(H_0 + \lambda V)$. The fictitious gases may, of course, be considered in their Hamiltonian evolution for all finite times, $-\infty < \tau < +\infty$. Even the quantum Hamiltonian $H(\tau)$ in either parametrization, (6.10.11) or (6.10.14), generally remains meaningful and faithful to the fictitious-gas dynamics when τ is so extended. The finite time span $0 \le \tau \le \tau_0$ suffices, however, to describe the single continuous transition of $H(\tau)$ from H_0 to V.

When attempting, analogously to Sects. 6.8 and 6.9, to recover the Gaussian ensembles of random matrices from the equilibrium statistical mechanics of a fictitious gas, one would, of course, rely on the gas with the confining potential. Moreover, it is well to recall from Sect. 6.7 that control parameter averages need to go over an interval not much larger than a collision time of the gas, i.e., an interval shrinking to zero as $\hbar \to 0$.

As was already anticipated in concluding Sect. 6.9, the modified gas with the confining potential allows derivation of the eigenvalue distribution of the Gaussian ensembles of random matrices (4.3.15) from a certain canonical ensemble. The reader will enjoy verifying this statement in solving Problem 6.17.

6.11 Level Curvature Statistics

Interest in the statistics of the curvature of energy levels was stimulated by *Thouless* [152] who suggested that the conductance of a disordered system should reflect the sensitivity of the spectrum to changes of boundary conditions. The discussion was revived when *Gaspard* and co-workers [153, 154] characterized spectral fluctuations of random matrices and of systems with classical chaos by the density of level curvatures

$$K_n = \frac{d^2 E_n}{d\lambda^2} \tag{6.11.1}$$

at some fixed value of a control parameter λ. Curvature statistics have in fact gained considerable popularity since, both for disordered systems and in the quantum chaos community. They have been studied experimentally for the energy levels of atomic hydrogen [155] and for the eigenfrequencies of microwave resonators [156]. The relevance for electric and magnetic properties of mesoscopic probes is beautifully highlighted in a connection between the disorder-averaged conductance $\langle g \rangle$ and the spectral average of the modulus $\langle |K_n| \rangle$ of a dimensionless curvature [157],

$$\langle g \rangle = \langle |K_n| \rangle \frac{\pi}{\Delta} \frac{e^2}{h} \tag{6.11.2}$$

where Δ is the mean level spacing and the control parameter with respect to which the second derivative is taken is the magnetic flux Φ through the probe, normalized to the flux quantum $\Phi_0 = h/e$, as $\lambda = 2\pi\Phi/\Phi_0$.

The curvatures of two neighboring levels become large when the levels have their closest approach in an avoided crossing. Therefore, we can expect the large-K behavior of the curvature distribution $W(K)$ to be uniquely related to the small-S behavior of the spacing distribution $P(S)$. To establish that relationship for quasi-energies, we employ level dynamics from (6.2.15, 16) – second-order perturbation theory would do as well – to write the curvature of a level as

$$K_m = \ddot{\varphi}_m = \sum_{n(\neq m)} V_{mn}V_{nm}\frac{\cos[(\varphi_m - \varphi_n)]}{\sin[(\varphi_m - \varphi_n)]}. \tag{6.11.3}$$

In a close encounter of two levels, thus, indeed, $|K| \propto \frac{1}{S}$. For a quick guess at the asymptotic form of the density, we may use $P(S) \propto S^\beta$, assume the matrix elements V_{mn} statistically independent of the eigenphases, and so find

$$W(K) = \int dS\,\delta\big(|K| - \frac{1}{S}\big)P(S) \propto K^{-(\beta+2)}. \tag{6.11.4}$$

Needless to say, the degree of level repulsion β takes on the values 1, 2, and 4, respectively, for the orthogonal, unitary, and symplectic universality classes. The same result obtains for energy levels if we work with the Hamiltonian[6]

$$H = H_0 \cos \lambda + H_1 \sin \lambda \tag{6.11.5}$$

and its Pechukas–Yukawa partner (6.10.13) which has the levels harmonically confined according to $K_m = \ddot{E}_m|_{\lambda=0} = -E_m + 2\sum_{n(\neq m)} H_{mn}H_{1nm}/(E_m - E_n)$. But what right do we have to assume statistical independence of eigenvalues and matrix elements or, what amounts to the same, eigenvectors? Here are three good reasons: First, there is numerical evidence of such independence for dynamic systems faithful to random-matrix theory in their spectral fluctuations; second, random-matrix theory itself predicts precisely that independence, see Sect. 4.3; finally, we have just seen, in Sects. 6.7–6.9, in which sense level dynamics implies random-matrix theory.

The asymptotic power law (6.11.4) was confirmed numerically for random matrices as well as dynamic systems in [153, 154] and for disordered systems in [158]. An illustration is provided by Fig. 6.6. *Zakrzewski* and *Delande* [159] went one step further and, on the basis of their numerical investigations, conjectured the full density

$$W(K/K_0) = C_\beta(1 + (K/K_0)^2)^{-(1+\beta/2)} \tag{6.11.6}$$

with K_0 a suitable curvature unit and C_β a normalization constant. Eventually, *von Oppen* [160] and *Fyodorov* and *Sommers* [161] showed that this was indeed

---

[6] Note that the "time" τ of Sect. 6.10 is renamed as λ such that $g = \cos\lambda$, and now the previous parameter $\lambda = \tan\tau$ appears as $\tan\lambda$; moreover, for the reader's convenience, we rename the perturbation V as H_1.

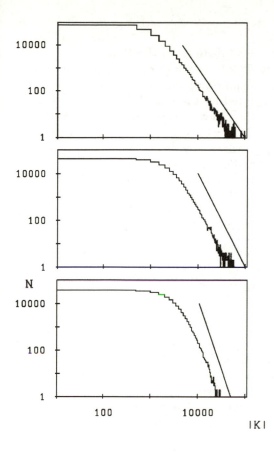

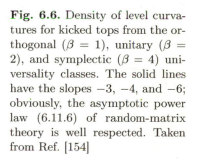

Fig. 6.6. Density of level curvatures for kicked tops from the orthogonal ($\beta = 1$), unitary ($\beta = 2$), and symplectic ($\beta = 4$) universality classes. The solid lines have the slopes -3, -4, and -6; obviously, the asymptotic power law (6.11.6) of random-matrix theory is well respected. Taken from Ref. [154]

correct, for the Gaussian ensembles of random matrices as well as for disordered systems.

Now, I shall proceed to derive the distribution (6.11.5) for the Gaussian orthogonal ensemble along the lines of Ref. [161]. The λ-dependent Hamiltonian is taken from the form (6.11.5) with H_0 and H_1 statistically independent but from the orthogonal universality class such that their joint density reads

$$P(H_0, H_1) \propto \exp\left(-\frac{1}{4J_0^2}\mathrm{Tr}H_0^2 - \frac{1}{4J_1^2}\mathrm{Tr}H_1^2\right) . \tag{6.11.7}$$

Eventually but not right away, we shall set $J_0 = J_1 = J$. Then, the semicircle law for the density takes the form

$$\bar{\varrho}(E) = \langle\frac{1}{N}\mathrm{Tr}\,\delta(E - H)\rangle = \frac{1}{2\pi J^2 N}\sqrt{4NJ^2 - E^2} \tag{6.11.8}$$

and the variances of the various matrix elements implied by (6.11.7) are given by

$$\langle H_{ii}^2\rangle = 2\langle|H_{ij}|^2\rangle = 2J^2 = \mathcal{O}(\frac{1}{N}) . \tag{6.11.9}$$

We choose $J^2 = \mathcal{O}(\frac{1}{N})$ to have a finite radius $2J\sqrt{N}$ of the semicircle law, i.e., a finite range for the eigenvalues E_n.

It is well to emphasize that the model defined by the Hamiltonian (6.11.5) and the ensemble (6.11.7) cannot be blindly applied to all classically chaotic systems. Quantum localization would require different treatment. Moreover, for perturbations to be reasonably represented by the model, they must "globally" affect all wave functions throughout the relevant configuration space and the spectral range of H_0. A counterexample to which we shall have to return to in the next section is a chaotic billiard perturbed by an antenna smaller in linear dimension than the wavelength range under investigation.

To avoid any distinction of the initial value $\lambda = 0$, we rewrite the level dynamics (6.10.3) with the help of $H_1 = H \sin \lambda + \dot{H} \cos \lambda$ as

$$K_m = \ddot{E}_m = -E_m + 2 \sum_{n(\neq m)} \frac{|\langle m|\dot{H}|n\rangle|^2}{E_m - E_n}. \tag{6.11.10}$$

The statistical independence of H_0 and H_1 entails

$$\langle H\dot{H}\rangle = (-\langle H_0^2\rangle + \langle V^2\rangle) \cos \lambda \sin \lambda \tag{6.11.11}$$

such that H and $\dot{H}$ are statisically independent provided $\langle H_0^2\rangle = \langle H_1^2\rangle$, which latter equality holds only once we choose $J_0 = J_1 = J$.

Even though we want to establish our final result for arbitrary values of the control parameter λ, it is convenient to consider $\lambda = 0$ first. Looking at the ensemble mean of the curvature of the mth level $E_m = E$ will provide a good warm-up. Exploiting the statistical independence of eigenvalues and eigenvectors of matrices in the Gaussian ensembles,

$$\langle K_m\rangle = -E + 2J_1^2 \left\langle \sum_{n(\neq m)} \frac{1}{E_m - E_n} \right\rangle \Bigg|_{E_m=E}. \tag{6.11.12}$$

For $\lambda = 0$, the eigenvalues E_n are those of H_0. The average on the r. h. s. can be related to the Green function of H_0 if we replace $E = E_m$ by a complex variable $\mathcal{E}$; then $\langle \sum_{n(\neq m)} (\mathcal{E} - E_n)^{-1}\rangle = \langle \text{Tr}\,(\mathcal{E} - H_0)^{-1}\rangle (1 + \mathcal{O}(\frac{1}{N}))$. Now with $\mathcal{E} = E - \mathrm{i}0^+$, from (4.6.4),

$$\left\langle \text{Tr} \frac{1}{E + \mathrm{i}0^+ - H_0} \right\rangle = \frac{1}{2J_0^2} \left(E - \mathrm{i}\sqrt{4NJ_0^2 - E^2} \right) = \frac{E}{2J_0^2} - \mathrm{i}\pi N \bar{\varrho}(E). \tag{6.11.13}$$

Thereore, the mean curvature

$$\langle K\rangle = -E \left(1 - \frac{J_1^2}{J_0^2} \right) \tag{6.11.14}$$

vanishes when $J_0 = J_1$ for which we now settle.

But now on to the fluctuations of the curvature which are fully described by the distribution

$$P(K,E) = \frac{1}{N\bar{\varrho}(E)} \left\langle \sum_{m=1}^{N} \delta(K - K_m)\delta(E - E_m) \right\rangle$$

$$= \frac{1}{N\bar{\varrho}(E)} \sum_{m=1}^{N} \int_{-\infty}^{\infty} d\omega\, e^{-i\omega(K+E)} \qquad (6.11.15)$$

$$\times \left\langle \delta(E - E_m) \exp\left\{ i2\omega \sum_{n(\neq m)} \frac{|H_{1mn}|^2}{E - E_n} \right\} \right\rangle .$$

As in (6.11.12), the eigenvalues appearing here are those of H_0, and we can first do the average over H_1. The density (6.11.7) yields

$$P(K,E) = \frac{1}{N\bar{\varrho}(E)} \sum_{m=1}^{N} \int_{-\infty}^{\infty} d\omega\, e^{-i\omega(K+E)} \qquad (6.11.16)$$

$$\times \left\langle \delta(E - E_m) \prod_{n(\neq m)}^{1...N} \left[1 - \frac{4i\omega J^2}{(E - E_n)} \right]^{-1/2} \right\rangle .$$

The remaining average over the N eigenvalues of H_0 can be done with their joint density (4.3.15). If we integrate out the delta function $\delta(E - E_m)$, setting $E = E_m \equiv E_1$ everywhere else, we arrive at

$$P(K,E) \propto \frac{1}{\bar{\varrho}(E)} \sum_{m=1}^{N} \int_{-\infty}^{\infty} d\omega\, e^{-i\omega(K+E)-\frac{E^2}{4J^2}} \qquad (6.11.17)$$

$$\times \left\langle \prod_{n=2}^{N} \left(\frac{|E - E_n|}{\sqrt{1 - \frac{4i\omega J^2}{(E-E_n)}}} \right) \right\rangle_{N-1} .$$

The subscript $N - 1$ at the rightmost bracket demands averaging with the joint density of the $N-1$ eigenvalues $E_2 \dots E_N$; the average over E_1 has already taken been care of by the delta function just mentioned. I shall omit that subscript in the sequel and simply imagine that the $N - 1$ eigenvalues pertain to a random $(N-1) \times (N-1)$ matrix H and do the average over the corresponding ensemble; accepting an (irrelevant) error of relative weight $1/N$, we need not even bother to change the width J. The square root in the foregoing expression (as well as below) is meant with its main branch, i.e., with a nonnegative real part. With the latter remark in mind, we may represent the fraction inside the round brackets above as

$$\frac{|E - E_n|}{\sqrt{1 - \frac{4i\omega J^2}{(E-E_n)}}} = \frac{(E - E_n)^2}{\sqrt{(E - E_n)^2 - 4i\omega J^2(E - E_n)}} . \qquad (6.11.18)$$

To save space, we shall look at the center of the spectrum, $E = 0$, reserving for later the extension to $E \neq 0$. Thus,

$$P(K,0) \equiv P(K) = \int_{-\infty}^{\infty} d\omega e^{-i\omega K} \left\langle \frac{\det H^2}{\det \sqrt{H^2 + 4i\omega J^2 H}} \right\rangle. \tag{6.11.19}$$

When writing a product $\prod_{n=2}^{N} f(E_n)$ as an $(N-1) \times (N-1)$ determinant $\det f(H)$, one has in mind replacing the average over eigenvalues by one over the full matrix H. That may look like a step in the wrong direction since the number of integrations increases by a factor of the order N. But progress is indeed possible since we can write the denominator as a Gaussian integral,

$$\frac{1}{\sqrt{\det f(H)}} \propto \int d^{N-1}z \, e^{-z^\dagger f(H)z} \tag{6.11.20}$$

with $f(H) = H^2 + 4i\omega J^2 H$ and z a real $(N-1)$ component vector. The integration measure is denoted as $d^{N-1}z$ to remind the reader of the number of independent real parameters in z.

Due to the symmetry of the GOE under orthogonal transformations, we can rotate the vector z with such a transformation to assign any desired direction to z. The orientation of z along any of the "axes" is desirable, say the first one which we had labelled as "2", since then the matrix $f(H)$ enters the exponent of the above Gaussian integral only with the single element $(f(H))_{22}$,

$$z^\dagger f(H)z = |z|^2 \left[\sum_{j=2}^{N} |H_{2j}|^2 + (4i\omega J^2)H_{22} \right], \tag{6.11.21}$$

where $|z_2|^2 = |z|^2$ is the squared length of the vector z. It follows that the GOE average $\langle \exp(-z^\dagger f(H)z) \rangle$ depends on z only through the squared length $|z|^2$.

The ω integral in (6.11.19) obviously gives a delta function, and we are left with the z integral and the GOE average,

$$P(K) = \int d^{N-1}z \left\langle (\det H)^2 \, \delta\big(K + 4J^2|z|^2 H_{22}\big) e^{-|z|^2 \sum_{j=2}^{N} |H_{2j}|^2} \right\rangle. \tag{6.11.22}$$

Turning to the average first, we immediately do the integral over H_{22} by setting

$$H_{22} \to \hat{H}_{22} = -K/4J^2|z|^2. \tag{6.11.23}$$

The integrals over the off-diagonal elements H_{2j} with $j > 2$ may be read as moments of the Gaussian $\exp\{-(\frac{1}{2J^2} + |z|^2)|H_{2j}|^2\}$ whose z-dependent width is smaller than that for the original Gaussian ensemble (6.11.7 with $J_1 = J_2 = J$); the basic moments are

$$\overline{H_{2j}} = 0, \quad \overline{|H_{2j}|^2} = \big(J^{-2} + 2|z|^2\big)^{-1} \tag{6.11.24}$$

which are denoted by the overbar rather than the angular brackets to emphasize the change of the Gaussian densities. When doing these integrations, we get the

z-dependent prefactor $(1 + 2J^2|z|^2)^{-(N-2)/2}$. As a further preparation to writing out an elevated form of the density $P(K)$ in search, we write the $(N-1) \times (N-1)$ matrix H in the form

$$H = \begin{pmatrix} H_{22} & H_{2j} \\ H_{k2} & V \end{pmatrix} \tag{6.11.25}$$

where V summarily denotes the lower right $(N-2) \times (N-2)$ block of H. Then, the determinant of H can be written as

$$\det H = (H_{22} - \sum_{kj=3}^{N} H_{2j} V_{jk}^{-1} H_{k2}) \det V . \tag{6.11.26}$$

Finally, we employ spherical coordinates with $|z| \equiv r$ for the z integral to get $d^{N-1}z \propto dr\, r^{N-2}$. Putting together the various pieces mentioned,

$$P(K) \propto \int_0^\infty dr\, r^{N-4} (1 + 2J^2 r^2)^{-(N-2)/2} e^{-\left(\frac{1}{4J^2} + r^2\right) \hat{H}_{22}^2}$$
$$\times \left\langle \det V^2 \overline{\left(\hat{H}_{22} - \sum_{kj>2} H_{2j} V_{jk}^{-1} H_{k2}\right)^2} \right\rangle_V$$

or, after inserting $\hat{H}_{22}$ from (6.11.23) and a little rearrangement which will facilitate the further discussion,

$$P(K) \propto \int_0^\infty dr\, r^{-4} (1 + 2J^2 r^2) \exp\left[-\frac{N}{2} \ln\left(1 + \frac{1}{2J^2 r^2}\right) - \left(\frac{1}{4J^2} + r^2\right) \frac{K^2}{16 J^4 r^4}\right]$$
$$\times \langle \det V^2 \overline{(\ldots)^2} \rangle_V ; \tag{6.11.27}$$

here the suffix V at the angular bracket demands averaging over V with the Gaussian distribution $\propto e^{-\frac{1}{4J^2} \mathrm{Tr} V^2}$. That average leaves us with only even-order terms in V,

$$\left\langle \det V^2 \overline{(\ldots)^2} \right\rangle_V = \left\langle \det V^2 \left\{ \hat{H}_{22}^2 + \overline{\left(\sum_{jk>2} H_{2j} V_{jk}^{-1} H_{k2}\right)^2} \right\} \right\rangle_V \tag{6.11.28}$$

$$\propto \left(\frac{K}{4J^2 r^2}\right)^2 + \left(\frac{1}{J^{-2} + 2r^2}\right)^2 \frac{\langle \det V^2 ((\mathrm{Tr} V^{-1})^2 + 2\mathrm{Tr} V^{-2}) \rangle_{N-2}}{\langle \det V^2 \rangle_{N-2}} .$$

The inconsequential overall factor $\langle \det V^2 \rangle$ has been dropped in the last expression. Moreover, now that the only remaining average is that over the $(N-2) \times (N-2)$ matrices V, the subscript V to the angular brackets has completed its short service and was replaced, with foresight, by a reminder of the dimension $N-2$ of the matrices V involved.

We should scrutinize the r dependence of the integrand in (6.11.27). The exponential grows monotonically from 0 to 1 as r grows from 0 to ∞. Near $r = 0$, the exponential is small, $\propto e^{-\mathrm{const}/r^4}$, and wrestles down the power-law blow-up, $\propto r^{-8}$, of the prefactor. For large r, on the other hand, convergence of the integral is provided by the power-law fall-off $\propto r^{-6}$ of the prefactor. These statements hold true for any N and must remain true for $N \to \infty$, for which latter limit we must recall $J^2 = \mathcal{O}(1/N)$. It follows that the correct large-N behavior of the integral is brought about by rescaling the integration variable as $r = NR$. Dropping corrections of relative weight $1/N$ everywhere, we can simplify the integral as

$$\int_0^\infty dR R^{-6} e^{-\frac{K^2 + 4J^2 N}{16 J^4 N^2 R^2}} \propto (K^2 + 4J^2 N)^{-\frac{5}{2}} \tag{6.11.29}$$

and thus find the curvature distribution as

$$P(K) \propto (K^2 + 4J^2 N)^{-\frac{5}{2}} \left\{ K^2 + 4J^4 \frac{\langle \det V^2 ((\mathrm{Tr} V^{-1})^2 + 2\mathrm{Tr} V^{-2}) \rangle_{N-2}}{\langle \det V^2 \rangle_{N-2}} \right\}.$$

Our final step is to show that the curly and round brackets in the foregoing expression coincide for large N. To that end, we consider the Gaussian ensemble of real symmetric $(N-1) \times (N-1)$ matrices with density $\propto e^{-\frac{1}{4J^2} \mathrm{Tr} V^2}$ and consider the mean value $\langle \det V^2 \rangle_{N-1}$. Once more employing (6.11.26), this time for the $(N-1) \times (N-1)$ matrix V and its lower right $(N-2) \times (N-2)$ submatrix, we average over the elements of the the first row and column in analogy with (6.11.28) but here with just the density specified and get the identity

$$\frac{\langle \det V^2 \rangle_{N-1}}{\langle \det V^2 \rangle_{N-2}} = 2J^2 + J^4 \frac{\langle \det V^2 ((\mathrm{Tr} V^{-1})^2 + 2\mathrm{Tr} V^{-2}) \rangle_{N-2}}{\langle \det V^2 \rangle_{N-2}}. \tag{6.11.30}$$

But the ratio on the l. h. s. must be finite for $N \to \infty$ and for dimensional reasons can thus differ from $J^2 N$ at most by a numerical factor of order unity. It follows that the first term on the r. h. s. is negligible and that the curly bracket under investigation reads $\{K^2 + 4J^2 N(?)\}$, with (?) the numerical factor just mentioned. That factor becomes in fact equal to unity when $N \to \infty$, as follows from the rigorous identity (valid for any N) [162]

$$\langle \det V^2 \rangle_N = \left(-J^2 \frac{\partial}{\partial \omega} \right)^N \frac{1}{(1 + \omega^2)\sqrt{1 - \omega^2}} \bigg|_{\omega=0}. \tag{6.11.31}$$

The function of ω to be differentiated here may be expanded in powers of ω, and the radius of convergence of that Taylor expansion obviously equals unity. The ratio a_{n+1}/a_n of neighboring Taylor coefficients must therefore approach unity as n grows indefinitely. By the required differentiation, the coefficient a_N acquires the factor $(-J^2)^N N!$ such that indeed

$$\frac{\langle \det V^2 \rangle_{N-1}}{\langle \det V^2 \rangle_{N-2}} \xrightarrow{N \to \infty} J^2 N. \tag{6.11.32}$$

The identity (6.11.32) can be proven by representing $\det V$ by a Grassmannian Gaussian integral as in (4.12.1) but I assume the reader will be just as happy without seeing the proof which would fill several more pages.

At any rate, we have established the Zakrzewski–Delande curvature distribution for the GOE, $P(K) \propto (K^2 + K_0^2)^{-3/2}$, i.e., (6.11.6) for $\beta = 1$. Before leaving level curvatures, I should at least mention that lifting the restrictions to $E = 0$ and $\lambda = 0$ is not an overly difficult task and yields $P(K, E) \propto [1/\bar{\varrho}(E)](K^2 + 4\pi^2\bar{\varrho}(E)^2)$. I shall not bother to work out that generalization here. Comparisons with numerical data for dynamical systems [159] require, as for all measures of spectral fluctuations, an unfolding of the spectrum after which the scaling of the curvature with the mean level spacing disappears anyway. Nevertheless, the interested reader may want to consult Ref. [161] for such details as well as for a parallel treatment of all three Gaussian ensembles ($\beta = 1, 2, 4$) and even ensembles of sparse matrices.

6.12 Level Velocity Statistics

The level velocity

$$v_n = \frac{dE_n}{d\lambda} \tag{6.12.1}$$

and its statistical properties enjoy a popularity comparable to that of the level curvature in both the quantum chaos and disordered systems communities. For disordered systems, the interest derives from the fact that the spectral and disorder average of the squared level velocity, $\langle v^2 \rangle$, where v is defined as the response to a magnetic field threading a mesoscopic probe, can be linked to the Kubo conductance $\langle g \rangle$ [163]. In particular, a simple relationship exists between $\langle v^2 \rangle$ and the mean curvature modulus [158, 161],

$$\Delta\langle|K|\rangle = 2\pi\langle v^2\rangle \propto \langle g\rangle. \tag{6.12.2}$$

Thus, a fruitful new perspective on the Thouless conjecture mentioned in the beginning of the preceding section is opened. Fyodorov and Mirlin [164] have shown that the distribution of level velocities is directly related to the wave function statistics.

For a quick orientation to theory, I shall here again consider level dynamics according to a Hamiltonian of the form (6.11.5), $H(\lambda) = H_0 \cos\lambda + H_1 \sin\lambda$, with H_0 and H_1 statistically independent random $N \times N$ (or for the symplectic ensemble, $2N \times 2N$) matrices, drawn from one of the classic Gaussian ensembles. The densities will be taken as

$$P_\beta(H_0, H_1) \propto \exp\left(-\frac{\beta}{4J^2}\mathrm{Tr}H_0^2 - \frac{\beta}{4J^2}\mathrm{Tr}H_1^2\right) \tag{6.12.3}$$

whereupon the semicircular law (6.11.8) holds for the three classes with the same radius $\sqrt{4NJ^2}$.

The first and simplest item to deal with is the stationary velocity distribution predicted by random-matrix theory itself and its applicability to dynamic systems. Inasmuch as we let H_0 and H_1 be random matrices, there is practically no distinction of the "initial moment" $\lambda = 0$, and therefore I shall not hesitate to look at the velocity distribution for that situation. But then, v_n is nothing but the momentum of the nth "particle" in the Pechukas–Yukawa gas,

$$v_n = p_n = H_{1nn},\tag{6.12.4}$$

with the state $|n\rangle$ an eigenstate of H_0. The distribution of the diagonal elements of the random matrix H_1 is the Gaussian implied by (6.12.3),

$$P_\beta(v) = \sqrt{\frac{\beta}{4\pi J^2}}\,e^{-\frac{\beta}{4J^2}v^2}.\tag{6.12.5}$$

There is good evidence (see Fig. 6.7) that quantum or wave chaotic systems (without localization) are faithful to that Gaussian, both from experiments on microwave billiards [156] and numerical data on the Sinai billiard [165], provided the perturbation is "globally" effective like a change of the linear dimension and unlike the displacement of a small antenna [167]. Such applicability of random-matrix theory is not much of a surprise from the point of view of level dynamics and its statistical mechanics: The distributions of level spacings, velocities, and curvatures should all be self-averaging in the same amount.

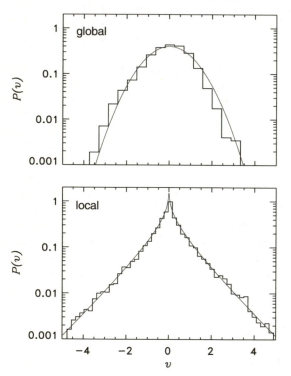

Fig. 6.7. Level velocity distributions measured for microwave resonators [166]. The Gaussian behavior in the upper part was obtained by moving a wall comparable in size to the linear dimension of the resonator [156]; the lower part was found by moving a comparatively small antenna [167] and is faithful to (6.12.6). Courtesy of Stöckmann

The exception of local perturbations by displacing an antenna small in its linear dimension compared to the mean wavelength under consideration was studied experimentally and theoretically by *Barth, Kuhl,* and *Stöckmann* [167]. Such a perturbation is always small, independently of the location of the antenna, and just probes the eigenfunctions of the unperturbed "Hamiltonian". Stöckmann and co-workers argue that the rate of change of a level with a spatial coordinate r of the antenna is given by $v = \partial E_n/\partial r = \alpha\partial|\psi_n|^2/\partial r$, where α is a geometry-dependent constant. According to a conjecture by *Berry* [168, 169] and to random-matrix theory (see Chap. 4.8.1), the local wave has Gaussian amplitude statistics. Assuming statistical independence of $\psi(r)$ and $\psi(r + \delta r) = \psi(r) + \psi'(r)\delta r$, we have statistical independence of the wave-function and its gradient, while the Gaussian distributions $P(\psi) \propto e^{-A|\psi|^2}$ at both points imply the Gaussian distribution $P(\psi') \propto e^{-\frac{1}{2}A(\delta r)^2|\psi'|^2}$ for the gradient. As to the length-scale parameter δr, there is no other natural choice than the mean wavelength of the spectral range considered. At any rate, the foregoing ideas uniquely yield the velocity distribution (for real ψ, i.e., situations with time-reversal invariance)

$$
\begin{aligned}
P(v) &= \langle\delta(v - 2\alpha\psi\psi')\rangle \propto \int d\psi d\psi' e^{-A(\psi^2 + (\delta r)^2\psi'^2/2)} \\
&= \frac{\beta}{\pi}K_0(\beta|v|),
\end{aligned}
\tag{6.12.6}
$$

where $\beta = A\delta r/\alpha\sqrt{2}$ and K_0 is a modified Bessel function. The lower part of Fig. 6.7 depicts velocity distributions experimentally obtained [167] for microwave billiards: local perturbations by small antennas yield good agreement with the Bessel function (6.12.6) while global perturbations give the expected Gaussian (6.12.5).

Quite a different spectral characteristic was suggested by *Yang* and *Burgdörfer* [170], the autocorrelation function of a level for different values of the control parameter alias time λ. In a stationary situation, that function should depend only on the "time" difference, $\langle v_n(\lambda)v_n(\lambda')\rangle = \langle v_n(\lambda - \lambda')v_n(0)\rangle$; the average may be imagined as taken over all levels of an (unfolded) spectrum. One defines the normalized autocorrelator as

$$
C(\lambda) = \frac{\langle v_n(\lambda - \lambda')v_n(0)\rangle}{\langle v_n(0)^2\rangle}
\tag{6.12.7}
$$

and would like to compare its single-system form, determined experimentally or numerically, with the prediction of random-matrix theory. Unfortunately, no closed form of $C(\lambda)$ is known from random-matrix theory and in particular none for the physically most interesting case where H_0 is time-reversal invariant but H_1 is not. The latter case is realized in disordered systems when λ is a magnetic field. *Simons* and *Altshuler* [171] have come closest to an explicit result with a superanalytic treatment of a correlation function similar to the above $C(\lambda)$ in which the energy is involved as a second independent variable. These authors also find universal behavior of $C(\lambda)$ after unfolding the spectrum to uniform mean level density and also employing the rescaled control parameter

$$
\tau = \lambda\pi\sqrt{\langle v^2\rangle/\Delta^2}
\tag{6.12.8}
$$

where $\Delta = (N\bar{\varrho})^{-1}$ is the mean level spacing. In particular, the large-τ behavior comes out as

$$C(\lambda) \to -\frac{2}{\beta\tau^2} \qquad \text{for} \qquad \tau \gg 1. \tag{6.12.9}$$

The small-τ behavior was found, for $\beta = 2$ in Ref. [171] and for $\beta = 4$ in Ref. [172], as

$$C_{\text{GUE}}(\lambda) \to 1 - 2\tau^2$$
$$C_{\text{GSE}}(\lambda) \to 1 - \tfrac{8}{3}\tau^2. \tag{6.12.10}$$

Numerical data are available to interpolate between the positive correlation at small τ and the negative one at large τ; a minimum value arises for a value of τ of order unity. The change of the relative sign of the level velocity is clearly related to an avoided crossing and defines, in the fictitious-gas language, a typical collision time $\tau_{\text{coll}} \propto \lambda_{\text{coll}}$ for neighboring particles. The N dependence of λ_{coll} results from $\tau_{\text{coll}} = \mathcal{O}(N^0)$ with the help of (6.12.3) or $\langle|H_{1nn}|^2\rangle = \frac{2J^2}{\beta} \propto \frac{1}{N}$ as

$$\lambda_{\text{coll}} = \mathcal{O}(\frac{1}{N}), \tag{6.12.11}$$

a result already anticipated in Sect. 6.7.

I shall present here a simple argument due to *Zakrzewski* [172] that produces the small-τ behavior (6.12.11) and at the same time reveals that for the GOE, the behavior of $C_{\text{GOE}}(\lambda)$ at $\lambda = 0$ is too singular for an $\mathcal{O}(\tau^2)$ term to exist. By stationarity, $\langle v_n(\lambda)v_n(0)\rangle = \langle v_n(\lambda/2)v_n(-\lambda/2)\rangle$; upon expanding both sides of this identity in powers of λ as $v_n(\lambda) = v_n(0) + K_n(0)\lambda + \frac{1}{2}K'_n(0)\lambda^2 + \ldots$, we conclude that $\langle v_n(0)K_n(0)\rangle = 0$ and $\langle v_n(0)K'_n(0)\rangle = -\langle K_n(0)^2\rangle$ and thus

$$\langle v_n(\lambda)v_n(0)\rangle = \langle v_n(0)^2\rangle - \frac{1}{2}\langle K_n(0)^2\rangle\lambda^2 + \ldots. \tag{6.12.12}$$

The curvature variance $\langle K_n(0)^2\rangle$ can be computed from the distribution (6.11.6). It does not exist for the GOE and in the other two cases comes out, with $K_0 = 4J^2N$, as

$$\langle K_n(0)^2\rangle = 4J^2N \times \begin{cases} 2 \\ \frac{1}{3} \end{cases} \qquad \text{for} \qquad \beta = \begin{cases} 2 \\ 4. \end{cases} \tag{6.12.13}$$

After rescaling the control parameter as in (6.12.6), we establish the small-τ behavior claimed in (6.12.9). A more thorough analysis [173] reveals singularities at $\tau = 0$ for all three ensembles; these are of higher order in the unitary ($\propto |\tau|^3$) and symplectic ($\propto |\tau|^5$) cases while the orthogonal ensemble produces a departure from 1 with $\tau^2 \ln|\tau|$.

6.13 Level Equilibration for Classically Integrable Systems

A given Hamiltonian or Floquet operator need not display generic level fluctuations. In fact, the simplest classically integrable systems such as the harmonic oscillator, a magnetic moment precessing in a magnetic field, and the hydrogen atom have untypical spectra: there are no fluctuations at all since the levels are equidistant or become so after unfolding. However, these systems may acquire typical spectral fluctuations when a suitable perturbation is switched on. Now, I shall show, first for classically integrable systems, that such a transition of the level spacing distribution from exceptional to generic corresponds to an equilibration process in the associated fictitious N-particle system. For convenience, we work again with a periodically kicked system, i.e., we consider a Floquet operator of the form

$$F = e^{-i\lambda V} e^{-iH_0} . \tag{6.13.1}$$

To ensure integrability of the area-preserving map arising from F in the classical limit, we assume that $[H_0, V] = 0$ and that $H_0 + \lambda V$ is integrable.

Before embarking on technicalities, it will help to think briefly how an integrable classical N-particle system like our level dynamics can display any kind of equilibration, while moving on an N-torus in phase space. Working with a cloud of phase-space points, i.e., with an ensemble of replicas of the N-particle systems, promises no help: Such a cloud maps out a bundle of trajectories on the torus and never changes shape; therefore at no instant will it uniformly cover the torus unless it does so to begin with. An intuitive understanding of the possibility of equilibration is in fact easier to achieve by looking, not at phase space, but at the one-dimensional "real" space (here the unit circle) on which the N particles exist. Here the state of the system is represented by N points. Their N nearest neighbor spacings constitute, in the limit of large N, a mass of data open to statistical analysis. Moreover, the motion of the phase-space point might well entail a spacing distribution tending to become self-averaging and time-independent once the phase-space point has departed from a possibly exceptional initial location and is wandering in typical terrain on the N-torus. To sum up this qualitative discussion, it is not the phase-space density itself which relaxes towards equilibrium. Only certain observables like the spacing distribution can be expected to equilibrate. The stationary values eventually reached by these observables can be calculated as if the phase-space density had relaxed to uniform coverage of the N-torus since the phase-space motion is ergodic. Needless to say, what has just been sketched is traditional statistical lore about many-particle systems.

A simple example is provided by a Floquet operator with

$$H_0 = \omega J_z , \quad \lambda V = \lambda \frac{J_z^2}{2j+1} \tag{6.13.2}$$

where J_z is the z component of an angular momentum vector and $2j + 1 = N$ is the dimension of the Hilbert space considered. Due to the commutativity of H_0 and V, $l_{mn} = 0$, and thus there is free uniform motion of the N fictitious particles around the unit circle. The initial coordinates at $\lambda = 0$ are $\phi_m(0) = \omega m$ (modulo 2π), and they may form a highly nongeneric configuration; for instance, if ω were chosen as $\omega = 2\pi/N$, the particles would occupy equidistant positions spaced by

$2\pi/N$. At later "times" λ, the particle trajectories read

$$\phi_m(\lambda) = \left(\omega m + \frac{\lambda m^2}{2j+1} \right) \bmod 2\pi , \quad -j \leq m \leq +j . \tag{6.13.3}$$

For $j \gg 1$, these $\phi_m(\lambda)$ pseudorandomize on the unit circle rather rapidly. One would expect the equilibration to be complete after a time such that the $2j + 1$ numbers $\omega m + \lambda m^2/j$ span a range several times 2π since the ϕ_m then form several quasi-independent ladders, each of which goes around the unit circle once. Figure 6.8 confirms that expectation.

To achieve a closed-form analytic description, one must slightly modify the dynamics (6.13.1, 2). We still assume that $[H_0 V] = 0$ such that the fictitious particles remain noninteracting. The momenta $p_m = V_{mm}$ are taken to be independent random numbers and the initial coordinates $\phi_m(0)$ coincident, $\phi_m(0) = 0$. Again, the coordinates $\phi_m(\lambda)$, in the limit $N \to \infty$, will approach a uniform density and an exponential spacing distribution. We shall calculate the spacing distribution

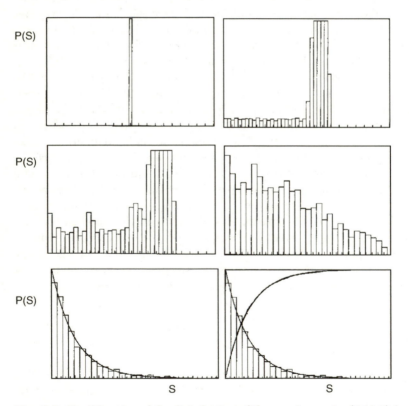

Fig. 6.8. Equilibration of the distribution of the quasi-energies (6.13.3) for an integrable top. Parameters were chosen as $\omega = 0$, $j = 1000$, and the sequence of "times" λ is 0, 1, 2, 10, and 20 000. The last picture shows the spacing staircase as hardly resolvable from that of a Poissonian process

$P(S, \lambda)$ as the second derivative of the gap probability $E(S)$ with respect to S (Sects. 4.11 and 5.6),

$$P(S, \lambda) = E''(S, \lambda) . \tag{6.13.4}$$

The quantity S is the particle spacing expressed in units of the mean spacing $2\pi/N$. The probability $E(S, \lambda)$ may be written in the following form [133]:

$$E(S, \lambda) = \int_0^{2\pi} \frac{d\alpha}{2\pi} \prod_{n=1}^{N} \left[\left(\int_0^{\alpha - \frac{\pi S}{N}} + \int_{\alpha + \frac{\pi S}{N}}^{2\pi} \right) d\phi_n \left\langle \delta \left(\phi_n - \phi_n(\lambda) \right) \right\rangle \right] . \tag{6.13.5}$$

The integral over α amounts here to an average over all intervals of length $2\pi S/N$ while the brackets $\langle \ldots \rangle$ denote an average over the independent random momenta p_n. Without the latter average, the density $P(S, \lambda)$ would be a sum of N delta functions, one for each spacing, and the weight of each such peak would be of the order $1/N$. The density of these peaks along the S-axis becomes more and more faithful to the distribution $\varrho(p)$ of the random momenta as N grows. In the limit $N \to \infty$, the spacing distribution $P(S, \lambda)$ becomes effectively smooth and is given by (6.13.5).

The evaluation of $E(S, \lambda)$ from (6.13.5) is facilitated if the delta functions are represented by Fourier series as

$$\delta(\phi - \phi') = \frac{1}{2\pi} \sum_{k=-\infty}^{+\infty} e^{ik(\phi - \phi')} . \tag{6.13.6}$$

This Poisson summation formula even manifests the 2π-periodicity of the delta functions tacitly assumed in (6.13.5). Now, the probability $E(S, \lambda)$ reads

$$E(S, \lambda) = \int_0^{2\pi} \frac{d\alpha}{2\pi} \prod_{n=1}^{N} \left(\int^* \frac{d\phi_n}{2\pi} \sum_{k_n} e^{ik_n \phi_n} \left\langle e^{-ik_n p_n \lambda} \right\rangle \right) , \tag{6.13.7}$$

where

$$\int^* d\phi = \left(\int_0^{\alpha - \pi S/N} + \int_{\alpha + \pi S/N}^{2\pi} \right) d\phi .$$

The average

$$\left\langle e^{-ik_n p_n \lambda} \right\rangle = \int_{-\infty}^{+\infty} dp_n e^{-ip_n k\lambda} \varrho(p_n) = \tilde{\varrho}(k\lambda) \tag{6.13.8}$$

arising here is nothing other than the Fourier transform of the probability density $\varrho(p)$ of the random momenta.

For all momentum distributions whose Fourier transforms $\tilde{\varrho}(k\lambda)$ decay to zero for $k\lambda \to \infty$ [i.e., roughly speaking, for all smooth $\varrho(p)$], the equilibrium form of $E(S, \infty)$ is immediately accessible since the N Fourier sums in (6.13.7) contribute only via their $k = 0$ terms,

$$E(S, \infty) = \int_0^{2\pi} \frac{d\alpha}{2\pi} \prod_{n=1}^{N} \int^* \frac{d\phi_n}{2\pi} = \int_0^{2\pi} \frac{d\alpha}{2\pi} \left(1 - \frac{S}{N} \right)^N . \tag{6.13.9}$$

Clearly, $E(S, \infty)$ approaches e^{-S} as $N \to \infty$, as does the spacing distribution

$$P(S, \infty) \to e^{-S} \quad \text{for} \quad N \to \infty \,. \tag{6.13.10}$$

With the stationary form of the spacing distribution established, it remains to study the time evolution toward equilibrium. Carrying out the ϕ_n integral in (6.13.7) yields

$$E(S, \lambda) = \int_0^{2\pi} \frac{d\alpha}{2\pi} \left[1 - \frac{S}{N} - \sum_{k \neq 0} e^{ik\alpha} \frac{\sin(\pi k S/N)}{\pi k} \left\langle e^{-ikp\lambda} \right\rangle \right]^N \,. \tag{6.13.11}$$

For times λ not too close to zero and S of order unity, the Fourier transformed momentum density cuts off the sum over k before $|k|S/N$ grows to order unity. Thus, one can use the approximation $\sin(\pi k S/N) \to \pi k S/N$. Then, the term S/N in the integrand can be included as the $k = 0$ contribution to the sum such that in the limit as $N \to \infty$

$$E(S, \lambda) = \int_0^{2\pi} \frac{d\alpha}{2\pi} \left[1 - \frac{S}{N\sigma(\alpha, \lambda)} \right]^N = \int_0^{2\pi} \frac{d\alpha}{2\pi} e^{-S/\sigma(\alpha, \lambda)} \,, \tag{6.13.12}$$

$$\frac{1}{\sigma(\alpha, \lambda)} = \sum_k e^{ik\alpha} \left\langle e^{-ikp\lambda} \right\rangle \,. \tag{6.13.13}$$

It should be emphasized that this is an asymptotic result which cannot be extrapolated back to $\lambda = 0$. Roughly speaking, for N large but still finite, $S/N\lambda$ must still be small compared to the width of the momentum distribution $\varrho(p)$ for the simplification of $\sin(\pi k S/N)$ in (6.13.11) to be permissible. Since values of S much larger than unity are of no interest, the lower bound for λ is of order $1/N$. With these words of caution in mind, the spacing distribution can be written as

$$P(S, \lambda) = \int_0^{2\pi} \frac{d\alpha}{2\pi} \frac{1}{\sigma(\alpha, \lambda)^2} e^{-S/\sigma(\alpha, \lambda)} \,. \tag{6.13.14}$$

It is noteworthy that this distribution is correctly normalized,

$$\int_0^\infty dS \, P(S, \lambda) = \int_0^{2\pi} \frac{d\alpha}{2\pi} \frac{1}{\sigma(\alpha, \lambda)} = 1 \,, \tag{6.13.15}$$

since only the $k = 0$ term in (6.13.13) contributes to the α integral in (6.13.15). Moreover, $P(S, \lambda)$ implies that the mean spacing is unity at all times λ.

To obtain a fully explicit result for the time-dependent distribution $P(S, \lambda)$ one needs to specify the momentum distribution. Various choices of $\varrho(p)$ allow evaluating the width $\sigma(\alpha, \lambda)$. The typical behavior is that $\sigma(\alpha, \lambda) - 1$ and thus $P(S, \lambda) - e^{-S}$ decay to zero like an inverse power of the "time" λ. A nice example is provided by the box distribution

$$\varrho(p) = \frac{1}{\Omega} \Theta \left(\frac{\Omega}{2} - |p| \right) \tag{6.13.16}$$

whose Fourier transform reads

$$\tilde{\varrho}(k, \lambda) = \left(\frac{2}{k\Omega\lambda}\right) \sin\left(\frac{k\Omega\lambda}{2}\right) . \tag{6.13.17}$$

Upon inserting this into (6.13.13) and using the Fourier series representation of the unit staircase [the integral of the periodic delta function (6.13.6)],

$$\sum_{n=1}^{\infty} \Theta(\phi - n2\pi) = \frac{1}{2\pi i} \sum_{k=-\infty}^{+\infty} \frac{1}{k} \left(e^{ik\phi} - 1\right) , \tag{6.13.18}$$

the width can be expressed as

$$\frac{1}{\sigma(\alpha, \lambda)} = \frac{2\pi}{\Omega\lambda} \left[\sum_{n=1}^{\infty} \Theta\left(\frac{\Omega\lambda}{2} + \alpha - n2\pi\right)\right.$$
$$\left. + \sum_{n=1}^{\infty} \Theta\left(\frac{\Omega\lambda}{2} - \alpha - n2\pi\right)\right] . \tag{6.13.19}$$

It follows that $\sigma(\alpha, \lambda) = 1$ at the equidistant instants $\lambda = \nu 2\pi/\Omega$, $\nu = 1, 2, 3, \ldots$. Between these special instants, one obviously has the decay

$$\frac{1}{\sigma(\alpha, \lambda)} - 1 \sim \frac{1}{\Omega\lambda} \quad \text{for} \quad \Omega\lambda \gg 1 . \tag{6.13.20}$$

To conclude, N particles moving freely on a circle with random velocities do display an equilibrating spacing distribution for $N \to \infty$ and so do the quasi-energies of a Floquet operator with an integrable classical limit.

6.14 Level Equilibration for (a Caricature of) Classically Chaotic Systems

A particularly interesting transition is the change from level clustering to level repulsion, a frequently observed quantum signature of the development of chaos in the classical limit. A typical example is provided by the Floquet operator of a kicked top,

$$F = e^{-i\lambda V} e^{-iH_0} \tag{6.14.1}$$

with noncommuting operators H_0 and V both of which are functions of the components of an angular momentum $\boldsymbol{J}$. When at least one of these functions is nonlinear in $\boldsymbol{J}$, the Floquet operator is in general nonintegrable in the classical limit. (When both H_0 and V are linear, F is a rotation and is thus integrable; the caution "generically" excludes trivial nonlinearities like $\boldsymbol{J}^2 J_z$). The well-investigated special case [129, 149]

$$H_0 = pJ_y , \quad V = \frac{J_z^2}{2j+1} \tag{6.14.2}$$

has a classical limit dominated by regular motion for $p \approx \pi/2$, $\lambda \approx 1$ while chaos is fully developed for $\lambda \gtrsim 5$. Correspondingly, the spacing distributions are Poissonian and COE-like.

The fictitious N-particle system corresponding to (6.14.1) has nonvanishing pair interactions because of $[V, F] \neq 0$. Now, interparticle collisions dominate the dynamics, and for sufficiently large "times" λ, the repulsion tends to forbid encounters of pairs with separations much smaller than the mean spacing $2\pi/N$. A typical λ dependence of the coordinates $\phi_m(\lambda)$ is displayed in Fig. 6.3. Clearly, the collisions do not hinder the equilibration of the spacing distribution.

As regards the quantitative description of the equilibration in question it is unfortunate that there is no classically nonintegrable F for which exact closed-form solutions for the $\phi_m(\lambda)$ are known. To illustrate the capability of our re-pelling particles to display equilibration of their spacing distribution, we may resort to model "trajectories" $\phi_m(\lambda)$ that simulate the essential characteristics of the solutions of (6.2.16).

A simple but not absurd caricature is provided by N "trajectories" oscillating sinusoidally about equidistant centers [133],

$$\phi_m(\lambda) = m\frac{2\pi}{N} + a_m \sin \omega_m \lambda \, ,$$

$$|a_m| \leq \tfrac{\pi}{N} \, , \qquad\qquad\qquad\qquad\qquad (6.14.3)$$

$$m = 1, 2, \ldots, N \, .$$

The restriction on the amplitudes a_m serves to forbid crossings and the frequencies ω_n are again taken as independent random numbers. Clearly, the lack of correlation even between neighboring particles implies a drastic oversimplification of the true N-particle dynamics (6.2.16). Due to the absence of crossings and the oscillatory character, the $\phi_m(\lambda)$ in (6.14.3) do bear a certain superficial similarity to the λ dependence of real quasi-energies shown in Fig. 6.9. The simplest version of the caricature (6.14.3) worthy of consideration has all amplitudes equal, $a_m = \pi/N$, and the frequencies distributed according to the box density

$$\varrho(\omega) = \begin{cases} 1/\Omega & \text{for } 0 \leq \omega \leq \Omega \\ 0 & \text{for } \Omega < \omega \, . \end{cases} \qquad\qquad (6.14.4)$$

Then, the distribution of the normalized spacings can be obtained as

$$P(S, \lambda) = \frac{1}{N} \sum_{i=1}^{N} \left\langle \delta\left(S - 1 - \frac{1}{2}\sin \omega_{i+1}\lambda + \frac{1}{2}\sin \omega_i\lambda\right)\right\rangle \qquad (6.14.5)$$

where the angular brackets denote the frequency average in which each ω_i is averaged with its own weight $\varrho(\omega_i)$. That average is justified similarly to the momentum average encountered in the last section. It is noteworthy that the prohibition of crossings in (6.14.3) and the confinement of each $\phi(\lambda)$ to its own fixed interval make possible the simple formula (6.14.5) for $P(S, \lambda)$. One may even consider the delta function in (6.14.5) as ordinary nonperiodic since none of the particles ever leaves the interval $[0, 2\pi]$.

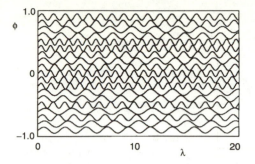

Fig. 6.9. Caricature of the control parameter dependence of quasi-energy spectrum according to (6.8.3) with $a_m = \pi/N$ and randomly chosen ω

To evaluate the average in (6.14.5), we represent the delta function by a Fourier integral,

$$P(S,\lambda) = \frac{1}{N} \sum_{i=1}^{N} \int \frac{dk}{2\pi} e^{ik(S-1)} \left\langle e^{-ik/2 \, \sin \, \omega_{i+1}\lambda} \right\rangle \left\langle e^{+ik/2 \, \sin \, \omega_i\lambda} \right\rangle .$$

(6.14.6)

Finally, invoking the distribution (6.14.4) for the two independent averages over ω_i and ω_{i+1}, one arrives at

$$P(S,\lambda) = \int \frac{dk}{2\pi} e^{ik(S-1)} \left| \frac{1}{\Omega\lambda} \int_0^{\Omega\lambda} d\Theta e^{-i(k/2) \, \sin \, \Theta} \right|^2 .$$

(6.14.7)

Due to the 2π-periodicity of $\sin \Theta$, the distribution $P(S,\lambda)$ already takes on its stationary form $P(S,\infty)$ at the equidistant points of "time" $\lambda = 2\pi\nu/\Omega$, $\nu = 1, 2, 3, \ldots$. At these instants, the Θ integral is just the Bessel function $J_0(k/2)$, the square of which remains to be Fourier transformed,

$$P(S,\infty) = \int \frac{dk}{2\pi} e^{ik(S-1)} \left[J_0 \left(\frac{k}{2} \right) \right]^2 .$$

(6.14.8)

This transformation can be carried out to yield the complete elliptic integral of the first kind [150],

$$P(S,\infty) = \frac{2}{\pi^2} K \left(\sqrt{S(2-S)} \right) .$$

(6.14.9)

Figure 6.10 depicts this distribution. Incidentally, as was the case for the integrable Floquet operator treated in Sect. 6.13, the stationary spacing distribution is independent of the frequency density $\varrho(\omega)$ provided only that the latter function is sufficiently smooth (Problem 6.7). Well-known properties of the elliptic integral imply that $P(S,\infty)$

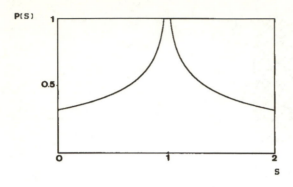

P(S)

Fig. 6.10. Caricature of the stationary level-spacing distribution according to (6.8.3, 4, 9)

(i) is properly normalized,

$$\int_0^2 dS\, P(S, \infty) = 1 \; ; \tag{6.14.10}$$

(ii) has unit mean spacing,

$$\int_0^2 dS\, P(S, \infty)S = 1 \; ; \tag{6.14.11}$$

(iii) is minimal at $S = 0$ and $S = 2$,

$$P(0, \infty) = P(2, \infty) = \frac{1}{\pi} \; ; \tag{6.14.12}$$

(iv) is maximal and in fact logarithmically divergent at $S = 1$; and

(v) is symmetric around $S = 1$. A level repulsion of sorts is indicated by $P(S, \infty)$ $\geq P(0, \infty) = 1/\pi$, but the caricature used for the particle trajectories is too crude to allow one to distinguish the three universality classes of level repulsion.

The main purpose of the caricature (6.14.3) was of course to illustrate the relaxation of $P(S, \lambda)$ to some equilibrium form $P(S, \infty)$. The detailed course of that relaxation is open to inspection from (6.14.7). Obviously, the deviation $P(S, \lambda) - P(S, \infty)$ in between the instants $\lambda = 2\pi\nu/\Omega$ decays to zero in proportion to $1/\lambda$. That decay is nicely illustrated by the mean squared deviation of the spacing from its mean,

$$\langle S(\lambda)^2 \rangle - 1 = \frac{1}{4} - \frac{\sin 2\Omega\lambda}{8\Omega\lambda} - \frac{1}{2}\left(\frac{1 - \cos \Omega\lambda}{\Omega\lambda}\right)^2 \; . \tag{6.14.13}$$

6.15 Dyson's Brownian-Motion Model

Now, I propose to consider a classically nonintegrable autonomous quantum system, whose Hamiltonian $H_0 + \lambda V$ has H_0 time-reversal invariant but V not so restricted. Without loss of generality, H_0 and V may be assumed to have the same mean level spacing. As λ increases from zero to infinity, the spectrum of H should undergo a transition from linear to quadratic level repulsion. This transition is described by the Hamiltonian level dynamics (6.2.27) but should actually look like an irreversible relaxation into equilibrium with respect, e.g., to the level spacing distribution $P(S, \lambda)$, provided $N \gg 1$.

Inasmuch as $P(S, \lambda)$ or other characteristics of spectral fluctuations are ergodic in Pandey's sense (Sect. 4.9), they may be derived from the whole family of similar Hamiltonians represented by the density, instead of from an individual Hamiltonian $H_0 + \lambda V$ [151, 174, 175]

$$\tilde{P}(H, \lambda) = \langle \delta(H - H_0 - \lambda V) \rangle \ . \tag{6.15.1}$$

The average in (6.15.1) is over the GOE and over the GUE for H_0 and V, respectively, such that $\tilde{P}(H, \lambda)$ takes the form of a convolution,

$$\tilde{P}(H, \lambda) = \int dH_0 P_{\mathrm{GOE}}(H_0) P_{\mathrm{GUE}} \left(\frac{H - H_0}{\lambda} \right) \lambda^{-N^2} \ . \tag{6.15.2}$$

The distributions $P_{\mathrm{GOE}}(X)$ and $P_{\mathrm{GUE}}(X)$ of $N \times N$ matrices X are special cases [see (4.2.20)] of

$$
\begin{aligned}
P_\beta(X) &= \mathcal{N}_\beta^{-1} e^{-(\beta/2)\,\mathrm{Tr}\,\{X^2\}} \\
&= \mathcal{N}_\beta^{-1} \left(\prod_{i}^{1\,\ldots\,N} e^{-(\beta/2)X_{ii}^2} \right) \left(\prod_{i<j}^{1\,\ldots\,N} e^{-\beta|X_{ij}|^2} \right) ,
\end{aligned}
$$

$$\mathcal{N}_\beta^{-1} = \left(\sqrt{\frac{\beta}{2\pi}} \right)^{N} \left(\sqrt{\frac{\beta}{2\pi}} \right)^{N(N-1)\beta/2} , \tag{6.15.3}$$

$$\beta = \begin{cases} 1 & \text{GOE} \\ 2 & \text{GUE.} \end{cases}$$

Of course, the number of independent matrix elements of X,

$$p_\beta = \tfrac{1}{2}N(N+1) + \tfrac{1}{2}N(N-1)(\beta - 1) \ , \tag{6.15.4}$$

is different in the two cases since X is real symmetric for the GOE but complex Hermitian for the GUE. Thus, the remaining integral in (6.15.2) is over $p_1 = N(N+1)/2$ real matrix elements.

With the GUE average done, the distribution $\tilde{P}(H, \lambda)$ given in (6.15.2) allows a most interesting and perhaps surprising interpretation: It appears to be the solution of an N^2-dimensional diffusion with λ^2 as the time; for each of the N diagonal elements H_{ii}, the diffusion constant is $1/2$ while the $N^2 - N$ independent real elements in $H_{ij} = \mathrm{Re}\,\{H_{ij}\} + \mathrm{i}\,\mathrm{Im}\,\{H_{ij}\}$ for $i < j$ have the diffusion constants $1/4$. For $\lambda^2 \to 0$, the Green's function in (6.15.2), $\lambda^{-N^2} P_{\mathrm{GUE}}((H - H_0)/\lambda)$, indeed shrinks to the N^2-dimensional delta function $\delta(H - H_0)$ (which actually sets $\text{Im}\{H\} = 0$ and $\text{Re}\{H\} = H_0$). For $\lambda^2 \to \infty$, on the other hand, the Green's function seems a little sick, as it stands, since it does not tend to assign to H the desired GUE distribution.

The sickness just mentioned is due to a certain negligence in the definition (6.15.1). While H_0 and V were assumed to have the same energy scale, i.e., the same mean level spacings in their separate spectra, the definition (6.15.1) fails to impose that condition on H. Rather, the mean level spacing of $H_0 + \lambda V$ grows in proportion to λ for $\lambda \to \infty$ and therefore the Green's function $\lambda^{-N^2} P_{\text{GUE}}((H - H_0)/\lambda)$ cannot approach the desired limit. Hurrying to repair the defect $\tilde{P}(H, \lambda)$, we introduce a scaling factor f,

$$
\begin{aligned}
P(H) &= \left\langle \delta\left(H - \sqrt{f}(H_0 + \lambda V)\right)\right\rangle \\
&= \left(\lambda\sqrt{f}\right)^{-N^2} \int dH_0\, P_{\text{GOE}}(H_0) P_{\text{GUE}}\left(\frac{H - \sqrt{f}H_0}{\lambda\sqrt{f}}\right). \quad (6.15.5)
\end{aligned}
$$

When

$$
\lambda\sqrt{f} \to \begin{cases} 0 & \text{for } \lambda \to \\ 1 & \text{for } \lambda \to \infty, \end{cases} \quad (6.15.6)
$$

the kernel $(\lambda\sqrt{f})^{-N^2} P_{\text{GUE}}((H - \sqrt{f}H_0)/\lambda\sqrt{f})$ retains its initial delta function shape and approaches $P_{\text{GUE}}(H)$ for $\lambda \to \infty$. Of course, the "boundary" condition (6.15.6) does not uniquely determine the factor f. To fix the variation of f with the coupling constant λ, we require that H has the same mean level spacing as H_0 and V. Actually, that specification of f is hard to implement and also a bit stronger than necessary. It suffices that the mean level spacings of H, H_0, and V all have the same scaling with N, i.e., are proportional to $1/\sqrt{N}$ [see Sect. 4.6, especially the remark following (4.6.3)]. The latter behavior will be ensured by the choice of f to be adopted presently.

It is worth noting that the Green's function of the N^2-dimensional free diffusion, $\lambda^{-N^2} P_{\text{GUE}}((H - H_0)/\lambda)$, obeys the Chapman–Kolmogorov equation [176], a property characteristic of Markovian stochastic processes. Any nontrivial rescaling will change the free diffusion into some other N^2-dimensional stochastic process. The question naturally arises whether there is a particular rescaling that leaves the resulting process Markovian in character. To answer that question, we look for two functions $\lambda(\tau)$, $f(\tau)$ of a rescaled "time" τ such that

$$
P(H, \tau|H_0) \equiv \left[\lambda(\tau)\sqrt{f(\tau)}\right]^{-N^2} P_{\text{GUE}}\left(\frac{H - \sqrt{f(\tau)}H_0}{\lambda(\tau)\sqrt{f(\tau)}}\right) \quad (6.15.7)
$$

again obeys the Chapman–Kolmogorov equation

$$
P(H, \tau + \tau'|H') = \int dH'' P(H, \tau|H'') P(H'', \tau'|H') \quad (6.15.8)
$$

with $\tau, \tau' \geq 0$ and H, H'', H' all complex Hermitian.

By performing the N^2 Gaussian integrals on the right-hand side of (6.15.8), one easily finds two functional equations for $\lambda(\tau)$ and $f(\tau)$,

$$f(\tau)f(\tau') = f(\tau + \tau')$$
$$\lambda(\tau')^2 f(\tau')f(\tau) + \lambda(\tau)^2 f(\tau) = \lambda(\tau + \tau')^2 f(\tau + \tau') \, . \tag{6.15.9}$$

The first of these implies $f(\tau) = e^{-2\Gamma\tau}$ with arbitrary Γ, and the second then yields $\lambda^2(\tau) = a(e^{2\Gamma\tau} - 1)$ with an arbitrary constant a. The boundary condition (6.15.6) for $\lambda \to \infty$ entails $a = 1$. As a matter of convenience, we choose $\Gamma > 0$ so that $\lambda \to \infty$ corresponds to $\tau \to \infty$, whereupon Γ can be absorbed in the scale of the "time" τ. Within the latter freedom for the direction and the unit of τ, the solution of (6.15.6, 6.15.9) is unique and reads

$$\lambda(\tau)^2 = e^{2\tau} - 1 \, , \quad f(\tau) = e^{-2\tau} \, , \quad f = \frac{1}{1 + \lambda^2} \, . \tag{6.15.10}$$

The resulting conditional probability density (6.15.7) turns out to be the Green's function of the N^2-dimensional Ornstein–Uhlenbeck process [176]. It obeys the Fokker–Planck equation

$$\frac{\partial}{\partial\tau}P(H,\tau) = \sum_\mu \left(\frac{\partial}{\partial H_\mu}H_\mu + \frac{1}{2}D_\mu\frac{\partial^2}{\partial H_\mu^2} \right) P(H,\tau) \, , \tag{6.15.11}$$

where the shorthand H_μ denotes the N^2 independent variables, i.e., H_{ii} and, for $i < j$, $\text{Re}\{H_{ij}\}$ and $\text{Im}\{H_{ij}\}$. The diffusion constants D_μ are the same as for the free diffusion encountered before the rescaling, i.e., $D_\mu = 1/2$ for the diagonal elements H_i, and $D_\mu = 1/4$ for the off-diagonal elements $\text{Re}\{H_{ij}\}$ and $\text{Im}\{H_{ij}\}$. The Fokker–Planck equation (6.15.11) is invariant under unitary transformations of H; thus the representation of H may be chosen at will.

The distribution $P(H,\tau)$ interpolating between $P_{\text{GOE}}(H) = P(H,0)$ and $P_{\text{GUE}}(H) = P(H,\infty)$ can also be obtained explicitly by doing the $N(N+1)/2$ Gaussian integrals in (6.15.5),

$$
\begin{aligned}
P(H,\tau) \;=\; & \prod_i \frac{\exp\left[-H_{ii}^2/(1 + e^{-2\tau})\right]}{\sqrt{\pi(1 + e^{-2\tau})}} \\
& \times \prod_{i<j} \left(\frac{\exp\left[-(\text{Re}\{H_{ij}\})^2 2/(1 + e^{-2\tau})\right]}{\sqrt{(\pi/2)(1 + e^{-2\tau})}} \right. \\
& \left. \times \frac{\exp\left[-(\text{Im}\{H_{ij}\})^2 2/(1 - e^{-2\tau})\right]}{\sqrt{(\pi/2)(1 - e^{-2\tau})}} \right) \, .
\end{aligned}
\tag{6.15.12}
$$

It is interesting to note the invariance of $P(H,\tau)$ under orthogonal transformations for $0 \le \tau < \infty$ while in the limit $\tau \to \infty$, the full symmetry of the GUE is established.

The Ornstein–Uhlenbeck process for the matrix elements of a Hamiltonian is known in random-matrix theory as Dyson's Brownian-motion model [134]. It was suggested by Dyson as the simplest random process displaying relaxation into a stationary regime described by one of the three Gaussian random-matrix ensembles. The foregoing arguments elevate the Brownian-motion model to a rigorous description of the transition of a Hamiltonian from a typical member of the GOE to a typical member of the GUE when an additive term λV breaking time-reversal invariance is switched on. Evidently, the considerations outlined above for the transition GOE $\rightarrow$ GUE can be extended immediately to the other transitions between pairs of universality classes. The significance of this result lies in its apparent universality. Indeed, inasmuch as the time-reversal invariant Hamiltonian H_0 of a dynamic system generically displays GOE fluctuations in its spectrum, while, similarly, a perturbation V without antiunitary symmetries has GUE fluctuations, the perturbed Hamiltonian $\sqrt{\bar{f}}(H_0 + \lambda V)$ should be faithful to (6.15.12).

A further comment about the rescaling (6.15.10) is in order. According to Wigner's semicircle law [see (4.6.3) and the remark following that equation] both H_0 and V have a mean level spacing proportional to $1/\sqrt{N}$ in the middle of their spectra. The family of Hamiltonians described by (6.15.12) should share that property since the widths of the distributions of the matrix elements of H interpolate monotonically between the two limits at $\tau = 0$ and $\tau \rightarrow \infty$. Indeed, by using the fact that $\langle H_{ij} H_{ji} \rangle = 1$ independent of τ for the distribution (6.15.12), it is easy to see that the arguments of Sect. 4.6 yield the same semicircle law as for the GOE and the GUE (Problem 6.10).

The Ornstein–Uhlenbeck process (6.15.11) contains a subset of N stochastic variables, namely, the eigenvalues E_i of the matrix H, which itself undergoes a separate Markovian process. The joint probability density of the E_i obeys a Fokker–Planck equation of the form

$$\frac{\partial}{\partial \tau} \varrho(E, \tau) = \left[- \sum_{i}^{1 \ldots N} \frac{\partial}{\partial E_i} D_i(E) + \frac{1}{2} \sum_{ij}^{1 \ldots N} \frac{\partial^2}{\partial E_i \partial E_j} D_{ij}(E) \right] \varrho(E, \tau) .$$

(6.15.13)

Different paths may be followed to prove the validity of (6.15.13) and to determine the drift coefficients $D_i(E)$ and the diffusion matrix $D_{ij}(E)$. In the fashion of Sect. 4.3, a transformation of variables may be performed from the N^2 matrix elements H_{ij} to the N eigenvalues E_i and the $N^2 - N$ angles that specify the unitary matrix diagonalizing H. When these angles are integrated out of the Fokker–Planck equation (6.15.11), the reduced Fokker–Planck equation (6.15.13) results. A more convenient route is opened by the observation that the Green's function $\varrho(E + \delta E, \tau + \delta \tau | E, \tau)$ of (6.15.13), for an infinitesimal time increment $\delta \tau$, has the mean and mean-squared eigenvalue increments [176]

$$\langle \delta E_i \rangle = \langle E_i(\tau + \delta \tau) \rangle - E_i = D_i(E) \delta \tau + \mathcal{O}(\delta \tau^2)$$
$$\langle \delta E_i \delta E_j \rangle = D_{ij}(E) \delta \tau + \mathcal{O}(\delta \tau^2) .$$

(6.15.14)

These identities can in fact be obtained immediately from (6.15.13). On the other hand, they may be derived from the short-time version of the Green function

(6.15.7) of the Ornstein–Uhlenbeck process,

$$P\left(H + \delta H, \tau + \delta\tau | H, \tau\right) = \left(\frac{1}{2\delta\tau}\right)^{N^2/2} P_{\text{GUE}}\left(\frac{\delta H + H\delta\tau}{\sqrt{2\delta\tau}}\right) , \tag{6.15.15}$$

and the perturbative relationship

$$\delta E_i = \delta H_{ii} + \sum_{j(\neq i)} \frac{\delta H_{ij}\delta H_{ji}}{E_i - E_j} + \ldots \tag{6.15.16}$$

between the matrix increment δH and the eigenvalue increment δE_i. Obviously, the relation (6.15.16) refers to the eigenrepresentation of H, which, according to the "unitary" invariance of the Fokker–Planck equation (6.15.11) of the Ornstein–Uhlenbeck process, we are free to adopt at time τ. If the Green's function (6.15.15) is also taken in that representation, it has the moments

$$\langle \delta H_{ii} \rangle = -E_i \delta\tau$$
$$\langle (\delta H_{ii})^2 \rangle = \delta\tau \tag{6.15.17}$$
$$\langle (\text{Re}\,\{\delta H_{ij}\})^2 \rangle = \langle (\text{Im}\,\{\delta H_{ij}\})^2 \rangle = \delta\tau/2 , \quad i < j ,$$

and these, together with (6.15.16), imply

$$\langle \delta E_i \rangle = \left(-E_i + \sum_{j(\neq i)} \frac{1}{E_i - E_j}\right) \delta\tau$$
$$\langle \delta E_i \delta E_j \rangle = \delta_{ij}\delta\tau . \tag{6.15.18}$$

In all increment moments (6.15.14, 17, 18) powers of $\delta\tau$ higher than the first have been neglected. To that accuracy, higher-than-second orders in δH could be neglected in the perturbation expansion (6.15.16). Comparison of the moments in (6.15.18) and (6.15.14) now yields the drift and diffusion coefficients for the Brownian motion of the eigenvalues

$$D_i(E) = -E_i + \sum_{j(\neq i)} \frac{1}{E_i - E_j} , \quad D_{ij} = \delta_{ij} . \tag{6.15.19}$$

Like the Ornstein–Uhlenbeck process of the H_{ij} the Brownian motion of the E_i has a harmonic binding force in the drift. The additional two-body interaction term in $D_i(E)$ can be interpreted as a repulsive coulombic force between charged particles in two spatial dimensions; the interaction energy for a pair of particles varies logarithmically with the distance. In view of this interaction the process is often referred to as to Dyson's stochastic Coulomb gas model [134]. While the repulsive character reflects level repulsion, the harmonic binding follows from the energy rescaling f specified in (6.15.5, 10): the purpose of that rescaling is to prevent the "particles" from flying apart indefinitely for $\lambda \to \infty$.

Interestingly, the (logarithmic) Coulomb gas interaction differs from the (inverse-square) two-body interaction in Pechukas' level dynamics (6.2.27). See Problem 6.13.

Since the Fokker–Planck equation (6.15.13, 19) of the Coulomb gas obeys the condition of detailed balance [176], the stationary solution $\varrho(E,\infty)$ can be given immediately in closed form. As it must, $\varrho(E,\infty)$ comes out as the joint probability density of the eigenvalues for the GUE [see (4.3.15) with $n=2$ or (6.15.22) below]. The time-dependent solution $\varrho(E,\tau)$ originating from the joint eigenvalue density of the GOE at $\tau=0$ is also known. It was determined by *Pandey* and *Mehta* by integrating $P(H,\tau)$ as given in (6.15.12) over the N^2-N "angles" alluded to above, that "integration over the unitary group" yields [177, 178], for even N,

$$
\varrho(E,\tau) \;=\; \mathcal{N}(\tau)^{-1}\left[\prod_{i<j}(E_i-E_j)\mathrm{Pf}\left(\mathrm{erf}\,\frac{E_i-E_j}{\sqrt{e^{2\tau}-e^{-2\tau}}}\right)\right]
$$

$$
\times \exp\left[-\sum_i \frac{E_i^2}{1+e^{-2\tau}}\right]\,, \tag{6.15.20}
$$

$$
\mathcal{N}(\tau)^{-1} = 2^{3N/2}e^{-(N^2-N)\tau/2}\left[\prod_j \Gamma\left(1+\frac{j}{2}\right)\right]\,.
$$

Here, we encounter the Pfaffian $\mathrm{Pf}\,(a_{ij}) = \sqrt{\det(a_{ij})}$ of an antisymmetric $N\times N$ matrix a_{ij} with even N introduced in (4.12.13). For the case of odd N, the reader may consult Ref. [177, 178].

To check on the limiting form of $\varrho(E,\tau)$ for $\tau\to\infty$, both the Pfaffian and the normalization factor $\mathcal{N}(\tau)$ must be inspected for their time dependence. The argument a_{ij} of the Pfaffian simplifies as $a_{ij}\to\mathrm{erf}\,[e^{-\tau}(E_i-E_j)]$ and may be expanded in powers of $e^{-\tau}(E_i-E_j)$. The exponential blowup of $\mathcal{N}(\tau)$ can be compensated for only when the $N/2$ factors a_{ij} in each additive term of the Pfaffian build up $(N^2-N)/2$ factors $e^{-\tau}(E_i-E_j)$. Such contributions arise when all the a_{ij} join the $(N-1)$th-order term of their Taylor expansions. At any rate, since the Pfaffian is totally antisymmetric in the variables E_i, it must have the structure

$$
\mathrm{Pf}\left\{\mathrm{erf}\left[e^{-\tau}(E_i-E_j)\right]\right\}
$$

$$
= e^{-(N^2-N)\tau/2}\left[\prod_{i<j}(E_i-E_j)\right]F_{\mathrm{sym}}\left(\{e^{-\tau}E_j\}\right) \tag{6.15.21}
$$

where F_{sym} is totally symmetric in its arguments. The stationary form of $\varrho(E,\tau)$ follows as

$$
\varrho(E,\infty) \;=\; 2^{-3N/2}\left[\prod_j \Gamma\left(1+\frac{j}{2}\right)\right]^{-1}F_{\mathrm{sym}}\left(\{0\}\right)
$$

$$
\times \left[\prod_{i<j}(E_i-E_j)^2\right]e^{-\sum_i E_i^2}
$$

$$
= \varrho_{\mathrm{GUE}}(E)\,. \tag{6.15.22}
$$

The constant factor $F_{\text{sym}}(\{0\})$ may either be constructed from the Pfaffian or simply determined by normalization.

Verifying the correct initial behavior of the solution (6.15.20) is also instructive, starting from $a_{ij}(\tau) \to (E_i - E_j)/|E_i - E_j|$. Then, the Pfaffian has the value $(-1)^{\mathcal{Q}}$ where $\mathcal{Q}$ is the permutation $i_1 i_2 \ldots i_N$ of $1\,2 \ldots N$ corresponding to $E_{i_1} > E_{i_2} > \ldots > E_{i_N}$. But the same permutation $\mathcal{Q}$ determines the sign of the Vandermonde determinant $\prod_{i<j}(E_i - E_j)$ as $(-1)^{\mathcal{Q}}$. Thus, the Pfaffian may be replaced by the modulus operation on the Vandermonde determinant, whereupon $\varrho(E, 0) = \varrho_{\text{GOE}}(E)$ is established.

An important conclusion may be drawn from the solution (6.15.20) for the time scale of the transition from GOE to GUE behavior. Clearly, the Gaussian factors in $\varrho(E, \tau)$ are irrelevant for the change in the degree of level repulsion; their principle role is to keep the spectrum confined, roughly, to $-\sqrt{N} \lesssim E \lesssim +\sqrt{N}$ according to the semicircle law. It is actually the product of the Pfaffian and the Vandermonde determinant that signals the transition. As long as the arguments $(E_i - E_j)/\sqrt{e^{2\tau} - e^{-2\tau}}$ of the error functions in the Pfaffian, for a typical spectrum $\{E_i\}$, are all large, GOE behavior still prevails; the transition to the GUE is more or less complete once these arguments have all become small. Again taking the typical spacing $E_i - E_j \approx 1/\sqrt{N}$ from the semicircle law, the time scale $\hat{\tau}$ of the transition may be estimated as $N \sinh 2\hat{\tau} \approx 1$, i.e.,

$$\hat{\tau} \approx \frac{1}{N} \Leftrightarrow \hat{\lambda} \approx \frac{1}{\sqrt{N}} \,. \tag{6.15.23}$$

For $\tau \ll \hat{\tau}$ or $\lambda \ll \hat{\lambda}$, the degree of level repulsion still appears to be unity on a scale given by the mean level spacing. It follows that in the limit of large N a rather small symmetry-breaking addition λV to the time-reversal invariant H_0 brings about quadratic level repulsion. It also follows that the crossover in the inverse transition, GUE $\to$ GOE, in $H = \sqrt{f\lambda^2}(V + H_0/\lambda) \equiv \sqrt{\tilde{f}}(V + \tilde{\lambda}H_0)$ takes place at $\tilde{\lambda} \approx \sqrt{N}$; a comparatively large time-reversal-invariant addition is needed to overwhelm a Hamiltonian V without antiunitary symmetries. Figure 6.11 illustrates the transition.

Incidentally, the "time" scale (6.15.23) for the transition GOE $\to$ GUE can also be obtained from the matrix density (6.15.12). To this end, we consider the imaginary part of H_{ij} as a small perturbation which shifts the ith level by

$$\delta E_i = \sum_{j(\neq i)} \frac{(\text{Im}\,\{H_{ij}\})^2}{E_i - E_j} \,. \tag{6.15.24}$$

In the average over the ensemble (6.15.12),

$$\langle (\text{Im}\,\{H_{ij}\})^2 \rangle = \tfrac{1}{4}(1 - e^{-2\tau}) \approx \tfrac{1}{2}\tau \,. \tag{6.15.25}$$

The N dependence of the sum on the r.h.s. in (6.15.24) can be estimated from the mean level spacing, $\Delta E \sim 1/\sqrt{N}$. Finally, the transition will be more or less completed once the mean shift $\langle \delta E_i \rangle$ has grown to the order of magnitude of the

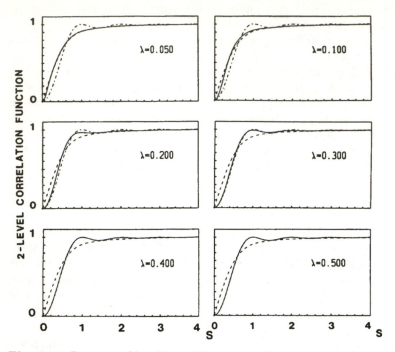

Fig. 6.11. Progress of breaking of time-reversal invariance in the two-level correlation function [normalized density of two eigenvalues $P(E_1, E_2)/P(E_1)P(E_2)$ as implied by (6.9.20). The full curves pertain to the limit $N \to \infty$ which becomes accessible after referring the spacing $E_1 - E_2$ to the mean spacing and the control parameter to the scale (6.9.24)

mean spacing. Putting all of these estimates into (6.15.24), one indeed recovers the result (6.15.23) for the transition time $\hat{\tau}$.

While the foregoing discussion was confined to transitions between the GOE and the GUE, the other transitions between universality classes, GOE $\leftrightarrow$ GSE and GUE $\leftrightarrow$ GSE, can be treated similarly, at least as far as the interpolating matrix densities $P(H, \tau)$ are concerned. Even though the interpolating reduced densities of the eigenvalues are not known in closed form for the latter transitions, the transition times can be determined using the above perturbative argument. The result again is $\hat{\lambda} \sim 1/\sqrt{N}$ for the easy direction (symmetry breaking) and $\hat{\lambda} \sim \sqrt{N}$ for the inverse transition (symmetry restoring). It is worth emphasizing once more that a given transition (with fixed sense) may be achieved either through breaking or restoring a symmetry. For instance, GSE $\to$ GUE takes place when (1) a time-reversal invariance of H_0 with $T^2 = -1$ is broken in the absence of any geometric symmetry or (2) both H_0 and V are invariant under T with $T^2 = -1$ while V introduces a parity not shared by H_0 (Sect. 2.7); clearly, the first transition is easy, and the second one hard.

A particularly interesting crossover arises when H_0 is classically integrable but $H_0 + \lambda V$ with $\lambda > 0$ nonintegrable, such that the classical phase space becomes dominated by chaos once λ is sufficiently large. As a quantum parallel to this spreading of classical chaos, one expects that the level clustering typically present for $\lambda = 0$ eventually gives way to level repulsion. A convenient random-matrix model for such a case assumes a diagonal $N \times N$ matrix H_0 whose eigenvalues are independent random numbers with a Gaussian density of zero mean and variance $\langle (H_{0ii})^2 \rangle = N$; the latter normalization implies an average eigenvalue spacing of the order $1/\sqrt{N}$, in coincidence with that assumed for V. For the sake of concreteness, taking V as an $N \times N$ matrix without antiunitary symmetry, one is led to consider, in analogy with (6.15.5), the matrix ensemble

$$P(H, \tau) = \left(\lambda \sqrt{f} \right)^{-N^2} \int dH_0 P_P(H_0) P_{\text{GUE}} \left(\frac{H - \sqrt{f} H_0}{\lambda \sqrt{f}} \right) , \qquad (6.15.26)$$

where the integral is over the N diagonal elements (i.e. eigenvalues) of H_0. The density (6.15.26) interpolates between the initial "Poissonian" ensemble (PE)

$$P_P(H_0) = \prod_{i=1}^{N} (2\pi N)^{-1/2} e^{-(H_{0ii})^2 / 2N} \qquad (6.15.27)$$

and the final GUE. The N-fold integral in (6.15.26) is once more Gaussian and thus assigns to the interpolating density a Gaussian form as well; all matrix elements are distributed independently of one another; they have zero means and the variances

$$\langle (H_{ii})^2 \rangle = \tfrac{1}{2} \left[1 + (2N - 1) e^{-2\tau} \right] ,$$

$$\langle (\text{Re}\{H_{ij}\})^2 \rangle \langle (\text{Im}\{H_{ij}\})^2 \rangle = \tfrac{1}{4} (1 - e^{-2\tau}) , \quad i \neq j . \qquad (6.15.28)$$

One would like to extract from the Gaussian matrix density defined by (6.15.28) an explicit expression for, e.g., the spacing distribution $P(S, \lambda)$ which interpolates between $P(S, 0) = e^{-S}$ and $P(S, \infty) = P_{\text{GUE}}(S)$. Unfortunately, the integrals involved have thus far resisted all attempts at evaluation. See, however, Problem 6.19 and [187, 188].

The discussions of this section may be looked upon from a different perspective. Once the functions $g(\tau)$ and $\lambda(\tau)$ are specified, the Hamiltonian $g(H_0 + \lambda V)$ moves deterministically along a line in the space of complex Hermitian $N \times N$ matrices. The family of Hamiltonians described by Green's function (6.15.7, 6.15.10) corresponds to a bundle of such trajectories, all of which originate from a single point at $\tau = 0$. At each moment $\tau > 0$, the bundle of trajectories defines a cloud of points. The bundle is made unique by the cloud reached at $\tau \to \infty$, and that final cloud represents the GUE. Thus, the matrix-valued Ornstein–Uhlenbeck process may be thought of as determined by a teleological average over a bundle of deterministic trajectories.

6.16 Local and Global Equilibrium in Spectra

As explained in Sect. 4.7, the notion of an average density of levels $\varrho(E) = \Delta N / \Delta E$ for a given Hamiltonian would not make sense if $\varrho(E)$ were not smooth, and in fact constant, on the energy scale of a mean level spacing. The scale on which $\varrho(E)$ is allowed to vary in the limit of large N is a "global" one on which fluctuations in the level sequence tend not to be noticeable. For instance, if the Hamiltonian H is a typical member of any of the Gaussian ensembles of random matrices, $\varrho(E)$ obeys the semicircle law and thus has the radius of the semicircle as its natural scale of energy. To obtain a well-defined limiting form of the density of levels for $N \to \infty$, one must therefore refer the energy variable to a global unit (such that, e.g., the radius of the semicircle law becomes independent of N).

The energy scale on which fluctuations in the sequence of levels become manifest is smaller by a factor of the order N, however. Indeed, the natural energy unit for quantities describing such fluctuations is the local mean spacing $1/\varrho(E)$. In discussing fluctuations of the density of levels in Sect. 4.9, I used the unfolded energy $e = E\varrho(E)$ and reported Dyson's results (4.9.15) for the two-level cluster functions of the Gaussian ensembles of random matrices; these results indeed imply that density fluctuations at different energies become statistically independent when the energies differ by many mean spacings; more quantitatively speaking, the two-point correlation function approaches the product of two mean densities as $1/(e - e')^2$. Similar results hold for the matrix ensemble (6.15.12, 20) which smoothly interpolates between the GOE and the GUE, as explicitly verified by *Pandey* and *Mehta* [177, 178] who calculated all n-point correlation functions by integrating the joint density (6.15.20) of all N eigenvalues over $N - n$ of its arguments.

Intimately related to the energy scale separation just described is a separation of "time" or control parameter scales for "processes" in which a spectrum changes its global and local properties as a control parameter λ in the Hamiltonian $H = H_0 + \lambda V$ or the Floquet operator $F = e^{-i\lambda V}e^{-iH_0}$ is varied: The level density $\varrho(E, \lambda)$ changes much more slowly with λ than any of the quantities describing local fluctuations. For instance, the transitions between the Gaussian ensembles of random matrices discussed in Sect. 6.15 have a level density independent of the "time" $\tau = \ln \sqrt{1 + \lambda^2}$ while the spacing distribution and low-order correlation functions of density fluctuations have transition "time" scales of the order $1/N$ [see (6.15.23)].

The transition from the Poissonian ensemble to the GUE briefly treated at the end of the previous section does not display a strictly vanishing time scale ratio for the density fluctuations and the mean density but still one of order $1/N$. Indeed, the characteristic time for the density of levels in that case can be revealed to be of order unity by looking at the mean-squared matrix elements (6.15.28) and recalling the derivation of the semicircle law in Sect. 4.6. Initially, at $\tau = 0$, the Gaussian level density is governed by the diagonal elements H_{ii}; at $\tau \to \infty$, as is clear from Sect. 4.6, the validity of the semicircle law hinges on the negligibility of the diagonal elements in the sense $\sum_i \langle H_{ii}^2 \rangle / \sum_{i \neq j} \langle |H_{ij}|^2 \rangle = \mathcal{O}(1/N)$. Since the latter ratio has the order N during the early stages during which $\tau = \mathcal{O}(1/N)$, the transition from the Poissonian to the final behavior may roughly be located at times of order unity, when the ratio in consideration itself has the intermediate order unity. Note that according to the mean-squared matrix elements (6.15.28), the ratio falls to $\mathcal{O}(1/N)$ at $\tau = \mathcal{O}(\ln N)$. However, local equilibrium within clusters of levels extending over a few mean spacings should be attained more rapidly on a time scale $\sim 1/N$, just as for the transition, say, from the GOE to the GUE. Unfortunately, the latter expected behavior cannot be considered obvious from the interpolating matrix density defined by (6.15.28) since the corresponding level spacing distribution and low-order correlation functions have not yet been evaluated. Lacking such explicit rigorous results, perturbative arguments have been invoked in the literature [179] to indicate the proportionality of the characteristic time of local fluctuations to $1/N$.

The separations of time and energy scales just described is certainly expected by analogy with ordinary many-particle systems. There, too, the approach to global equilibrium is characterized by long "hydrodynamic" times while local equilibrium across microscopically correlated clusters is established within a few collision times.

The remainder of this section is devoted to a more systematic discussion of the scaling properties of the correlation functions

$$\varrho_n(E_1 \ldots E_n, \tau) = \frac{1}{(N-n)!} \int dE_{n+1} \ldots dE_N \varrho_N(E_1 \ldots E_N, \tau) \,,$$

$$(6.16.1)$$

which are normalized as

$$\int dE_1 \ldots dE_n \varrho_n(E_1 \ldots E_n, \tau) = \frac{N!}{(N-n)!} \,. \qquad (6.16.2)$$

Note that the correlation function $\varrho_N(E_1 \ldots E_N, \tau)$ of all N levels differs from the joint probability density $P(E_1 \ldots E_N, \tau)$ of all N levels by normalization, $\varrho_N(E_1 \ldots E_N, \tau) = N! P(E_1 \ldots E_N, \tau)$. These correlation functions arise naturally as the following ensemble averages,

$$\varrho_1(E_1, \tau) = \left\langle \sum_i^{1 \ldots N} \delta(E - E_i) \right\rangle$$

$$\varrho_n(E_1 \ldots E_n, \tau) = \left\langle \sum_{i_1 \neq i_2 \neq \ldots \neq i_n}^{1 \ldots N} \delta(E_1 - E_{i_1}) \ldots \delta(E_n - E_{i_n}) \right\rangle \,,$$

$$(6.16.3)$$

the averages are defined with $P(E_1 \ldots E_N, \tau)$ as the weight. Of course, $\varrho_1(E) \equiv \varrho(E)$ is the average density of levels referred to above. The reader is invited to establish the relationship of $\varrho_2(E_1, E_2)$ to the "structure factor" $S(E_1, E_2)$ employed in Sect. 4.9.

By integrating the Fokker–Planck equation (6.15.13, 19) over $N - n$ energy variables, one obtains the following set [180] of integrodifferential equations for the correlation functions ϱ_n:

$$\frac{\partial}{\partial \tau} \varrho_n(E_1 \ldots E_n, \tau)$$

$$= \sum_i^{1 \ldots n} \frac{\partial}{\partial E_i} \left(E_i - \sum_{j(\neq i)}^{1 \ldots n} \frac{1}{E_i - E_j} + \frac{1}{2} \frac{\partial}{\partial E_i} \right) \varrho_n(E_1 \ldots E_n, \tau)$$

$$- \sum_i^{1 \ldots n} \frac{\partial}{\partial E_i} \mathcal{P} \int dE_{n+1} \frac{1}{E_i - E_{n+1}} \varrho_{n+1}((E_1 \ldots E_{n+1}, \tau) , \qquad (6.16.4)$$

which in the context of ordinary many-particle theory is called the BBGKY hierarchy. Now, I propose to show that this hierarchy is consistent with the separation of time and energy scales discussed above. Let us first consider (6.16.4) for $n = 1$ in the form

$$\frac{\partial}{\partial \tau} \varrho_1(E, \tau) = \frac{\partial}{\partial E} \left(E - \mathcal{P} \int dE' \frac{\varrho_1(E', \tau)}{E - E'} + \frac{1}{2} \frac{\partial}{\partial E} \right) \varrho_1(E, \tau)$$

$$+ \frac{\partial}{\partial E} \mathcal{P} \int dE' \frac{1}{E - E'} [\varrho_1(E, \tau) \varrho_1(E', \tau)$$

$$- \varrho_2(E, E', \tau)] . \qquad (6.16.5)$$

Note that I have added and subtracted a piece bilinear in ϱ_1 so that Dyson's two-level cluster function $\varrho_1(E, \tau) \varrho_1(E', \tau) - \varrho_2(E, E', \tau)$ appears in the last integral; the latter function was already met in Sect. 4.9 (4.9.14), albeit expressed in different units.

Now, I shall argue that the last integral in (6.16.5) vanishes. To that end, I assume, subject to later proof of consistency with (6.16.4) for $n > 1$, that the two-level cluster function has the scaling properties announced before: it approaches an adiabatic equilibrium, contingent on the local and current value of the density $\varrho_1(E, \tau)$, roughly N times faster than ϱ_1 relaxes to its final semicircular form. The assumed time-scale separation entails, as will become clear presently, that the two-level cluster function makes negligible contributions to the last integral in (6.16.5) once E and E' are further apart than $\Delta E = N^\varepsilon$ with $-1/2 < \varepsilon < +1/2$. To appreciate the required N dependence of the interval ΔE, the reader must recall that in the units chosen the final semicircle law has a radius $\sim \sqrt{N}$ while the mean level spacing is of the order $1/\sqrt{N}$.

Due to the assumed adiabaticity of the temporal variation of $\varrho_1(E, \tau)$ with respect to the two-level cluster function, the latter may, under the last integral in (6.16.5), be taken to be in adiabatic equilibrium during the whole relaxation of $\varrho_1(E, \tau)$ toward the semicircle law,

$$\varrho_1(E, \tau) \varrho_1(E', \tau) - \varrho_2(E, E', \tau) = \varrho_1(E, \tau) \varrho_1(E', \tau) Y_{\mathrm{GUE}}(e - e') \qquad (6.16.6)$$

with

$$e - e' = E \varrho_1(E, \tau) - E' \varrho_1(E', \tau)$$

and with $Y_{\mathrm{GUE}}(e)$ given in (4.9.15). Then, consider the contribution to the last term in (6.16.5) from the integration range outside the interval $\Delta E \sim N^\varepsilon$

around E,

$$\frac{\partial}{\partial E}\varrho_1(E,\tau)\left(\int_{-\infty}^{E-\Delta E/2}+\int_{E+\Delta E/2}^{\infty}\right)dE'\frac{\varrho_1(E',\tau)}{E-E'}Y_{\text{GUE}}(e-e') . \quad (6.16.7)$$

By order of magnitude, we may put $\varrho_1(E',\tau)\approx\sqrt{N}$ and $Y_{\text{GUE}}(e)\approx 1/e^2$ such that the term (6.16.7) turns out to have the weight

$$\frac{\partial}{\partial E}\varrho_1(E,\tau)\sqrt{N}\frac{4}{N(\Delta E)^2}\approx\frac{\partial}{\partial E}\varrho_1(E,\tau)\sqrt{N}4N^{-1-2\varepsilon} , \quad (6.16.8)$$

i.e., smaller than the weight of the linear drift term $\frac{\partial}{\partial E}E\varrho_1(E,\tau)$ in (6.16.5) by a factor $N^{-1-2\varepsilon}$ and thus indeed negligible in the limit $N\to\infty$ for $-1/2<\varepsilon$.

It remains to discuss the contribution to the last term in (6.16.5) from the integration interval $\Delta E=N^\varepsilon$ around E. Instead of (6.16.7), we are facing

$$\frac{\partial}{\partial E}\varrho_1(E,\tau)\int_{E-\Delta E/2}^{E+\Delta E/2}dE'\frac{\varrho_1(E',\tau)}{E-E'}Y_{\text{GUE}}(e-e') . \quad (6.16.9)$$

Assuming once more that the variation in energy of $\varrho_1(E',\tau)$ is characterized by the scale $\sqrt{N}$ given by the final semicircle law and now requiring $\varepsilon<1/2$ we conclude that $\varrho_1(E',\tau)$ cannot vary noticeably across the interval ΔE; thus the integral in (6.16.9) is seen to vanish since $Y(e)$ is an even function of e.

With the whole last term in (6.16.5) now having disappeared, the integrodifferential equation for the level density is decoupled from the rest of the hierarchy (6.16.4) with $1<n\le N$. One further simplification actually arises in the equation (6.16.5) for $\varrho_1(E,\tau)$. Inasmuch as $\varrho_1(E,\tau)$ indeed varies on a global energy scale $\sim\sqrt{N}$, a well-defined limiting form of the density for $N\to\infty$ becomes accessible only through the change of units

$$\varrho_1(E,\tau)=\sqrt{N}\tilde{\varrho}_1(\tilde{E},\tau) ; \quad \tilde{E}=\frac{E}{\sqrt{N}} . \quad (6.16.10)$$

The radius of the final semicircle is thereby assigned a value independent of N, while the mean spacing of neighboring levels appears to be of the order $1/N$, and the new density $\tilde{\varrho}_1$ is normalized as a probability, $\int d\tilde{E}\tilde{\varrho}_1(\tilde{E},\tau)=1$. When (6.16.5) is expressed in these new units, the diffusion term takes the form $(1/2N)\partial^2/\partial\tilde{E}^2$; clearly, such a term cannot affect the $\tilde{E}$ dependence of $\tilde{\varrho}_1(\tilde{E},\tau)$ in the limit $N\to\infty$. Thus, the final form of the evolution equation, derived by *Dyson* [181] and independently by *Pastur* [182], reads

$$\frac{\partial\tilde{\varrho}_1(\tilde{E},\tau)}{\partial\tau}=\frac{\partial}{\partial\tilde{E}}\left(\tilde{E}-\mathcal{P}\int dE'\frac{\tilde{\varrho}_1(E',\tau)}{\tilde{E}-E'}\right)\tilde{\varrho}_1(\tilde{E},\tau) . \quad (6.16.11)$$

It is easy to check that this has the stationary solution $\tilde{\varrho}_1(\tilde{E},\infty)=(1/\pi)\sqrt{2-\tilde{E}^2}$.

When turning to the hierarchy (6.16.4) with $n>1$, the most interesting situation to discuss arises when all n energy arguments in $\varrho_n(E_1\ldots E_n,\tau)$ lie in a single cluster of extension $\Delta E \sim 1/\sqrt{N}$; otherwise, if the $\{E_i\}$ form several such clusters and the intercluster distances are noticeable on the global energy scale $\sqrt{N}$, ϱ_n would factor into a product in which each local subcluster would be represented by its own correlation function. Assuming that ϱ_n describes a single local cluster, it is helpful to start the discussion of (6.16.4) by noting the order of magnitude of the various terms. We have $\mathcal{O}(\varrho_n) = \mathcal{O}(\varrho_1^n) = N^{n/2}$. In contrast to the situation discussed above, the energy scale for all ϱ_n with $n > 1$ is of the order of a mean spacing, i.e., $\sim 1/\sqrt{N}$, such that formally $\mathcal{O}((\partial/\partial E_i)\varrho_n) = N^{(n+1)/2}$; however, a slight complication is of importance: the n energy arguments of ϱ_n can be reorganized into one "center-of-energy" variable and $n - 1$ relative energies; only the latter can be expected to have a range $\sim 1/\sqrt{N}$. The dependence of ϱ_n on the center-of-energy variable, on the other hand, must certainly be weaker, i.e., must have the global scale $\sqrt{N}$; otherwise, a conflict would arise with the previous assumption that the density of levels $\varrho_1(E)$ is constant across a mean spacing $1/\varrho_1 \sim 1/\sqrt{N}$ and varies only on the global scale $\sqrt{N}$. For instance, the definition (6.16.1) implies that $\int dE_2 \varrho_2(E_1, E_2) = (N - 1)\varrho_1(E_1)$ and therefore $\varrho_1(E_1)$ would vary on the scale $1/\sqrt{N}$ if $\varrho_2(E_1, E_2)$ were allowed to do so independently in both of its arguments. It follows that $\mathcal{O}(\sum_{i=1}^{n}(\partial/\partial E_i)\varrho_n) = N^{(n-1)/2}$, a point to be returned to several times below.

The foregoing order-of-magnitude estimates imply that, of the three terms involving ϱ_n on the right-hand side in (6.16.4), the Coulomb drift and the diffusion overwhelm the linear drift $(\partial/\partial E_i)E_i\varrho_n$ by a factor N. Indeed, the linear restoring force, essential for the global behavior of the density of levels ϱ_1, is entirely negligible on the smaller energy scale of relevance for local fluctuations. Conversely, the diffusion was ineffective globally but is now as important as the drift caused by the Coulombic "interparticle" force. Incidentally, there are two types of Coulomb terms in (6.16.4) for $n > 1$, both of equal weight in the sense of the estimates in consideration: The first such term comprises all intracluster interactions while the second refers to pairs of "particles" with one member particle within the cluster and the other outside; the second term involves the $(n + 1)$-level correlation function ϱ_{n+1}.

Having commented on all terms on the right-hand side of (6.16.4) for $n > 1$, we should also look at the left-hand member, $(\partial/\partial\tau)\varrho_n$. Its formal weight is $\mathcal{O}(N^{n/2}/\tau)$, and the equation therefore requires that $\mathcal{O}(\tau) = 1/N$. It is at this point that we get back, self-consistently, the time-scale separation for local fluctuations and global density equilibration.

A convenient implementation of the scaling with N just discussed employs new units, $\varrho_n(E_1, \ldots E_n, \tau) \rightarrow \tilde{\varrho}_n(e_1, \ldots e_n, \tilde{\tau})$, such that the latter function has a well-defined limiting form for $N \rightarrow \infty$. Following *Pandey* [180], I choose

$$e_i = \int_{E}^{E_i} dE' \varrho_1(E', \tau) \approx (E_i - E)\varrho_1(E, \tau) \,,$$

$$\tilde{\tau} = \tau \varrho_1(E, \tau)^2 \sim \tau N \,, \qquad\qquad (6.16.12)$$

$$\tilde{\varrho}_n(E_1, \ldots E_n, \tau) = \varrho_1(E, \tau)^n \tilde{\varrho}_n(e_1, \ldots e_n, \tilde{\tau}) \,.$$

The introduction of the e_i is equivalent to the familiar unfolding of the spectrum to unit mean spacing (see Sects. 4.7, 4.9). It is important to realize that the e_i are effectively independent of τ on the time scale of relevance to local fluctuations. The reference energy E specifies the location of the cluster. The new time $\tilde{\tau}$ is effectively linear in τ and independent of E throughout the cluster. The reader should note the difference between the scalings (6.16.10) for $n = 1$ and (6.16.12) for $n > 1$.

In the limit $N \to \infty$, the new n-point correlation function must be restricted so as to become independent of the center of energy. To see this, consider a displacement of the original energies

$$E_i \;\to\; E_i + \frac{\varepsilon}{\varrho_1} = E + \frac{\varepsilon}{\varrho_1} + \frac{e_i}{\varrho_1} \tag{6.16.13}$$

and the corresponding correlation function ϱ_n. Displaying only variables of relevance for the argument, I have the identity

$$\varrho_n\left(\left\{E_i + \frac{\varepsilon}{\varrho_1}\right\}\right) \;=\; \varrho_1\left(E + \frac{\varepsilon}{\varrho_1}\right)^n \tilde{\varrho}_n\left(\{e_i\}\right)$$

$$=\; \varrho_1(E)^n \tilde{\varrho}_n\left(\{e_i + \varepsilon\}\right) . \tag{6.16.14}$$

To first order in ε this yields

$$\frac{1}{\tilde{\varrho}_n(\{e_i\})} \sum_{j=1}^{n} \frac{\partial}{\partial e_j} \tilde{\varrho}_n\left(\{e_i\}\right) = \frac{n}{\varrho_1^2} \frac{\partial}{\partial E} \varrho_1 . \tag{6.16.15}$$

As long as n remains finite for $N \to \infty$, the right-hand side in (6.16.15) is of order $1/N$, and thus

$$\mathcal{O}\left(\sum_{j=1}^{n} \frac{\partial}{\partial e_j} \tilde{\varrho}_n\right) = \frac{1}{N} \to 0 \ \text{ for } \ N \to \infty . \tag{6.16.16}$$

Of course, this restriction on ϱ_n and $\tilde{\varrho}_n$ can be imposed only at some initial time, e.g., at $\tau = 0$. We shall see below, however, that the dynamics will preserve local homogeneity at later times.

When inserting the rescaling (6.16.12) into (6.16.4) it is once more important to use the effective constancy of the density $\varrho_1(E_i, \tau)$ across the cluster,

$$\frac{\partial}{\partial E_i} = \varrho_1(E_i, \tau) \frac{\partial}{\partial e_i} \approx \varrho_1(E, \tau) \frac{\partial}{\partial e_i}$$

$$\frac{\partial^2}{\partial E_i^2} \approx \varrho_1(E, \tau)^2 \frac{\partial^2}{\partial e_i^2} , \tag{6.16.17}$$

thereby neglecting corrections vanishing in the limit $N \to \infty$. Similarly, in the denominator of the intracluster Coulomb drift, we may put

$$e_i - e_j = \int_{E_j}^{E_i} dE' \varrho_1(E', \tau) \approx (E_i - E_j)\varrho_1(E, \tau) . \tag{6.16.18}$$

Dropping the linear-drift term, I obtain the rescaled hierarchy as

$$
\frac{\partial}{\partial \tilde{\tau}} \tilde{\varrho}_n(e_1, \ldots e_n, \tilde{\tau}) = \sum_i^{1 \ldots n} \frac{\partial}{\partial e_i} \left(- \sum_{j(\neq i)}^{1 \ldots n} \frac{1}{e_i - e_j} + \frac{1}{2} \frac{\partial}{\partial e_i} \right)
$$

$$
\tilde{\varrho}_n(e_1, \ldots, e_n, \tilde{\tau})
$$

$$
- \varrho_1(E, \tau)^{-(n+2)} \sum_i^{1 \ldots n} \frac{\partial}{\partial E_i} \int dE_{n+1} \frac{1}{E_i - E_{n+1}}
$$

$$
\varrho_{n+1}(E_1, \ldots, E_{n+1}, \tau) .
$$

$$(6.16.19)$$

It remains to discuss the last term in (6.16.19) which describes the Coulomb drift of levels within the cluster caused by levels outside. As was done in treating the analogous term for $n = 1$, I split the integral into an integral over an interval $\Delta E \sim N^\varepsilon$ with $-1/2 < \varepsilon < 1/2$ around the center of the cluster and a remainder. Note that ΔE is chosen larger than the extension of the cluster such that the contribution to the integral from outside ΔE, the $(n+1)$-level correlation function, factors: $\varrho_{n+1} \to \varrho_n \varrho_1$, is

$$
\varrho_1^{-(n+2)} \sum_i^{1 \ldots n} \frac{\partial}{\partial E_i} \varrho_n(E_1, \ldots E_n, \tau) \int_{\substack{\text{outside} \\ \Delta E}} dE_{n+1} \frac{1}{E_i - E_{n+1}} \varrho_1(E_{n+1}, \tau) .
$$

$$(6.16.20)$$

This "remainder" is easily seen to vanish asymptotically: since $\varepsilon > -1/2$ the intracluster positions E_i in the Coulomb-force denominator can be replaced by the center energy E of the cluster; after inserting the rescaling (6.16.12, 13) for the n-level correlation function, the local homogeneity (6.16.15, 16) can be invoked. Finally, the integral over the interval $\Delta E \sim N^\varepsilon$ can, using $\varepsilon > -1/2$ once more, be extended over the whole energy axis such that the hierarchy takes the form

$$
\frac{\partial}{\partial \tilde{\tau}} \tilde{\varrho}_n = \sum_i^{1 \ldots n} \frac{\partial}{\partial e_i} \left(- \sum_{j(\neq i)}^{1 \ldots n} \frac{1}{e_i - e_j} + \frac{1}{2} \frac{\partial}{\partial e_i} \right) \tilde{\varrho}_n
$$

$$
- \sum_i^{1 \ldots n} \frac{\partial}{\partial e_i} \mathcal{P} \int de_{n+1} \frac{\tilde{\varrho}_{n+1}}{e_i - e_{n+1}} , \quad n > 1 . \qquad (6.16.21)
$$

Now, the correlation function $\tilde{\varrho}_{n+1}$ under the integral represents a local cluster of $n+1$ levels. Remarkably enough, apart from the absence of the linear drift term from (6.16.21), the original hierarchy (6.16.4) and the rescaled one have identical structures.

Having invoked local homogeneity for the correlation functions $\tilde{\varrho}_n$ of local clusters in showing the "locality" of the integral term in (6.16.21), it is imperative to demonstrate the consistency of the homogeneity property (6.16.16) with the hierarchy (6.16.21). Indeed, by differentiating (6.16.21) and integrating by parts in the last term, it is easy to check that the quantities $D_n = \sum_i^{1 \ldots n} (\partial/\partial e_i) \tilde{\varrho}_n$

obey the same hierarchy of integrodifferential equations (6.16.21) as the correlation functions $\tilde{\varrho}_n$. Consequently, if all D_n are of order $1/N$ initially, none of them can grow enough to exceed that order of magnitude. To appreciate this statement, the reader should realize that both the diffusion and the Coulomb repulsion tend to make $\tilde{\varrho}_n$ smooth and thus D_n small.

The separation of energy and time scales for local fluctuations and global relaxation of the density must hold for any reasonable initial condition. A particularly interesting example is the transition from the GOE to the GUE treated in the preceding section. The transition from the Poissonian ensemble to the GUE, also briefly mentioned in Sect. 6.15, must abide by that separation as well. It follows that the latter transition cannot be considered the quantum parallel of classical crossovers from predominantly regular motion to global chaos, except when that classical transition is abrupt. In fact, I know of no classical system with a Hamiltonian $H_0 + \lambda V$ or a Floquet operator $\mathrm{e}^{-\mathrm{i}\lambda V}\mathrm{e}^{-\mathrm{i}H_0}$ which, while integrable for $\lambda = 0$, is globally chaotic for *all* non-zero values of λ. Such abrupt classical crossovers can happen for billiards upon changes of the boundaries [183]. In that case, however, level dynamics does not apply: Even a small deformation of the shape of a billiard boundary amounts to strong perturbations for sufficiently high degrees of excitation.

Neither the Gaussian ensembles of Hermitian random matrices nor their Brownian-motion dynamizations could yield good models of Hamiltonians of quantum systems with globally chaotic classical limits, were it not for the disparity of scales under discussion. In fact, concrete dynamic systems do not in general obey the semicircle law for their level densities; it is only the local fluctuations in the spectra which tend to be faithful to random-matrix theory. Were there not a complete decoupling of local fluctuations and global variations of the density, provided by the scale separation, the predictions of random-matrix theory would be as unreliable locally as they in fact are globally [181, 184].

6.17 Adiabatic Invariance of Quantum Maps

According to a well-known adiabatic-invariance law, an eigenstate of a Hamiltonian H_0 can be continuously transformed into an eigenstate of another Hamiltonian H pertaining to the same Hilbert space. The transformation needed is a time evolution generated by a time-dependent Hamiltonian $H(t)$ which interpolates adiabatically, i.e., sufficiently slowly, between the initial H_0 and the final H. The level dynamics (6.2.27) for Hamiltonians can be used in an elegant proof of the adiabatic invariance (Problem 6.9). There is an interesting extension of the traditional theorem just mentioned to the Floquet operators of periodically kicked systems [185]: For given fixed H_0 and V, one considers the two Floquet operators $F(\lambda) = \mathrm{e}^{-\mathrm{i}\lambda V}F(0)$ and $F(\lambda + \delta\lambda) = \mathrm{e}^{-\mathrm{i}(\lambda + \delta\lambda)V}F(0)$; the eigenstate $|m, \lambda\rangle$ of $F(\lambda)$ can be adiabatically transformed into the eigenstate $|m, \lambda + \delta\lambda\rangle$, and the transformation needed is equivalent to the level dynamics discussed in Sect. 6.2.

To prove the theorem, we first consider an infinitesimal interval $\delta\lambda$ and even subdivide $\delta\lambda$ into N equal parts. Then, a sequence of N infinitesimally close Floquet operators $F(\lambda + s\delta\lambda/N)$ with $s = 1, 2, \ldots N$ can be defined and allowed to act consecutively on the eigenstate $|m, \lambda\rangle$ of $F(\lambda)$,

$$
\begin{aligned}
|\rangle &= \left[\prod_{s=1}^{N} F\left(\lambda + \frac{s\delta\lambda}{N}\right)\right]_{+} F(\lambda)^{-N} |n, \lambda\rangle \\
&= e^{iN\phi_n(\lambda)} \left[\prod_{s=1}^{N} F\left(\lambda + \frac{s\delta\lambda}{N}\right)\right]_{+} |n, \lambda\rangle .
\end{aligned}
\tag{6.17.1}
$$

Here the index $+$ means positive "time" order for the sequence of N Floquet operators. It is for convenience and not essential for the subsequent argument that the phase factor $e^{iN\phi_n(\lambda)}$ is included in the definition of the state $|\rangle$ which in the limit of large N may be said to arise adiabatically from the state $|n, \lambda\rangle$. Now, we proceed to show that this state is indeed an eigenstate $|m, \lambda + \delta\lambda\rangle$ of $F(\lambda + \delta\lambda)$. To that end, we expand all N operators $F(\lambda + s\delta\lambda/N)$ in powers of $\delta\lambda$, drop second- and higher order terms, and write F for $F(\lambda)$

$$
\begin{aligned}
|\rangle &= e^{iN\phi_n(\lambda)} (1 - i\delta\lambda V) F \left(1 - i\delta\lambda \frac{N-1}{N} V\right) F \\
&\quad \ldots \left(1 - i\delta\lambda \frac{2}{N} V\right) F \left(1 - i\delta\lambda \frac{1}{N} V\right) F |n, \lambda\rangle \\
&= \left(1 - i\delta\lambda \sum_{s=1}^{N} \frac{s}{N} F^{N-s} V F^{s-N}\right) |n, \lambda\rangle .
\end{aligned}
\tag{6.17.2}
$$

After taking the scalar product with $\langle m, \lambda|$, we can carry out the sum over s to obtain

$$
\langle n, \lambda |\rangle = 1 - i\delta\lambda V_{nn} \frac{N-1}{2} = e^{-i\delta\lambda V_{nn}(N-1)/2} + \mathcal{O}(\delta\lambda^2)
\tag{6.17.3}
$$

and, for $m \neq n$ and $\phi_m \neq \phi_n$ with $N \to \infty$,

$$
\langle m, \lambda |\rangle = \frac{i\delta\lambda V_{mn}}{e^{-i\phi_{mn}} - 1} .
\tag{6.17.4}
$$

The scalar product in (6.17.3) is a phase factor which could be absorbed into the definition of the state $|\rangle$; this result is in accordance with the statement in Sect. 6.2 that the infinitesimal increment $\langle m, \lambda|m, \lambda\rangle\,\delta\lambda$ can be made zero by properly choosing a phase factor for $|m, \lambda\rangle$. The scalar product (6.17.4), on the other hand, is just the level dynamics result (6.2.6). It follows that the state $|\rangle$ is indeed, to within a physically irrelevant phase factor, the eigenstate of $F(\lambda + \delta\lambda)$ which is generated from the eigenstate $|m, \lambda\rangle$ of $F(\lambda)$ by the level dynamics of Sect. 6.2.

6.18 Problems

6.1. Determine the number of independent dynamic variables of the fictitious N-particle system when $O(N)$, $U(N)$, and $Sp(N)$ are the groups of canonical transformations.

6.2. Show that the angular momentum matrix l fulfills $\tilde{l} = -l$, $l^\dagger = -l$, and $\tilde{l}Z = -Zl$ when the Floquet operator has $O(N)$, $U(N)$, and $Sp(N)$, respectively, as its canonical group.

6.3. Having done this problem you will understand why the variables l_{mn} are often called angular momenta and why their Poisson brackets define the Lie algebras $o(N), u(N)$, and $sp(N)$ in the orthogonal, unitary, and symplectic cases, respectively. Imagine an N-dimensional real configuration space with Cartesian coordinates $x_1 \ldots x_N$. A $2N$-dimensional phase space arises by associating N conjugate momenta $p_1 \ldots p_N$. Rotations in the configuration space are generated by the angular momenta $l_{mn} = \frac{1}{2}(x_m p_n - x_n p_m)$. The Poisson brackets (6.2.23) then follow from $\{p_m, x_n\} = \delta_{mn}$. Think about how to generalize to complex x, p to get the Poisson brackets (6.2.25) for the unitary case where the l_{mn} generate rotations in an N-dimensional complex vector space. In what vector space do the l_{mn} generate rotations for the symplectic case? See also Sect. 9.10 and Problem 9.19.

6.4. Verify

$$\dot{x}_m = p_m$$

$$\dot{p}_m = -\sum_{l(\neq m)} \frac{2l_{ml}l_{lm}}{(x_m - x_l)^3}$$

$$\dot{l}_{mn} = \sum_{l(\neq m,n)} l_{ml}l_{ln}\left[(x_m - x_l)^{-2} - (x_l - x_n)^{-2}\right]$$

as the level dynamics for time-independent Hamiltonians $H = H_0 + \lambda V$.

6.5. Show that $a_{mm} = 0$ for the Lax generator a can be achieved by properly choosing phase factors for the eigenvectors of the Floquet operator (or the Hamiltonian).

6.6. Prove (6.6.7) by showing that $l_{1\bar{1}} = l_{2\bar{2}} = 0$ and $|l_{12}|^2 + |l_{1\bar{2}}|^2 = $ const. Use the T-invariance of F with $T^2 = -1$ and the ensuing identity for TVT^{-1}.

6.7. Show that the stationary spacing distribution (6.14.9) holds for all sufficiently smooth frequency distributions $\varrho(\omega)$.

6.8. Starting from (6.14.5), show that in the limit $N \to \infty$ with a smooth density $\varrho(\omega)$, the mean spacing is independent of λ, $\langle S(\lambda)\rangle = 1$. Verify the inverse-power decay of $\langle S(\lambda)^2\rangle$ for various choices of $\varrho(\omega)$: Lorentzian, Gaussian, Wigner distribution.

6.9. Prove the adiabatic invariance law for Hamiltonians using the level dynamics (6.2.27).

6.10. Using the arguments of Sect. 4.6 show that the ensemble (6.15.12) implies Wigner's semicircle law for the mean level density in the limit $N \to \infty$.

6.11. Show that the free diffusion described by (6.15.2) with λ^2 as a "time" implies a Brownian motion with Coulomb gas interaction for the eigenvalues of H. Discuss how this differs from (6.15.13, 19).

6.12. Study a two-body collision in the Coulomb gas model (6.15.13, 19).

6.13. Write the Fokker–Planck equation (6.15.13, 19) in the compact form $\dot{\varrho} = L\varrho$, $L = \sum_i \partial_i \varrho_{\mathrm{GUE}} \partial_i \varrho_{\mathrm{GUE}}^{-1}$. Transform as $\varrho = \sqrt{\varrho_{\mathrm{GUE}}}\,\tilde{\varrho}$, and show that the transformed generator $\tilde{L}$ is a Hermitian differential operator. Furthermore, show that $\tilde{L}$ contains pair interactions inversely proportional to the squared distance between the "particles", as does the Hamiltonian level dynamics (6.2.27). Appreciate the conceptual difference between the latter and the motion generated by $\tilde{L}$.

6.14. Derive the Hamiltonian flows generated by the Hamiltonians (6.16.13, 15) by inserting the transformation $E_m(\tau) = g(\tau) \cdot x_m(\lambda(\tau))$ into Pechukas' dynamics (6.2.27).

6.15. Proceeding as in Sect. 6.3, find the Lax form of the Pechukas equations (6.2.27). Hint: The Pechukas equations follow from (6.2.16) by linearizing as $\cos x \to 1$, $\sin x \to x$.

6.16. Find the Lax form of the modified Pechukas equations pertaining to the Hamiltonian function (6.16.15). Hint: Proceed as in Sect. 6.3 but use (6.16.6).

6.17. Establish the most general canonical ensemble for the modified Pechukas–Yukawa gas (with the confining potential) which rigorously gives the joint distribution of eigenvalues (4.3.15) of the Gaussian ensembles of random matrices. Use the results of Problem 6.16, and argue as in Sect. 6.5. Where does the Gaussian factor come from?

6.18. Find the spacing distribution $P(S, \lambda)$ interpolating between the Gaussian orthogonal and unitary ensemble of 2×2 matrices. Use the method of Sect. 4.4, but start from the matrix density (6.15.12), taking the latter for $N = 2$. The result is [187]

$$P(S, \lambda) = \sqrt{1 + \lambda^2/2}\, S\phi(\lambda)^2 e^{-S^2\phi(\lambda)^2/2}\, \mathrm{erf}\,(S\phi(\lambda)/\lambda)$$

where

$$\phi(\lambda) = \sqrt{1 + \lambda^2/2}\sqrt{\pi/2}\left[1 - \frac{2}{\pi}\left(\arctan\frac{\lambda}{\sqrt{2}} - \frac{\lambda\sqrt{2}}{2 + \lambda^2}\right)\right]$$

and $\lambda^2 = e^{2\tau} - 1$.

6.19. Repeat Problem 6.18 but for the transition of Poisson to Wigner. Start from the Gaussian defined by (6.15.28) to find

$$P(S, \lambda) = S\lambda\psi(\lambda)^2 e^{-S^2\psi(\lambda)^2/4} \int_0^\infty d\xi e^{-\xi^2 - 2\xi\lambda} I_0\left[S\psi(\lambda)\right] .$$

By requiring $\bar{S} = 1$, the scale factor $\psi(\lambda)$ can be determined and is related to the Kummer function [187, 188].

7. Quantum Localization

7.1 Preliminaries

This chapter will focus mainly on the kicked rotator that displays global classical chaos in its cylindrical phase space for sufficiently strong kicking. The chaotic behavior takes the form of "rapid" quasi-random jumps of the phase variable around the cylinder and "slow" diffusion of the conjugate angular momentum p along the cylinder. A quantum parallel of this time-scale separation of the motions along the two principal directions of the classical phase space is the phenomenon of quantum localization: the quasi-energy eigenfunctions of the kicked rotator turn out to be localized in the angular momentum representation [189]–[191]. (We shall be concerned with the generic case of irrational values of a certain dimensionless version of Planck's constant – for rational values extended eigenfunctions arise [192].)

The difference between classical and quantum predictions for the rotator is nothing like a small quantum correction, even if states with large angular-momentum quantum numbers are involved. Classical behavior would be temporally unlimited diffusive growth of the kinetic energy, $\langle p^2 \rangle \propto t$. Quantum localization must set an end to any such growth starting from a finite range of initially populated angular-momentum states, since only those Floquet states can be excited that overlap the initial angular momentum interval.

Given localization of all quasi-energy eigenfunctions, it is clear that classical chaos cannot be accompanied by repulsion of all quasi-energies: Indeed, eigenfunctions without overlap in the angular momentum basis have no reason to keep their quasi-energies apart. On the other hand, if a level spacing distribution is established by including only eigenfunctions with their supports within one localization length, one expects, and does indeed find, a tendency to level repulsion again [193].

From the point of view of random-matrix theory, the Floquet operator F of the kicked rotator appears as nongeneric in other respects as well. In the angular momentum representation, F takes the form of a narrowband matrix whose off-diagonal elements decrease rapidly as their distance from the main diagonal grows. In appreciating this statement, the reader should realize that the angular momentum representation is the eigenrepresentation of a natural observable of the kicked rotator, not merely a representation arrived at through an incomplete diagonalization. Banded random matrices will be considered separately in Chap. 10.

It is tempting but perhaps a bit speculative to associate the band structure of F (a near diagonality) with a weakly broken symmetry, rotational invariance. For vanishing kick strength, full rotational symmetry reigns and implies conservation of angular momentum. With strong kicking present, i.e., under conditions of global classical chaos, rotational invariance holds in a weaker sense. The phase variable, effective only modulo 2π, varies randomly from kick to kick, does not tend to display kick-to-kick correlations, and shows no preference for any particular value.

In the following I shall discuss some peculiarities of the dynamics of the kicked rotator related to localization. Particular emphasis will be placed on the analogy of the kicked rotator to Anderson's hopping model in one dimension. Moreover, I shall contrast the kicked rotator with the kicked top. The latter system does not display localization except in a very particular limit in which the top actually reduces to the rotator. The absence of quantum localization from the generic top, it will be seen, is paralleled by the absence of any time-scale separation for its two classical phase-space coordinates. Correspondingly, there is no slow, nearly conserved variable.

No justice can be done to the huge literature on localization, not even as regards spectral statistics of systems with localization. The interested reader can pick up threads not followed here in Refs. [194–196] and in *Efetov's* and *Imry's* recent books [197, 198].

Before diving into our theoretical discussions, it is appropriate to mention that quantum localization has been observed in rather different types of experiments. Microwave ionization of Rydberg states was investigated by *Bayfield, Koch*, and co-workers for atomic hydrogen [199, 204] and by *Walther* and co-workers for rubidium, as reviewed in Ref. [205]. The degree of atomic excitation in these experiments was so high that one might expect and in some respects does find effectively classical behavior of the valence electron. Quasi-classical behavior would, roughly speaking, amount to diffusive growth of the energy toward the ionization threshold. What is actually found is a certain resistance to ionization relative to such classical expectation. Quantitative analysis reveals an analogy to the localization-imposed break of diffusion for the kicked rotator. For reviews on this topic, the reader is referred to Refs. [206, 207].

A second class of experiments realizing kicked-rotator behavior was suggested by *Graham, Schlautmann* and *Zoller* [208] and conducted by *Raizen* and co-workers, on cold (some micro-Kelvins) sodium atoms exposed to a standing-wave light field whose time dependence is monochromatic, apart from a periodic amplitude modulation [209–211]. The laser frequency is chosen sufficiently close to resonance with a single pair of atomic levels so that all other levels are irrelevant. The selected pair gives rise to a dipole interaction with the electric field of the standing laser wave. Thus, the translational motion of the atom takes place in a light-induced potential $U(x,t) = -K\cos(x - \lambda\sin\omega t)$ where ω is the modulation frequency, λ is a measure of the modulation amplitude, and K is proportional to the squared dipole matrix element. The periodically driven motion in that potential is again characterized by localization-limited diffusion of the momentum p.

An account of the physics involved in the quantum optical experiments on localization is provided by *Stöckmann's* recent book [212]. Experiments and the theory of localization in disordered electronic systems are reviewed in Ref. [213].

7.2 Localization in Anderson's Hopping Model

In preparation for treating the quantum mechanics of the periodically kicked rotator, we now discuss Anderson's model of a particle whose possible locations are the equidistant sites of a one-dimensional chain. At each site, a random potential T_m acts and hopping of the particle from one site to its rth neighbor is described by a hopping amplitude W_r. The probability amplitude u_m for finding the particle on the mth site obeys the Schrödinger equation [214–216]

$$T_m u_m + \sum_r W_r u_{m+r} = E u_m \ . \tag{7.2.1}$$

If the potential T_m were periodic along the chain with a finite period q such that $T_{m+q} = T_m$, the solutions of (7.2.1) would, as is well known, be Bloch functions, and the energy eigenvalues would form a sequence of continuous bands. However, of prime interest with respect to the kicked rotator is the case in which the T_m are random numbers, uncorrelated from site to site and distributed with a density $\varrho(T_m)$. The hopping amplitudes W_r, in contrast, will be taken to be non-random and will decrease fast for hops of increasing length r.

In the situation just described, the eigenstates of the Schrödinger equation (7.2.1) are exponentially localized. They may be labelled by their center ν such that

$$u_m^\nu \sim \mathrm{e}^{-|\nu-m|/l} \text{ for } |\nu - m| \to \infty \ . \tag{7.2.2}$$

The so-called localization length l is a function of the energy E and may also depend on other parameters of the model such as the hopping range and the density $\varrho(T_m)$. For the special case of Lloyd's model, defined by a Lorentzian density $\varrho(T_m)$ and hoppings restricted to nearest-neighbor sites, an exact expression for l will be derived in Sect. 7.4. Typical numerical results for the eigenfunctions u_m^ν are depicted in Fig. 7.1.

Numerical results for the energy eigenvalues E_ν indicate that in the limit of a long chain of sites ($N \gg 1$), a smooth average level density arises. Moreover, the level spacings turn out to obey an exponential distribution just as if the levels were statistically independent. It is easy to see, in fact, that the model must display level clustering rather than level repulsion due to the exponential localization of the eigenfunctions: two eigenfunctions with their centers separated by more than a localization length l have an exponentially small matrix element of the Hamiltonian and thus no reason to keep their energies apart.

A further aspect of localization is worthy of mention. The Hamiltonian matrix H_{mn} implicitly defined by (7.2.1) is a random matrix with independent random

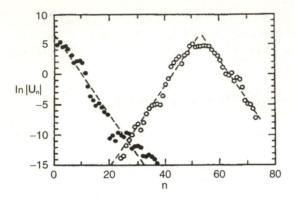

Fig. 7.1. Exponential localization of quasi-energy eigenfunctions of kicked rotator From [189]

diagonal elements $H_{mm} = T_m$ and with (noticeably) non-zero off-diagonal elements $H_{m,m+r} = W_r$ only in a band around the diagonal; the "bandwidth" is small compared to the dimension N of the matrix. Upon numerically diagonalizing banded random matrices, one typically finds localized eigenvectors and eigenvalues with Poissonian statistics. Both the small bandwidth and the randomness of H are essential for that statement to hold true: As we have seen in Chap. 4, full random matrices whose elements are independent to within the constraints of symmetry typically display level repulsion. On the other hand, any Hermitian matrix with repelling eigenvalues can be unitarily transformed so as to take a banded and even diagonal form, but then the matrix elements must bear correlations. We shall return to banded random matrices in Chap. 10 and show there that localization takes place if the bandwidth b is small in the sense $b^2/N < 1$.

An interesting argument for Anderson localization can be based on a theorem due to *Furstenberg* [217, 218]. This theorem deals with unimodular random matrices M_i (i.e., random matrices with determinants of unit modulus) and states that under rather general conditions

$$\lim_{Q \to \infty} \frac{1}{Q} \ln \text{Tr} \{M_Q M_{Q-1} \ldots M_2 M_1\} \equiv \gamma > 0 \,. \tag{7.2.3}$$

To apply (7.2.3) in the present context, we restrict the discussion to nearest neighbor hops ($W_{\pm 1} = W$, $W_r = 0$ for $|r| > 1$) and rewrite the Schrödinger equation (7.2.1) in the form of a map for a two-component vector,

$$\begin{pmatrix} u_{m+1} \\ u_m \end{pmatrix} = \begin{pmatrix} (E - T_m)/W & -1 \\ 1 & 0 \end{pmatrix} \begin{pmatrix} u_m \\ u_{m-1} \end{pmatrix} \equiv M_m \begin{pmatrix} u_m \\ u_{m-1} \end{pmatrix} \,. \tag{7.2.4}$$

The 2×2 matrix in this map is indeed unimodular, $\det M_n = 1$, and random due to the randomness of the potential T_m. Starting from the values of the wave function on two neighboring sites, say 0 and 1, one finds the wave function further to the left or to the right by appropriately iterating the map (7.2.4),

$$\begin{pmatrix} u_{m+1} \\ u_m \end{pmatrix} = M_m M_{m-1} \ldots M_2 M_1 \begin{pmatrix} u_1 \\ u_0 \end{pmatrix} \,. \tag{7.2.5}$$

Far away from the starting sites, i.e., for $|m| \to \infty$, Furstenberg's theorem applies and means that the unimodular matrix $M_m M_{m-1} \ldots M_2 M_1$ has the two eigenvalues $\exp(\pm m\gamma)$. It follows that almost all "initial" vectors u_1, u_0 will give rise to wave functions growing exponentially both to the left and to the right. Thus, the parameter γ is appropriately called the Lyapunov exponent of the random map under consideration. By special choices of u_0 and u_1, it is possible to generate wave functions growing only in one direction and decaying in the other. However, to enforce decay in both directions, special values of the energy E must be chosen. In other words, the eigenenergies of the one-dimensional Anderson model cannot form a continuum and in fact typically constitute a discrete spectrum. The corresponding eigenfunctions are exponentially localized and the Lyapunov exponent turns out to be the inverse localization length (we disregard the untypical exception of so-called singular continuous spectra).

An interesting consequence arises for the temporal behavior of wave packets. An initially localized packet is described by an effectively finite number of the localized eigenstates. As a sum of an effectively finite number of harmonically oscillating terms, the wave packet must be quasi-periodic in time. At any site m, the occupation amplitude $u_m(T)$ will get close to its initial value again and again as time elapses. A particle initially placed at, say, $m = 0$ such that $u_m(0) = \delta_{m,0}$ may, while hopping to the left and right in the random potential, display diffusive behavior $\langle (\Delta m)^2 \rangle \sim t$. As soon as a range of the order of the localization length has been explored, however, the diffusive growth of the spread $\langle (\Delta m)^2 \rangle$ must give way to manifestly quasi-periodic behavior.

7.3 The Kicked Rotator as a Variant of Anderson's Model

The periodically kicked rotator, one of the simplest and best investigated models capable of displaying classical chaos, has a single pair of phase-space variables, an angle Θ and an angular momentum p. As quantum operators, these variables obey the canonical commutation rule

$$[p, \Theta] = \frac{\hbar}{i} \; . \tag{7.3.1}$$

Their dynamics is generated by the Hamiltonian [219]

$$H(t) = \frac{p^2}{2I} + \lambda \frac{I}{\tau} V(\Theta) \sum_{n=-\infty}^{+\infty} \delta(t - n\tau) \; , \tag{7.3.2}$$

where I is the moment of inertia, τ the kicking period, and λ a dimensionless kicking strength. The dimensionless potential $V(\Theta)$ must be 2π-periodic in Θ. We shall deal for the most part with the special case

$$V(\Theta) = \cos \Theta \; . \tag{7.3.3}$$

Due to the periodic kicking, a stroboscopic description is appropriate with the Floquet operator

$$F = e^{-i\lambda(I/\tau\hbar)V} e^{-ip^2\tau/2I\hbar} \tag{7.3.4}$$

which accounts for the evolution from immediately after one kick to immediately after the next. The wave vector at successive instants of this kind obeys the stroboscopic map

$$\Psi(t+1) = F\Psi(t) . \tag{7.3.5}$$

Equivalently, the Heisenberg-picture operators obey

$$X_{t+1} = F^\dagger X_t F . \tag{7.3.6}$$

In (7.3.5, 6) and throughout the remainder of this chapter, the time variable t is a dimensionless integer counting the number of periods executed. Special cases of (7.3.6) are the discrete-time Heisenberg equations

$$p_{t+1} = p_t - \lambda \frac{I}{\tau} V'(\Theta_{t+1}) ,$$
$$\Theta_{t+1} = \Theta_t + \frac{\tau}{I} p_t , \tag{7.3.7}$$

which follow from the commutator relation (7.3.1) (Problem 7.1). Due to the absence of products of p and Θ, the recursion relations (7.3.7) also hold classically as discrete-time Hamiltonian equations.

A word about parameters and units is now in order. The parameter I/τ has the dimension of an angular momentum. Upon introducing a dimensionless angular momentum $P = p\tau/I$, the stroboscopic equations of motion (7.3.7) simplify to reveal the dimensionless kicking strength λ as the only control parameter of classical dynamics. On the other hand, in the commutator relation $[P, \Theta] = \hbar\tau/I$i and also in the Floquet operator F, the dimensionless version of Planck's constant $\hbar\tau/I$ appears as a second parameter controlling the effective strength of quantum fluctuations. With these remarks in mind, we may simplify the notation and henceforth set τ and I equal to unity.

It is appropriate first to comment on the classical behavior of the rotator. The period-to-period increment of angle Θ becomes greater, the larger the chosen kick strength λ. For sufficiently large λ, the angle may in fact jump by several 2π, without $\Theta_{t+1}(\mathrm{mod}\, 2\pi)$ showing any obvious preference for any part of the interval $[0, 2\pi]$. It follows that the momentum (which reacts only to $\Theta_{t+1}(\mathrm{mod}\, 2\pi)$) tends to undergo a random motion for strong kicking. In a rough idealization, we may assume that a series of successive values of Θ uniformly fills the interval $[0, 2\pi]$ and keeps no kick-to-kick memory; thus, the force $-V'(\Theta)$ in (7.3.7) appears as noise with no kick-to-kick correlation. Then, the map (7.3.7) can be approximated by

$$p_t = p_0 - \sum_{\nu=1}^{t} \lambda V'(\Theta_\nu) \tag{7.3.8}$$

and yields a constant mean momentum,

$$\overline{p_t} = \overline{p_0} - \sum_{\nu=1}^{t} \lambda \overline{V'(\Theta_\nu)} = \overline{p_0} \, , \tag{7.3.9}$$

since the mean force

$$\overline{V'(\Theta)} = \frac{1}{2\pi} \int_0^{2\pi} d\Theta \, V'(\Theta) = 0 \tag{7.3.10}$$

vanishes for a periodic potential. Similarly, the mean-squared momentum,

$$\begin{aligned}
\overline{p_t^2} &= \overline{p_0^2} + \sum_{\mu,\nu=1}^{t} \lambda^2 \overline{V'(\Theta_\mu)V'(\Theta_\nu)} \\
&= \overline{p_0^2} + \left(\lambda^2 \frac{1}{2\pi} \int_0^{2\pi} d\Theta \, V'(\Theta)^2 \right) t \, ,
\end{aligned} \tag{7.3.11}$$

grows linearly with time inasmuch as the force displays no kick-to-kick correlations,

$$\overline{V'(\Theta_\mu)V'(\Theta_\nu)} = \delta_{\mu\nu} \overline{V'(\Theta)^2} \, . \tag{7.3.12}$$

Evidently, the crude idealization of the dynamics (7.3.7) for $\lambda \gg 1$ by a random process suggests diffusive behavior of the (angular) momentum with the diffusion constant

$$D = \frac{\lambda^2}{2\pi} \int_0^{2\pi} d\Theta \, V'(\Theta)^2 \, . \tag{7.3.13}$$

In particular, for $V(\Theta) = \cos \Theta$,

$$D = \frac{\lambda^2}{2} \, . \tag{7.3.14}$$

In fact, by numerically iterating Chirikov's standard map [219]

$$\begin{aligned}
p_{t+1} &= p_t + \lambda \sin \Theta_{t+1} \\
\Theta_{t+1} &= (\Theta_t + p_t) \bmod (2\pi)
\end{aligned} \tag{7.3.15}$$

for an ensemble of initial angles Θ_0 and vanishing initial momentum, one does find the expected diffusive behavior if λ is chosen sufficiently large, $\lambda \gtrsim 1.5$. Figure 7.2 depicts this numerical evidence.

The effectively diffusive behavior of the momentum is tantamount to chaos, as follows from the linearized version of the standard map that describes the fate of infinitesimally close phase-space points under iteration,

$$\begin{pmatrix} \delta p_{t+1} \\ \delta \Theta_{t+1} \end{pmatrix} = \begin{pmatrix} 1 + \lambda \cos \Theta_{t+1} & \lambda \cos \Theta_{t+1} \\ 1 & 1 \end{pmatrix} \begin{pmatrix} \delta p_t \\ \delta \Theta_t \end{pmatrix} \equiv M_t \begin{pmatrix} \delta p_t \\ \delta \Theta_t \end{pmatrix} . \tag{7.3.16}$$

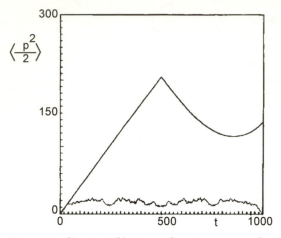

Fig. 7.2. Classical (*full curve*) and quantum (*dotted*) mean kinetic energy of the periodically kicked rotator, determined by numerical iteration of the respective maps. The quantum mean follows the classical diffusion for times up to some "break" time and then begins to display quasi-periodic fluctuations. After 500 kicks, the direction of time was reserved; while the quantum mean accurately retraces its history, the classical mean reverts to diffusive growth thus revealing the extreme sensitivity of chaotic systems to tiny perturbations (here round-off errors). Courtesy of Dittrich and Graham [352]

Inasmuch as Θ_t is effectively random, M_t is a unimodular random matrix. Furstenberg's theorem again applies (now in a classical context, in contrast to the application in Sect. 7.2) and secures a positive Lyapunov exponent. Exponential separation of classical trajectories, i.e., chaos, is manifest.

Now, we turn to the quantum version of the kicked rotator. Slightly changing notation, we rewrite the quantum map (7.3.5) as

$$|\Psi^+(t+1)\rangle = F|\Psi^+(t)\rangle , \quad t = 0, 1, 2, \ldots . \tag{7.3.17}$$

Instead of looking at the wave vector $|\Psi^+(t)\rangle$ immediately after the tth kick, one may study the wave vector just before that kick

$$|\Psi^-(t)\rangle = e^{-iH_0/\hbar}|\Psi^+(t-1)\rangle , \quad H_0 = \frac{p^2}{2} . \tag{7.3.18}$$

To solve Schrödinger's equation for either $|\Psi^+(t)\rangle$ or $|\Psi^-(t)\rangle$, a representation must now be chosen.

The free motion in between kicks is most conveniently described in the basis constituted by the eigenvectors of p,

$$p|n\rangle = n\hbar|n\rangle ,$$
$$n = 0, \pm 1, \pm 2, \ldots , \tag{7.3.19}$$

since the Hamiltonian is diagonal in that representation:

$$H_0|n\rangle = \frac{\hbar^2}{2}n^2|n\rangle . \tag{7.3.20}$$

By expanding as

$$|\Psi^{\pm}(t)\rangle = \sum_n \Psi_n^{\pm}(t)|n\rangle \, , \qquad (7.3.21)$$

one finds that the free motion is described by

$$\Psi_n^-(t+1) = e^{-i\hbar n^2/2}\Psi_n^+(t) \, . \qquad (7.3.22)$$

An individual kick, on the other hand, is most easily dealt with in the Θ representation,

$$|\Psi^{\pm}(t)\rangle = \int_0^{2\pi} d\Theta \, \Psi^{\pm}(\Theta, t)|\Theta\rangle \, , \qquad (7.3.23)$$

since

$$\Psi^+(\Theta, t) = e^{-i\lambda V(\Theta)/\hbar}\Psi^-(\Theta, t) \, . \qquad (7.3.24)$$

Combining the two steps in the p representation, one encounters the map

$$\Psi_m^+(t+1) = \sum_{n=-\infty}^{+\infty} J_{m-n}e^{-i\hbar n^2/2}\Psi_n^+(t) \, ,$$
$$J_{m-n} = \frac{1}{2\pi}\int_0^{2\pi} d\Theta \, e^{i(m-n)\Theta}e^{-i\lambda V(\Theta)/\hbar} \, . \qquad (7.3.25)$$

For the special case $V = \cos\Theta$, the matrix element of the unitary kick operator becomes the Bessel function

$$J_n(z) = \frac{1}{\pi i^n}\int_0^{\pi} d\Theta \, e^{iz\cos\Theta}\cos n\Theta \, . \qquad (7.3.26)$$

To establish a relation between the kicked rotator and Anderson's model [189], we must consider the eigenvalue problem for the Floquet operator F,

$$F|u^+\rangle = e^{-i\phi}|u^+\rangle \, , \qquad (7.3.27)$$

which, in the momentum representation, reads

$$\sum_n J_{m-n}e^{-i\hbar n^2/2}u_n^+ = e^{-i\phi}u_m^+ \, . \qquad (7.3.28)$$

The corresponding problem with the order of kick and free motion reversed will also play a role; the corresponding eigenvectors $|u^-\rangle$ are most easily related to the $|u^+\rangle$ in the Θ representation,

$$u^-(\Theta) = e^{i\lambda V(\Theta)/\hbar}u^+(\Theta) \, , \qquad (7.3.29)$$

where $u^-(\Theta)$ and $u^+(\Theta)$ pertain to the same eigenphase ϕ of $e^{-iH_0/\hbar}e^{-i\lambda V/\hbar}$ and $F = e^{-i\lambda V/\hbar}e^{-iH_0/\hbar}$, respectively. With the help of (7.3.27), the relation (7.3.29) can be rewritten as

$$u^-(\Theta) = e^{i(\phi - H_0/\hbar)}u^+(\Theta) \qquad (7.3.30)$$

and thus in the p representation

$$u_m^- = e^{i(\phi - \hbar m^2/2)} u_m^+ . \tag{7.3.31}$$

Now, we represent the unitary kick operator in terms of a Hermitian operator W,

$$e^{-i\lambda V/\hbar} = \frac{1 + iW}{1 - iW} , \quad W = -\tan\frac{\lambda V}{2\hbar} , \tag{7.3.32}$$

and define the vector

$$|u\rangle = \tfrac{1}{2}\left(|u^+\rangle + |u^-\rangle\right) . \tag{7.3.33}$$

From (7.3.29), this vector is obtained as

$$u(\Theta) = \frac{u^+(\Theta)}{1 + iW(\Theta)} = \frac{u^-(\Theta)}{1 - iW(\Theta)} \tag{7.3.34}$$

and together with (7.3.31) yields an integral equation for the function $u(\Theta)$,

$$[1 - iW(\Theta)]\, u(\Theta) = e^{i(\phi - H_0/\hbar)}\, [1 + iW(\Theta)]\, u(\Theta) . \tag{7.3.35}$$

This equation assumes an algebraic appearance, however, when expressed in the p representation,

$$T_m u_m + \sum_{r(\neq 0)} W_r u_{m+r} = E u_m , \tag{7.3.36}$$

where we have introduced

$$T_m = i\frac{1 - e^{i(\phi - \hbar m^2/2)}}{1 + e^{i(\phi - \hbar m^2/2)}} = \tan\left(\frac{\phi - \hbar m^2/2}{2}\right) , \tag{7.3.37}$$

$$E = -W_0 .$$

Now, the desired relationship between Anderson's hopping model and the kicked rotator is established: the algebraic equation (7.3.36) for $u_m = \langle m|u\rangle$ does indeed have the form of the Schrödinger equation (7.2.1) for a particle on a lattice, where T_m is a single-site potential and W_r is a hopping amplitude. Interestingly, the quasi-energy ϕ of the rotator has become a parameter in the potential T_m while the zeroth Fourier component W_0 of the function $W(\Theta)$ has formally taken on the role of the energy of the hopping particle.

A subtle difference between the kicked rotator and the Anderson model deserves discussion. The amplitudes T_m defined in (7.3.37) are not strictly random but only pseudorandom numbers. In fact, since the quantities $(\phi - \hbar m^2/2)$ enter T_m as the argument of a tangent, they become effective modulo π. According to a theorem of Weyl's [220], the sequence $(\phi - \hbar m^2/2) \bmod \pi$ is ergodic in the interval $[0, \pi]$ and covers that interval with uniform density as $\pm m = 0, 1, 2, \ldots$.

It follows that the T_m have a density $W(T)dT = d\phi/\pi$. With $dT/d\phi = 1 + T^2$, one finds

$$W(T_m) = \frac{1}{\pi(1 + T_m^2)} ,$$
(7.3.38)

i.e., a Lorentzian distribution. Nevertheless, the T_m are certainly not strictly independent from one value of m to the next. Therefore, it may be appropriate to speak of the kicked rotator in terms of a pseudo-Anderson model or, actually, a pseudo-Lloyd model since the Anderson model with truly random T_m distributed according to a Lorentzian is known as Lloyd's model [221].

There is another difference between the simplest version of the Anderson model and the simplest rotator. Rather than allowing for hops to nearest neighbor sites, the kicking potential $V(\Theta) = \cos \Theta$ gives rise to

$$W(\Theta) = -\tan\left(\frac{\lambda \cos \Theta}{2\hbar}\right)$$
(7.3.39)

with Fourier components

$$W_r = -\int_0^{2\pi} \frac{d\Theta}{2\pi} e^{ir\Theta} \tan\left(\frac{\lambda \cos \Theta}{2\hbar}\right) .$$
(7.3.40)

These hopping amplitudes can be calculated in closed form for $\lambda < \pi$ and, it may be shown, fall off exponentially with increasing r.

Even though nobody has proved localization for the kicked rotator, the pseudo-randomness of the potential T_m and the finite range of the hopping amplitude W_r strongly suggest such behavior. Moreover, all numerical evidence available favors an effective equivalence of the Anderson model and the kicked rotator.

7.4 Lloyd's Model

It may be well to digress a little and add some hard facts to the localization lore. Anderson's Schrödinger equation allows an exact evaluation of the localization length l provided that the hopping amplitude is non-zero only for nearest neighbor hops and that the potential T_m is independent from site to site and distributed according to a Lorentzian [221],

$$T_m u_m + \frac{\kappa}{2}(u_{m-1} + u_{m+1}) = E u_m ,$$
(7.4.1)

$$\varrho(T_m) = \frac{1}{\pi(1 + T_m^2)} .$$
(7.4.2)

Now, we shall prove[1] the following well-known result [216, 222] for the ensemble-averaged localization constant $\overline{\gamma} \equiv 1/l$,

$$\cosh \overline{\gamma} = \frac{1}{2\kappa}\left[\sqrt{(E - \kappa)^2 + 1} + \sqrt{(E + \kappa)^2 + 1}\right]$$
(7.4.3)

which implies $\overline{\gamma} > 0$ for all values of the energy E.

---

[1] I am indebted to H.J. Sommers for showing this proof to me.

The starting point is the tridiagonal Hamiltonian matrix

$$H_{mn} = T_m \delta_{mn} + \frac{\kappa}{2} \left(\delta_{m,n+1} + \delta_{m,n-1} \right) \tag{7.4.4}$$

inserted in the Green function

$$G_{mn} = \frac{1}{N} \left(\frac{1}{E-H} \right)_{mn} = \frac{1}{N} (-1)^{m-n} \frac{\det_{nm}(E-H)}{\det(E-H)} . \tag{7.4.5}$$

The determinant $\det_{mn}(E-H)$ is obtained from $\det(E-H)$ by cancelling the nth row and the mth column. Assuming a finite number N of sites, the element G_{1N}, i.e., the one referring to the beginning and end of the chain of sites, takes an especially simple form since

$$\det_{1N}(E-H) = \left(-\frac{\kappa}{2} \right)^{N-1} . \tag{7.4.6}$$

It is an obvious result of the tridiagonality of H_{mn} that the element G_{1N} reads simply

$$G_{1N} = \frac{1}{N} \left(+\frac{\kappa}{2} \right)^{N-1} / \prod_{\nu=1}^{N} (E - E_\nu) . \tag{7.4.7}$$

On the other hand, by invoking the spectral representation of Green's function,

$$G_{mn} = \frac{1}{N} \sum_\nu \frac{u_m^\nu u_n^\nu}{E - E_\nu} , \tag{7.4.8}$$

and comparing the residues of the pole at $E = E_\nu$ in (7.4.7) and (7.4.8), one obtains

$$u_1^\nu u_N^\nu = \left(+\frac{\kappa}{2} \right)^{N-1} / \prod_{\mu(\neq\nu)} (E_\nu - E_\mu) . \tag{7.4.9}$$

This latter expression lends itself to a determination of the localization length for the νth eigenvector. If that eigenvector were extended like a Bloch state, one would have $u_1^\nu u_N^\nu \sim 1/N$, i.e., $|u_1^\nu u_N^\nu|^{1/N} \sim (1/N)^{1/N} \sim \exp\left(-(\ln N)/N \right) \to 1$. For an exponentially localized state, however, $u_1^\nu u_N^\nu \sim A e^{-\gamma_\nu N}$ and thus $|u_1^\nu u_N^\nu|^{1/N} \sim |A|^{1/N} e^{-\gamma_\nu} \to e^{-\gamma_\nu} < 1$. Therefore, the localization length $1/\gamma_\nu$ can be obtained as

$$\gamma_\nu = \frac{1}{N} \sum_{\mu(\neq\nu)} \ln|E_\nu - E_\mu| - \ln\frac{\kappa}{2} . \tag{7.4.10}$$

The limit $N \to \infty$ is implicitly understood, so we may replace the sum in (7.4.10) by an integral with the level density $\varrho(E)$ as a weight

$$\gamma(E) = \int dx\, \varrho(x) \ln|E - x| - \ln\frac{\kappa}{2} , \tag{7.4.11}$$

with the normalization of $\varrho(x)$ chosen as that of a probability density,

$$\int_{-\infty}^{+\infty} dx\, \varrho(x) = 1 \,. \tag{7.4.12}$$

More easily accessible than $\gamma(E)$ itself is its derivative

$$\gamma'(E) = \mathcal{P} \int dx\, \frac{\varrho(x)}{E-x} \,, \tag{7.4.13}$$

since the principal-value integral occurring here is easily related to Green's function. Indeed, the spectral representation (7.4.8) entails

$$\mathrm{Tr}\left\{G(E)\right\} = \frac{1}{N} \sum_\nu \frac{1}{E-E_\nu} \to \int dx\, \frac{\varrho(x)}{E-x} \,. \tag{7.4.14}$$

The continuum approximation endowed Green's function with a cut along the real axis in the complex energy plane, so different limiting values of G arise immediately above and below the real energy axis[2]

$$\mathrm{Tr}\left\{G(E \pm \mathrm{i}0^+)\right\} = \mathcal{P} \int dx\, \frac{\varrho(x)}{E-x} \mp \mathrm{i}\pi\, \varrho(E) \,. \tag{7.4.15}$$

Usually, one is interested in the imaginary part of this equation which yields a strategy for calculating the level density (Sect. 4.6). Here, however, we compare the real part with (7.4.13) and conclude that

$$\gamma'(E) = \mathrm{Re}\left\{\mathrm{Tr}\left\{G(E - \mathrm{i}0^+)\right\}\right\} \,. \tag{7.4.16}$$

The latter identity holds for any configuration of the potential T_m. Averaging over all configurations $\{T_m\}$ with the Lorentzian density (7.4.2) gives [222]

$$\overline{\gamma}'(E) = \mathrm{Re}\left\{\mathrm{Tr}\left\{\overline{G}(E \pm \mathrm{i}0^+)\right\}\right\} \,, \tag{7.4.17}$$

a result often referred to as *Thouless' formula*. Due to the assumed site-to-site independence,

$$\overline{(\ldots)} = \left(\prod_m \int_{-\infty}^{+\infty} dT_m\, \frac{1}{\pi(1+T_m^2)} \right) (\ldots) \,. \tag{7.4.18}$$

To carry out the average of $G(E)$, it is convenient to represent Green's function by a multiple Gaussian integral,

$$\left(\prod_{i=1}^N \prod_{\alpha=1}^n int_{-\infty}^{+\infty} dS_i^\alpha \right) S_p^1 S_q^1 \exp\left[-\mathrm{i} \sum_{ij\alpha} (E - H - \mathrm{i}0^+)_{ij} S_i^\alpha S_j^\alpha \right]$$

$$= \frac{N}{2} G_{pq}(E - \mathrm{i}0^+) \left[\frac{\pi^N}{\det(E - H - \mathrm{i}0^+)} \right]^{n/2} \,. \tag{7.4.19}$$

---

[2] Here and in the following 0^+ denotes an arbitrarily small positive number.

An especially useful identity arises from (7.4.19) when the parameter n, after performing the integral, is elevated from integer to continuous real and then sent to zero,

$$NG_{pq}(E - i0^+) = \lim_{n \to 0} 2 \left(\prod_{i=1}^{N} \prod_{\alpha=1}^{n} \int dS_i^\alpha \right) S_p^1 S_q^1$$

$$\times \exp\left[-i \sum_{ij\alpha} (E - H - i0^+)_{ij} S_i^\alpha S_j^\alpha \right] . \tag{7.4.20}$$

This representation of the inverse of the matrix $(E - H - i0^+)_{pq}$ by a multiple Gaussian integral with the subsequent manipulation of the parameter n is known in the theory of disordered spin systems as the "replica trick" [223]. Its virtue is that the random numbers T_m now appear in a product (of exponentials), each factor involving a single T_m. Therefore, the average (7.4.18) can be taken site by site according to

$$\overline{\exp\left(i \sum_j T_j S_j^\alpha S_j^\alpha \right)} = \prod_{j=1}^{N} \int dT_j \frac{\exp\left[iT_j S_j^\alpha S_j^\alpha \right]}{\pi(1 + T_j^2)}$$

$$= \exp\left[-\sum_j S_j^\alpha S_j^\alpha \right] . \tag{7.4.21}$$

The average Green function then takes the simple form

$$N\overline{G}_{pq}(E - i0^+) = \lim_{n \to 0} \left(2 \prod_i \prod_\alpha \int dS_i^\alpha \right) S_p^1 S_q^1$$

$$\times \exp\left[-i \sum_{ij\alpha} (E - \tilde{H} - i0^+)_{ij} S_i^\alpha S_j^\alpha \right] \tag{7.4.22}$$

with the matrix

$$\tilde{H}_{pq} = i\delta_{pq} + \frac{\kappa}{2} (\delta_{p,q+1} + \delta_{p,q-1}) . \tag{7.4.23}$$

It is remarkable that $\tilde{H}_{pq}$ differs from the original Hamiltonian only by the replacement $T_p \to i$. At any rate, one can now read the multiple integral (7.4.21) backwards as a representation of $(E - \tilde{H} - i0^+)^{-1}$, i.e.,

$$\left(N^{-1}\overline{G}^{-1} \right)_{pq} = (E - i)\delta_{pq} - \frac{\kappa}{2} (\delta_{p,q+1} + \delta_{p,q-1}) . \tag{7.4.24}$$

With the average over the random potential done, it remains to invert the tridiagonal matrix (7.4.23). To that end, we again employ (7.4.5) and write

$$\text{Tr}\left\{ N\overline{G} \right\} = \sum_m \frac{\det_{mm}(E - \tilde{H})}{\det (E - \tilde{H})} = \frac{\partial}{\partial E} \ln \det(E - \tilde{H}) \tag{7.4.25}$$

where in the last step we have exploited the fact that the energy E enters only the diagonal elements of $\det(E - \tilde{H})$. Indeed, simplifying the notation for a moment in an obvious manner and expanding $\det(E - \tilde{H})$ along the first row,

$$\partial \det = \det{}_{11} + (E - \mathrm{i})\partial \det{}_{11} - \frac{\kappa}{2}\partial \det{}_{12} = \det{}_{11} + \partial_1 \det \tag{7.4.26}$$

where the differentiation ∂_1 has to leave the first diagonal element untouched; similarly expanding along the second row gives $\partial_1 \det = \det_{22} + \partial_{12} \det$ where ∂_{12} now has to leave the first two diagonal elements of the determinant undifferentiated; after N such steps, one arrives at

$$\partial \det = \sum_{p=1}^{N} \det{}_{pp} \ .$$

Finally, we must evaluate the $N \times N$ determinant

$$D_N \equiv \det(E - \tilde{H}) \ . \tag{7.4.27}$$

That goal is most easily achieved recursively since

$$\begin{aligned} D_1 &= E - \mathrm{i} \\ D_2 &= (E - \mathrm{i})^2 - \frac{\kappa^2}{4} \\ D_n &= (E - \mathrm{i})D_{n-1} - \frac{\kappa^2}{4}D_{n-2} \ . \end{aligned} \tag{7.4.28}$$

These relations obviously allow the extension $D_0 = 1$. Since the recursion relation in (7.4.28) has coefficients independent of n, it can be solved by the ansatz $D_n \sim x^n$ where x is determined by the quadratic equation

$$x^2 - (E - \mathrm{i})x + \frac{\kappa^2}{4} = 0 \ . \tag{7.4.29}$$

The two solutions

$$x_\pm = \tfrac{1}{2}\left\{ E - \mathrm{i} \pm \sqrt{(E - \mathrm{i})^2 - \kappa^2} \right\} \tag{7.4.30}$$

together with the "initial" conditions $D_0 = 1$, $D_1 = (E - \mathrm{i})$ give

$$\det(E - \tilde{H}) = \frac{x_+^{N+1} - x_-^{N+1}}{x_+ - x_-} \ . \tag{7.4.31}$$

Combining (7.4.16, 7.4.25, 7.4.31),

$$\overline{\gamma'(E)} = \frac{\partial}{\partial E} \operatorname{Re}\left\{ \frac{1}{N} \ln\left(\frac{x_+^{N+1} - x_-^{N+1}}{x_+ - x_-} \right) \right\} \ . \tag{7.4.32}$$

This result simplifies considerably in the limit $N \to \infty$. Since $|x_+| > |x_-|$,

$$\overline{\gamma'(E)} = \frac{\partial}{\partial E} \ln |x_+| \ . \tag{7.4.33}$$

Integrating and fixing the integration constant by comparing with (7.4.10) for $E \to \infty$ gives

$$\overline{\gamma}(E) = \ln \left| \frac{E - \mathrm{i}}{\kappa} + \sqrt{\left(\frac{E - \mathrm{i}}{\kappa} \right)^2 - 1} \right| . \tag{7.4.34}$$

A little elementary algebra finally brings the localization constant to the form (7.4.3).

7.5 The Classical Diffusion Constant as the Quantum Localization Length

It was argued in Sect. 7.3 that the classical map

$$p_{t+1} = p_t + \lambda V'(\Theta_{t+1})$$
$$\Theta_{t+1} = (\Theta_t + p_t) \bmod (2\pi) \tag{7.5.1}$$

may, for sufficiently strong kicking λ, be simplified to

$$p_{t+1} = p_t + \text{noise} , \tag{7.5.2}$$

where the noise imparted to the momentum is uncorrelated from period to period. The ensuing diffusive behavior of the momentum,

$$\langle p_t^2 \rangle - \langle p_0^2 \rangle = Dt , \quad \langle p_t \rangle = \langle p_0 \rangle , \tag{7.5.3}$$

with the diffusion constant [219]

$$D = \frac{\lambda^2}{2\pi} \int_0^{2\pi} d\Theta \, V'(\Theta)^2 \tag{7.5.4}$$

is in fact in good agreement with numerical data obtained by following the bundle of solutions of (7.5.1) originating from a cloud of many initial points; see Fig. 7.2.

For the quantum rotator, however, diffusive growth of the squared momentum can prevail only until the spread of the wave packet along the momentum axis has reached the localization length l. Afterward, $\langle p_t^2 \rangle$ must display quasi-periodicity in time. The transition from diffusive to manifestly quasi-periodic behavior, depicted in Fig. 7.2, will take place at a "break time" t^*, whose order of magnitude may be estimated by

$$\langle (\Delta p_t)^2 \rangle = Dt^* = l^2 \hbar^2 . \tag{7.5.5}$$

An independent estimate for t^* or l is needed to complement (7.5.4) and (7.5.5) for a complete set of equations for D, t^*, and l. *Chirikov* et al. [224] have found a third relation that will be explained now. A quasi-periodic wave packet is spanned by about l basis states, for example, eigenstates of either the momentum

p or the Floquet operator F. The corresponding quasi-energies, roughly l in number, all lie in the interval $[0, 2\pi]$ and thus have a mean spacing of the order $2\pi/l$. The inverse of this spacing is the minimum time needed for the discreteness of the spectrum to manifest itself in the time dependence of the wave packet. Clearly, the time in question may be identified with the break time t^*, whereupon one has the important order-of-magnitude result

$$t^* \approx l \approx \hbar^{-2} D \approx \hbar^{-2} \frac{\lambda^2}{2\pi} \int_0^{2\pi} d\Theta\, V'(\Theta)^2 \ . \tag{7.5.6}$$

Shepelyansky [225] has numerically verified this result for the kicked rotator with $\lambda V(\Theta) = \lambda \cos \Theta$, and the kicking strength ranging over the interval $1.5 \leq \lambda \leq 29$.

While the above estimate of t^* certainly implies that $\langle p_t^2 \rangle$ must display quasi-periodicity for $t > t^*$, it does not explicitly suggest that $\langle p_t^2 \rangle$ precisely follows classical diffusion for $0 \leq t \lesssim t^*$. Clearly, for classical behavior to prevail at early times the, dimensionless version of $\hbar$ must be sufficiently small.

7.6 Absence of Localization for the Kicked Top

It should be pointed out that not all periodically kicked systems display localization analogous to that of Anderson's model. An interesting case is provided by kicked tops [226–229] which have already been alluded to several times and will be discussed now somewhat more systematically. Kicked tops have the three components of an angular momentum vector $\boldsymbol{J}$ as their only dynamic variables. In their Floquet operators $F = e^{-i\lambda V} e^{-iH_0}$ both H_0 and V are polynomials in $\boldsymbol{J}$ such that the squared angular momentum is a constant of the motion,[3]

$$\boldsymbol{J}^2 = j(j+1) \ , \quad j = \tfrac{1}{2}, 1, \tfrac{3}{2}, \ldots \ . \tag{7.6.1}$$

The classical limit is attained when the quantum number becomes large, $j \to \infty$.

The simplest model capable of chaotic motion in the classical limit is given by

$$F = e^{-i\lambda J_z^2/2j} e^{-i\alpha J_x} \ . \tag{7.6.2}$$

The second factor in this F clearly describes a linear precession of $\boldsymbol{J}$ around the x-direction by the angle α. Similarly, the first factor may be said to correspond to a nonuniform "rotation" around the z axis; instead of being a constant c number the rotational angle is itself proportional to J_z. The dimensionless coupling constant λ might now be called a torsion strength. The quantum number j appears in $V = J_z^2/2j$ to provide λV and $H_0 = \alpha J_z$ with the same asymptotic scaling with j when $j \to \infty$ for finite constant λ and α.

---

[3] While $\hbar\tau/I$ for the kicked rotator naturally arises as a dimensionless measure of Planck's constant, that role will be played by $1/j$ for the kicked top. Therefore, it is convenient to set $\hbar = 1$ in this section.

By using the angular momentum commutators

$$[J_i, J_j] = i\varepsilon_{ijk}J_k \, , \tag{7.6.3}$$

one finds the stroboscopic Heisenberg equations of motion $\boldsymbol{J}_{t+1} = F^\dagger \boldsymbol{J}_t F$ in the form

$$
\begin{aligned}
J_{x,t+1} &= \left\{ \tilde{J}_x, \cos\frac{\lambda}{j}\left(\tilde{J}_z - \frac{1}{2}\right) \right\} - \left\{ \tilde{J}_y, \sin \ldots \right\} \\
&\quad + \frac{i}{2}\left[\cos \ldots, \tilde{J}_y\right] + \frac{i}{2}\left[\sin \ldots, \tilde{J}_x\right] \, , \\
J_{y,t+1} &= \left\{ \tilde{J}_x, \sin\frac{\lambda}{j}\left(\tilde{J}_z - \frac{1}{2}\right) \right\} + \left\{ \tilde{J}_y, \cos \ldots \right\} \\
&\quad + \frac{i}{2}\left[\sin \ldots, \tilde{J}_y\right] - \frac{i}{2}\left[\cos \ldots, \tilde{J}_x\right] \, , \\
J_{z,t+1} &= \tilde{J}_z \, , \\
\tilde{J}_x &= J_{x,t} \, , \\
\tilde{J}_y &= J_{y,t}\cos\alpha - J_{z,t}\sin\alpha \, , \\
\tilde{J}_z &= J_{y,t}\sin\alpha + J_{z,t}\cos\alpha \, .
\end{aligned}
\tag{7.6.4}
$$

These clearly display the sequence $\boldsymbol{J}_t \to \tilde{\boldsymbol{J}}$, a linear precession by α around the x-axis and $\tilde{\boldsymbol{J}} \to \boldsymbol{J}_{t+1}$, a nonlinear precession around the z-direction. Since the latter rotation is by an angle involving the z component of the intermediate vector $\tilde{\boldsymbol{J}}$, symmetrized products as well as commutators of noncommuting operators occur in the corresponding equations in (7.6.4), $\{A, B\} = (AB + BA)/2$. In the classical limit, $j \to \infty$, the first two equations in (7.6.4) simplify inasmuch as the contribution $\lambda/2j$ to the rotational angle is negligible and the symmetrized operator products become ordinary products of the corresponding c number observables while the commutators disappear. Formally, the classical limit can be achieved by first rescaling the operators $\boldsymbol{J}$ as

$$\boldsymbol{X} = \frac{\boldsymbol{J}}{j} \tag{7.6.5}$$

whereupon the commutators (7.6.3) take the form

$$[X, Y] = \frac{i}{j}Z \, . \tag{7.6.6}$$

In the limit $j \to \infty$, the vector $\boldsymbol{X}$ tends to a unit vector with commuting components, and the stroboscopic equations of motion take the form already described above,

$$X_{t+1} = \tilde{X} \cos \lambda \tilde{Z} - \tilde{Y} \sin \lambda \tilde{Z}$$

$$Y_{t+1} = \tilde{X} \sin \lambda \tilde{Z} + \tilde{Y} \cos \lambda \tilde{Z}$$

$$Z_{t+1} = \tilde{Z}$$

$$\tilde{X} = X_t$$

$$\tilde{Y} = Y_t \cos \alpha - Z_t \sin \alpha$$

$$\tilde{Z} = Y_t \sin \alpha + Z_t \cos \alpha \,.$$

(7.6.7)

Due to the conservation law $\boldsymbol{X}^2 = 1$, the classical map (7.6.7) is two dimensional, expressible as two recursion relations for two angles defining the orientation of $\boldsymbol{X}$, for example,

$$X = \sin \Theta \cos \phi \,, \quad Y = \sin \Theta \sin \phi \,, \quad Z = \cos \Theta \,. \tag{7.6.8}$$

Actually, the surface of the unit sphere $\boldsymbol{X}^2 = 1$ is the phase space of the classical top with $Z = \cos \Theta$ and ϕ as canonical variables. This can be seen most easily by replacing the quantum commutators (7.6.6) with the classical Poisson brackets $\{X, Y\} = -Z$ and checking that the latter are equivalent to the canonical Poisson brackets $\{Z, \phi\} = 1$. Indeed, from $\{Z, \phi\} = 1$, one finds $\{f(Z), g(\phi)\} = f'(Z)g'(\phi)$, and this immediately gives $\{X, Y\} = -Z$ when the spherical representation (7.6.8) is used for X and Y.

The classical trajectories generated by the map (7.6.7) depend in their character on the values of the torsional strength λ and the precessional angle α. As shown in Fig. 7.3, the sphere $\boldsymbol{X}^2 = 1$ is dominated by regular motion for $\alpha = \pi/2$, $\lambda \lesssim 2.5$, whereas chaos prevails, at fixed $\alpha = \pi/2$, for $\lambda \gtrsim 3$. It is important to realize that as soon as chaos has become global with increasing torsional strength, the typical trajectory traverses the spherical phase space rapidly in time: a single kick suffices for the phase-space point to hop all around the sphere in any direction. Such stormy exploration is in blatant contrast to the behavior of the kicked rotator in its cylindrical phase space: While the cylinder may be surrounded in the direction of the angular coordinate once or even several times between two subsequent kicks, the momentum is capable of only slow quasi-random motion along the cylinder; the distance covered in the p-direction in a finite time is only a vanishing fraction of the infinite length of the cylinder.

Similarly striking is the difference in the quantum mechanical behavior of the top and of the rotator. The rotator displays localization along the one-dimensional angular momentum lattice, but the top, as we shall now proceed to show, does not. Rather, under conditions of classical chaos, wave packets pertaining to the top are in general limited in their spreads only by the finite size of the Hilbert space. To illustrate the typical behavior of the top, it is convenient to consider wave packets originating from coherent initial states.

Coherent states of an angular momentum [230, 231] with the quantum number j assign a direction to the observables $\boldsymbol{J}$ that can be characterized by angles

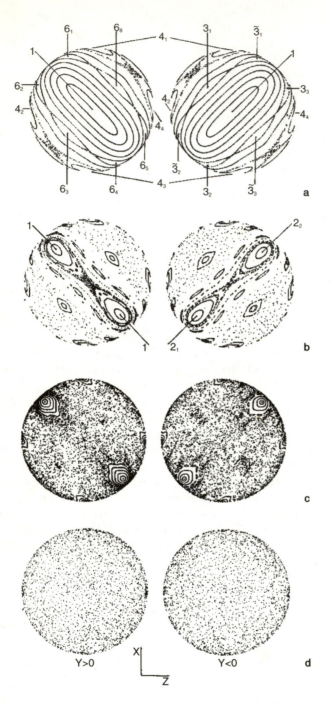

Fig. 7.3. Classical phase space portraits for the kicked top (7.6.7) with $\alpha = \pi/2$ and $\lambda = 2$ **(a)**; $\lambda = 2.5$ **(b)**; $\lambda = 3$ **(c)**; and $\lambda = 6$ **(d)**. Periodic orbits are marked by numerical labels

Θ and ϕ as

$$\langle\Theta\phi|J_z|\Theta\phi\rangle = j\cos\Theta$$
$$\langle\Theta\phi|J_x \pm \mathrm{i}J_y|\Theta\phi\rangle = je^{\pm\mathrm{i}\phi}\sin\Theta . \tag{7.6.9}$$

Clearly, the geometric meaning of these angles is the same as that of the angles used in (7.6.8) to specify the direction of the classical vector $\boldsymbol{X}$. However, due to the noncommutativity of the components J_i, it is with a finite precision only that the coherent state $|\Theta\phi\rangle$ defines an orientation. To find that precision, one may observe that two of the states $|\Theta\phi\rangle$ have $\Theta = 0$ and $\Theta = \pi$, i.e., $\langle J_z\rangle = \pm j$ and therefore can be identified with the joint eigenstates $|jm\rangle$ of $\boldsymbol{J}^2$ and J_z pertaining to $J_z = m = \pm j$ and $\boldsymbol{J}^2 = j(j+1)$. The relative variance of $\boldsymbol{J}$ with respect to these special coherent states is $(\langle\boldsymbol{J}^2\rangle - \langle\boldsymbol{J}\rangle^2)/j^2 = 1/j$. All other states $|jm\rangle$ with $m \neq \pm j$ have larger variances of $\boldsymbol{J}$ and so do all of their linear combinations (for fixed j) except the other coherent states $|\Theta\phi\rangle$. In fact, all coherent states $|\Theta\phi\rangle$ can be generated from the "polar" state $|\Theta = 0, \phi\rangle = |j,j\rangle$ by a rotation,[4]

$$|\Theta\phi\rangle = R(\Theta,\phi)|j,j\rangle , \tag{7.6.10}$$

$$\begin{aligned} R(\Theta,\phi) &= \exp\left[\mathrm{i}\Theta\left(J_x\sin\phi - J_y\cos\phi\right)\right] \\ &= e^{\gamma J_-}e^{-J_z\ln(1+\gamma\gamma^*)}e^{-\gamma J_+} , \end{aligned}$$

where $\gamma = e^{\mathrm{i}\phi}\tan(\Theta/2)$ and $J_\pm = J_x \pm \mathrm{i}J_y$. The variance of $\boldsymbol{J}$ remains unchanged under the rotation in question,

$$\left(\langle\Theta\phi|\boldsymbol{J}^2|\Theta\phi\rangle - \langle\Theta\phi|\boldsymbol{J}|\Theta\phi\rangle^2\right)/j^2 = \frac{1}{j} , \tag{7.6.11}$$

and indeed constitutes the minimum uncertainty of the orientation of $\boldsymbol{J}$ permitted by the angular-momentum commutators. A solid angle $\Delta\Omega = 1/j$ may be associated with the relative variance (7.6.11), and with respect to the classical unit sphere $\boldsymbol{X}^2 = 1$, a coherent state $|\Theta\phi\rangle$ may be represented by a spot of size $\Delta\Omega$ located at the point (Θ, ϕ). Such spots of the minimal size allowed quantum mechanically are often called Planck cells. In the classical limit $j \to \infty$, the spot in question shrinks to the classical phase-space point (Θ, ϕ).

Some further properties of the states $|\Theta\phi\rangle$ are worth mentioning for later reference. First, we should note the expansion in terms of the states $|jm\rangle$. It follows from (7.6.10) that

$$|\gamma\rangle \equiv |\Theta\phi\rangle = (1+\gamma\gamma^*)^{-j}e^{\gamma J_-}|j,j\rangle \tag{7.6.12}$$

and thus

$$\langle jm|\Theta\phi\rangle = (1+\gamma\gamma^*)^{-j}\gamma^{j-m}\sqrt{\binom{2j}{j-m}} . \tag{7.6.13}$$

---

[4] The reader will pardon the sloppiness of denoting the coherent state by $|\Theta\phi\rangle$ rather than $|j\Theta\phi\rangle$.

Therefore, the probability of finding $J_z = m$ in the coherent state $|\Theta\phi\rangle$ is given by the binomial distribution

$$|\langle jm|\Theta\phi\rangle|^2 = (1 + \gamma\gamma^*)^{-2j}(\gamma\gamma^*)^{j-m}\binom{2j}{j-m}, \tag{7.6.14}$$

which we shall use presently.

The expression (7.6.12) for the coherent state allows us to easily calculate expectation values like (7.6.9). For instance, obviously,

$$\langle\gamma|J_-|\gamma\rangle = (1 + \gamma^*\gamma)^{-2j}\frac{\partial}{\partial\gamma}(1 + \gamma^*\gamma)^{2j} = \frac{2j\gamma^*}{1 + \gamma^*\gamma}. \tag{7.6.15}$$

Next, by employing the easily checked identities

$$e^{-\gamma J_-}J_z e^{\gamma J_-} = J_z - \gamma J_-, \quad e^{-\gamma J_-}J_+ e^{\gamma J_-} = J_+ + 2\gamma J_z - \gamma^2 J_- \tag{7.6.16}$$

we get, similarly,

$$\langle\gamma|J_-|\gamma\rangle = \frac{2j\gamma^*}{1 + \gamma^*\gamma}, \quad \langle\gamma|J_z|\gamma\rangle = j\frac{1 - \gamma^*\gamma}{1 + \gamma^*\gamma},$$

$$\langle\gamma|J_+J_-|\gamma\rangle = |\langle\gamma|J_-|\gamma\rangle|^2 + \frac{2j}{(1 + \gamma^*\gamma)^2}. \tag{7.6.17}$$

A little trigonometry may finally be exercised to express these expectation values in terms of the angles Θ and ϕ recover (7.6.9).

A coherent state $|\Theta\phi\rangle$ does not in general remain coherent under the time evolution generated by the Floquet operator F. Figure 7.4 depicts the variance of $\boldsymbol{J}$ with respect to a state $F^t|\Theta\phi\rangle$ as the time grows. The behavior shown corresponds to the classical phase-space portrait of Fig. 7.3c, i.e., to $\alpha = \pi/2$ and $\lambda = 3$ and to initial angles well outside the islands of regular motion. The angular-momentum quantum number was chosen as $j = 100$ to make the spot size $\Delta\Omega = 1/j$ smaller than the solid angle range of the islands of regular motion. Several features of the full curve in Fig. 7.4 are worth noting. Figure 7.4 also displays the variance of the classical vector $\boldsymbol{X}_t$ for a bundle of phase-space trajectories originating from a cloud of 1000 initial points. The cloud was chosen to have uniform density and circular shape, and to be equal in location and size to the spot corresponding to the coherent initial state of the quantum top.

Due to classical chaos, one would expect the classical and the quantum variance to become markedly different for times of the order $\ln j$ (a typical quantum uncertainty $\sim 1/j$ is amplified to order unity within such a time by exponentially separating chaotic trajectories); this expectation is consistent with the behavior displayed in Fig. 7.4. A further qualitative difference between the two curves in Fig. 7.4 becomes manifest for times exceeding the inverse mean level spacing $(2j + 1)/2\pi$: While the classical curve becomes smooth, the quantum curve displays quasi-periodicity. Incidentally, the recurrent events are quite erratic in their sequence. Most importantly, there appears to be no limit to the spread of either the quantum wave packet or the classical cloud of phase-space points other than the finite size, respectively, of the Hilbert space and the phase space.

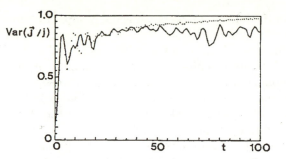

Fig. 7.4. Time-dependent variance $\langle (\Delta \boldsymbol{J}/j)^2 \rangle$ for the kicked top (7.6.2) ($\alpha = \pi/2$, $\lambda = 3$, $j = 100$) for a coherent initial state localized well outside classical islands of regular motion in Fig. 7.3c. The dotted line refers to the classical variance based on a bundle of 1000 classical trajectories

Further light may be shed on the problem of localization if coherent states $|\Theta\phi\rangle$ are represented as vectors with respect to the eigenstate of the Floquet operator F,

$$|\Theta\phi\rangle = \sum_{\mu=1}^{2j+1} C_\mu(\Theta, \phi)|\mu\rangle \ . \tag{7.6.18}$$

Taking the basis vectors $|\mu\rangle$ as normalized,

$$\sum_\mu |C_\mu|^2 = 1 \ . \tag{7.6.19}$$

Useful information can be gained by ordering the components C_μ according to decreasing modulus and truncating the representation (7.6.18) so as to include only the minimum number $N_{\min}$ of basis vectors necessary to exhaust the normalization of $|\Theta\phi\rangle$ to within, say, 1%. Numerical work [227] using $\alpha = \pi/2$, $\lambda = 3$ indicates that $N_{\min}$ scales quite differently with j, depending on whether the initial state lies in a region of classical chaos or in an island of classically regular motion,

$$N_{\min} \propto j^x$$

$$x = \begin{cases} 1 & \text{chaotic} \\ \frac{1}{2} & \text{regular} \end{cases} \tag{7.6.20}$$

The proportionality between $N_{\min}$ and j for coherent states located in the classically chaotic region supports the conclusion that the Floquet eigenstates related to classical chaos are not confined in their angular spread to a region of solid angle smaller than the range of classical chaos on the unit sphere depicted in Fig. 7.3c. Moreover, $N_{\min} \sim j$ is precisely the result that random-matrix theory would suggest [232].

Conversely, the value $1/2$ for the exponent x implies that Floquet eigenstates tend to be localized if they correspond to classically regular motion within islands around periodic orbits. Indeed, in the limit $j \to \infty$, a vanishingly small fraction $\sim 1/\sqrt{j}$ of the full set of eigenstates suffices to build a coherent state located in such an island. The proportionality $N_{\min} \sim \sqrt{j}$ is most easily understood when the island in question contains a fixed point of the classical map. A regular orbit surrounding such a fixed point must closely resemble linear precession of the vector $\boldsymbol{X}$ around the direction defined by the fixed point. Similarly, the eigenstates of the quantum dynamics must be well approximated by the states $|jm\rangle$ provided the z-axis is oriented toward the classical fixed point in question. Then, the weight $|\langle jm|\Theta\phi\rangle|^2$ of the approximate eigenstate $|jm\rangle$ in the coherent state $|\Theta\phi\rangle$ is given by the binomial distribution (7.6.14) and that distribution is easily seen to have a width of the order $\sqrt{2j}$ when j is large.

Numerical evidence for the scaling (7.6.20) is presented in Fig. 7.5, which again corresponds to the classical situation of Fig. 7.3c. The three curves pertain to $j = 200$, 400, and 500. The drop of $N_{\min}$ as the coherent state enters the island of classically regular motion is so pronounced that $N_{\min}$ might even be taken as an approximate quantum measure of the classical Lyapunov exponent. The plots are consistent with $N_{\min} \propto j^{1/2}$ in the classically regular region and support $N_{\min} \propto j$ quite convincingly whereas the coherent state ranges in the domain of classical chaos.

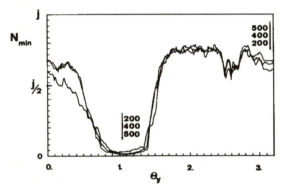

Fig. 7.5. Minimum number of eigenmodes of the kicked top (7.6.2) necessary to exhaust the normalization of coherent states to within 1 % versus the polar angle Θ_y (with respect to positive y-axis) at which the coherent state is located; the azimuthal angle is fixed at $\phi_y = \pi/4$. Coupling constants as in Fig. 7.3c. Classically, there is regular motion for $0.6 \lesssim \Theta_y \lesssim 1.4$ (the regular island in the lower right part of the $y > 0$ hemisphere in Fig. 7.3c). The curves follow the classical Lyapunov exponent better, the larger the j value (chosen here as 200, 400, and 500)

The "regular localization" and "chaotic delocalization" of the Floquet eigenvectors for the top[5] have interesting consequences for the time evolution of expectation values. In both cases quasi-periodicity must become manifest after a time of the order j, i.e., the inverse of the average quasi-energy spacing $2\pi/(2j + 1)$.

---

[5] The top does not, in general, display the phenomenon of quantum localization under conditions of classical chaos; from this fact it may be understood that the localization length is larger than $2j + 1$; see, however, the subsequent section.

However, as illustrated in Fig. 7.6, quasi-periodicity arises in two rather different varieties: a nearly periodic sequence of alternate collapses and revivals modulates oscillations of $\langle J_x \rangle$; those oscillations correspond to orbiting of the wave packet around the fixed point in one of the islands of Fig. 7.3c; this nearly periodic kind of quasi-periodicity accompanies $N_{\min} \propto j^{1/2}$ and may be interpreted as a quantum beat phenomenon dominated by a very small number of excited eigenmodes. When $N_{\min} \propto j$, on the other hand, the wave packet has neither a fixed point to orbit around nor are there any regular modulations; instead, recurrent events form a seemingly erratic sequence; this type of behavior might be expected for broadband excitation of eigenmodes.

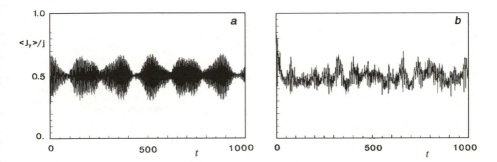

Fig. 7.6. Quasi-periodic behavior in time of $\langle J_x \rangle$ for the kicked top with the Floquet operator (7.6.2) for $\alpha = \pi/2$, $\lambda = 3$, $j = 100$. Initial states are coherent, see (7.6.10), localized within a classically regular island around a classical fixed point (see Fig. 7.3c) for curve (**a**) and in a classically chaotic region for curve (**b**). Note the near-periodic alternation of collapse and revival in the "regular" case (**a**) and the erratic variety of quasi-periodicity in the "irregular" case (**b**)

Yet another quantum distinction of regular and chaotic motion follows from the different localization properties of the Floquet eigenvectors just discussed [233]. The "few" eigenvectors localized in an island of regular motion around a fixed point (as shown in Fig. 7.3c) should not vary appreciably when a control parameter, say λ, is altered a little, as long as the classical fixed point and the surrounding island of regular motion are not noticeably changed. On the other hand, an eigenvector that is spread out in a large classically chaotic region might be expected to react sensitively to a small change of λ, since it must respect orthogonality to the "many" other eigenfunctions spread out over the same part of the classical phase space. To check on the expected different degree of sensitivity to small changes of the dynamics, one may consider the time-dependent overlap of two wave vectors, both of which originate from one and the same coherent initial state but evolve according to slightly different values of λ, for example,

$$\left| \langle \Theta, \phi | F(\lambda', \alpha)^{\dagger t} F(\lambda, \alpha)^t | \Theta \phi \rangle \right|^2 \quad \text{for } t = 0, 1, 2, \ldots . \tag{7.6.21}$$

Figure 7.7 displays that overlap for $\alpha = \pi/2$, $\lambda = 3$, $\lambda' = 3(1 + 10^{-4})$. The classical phase-space portraits for these two sets of control parameters are hardly

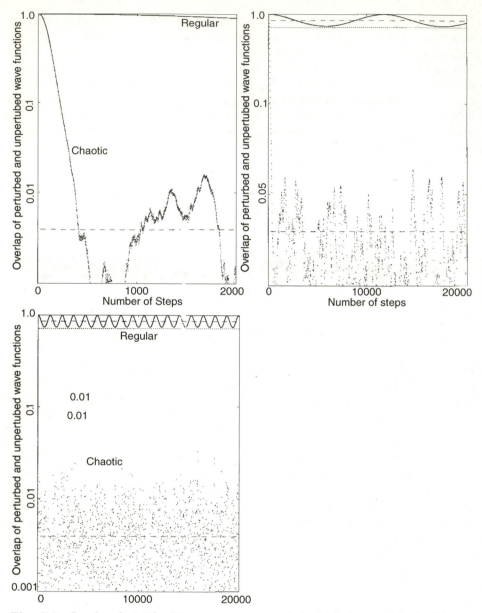

Fig. 7.7. Overlap (7.6.18) of two wave vectors of the kicked top with the Floquet operator (7.6.2) for $\alpha = \pi/2$, $\lambda = 3$, $\lambda' = 3(1 + 10^{-4})$, $j = 1600$. Initial coherent state as in Fig. 7.6a for upper curve (regular case) and as in Fig. 7.6b for lower curve. Note the extreme sensitivity to tiny changes of the dynamics in the irregular case. Courtesy of A. Peres [233]

distinguishable and appear as in Fig. 7.3c. For the upper curve in Fig. 7.7, the initial state was chosen that it lies within one of the regular islands of Fig. 7.3c; in accordance with the above expectation, the overlap (7.6.21) remains close to unity for all times in this "regular" case. The lower curve in Fig. 7.7 pertains to an initial state lying well within the chaotic part of the classical phase space shown in Fig. 7.3c; the overlap (7.6.21), it is seen, falls exponentially from its initial value of unity down to a level of order $1/j$. This highly interesting quantum criterion to distinguish regular and irregular dynamics, based on the overlap (7.6.21), was introduced by *Peres* [233]. Incidentally, the "regular" near-periodicity and "irregular" quasi-periodicity seen in Fig. 7.6 is again met in Fig. 7.7 on time scales of the order j.

To summarize, the top and the rotator display quite different localization behavior. Under conditions of classical chaos, the eigenvectors of the rotator dynamics localize while those for the top do not. Somewhat different in character is the localization described above for eigenvectors confined to islands of regular motion; the latter behavior, well investigated for the top only, must certainly be expected for the rotator and other simple quantum systems as well.

7.7 The Rotator as a Limiting Case of the Top

Now, I proceed to show that the top can be turned into the rotator by subjecting the torsional strength λ and the precessional angle α to a special limit [234]. From a classical point of view, that limit must confine the phase-space trajectory of the top to a certain equatorial "waistband" of the spherical phase space to render the latter effectively indistinguishable from a cylinder. The corresponding quantum mechanical restriction makes part of the $(2j + 1)$-dimensional Hilbert space inaccessible to the wave vector of the top: with the axis of quantization suitably chosen, the orientational quantum number m must be barred from the neighborhoods of the extremal values $\pm j$.

To establish the limit in question, one may look at the Floquet operators of the top

$$F_{\mathrm{T}} = \mathrm{e}^{-\mathrm{i}\lambda_{\mathrm{T}} J_z^2/2j\hbar^2} \mathrm{e}^{-\mathrm{i}\alpha J_x/\hbar} \tag{7.7.1}$$

and the rotator

$$F_{\mathrm{R}} = \mathrm{e}^{-\mathrm{i}p^2\tau/2I\hbar} \mathrm{e}^{-\mathrm{i}\lambda_{\mathrm{R}}(I/\tau\hbar)\cos\Theta} . \tag{7.7.2}$$

To avoid possible confusion, all factors $\hbar$, τ, and I are displayed here. The reader should note that the sequence of the unitary factors for free rotation and kick has been reversed with respect to the previous discussion of the rotator; the operator F_{T} defined in (7.7.2) describes the time evolution of the rotator from immediately before one kick to immediately before the next. This shift of reference is a matter of convenience for the present purpose.

By their very appearance, the two Floquet operators suggest the correspondence

$$\lambda_T = \frac{\hbar\tau}{I} j \ , \quad \alpha = \lambda_R \frac{I}{\tau\hbar} \frac{1}{j} \ . \tag{7.7.3}$$

For $j \to \infty$ with τ, I, and λ_R fixed, this clearly amounts to suppressing large excursions of $\boldsymbol{J}$ away from the "equatorial" region $J_z/j \approx 0$. To reveal the rotator-like behavior of the classical top in the limit (7.7.3), we call upon the classical recursion relations (7.6.7) of the top setting $J_z = p$, $J_x = \hbar j \cos \Theta$, and $J_y = \hbar j \sin \Theta$ and obtain

$$p' = p + \lambda_R \frac{I}{\tau} \sin \Theta$$
$$\phi' = \phi + \frac{\tau}{I} p' \ . \tag{7.7.4}$$

These are indeed the equations of motion for the rotator. In fact, to bring (7.7.4) into the form (7.3.7), we must shift back the reference of time to let the free precession precede the nonlinear kick, $\Theta = \Theta_{t+1}$, $p = p_t$.

The equivalence of the quantum mechanical Floquet operators (7.7.1) and (7.7.2) in the limit (7.7.3) becomes obvious when their matrix representations are considered with respect to eigenstates of J_z and p, respectively,

$$J_z|m\rangle = \hbar m|m\rangle \ , \quad -j \le m \le +j$$
$$p|m\rangle = \hbar m|m\rangle \ , \quad m = 0, \pm 1, \pm 2, \ \dots \ . \tag{7.7.5}$$

Then, the left-hand factors in F_T and F_R coincide once λ_T is replaced by $\hbar j\tau/I$ according to (7.7.3). The exponents in the right-hand factors of F_T and F_R read

$$\left\langle m \left| \frac{\alpha}{\hbar} J_x \right| m' \right\rangle = \frac{1}{2}\alpha\left\{ \sqrt{(j-m)(j+m+1)}\delta_{m',m+1} \right.$$
$$\left. + \sqrt{(j+m)(j-m+1)}\delta_{m',m-1} \right\} \ , \tag{7.7.6}$$

$$\left\langle m \left| \lambda_R \frac{I}{\tau\hbar} \cos \Theta \right| m' \right\rangle = \frac{1}{2}\lambda_R \frac{I}{\tau\hbar} \left\{ \delta_{m',m+1} + \delta_{m',m-1} \right\} \ .$$

These matrices become identical when the correspondence (7.7.3) is inserted and the limit $j \to \infty$ with $m/j \to 0$ is taken, the latter condition constituting the quantum mechanical analogue of a narrow equatorial waist-band of the classical sphere.

It is important to realize that quantum localization under conditions of classical chaos, impossible when α and λ_T remain finite for $j \to \infty$, arises in the limit (7.7.3). According to (7.5.6) the localization length can be estimated as $l \approx (\lambda I/\hbar\tau)^2/2$ and for localization to take place this length must be small compared to j, $l \ll j$. The limit (7.7.3) clearly allows that condition to be met.

7.8 Problems

7.1. Derive the discrete-time Heisenberg equations for the kicked rotator.

7.2. The random-phase approximation of Sect. 7.3 does not assign Gaussian behavior to the force $\xi = \lambda V(\Theta)$. Verify this statement by calculating $\langle \xi^n \rangle$ for $V = \cos \Theta$ and uniformly distributed phases Θ.

7.3. The result of Problem 7.3 notwithstanding, the moments $\langle p_t^{2n} \rangle$ tend to display Gaussian behavior for $t \to \infty$. Show this with the help of the random-phase approximation.

7.4. Rewrite the quantum map (7.3.17) in the Θ-representation.

7.5. Show that $V(\Theta) = -2 \arctan(\kappa \cos \Theta - E)$ for the kicked rotator corresponds to nearestneighbor hops in the equivalent pseudo-Anderson model.

7.6. Discuss the analyticity properties of Green's function $\overline{G}(E) = (E - \tilde{H})^{-1}$ with $\tilde{H}$ given by (7.3.23).

7.7. Generalize the simple Gaussian integrals

$$\int_{-\infty}^{+\infty} ds\, e^{-\alpha s^2} = \sqrt{\frac{\pi}{\alpha}}$$

$$\int_{-\infty}^{+\infty} ds\, s^2\, e^{-\alpha s^2} = -\frac{\partial}{\partial \alpha} \sqrt{\frac{\pi}{\alpha}} = \frac{1}{2\alpha} \sqrt{\frac{\pi}{\alpha}}$$

to the multiple integrals

$$\int_{-\infty}^{+\infty} d^N s\, e^{-\sum_{i,j} \alpha_{ij} S_i S_j} = \sqrt{\frac{\pi^N}{\det \alpha}}$$

$$\int_{-\infty}^{+\infty} d^N S\, S_1 S_2\, e^{-\sum_{i,j} \alpha_{ij} S_i S_j} = \left(\frac{1}{2\alpha}\right)_{1,2} \sqrt{\frac{\pi^\nu}{\det \alpha}}$$

for a positive symmetric matrix α. Use the eigenvectors and eigenvalues of α.

7.8. Diagonalize the matrix $V_{ij} = \kappa(\delta_{i,j+1} + \delta_{i,j-1})/2$.

7.9. Use (7.4.29, 7.4.30) to determine the eigenvalues of the matrix $V_{ij} = \kappa(\delta_{i,j+1} + \delta_{i,j-1})/2$. Locate the poles of Green's function given in (7.4.24). What happens to these poles in the limit $N \to \infty$?

7.10. Verify the stroboscopic Heisenberg equation (7.6.4) for the kicked top with the Floquet operator (7.6.2).

7.11. Determine the amplitude of the temporal fluctuations of var $(\boldsymbol{J})/j^2$ from the point of view of random-matrix theory.

8. Semiclassical Roles for Classical Orbits

8.1 Preliminaries

The precursor of quantum mechanics due to Bohr and Sommerfeld was already seen by Einstein [235] as applicable only to classically integrable dynamics. It has since been developed into a systematic semiclassical approximation, as sketched in Chap. 5. No prepotent guesses proved possible as to what quantum mechanics has to say about chaotic motion; the pertinent semiclassical approximations came into view with a delay of nearly fifty years in *Gutzwiller*'s pioneering work.

This chapter is devoted to Gutzwiller's periodic-orbit theory [236–239]. Doing justice to the enormous amount of knowledge gathered under that heading would require a separate book. I shall confine myself to a reasonably explicit derivation of the basic "trace formula" for both maps and autonomous flows, a discussion of the threat of divergence due to the infamous exponential proliferation of periodic orbits of increasing period, recent attempts at resummation by clever use of unitarity, and the impressive if still somewhat incomplete success of explaining the universality of spectral fluctuations in systems with global classical chaos. Along the way, somewhat playful encounters with Maslov theory and Lagrangian manifolds will provide fortification and entertainment.

By mentioning "divergence" and "attempts at resummation", I wish to warn the reader that not all is well mathematically here, and I do not intend to hide yet unresolved difficulties.

One invaluable source of inspiration for the development of periodic-orbit theory will be left out entirely, the statistics of the "nontrivial" zeros of Riemann's zeta function. For those, number theory provides a solid basis for the asymptotic treatment of the analogue of sums over periodic orbits. Such a basis is lacking for dynamic systems, and, conversely, the Riemann zeros still await the definitive word on dynamic interpretation.

The presentation aims to be self-contained and introductory in style. While readers with previous knowledge of classical nonlinear dynamics will have a head start on some issues, the uninitiated will find, I hope, sufficient background explanation here. Familiarity with Hamiltonian and Hamilton–Jacobi mechanics is an indispensable prerequisite, though. Therefore, I recommend that you start studying this chapter by doing Problems 8.1 and 8.2. Good luck, then!

8.2 Van Vleck Propagator

The Van Vleck propagator is a semiclassical approximant to the unitary time evolution operator $U(t) = \exp\left(-\frac{i}{\hbar}Ht\right)$ or $U(t) = \left\{\exp\left[-\frac{i}{\hbar}\int_0^t dt' H(t')\right]\right\}_+$ in, say, the coordinate representation; it expresses the transition amplitude $\langle q|U(t)|q'\rangle$ in terms of the action $S(q, q', t)$ of the classical path leading from q' to q during the time span t and a certain Morse index ν,

$$\langle q|U(t)|q'\rangle \approx \sum_\alpha \sqrt{\frac{1}{i2\pi\hbar}\left|\frac{\partial^2 S^\alpha(q, q', t)}{\partial q \partial q'}\right|}\, e^{i\{S^\alpha(q,q',t)/\hbar - \nu^\alpha\pi/2\}}\,; \qquad (8.2.1)$$

the sum over α picks up contributions from distinct classical paths, should there be such. The approximate-equality sign "$\approx$" indicating a semiclassical approximation will be with us throughout this chapter.

To derive this basic ingredient of all further semiclassical considerations, I start, for simplicity, with a particle of mass m moving along a straight line and assume that its unitary time evolution operator has the special product structure

$$U = e^{-iV(\hat{q})\tau/\hbar}\, e^{-i(\hat{p}^2/2m)\tau/\hbar}\,. \qquad (8.2.2)$$

We incur the following matrix elements in the position representation,[1]

$$\begin{aligned}
\langle q|U|q'\rangle &= \exp(-iV(q)\tau/\hbar)\sqrt{\frac{m}{i2\pi\hbar\tau}}\exp\{im(q-q')^2/2\tau\hbar\} \\
&\equiv \sqrt{\frac{m}{i2\pi\hbar\tau}}\, e^{iS(q,q')/\hbar}\,, \qquad (8.2.3) \\
S(q,q') &= \frac{m}{2\tau}(q-q')^2 - \tau V(q)\,.
\end{aligned}$$

The reader will recognize the solution of Schrödinger's equation of a free particle sharply localized at the point q' initially, as a factor in this probability amplitude for the transition $q' \to q$. I shall refer to the amplitude $\langle q|U|q'\rangle$ as the propagator.

The splitting of the evolution operator into two factors, one for free motion and the other involving only the potential energy $V(q)$, may be motivated in different ways. For one, we may take the time increment τ as a fraction of a larger time, $t = n\tau$ with integer n, and proceed to the nth power of U; then the limit $n \to \infty$ at fixed $t = n\tau$ yields the usual time evolution operator with the Hamiltonian $H = T+V$ and the finite time increment t through Trotter's product [240]

$$\lim_{n\to\infty}\left(e^{-iVt/n\hbar}\, e^{-iTt/n\hbar}\right)^n = e^{-i(T+V)t/\hbar}\,. \qquad (8.2.4)$$

Alternatively, if τ remains finite, U represents free motion during that time interval followed by an instantaneous kick according to the temporally delta-shaped potential $\tau\delta(t-\tau)V(\hat{q})$. Periodic repetition of that process invites a stroboscopic description with U as the Floquet operator.

---

[1] Throughout this chapter, $\sqrt{i} = e^{i\pi/4}$ will be understood.

8.2.1 Maps

I shall first adopt the second of the two interpretations just mentioned, i.e. take τ finite and thus treat quantum maps; the subsequent subsection will be devoted to infinitesimal τ, as appropriate for autonomous flows.

The quantity $S(q, q')$ in (8.2.2) is the action generating the classical transition from (q', p') to (q, p), in the time span from immediately after one kick to right after the next. Denoting the number of kicks passed by the label n, I write the classical kick-to-kick map as

$$p_{n-1} = -\frac{\partial S(q_n, q_{n-1})}{\partial q_{n-1}}, \quad p_n = \frac{\partial S(q_n, q_{n-1})}{\partial q_n} \tag{8.2.5}$$

or explicitly

$$q_n = q_{n-1} + (\tau/m)p_{n-1}, \quad p_n = p_{n-1} - \tau V'(q_n). \tag{8.2.6}$$

Noteworthy also is the second-order recurrence relation for the coordinate obtained by eliminating the momentum,

$$\frac{m}{\tau^2}(q_{n+1} - 2q_n + q_{n-1}) = -V'(q_n), \tag{8.2.7}$$

which not fortuitously looks like Newton's equation of motion in discretized time.

We may stroboscopically follow the classical phase-space path originating from, say, q_0, p_0 through n steps of the kick-to-kick map. Then, the n-fold composite map has the generating function

$$S^{(n)}(q_n, q_0) = \sum_{i=1}^{n} S(q_i, q_{i-1}) \tag{8.2.8}$$

where the intermediate coordinates $q_1, q_2, \ldots, q_{n-1}$ are to be considered functions of q_0 and q_n. Then, the analogue of (8.2.5) reads

$$p_0 = -\partial S^{(n)}/\partial q_0, \quad p_n = \partial S^{(n)}/\partial q_n. \tag{8.2.9}$$

Incidentally, there may be several such n-step paths from q_0 to q_n in which case they ought to be distinguished by a label like α on the intermediate coordinates, q_i^α with $i = 1, \ldots, n-1$, and on the generating function, $S^{(n,\alpha)}$.

With the foregoing classical lore in store for later semiclassical use, I turn back to quantum dynamics and propose to determine the matrix elements $\langle q|U^n|q'\rangle$ for $n = 2, 3, \ldots$. By just inserting the resolution of unity with position eigenstates, $\int_{-\infty}^{\infty} dx|x\rangle\langle x| = 1$, in between U's, we obtain the "path integral"

$$\langle q|U^n|q'\rangle = \int_{-\infty}^{+\infty} dx_1 \ldots \int_{-\infty}^{+\infty} dx_{n-1}\langle q|U|x_{n-1}\rangle \ldots \langle x_1|U|q'\rangle \tag{8.2.10}$$

$$= \left(\frac{m}{\mathrm{i}2\pi\hbar\tau}\right)^{\frac{n}{2}} \int_{-\infty}^{+\infty} dx_1 \ldots \int_{-\infty}^{+\infty} dx_{n-1} \exp\left[\mathrm{i}\sum_{i=1}^{n} S(x_i, x_{i-1})/\hbar\right];$$

the sum of all n single-step actions appears in the exponent in the second line but here, in contrast to the generating function (8.2.8) of the n-fold composite classical map, all $n-1$ intermediate coordinates serve as integration variables while the initial ($x_0 \equiv q'$) and final ones ($x_n \equiv q$) are fixed. As is typical for a Feynman-type path integral for a transition amplitude, we confront a superposition of amplitudes for arbitrary paths between fixed initial and final points rather than only from the particular path(s) admitted by classical mechanics.

Classically allowed paths do acquire distinction in quantum transition amplitudes as soon as one looks at the semiclassical limit. Then, a stationary-phase approximation to the above $(n-1)$-fold integral is called for; in its simplest implementation [236, 241] this amounts to expanding the generating function $\sum_i S(x_i, x_{i-1})$ about the (each, should there be several) classical path, keeping only terms up to second order in the deviation and thus approximating the integral by a Gaussian. The truncated expansion in powers of the deviation $\phi_i^\alpha = x_i - q_i^\alpha$ (but recall $\phi_0 = \phi_n = 0$) reads

$$\sum_{i=1}^{n} S(x_i, x_{i-1}) = S^{(n,\alpha)}(q_n, q_0) + \frac{1}{2}\sum_{i=1}^{n}\left[\frac{m}{\tau}(\phi_i^\alpha - \phi_{i-1}^\alpha)^2 - \tau V''(q_i^\alpha)(\phi_i^\alpha)^2\right]$$

$$\equiv S^{(n,\alpha)}(q_n, q_0) + \frac{1}{2}\delta^2 S^{(n,\alpha)}(\{\phi\}). \tag{8.2.11}$$

Prominence of the classical path is not just constituted by our choice of a reference for the expansion but is manifest in the absence of first-order terms: The latter vanish by virtue of the classical map (8.2.5). Indeed, the phase of the integrand of the path integral is stationary along the classical path. To do the resulting Gaussian integral, one may imagine the quadratic form in the truncated expansion (8.2.11) diagonalized. One is led to $n-1$ integrals of the Fresnel type,

$$\int_{-\infty}^{+\infty}\frac{dz}{\sqrt{i\pi}}e^{i\lambda z^2} = \frac{1}{\sqrt{|\lambda|}}\begin{cases}1 & \text{for }\lambda > 0 \\ e^{-i\pi/2} & \text{for }\lambda < 0\end{cases}. \tag{8.2.12}$$

Then, the propagator $\langle q|U^n|q'\rangle$ takes the semiclassical form of a sum of contributions from classical paths,

$$\langle q|U^n|q'\rangle = \sum_{\alpha} A^{(n,\alpha)}\,e^{iS^{(n,\alpha)}(q,q')/\hbar}\,e^{-i\nu^\alpha\pi/2}, \tag{8.2.13}$$

where the phase factor involves the generating function alias action for the αth classical path, coming from the first term in the expansion (8.2.11); the prefactor, a quantity much less wildly depending on Planck's constant $\hbar$, reads

$$A^{(n,\alpha)} = \sqrt{\frac{m}{i2\pi\hbar\tau|D_n^\alpha|}}. \tag{8.2.14}$$

Here $D_1^\alpha = 1$ due to (8.2.3) while otherwise $D_n^\alpha = \det G^{(n,\alpha)}$ is the determinant of the $(n-1)\times(n-1)$ matrix defining the second variation of the action in (8.2.11) as $\delta^2 S^{(n,\alpha)}(\{\phi\}) = \frac{m}{\tau} \sum_{ij} G_{ij}^{(n,\alpha)} \phi_i \phi_j$,

$$G^{(n,\alpha)} = \begin{pmatrix} d_1^\alpha & -1 & 0 & \cdots & & 0 \\ -1 & d_2^\alpha & -1 & 0 & \cdots & 0 \\ \vdots & & & & & \vdots \\ 0 & \cdots & 0 & -1 & d_{n-2}^\alpha & -1 \\ 0 & & \cdots & 0 & -1 & d_{n-1}^\alpha \end{pmatrix} , \quad d_i^\alpha = 2 - \frac{\tau^2}{m} V''(q_i^\alpha); \quad (8.2.15)$$

and ν^α, the so-called Morse index [240], is the number of negative eigenvalues of $G^{(n,\alpha)}$. That index emerges as the number of times the second line of (8.2.12) applies for the Fresnel integral while $\det G^{(n,\alpha)}$ enters through the product of eigenvalues of the matrix $G^{(n,\alpha)}$, from the $(n-1)$-fold usage of the Fresnel integral.[2] Needless to say, ν^α depends on q' and q as well as on the connecting path.

The determinant D_n^α obeys a recursion relationship easily established by expanding along the last row as

$$D_{i+1}^\alpha = d_i^\alpha D_i^\alpha - D_{i-1}^\alpha, \quad i = 1, 2, \ldots, n-1, \qquad (8.2.16)$$

where the initial conditions are $D_1^\alpha = 1, D_0^\alpha = 0$. In fact, the running index i may be extended beyond $n-1$ by imagining the path followed through any number of further steps, $q_0 \to q_1^\alpha \to q_2^\alpha \cdots \to q_n^\alpha \to q_{n+1}^\alpha \cdots$. The same recurrence relationship and initial conditions are obtained for $\partial q_i / \partial q_1^\alpha$, by considering the q_i^α for $i > 1$ as functions of q_1^α, q_0 and taking the partial derivative with respect to q_1^α (at constant q_0) in the equation of motion (8.2.7). We conclude that

$$D_n^\alpha = \det G^{(n,\alpha)} = \left.\frac{\partial q_n}{\partial q_1^\alpha}\right|_{q_0} = \left.\frac{m}{\tau} \frac{\partial q_n}{\partial p_0^\alpha}\right|_{q_0}; \qquad (8.2.17)$$

the last member of the foregoing chain results from $p_0^\alpha = (m/\tau)(q_1^\alpha - q_0)$. With the help of (8.2.5), now we may express the inverse determinant by a second derivative of the action as

$$\frac{\partial^2 S^{(n,\alpha)}}{\partial q_n \partial q_0} = -\frac{m}{\tau} \frac{\partial}{\partial q_n}(q_1^\alpha - q_0) = -\frac{m}{\tau D_n^\alpha}. \qquad (8.2.18)$$

Replacing the determinant D_n^α with the mixed second derivative of $S^{(n,\alpha)}$ in the semiclassical propagator (8.2.13, 14) we get the latter in the Van Vleck form[3]

$$\langle q|U^n|q'\rangle \approx \sum_\alpha \sqrt{\frac{1}{i2\pi\hbar} \left|\frac{\partial^2 S^{(n,\alpha)}}{\partial q \partial q'}\right|} \, e^{i\{S^{(n,\alpha)}(q,q')/\hbar - \nu^\alpha \pi/2\}}, \qquad (8.2.19)$$

first spelled out for quantum maps by *Tabor* [242].

---

[2] Formally, the contribution of the αth classical path can be seen as arising from a Gaussian integral of the type (4.12.22) after analytic continuation to purely imaginary exponents; that continuation, achieved by the Fresnel integral (8.2.12), brings about the modulus operation and the ν^α phase factors $e^{-i\pi/2}$ in (8.2.13).

[3] This is equally valid in continuous time; if we let $n \to \infty, \tau \to 0$ with $n\tau = t$ fixed, $S^{(n,\alpha)}(q,q') \to S^\alpha(q,q',t)$.

A simple prescription for determining the Morse index ν^α derives from the recursion relation (8.2.16) for the determinants D_i^α. A change of sign of D_{i+1}^α relative to D_i^α signals that the number of negative eigenvalues is larger by one for $G^{(i+1,\alpha)}$ than for $G^{(i,\alpha)}$. This follows from the fact that $G^{(i,\alpha)}$ is obtained from $G^{(i+1,\alpha)}$ by discarding the ith row and column. A well-known corollary of Courant's minimax theorem [244] then says[4] that the i eigenvalues of $G^{(i+1,\alpha)}$ form an alternating sequence with the $i-1$ eigenvalues of $G^{(i,\alpha)}$. Therefore, the number ν^α of negative eigenvalues of $G^{(n,\alpha)}$ must therefore equal the number of sign changes in the sequence of determinants D_i^α from D_1^α to D_n^α. Clearly, the prescription just explained relies on the kinetic energy to be a positive quadratic form in the momenta and distinguishes the coordinate representation; for more general situations see Ref. [245].

I leave to the reader as Problem 8.3 the generalization of the Van Vleck formula to f degrees of freedom. Actually, most applications of maps studied thus far are confined to $f = 1$, and reasonably so since periodically driven single-freedom systems can display chaos. It is only for autonomous dynamics (alias flows), to be looked at in Sect. 8.3.2, that nonintegrability requires $f > 1$.

The very appearance (8.2.19) of the Van Vleck propagator suggests validity beyond the limitations of its derivation given here. It is indeed not difficult to check that the simple product structure (8.2.2) of the Floquet operator can be abandoned in favor of arbitrary periodic driving, $U \to (\exp\{-i \int_0^\tau dt H(t)/\hbar\})_+$ where the time-dependent Hamiltonian need not even be the sum of a kinetic and a potential term.

According to its property (8.2.17), the determinant D_n^α characterizes the stability properties of the classical path $\{q_i^\alpha\}_{i=0,1,\dots,n}$ through $\delta q_n \approx D_n^\alpha \delta q_1$. We may define a Lyapunov exponent for the αth path between q_0 and q_n by

$$\lambda_\alpha = \lim_{n\to\infty} \frac{1}{n} \ln |D_n^\alpha| . \tag{8.2.20}$$

A path hopping about in a chaotic region will have positive λ_α such that the determinant in question will grow exponentially, $D_n^\alpha \sim e^{\lambda_\alpha n}$; a regular path, on the other hand, will have a vanishing Lyapunov exponent; its D_n^α is uncapable of exponential growth but might oscillate or grow like a power of n. For a thorough discussion of such different behaviors the reader may consult any textbook on nonlinear dynamics; here, I must confine myself to the few hints to follow.

For a rough qualitative characterization, we may replace all curvatures $V''(q_n)$ along the path by a suitable average $\overline{V''(q_n^\alpha)} = m\omega_\alpha^2$. Within this "harmonic" approximation $d_n^\alpha = 2 - \omega_\alpha^2 \tau^2 = d$ is independent of n, whereupon the recursion relation (8.2.16) allows for solution by exponentials a^n, whose bases are determined by the quadratic equation $a = d - 1/a$ as $a_\pm = 1 - \omega_\alpha^2\tau^2/2 \pm \sqrt{\omega_\alpha^2\tau^2(\omega_\alpha^2\tau^2 - 4)/4}$.

---

[4] When the eigenvalues of $G^{(i+1,\alpha)}$ are ordered as $g_1^{(i+1)} \leq g_2^{(i+1)} \leq \dots \leq g_i^{(i+1)}$ and those of $G^{(i,\alpha)}$ as $g_1^{(i)} \leq g_2^{(i)} \leq \dots \leq g_{i-1}^{(i)}$ the corollary to the minimax theorem in question yields $g_1^{(i+1)} \leq g_1^{(i)} \leq g_2^{(i+1)} \leq g_2^{(i)} \leq \dots g_{i-1}^{(i)} \leq g_i^{(i+1)}$.

Accounting for $D_0 = 0, D_1 = 1$, we get the solution

$$D_n^\alpha = \frac{a_+^n - a_-^n}{a_+ - a_-}. \tag{8.2.21}$$

Three cases arise:

(i) For $0 < \omega_\alpha^2 \tau^2 < 4$, the determinant oscillates as

$$D_n^\alpha = \frac{\sin n\phi_\alpha}{\sin \phi_\alpha} \quad \text{with} \quad \sin \frac{\phi_\alpha}{2} = \frac{1}{2}|\omega_\alpha \tau|, \quad 0 \le \phi_\alpha \le \pi. \tag{8.2.22}$$

The Morse index ν^α is the integer part of $n\phi_\alpha/\pi$, the number of sign changes in the sequence $\frac{\sin i\phi_\alpha}{\sin \phi_\alpha}$, $i = 1, 2, \ldots, n$. The boundary cases $\phi_\alpha = 0$ or π can be included here as limiting ones and yield the subexponential growth $D_n^\alpha = n$ with $\nu^\alpha = 0$. Paths of this type are stable and are called *elliptic*. They exist where the curvature of the potential $V(q)$ is positive and small.

(ii) For $\omega_\alpha^2 \tau^2 < 0$, the determinant grows exponentially,

$$D_n^\alpha = \frac{\sinh n\tilde{\lambda}_\alpha}{\sinh \tilde{\lambda}_\alpha} \quad \text{with} \quad \sinh \frac{\tilde{\lambda}_\alpha}{2} = \frac{1}{2}|\omega_\alpha \tau|, \quad \tilde{\lambda}_\alpha > 0. \tag{8.2.23}$$

The Morse index vanishes since D_n^α never changes sign. One speaks here of *hyperbolic* paths. They experience a potential with negative curvature.

(iii) Finally, for $\omega_\alpha^2 \tau^2 > 4$, the determinant grows exponentially in magnitude but alternates in sign from one n to the next,

$$D_n^\alpha = (-1)^{(n-1)} \frac{\sinh n\tilde{\lambda}_\alpha}{\sinh \tilde{\lambda}_\alpha} \quad \text{with} \quad \cosh \frac{\tilde{\lambda}_\alpha}{2} = \frac{1}{2}|\omega_\alpha \tau|, \quad \tilde{\lambda}_\alpha > 0.$$

$$\tag{8.2.24}$$

The Morse index is read off as $\nu^\alpha = n - 1$. Such *inverse hyperbolic* paths lie in regions where the potential $V(q)$ has a large positive curvature.

In the two unstable cases the parameter $\tilde{\lambda}_\alpha$ approximates the Lyapunov exponent.

The exponential growth of $|D_n^\alpha|$ with the stroboscopic time n entails the amplitude A_n^α in (8.2.13, 14) to decay like $e^{-\tilde{\lambda}_\alpha n/2}$ for unstable paths, while stable paths do not suffer such suppression. It would be wrong to conclude, though, that unstable paths have a lesser influence on the propagator than stable ones since the latter are outnumbered by the former, roughly in the ratio $e^{\lambda n}$ where λ is some average Lyapunov exponent. Such exponential proliferation even appears to threaten the semiclassical propagator (8.2.13, 19) with divergence as $n \to \infty$.

To appreciate the exponential proliferation, it is helpful to think momentarily of a compact two-dimensional phase space such as the spherical one of the kicked top. Consider two lines of constant q very close to one another and the stripe confined in between. Chaos and ergodicity assumed, repeated iterations of the map will not only deform that stripe but narrow and lengthen it in a self-avoiding way to keep the area constant. The ever narrowing, lengthening, and folding stripe will eventually visit every part of the phase space such that any line of constant q, contained or not in the original stripe, will be intersected a growing number of times as the number of iterations n grows. Inasmuch as an average Lyapunov exponent λ measures the lengthening of the stripe as $e^{\lambda n}$, one would indeed expect a similarly growing number of such crossings. This reasoning is not weakened much if we renounce the appeal to area conservation and compactness but insist on ergodicity for the fate of a single initial line of constant q under repeated iterations and its crossings with another line. On the other hand, an initial line pushed around by regular motion will not suffer foldings of ever increasing complexity, simply because it is not obliged to visit "everywhere" ergodically. Figure 8.1 illustrates the difference between chaotic and regular behavior just outlined for the kicked top. (No numerical work is needed to check on exponential proliferation for the baker's map; see Problem 8.4.)

8.2.2 Flows

As already mentioned, the Van Vleck propagator (8.2.19) also applies to the continuous time evolution generated by the Hamiltonian $H = T + V$: By taking the limit $n \to \infty, \tau \to 0$ with the product $t = n\tau$ fixed, we get the result anticipated in (8.2.1),

$$\langle q|e^{-iHt/\hbar}|q'\rangle \approx \sum_{\alpha} \sqrt{\frac{1}{i2\pi\hbar} \left| \frac{\partial^2 S^\alpha(q,q',t)}{\partial q \partial q'} \right|} \; e^{i\,[S^\alpha(q,q',t)/\hbar - \nu^\alpha \pi/2]}. \quad (8.2.25)$$

Here, the discrete time index n was replaced with the continuous time t as an independent argument of the action of the αth classical path going from q' to q during the time span t. As before, the action $S^\alpha(q, q', t)$ is the generating function for the classical transition in question.

To see what happens to the Morse index ν^α in the limit mentioned (nothing, really, except for a beautiful new interpretation as a certain classical property of the αth orbit), a little excursion back to the derivation of the Van Vleck propagator of Sect. 8.1 is necessary. To avoid any rewriting, I shall leave the time discrete, $t_i = i\tau$ with $i = 1, 2, \ldots, n$ and $t_n = t$, and imagine n so large and τ so small that the discrete classical path from $q_0 = q'$ to $q_n = q$ approximates the continuous path accurately. Recall that the Morse index was obtained as the number of negative eigenvalues of the matrix $G^{(n,\alpha)}$ in the second variation of the action about the classical path,

$$\delta^2 S^{(n,\alpha)}(\{\phi\}) = \frac{m}{\tau} \sum_{i,j=1}^{n-1} G_{ij}^{(n,\alpha)} \phi_i \phi_j. \quad (8.2.26)$$

Hamilton's principle characterizes the classically allowed paths as extremalizing the action. An interesting question, albeit quite irrelevant for the derivation of the equations of motion, concerns the character of the extremum. The interest in the present context hinges on its relevance for the value of the Morse index. Clearly, the action is minimal if the second variation $\delta^2 S^{(n,\alpha)}$ is positive, i.e., if

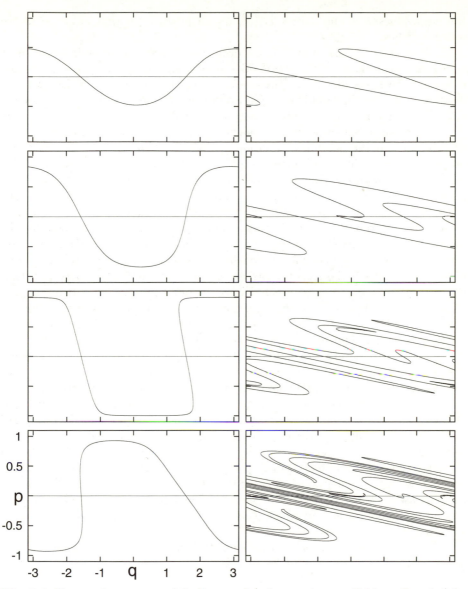

Fig. 8.1. Temporal successors of the line $p = 0$ (a Lagrangian manifold, see Sect. 8.4) for a kicked top under conditions of near integrability (left column) and global chaos (right column) after, from top to bottom, $n = 1, 2, 3, 4$ iterations of the classical map. The intersections with the line $p = 0$ define period-n orbits. The beginning of the exponential proliferation of such orbits is illustrated by the right column. Points q_c with $dp/dq = \infty$ give configuration-space caustics and, correspondingly, points p_c with $dq/dp = \infty$ give momentum-space caustics

the eigenvalues λ of the matrix $G_{ij}^{(n,\alpha)}$ are all positive; in that case the Morse index vanishes.

If we stick to Hamiltonians of the form $H = T + V$ and to sufficiently short time spans, minimality of the action does indeed hold, simply because locally any classical path looks like a straight line and resembles free motion. But for free motion, the eigenvalue problem in question reads $2\Phi_i - \Phi_{i-1} - \Phi_{i+1} = \lambda \Phi_i$ for $i = 1, 2, \ldots, n-1$ where the boundary conditions are $\Phi_0 = \Phi_n = 0$. The eigenvectors Φ can be constructed through an exponential ansatz for their components, $\Phi_i = a^i$, which yields the quadratic equation $a + \lambda - 2 + 1/a = 0$; the two roots $a_\pm$ obey $a_+ a_- = 1$; thus, each eigenvector is the sum of two exponentials, $\Phi_i = c_+ a_+^i + c_- a_-^i$ with coefficients $c_\pm$ restricted by the boundary conditions $c_+ + c_- = c_+ a_+^n + c_- a_-^n = 0$; nontrivial solutions $c_\pm \neq 0$ require $a_+^n = a_-^n$ and thus $a_+^{2n} = a_-^{2n} = 1$; now, both bases $a_\pm$ are revealed as unimodular, i.e., determined by a phase χ through $a_\pm = e^{\pm i\chi}$; then, the foregoing quadratic equation for the bases yields the eigenvalue $\lambda = 2(1 - \cos\chi)$ as bounded from below by zero; but zero is not admissible since a vanishing eigenvalue would entail a vanishing eigenvector; indeed, the eigenvectors are $\Phi_k \propto e^{ik\chi} - e^{-ik\chi} \propto \sin k\chi$ and thus $\Phi_k \equiv 0$ for $\chi = 0$. To ease possible worry about the limit $\chi \to 0$ in the last conclusion, one should check that the normalization constant is independent of χ; one easily finds $\Phi_k = (2/n)^{1/2} \sin k\chi$. As required for minimality of the action for free motion, all eigenvalues λ of G turn out positive.

But back to general classical paths! As we let the time t grow (always keeping a sufficiently fine gridding of the interval $[0, t]$), the action will sooner or later lose minimality (even though not extremality, of course). When this happens first, say for time $t_n = t^c$, an eigenvalue of G reaches zero and subsequently turns negative. The corresponding point $q(t^c)$ on the classical path is called "conjugate" to the initial one, $q(0) = q'$. More such conjugate points may and in general will follow later. At any conjugate point the number of negative eigenvalues of G will change by one. A theorem of *Morse's* [246] makes the stronger statement that the number of negative eigenvalues of G for Hamiltonians of the structure $H = T + V$ actually always keeps increasing by one as the configuration space path traverses a point conjugate to the initial one. *Therefore, the Morse index equals the number of conjugate points passed.*

To locate a conjugate point, one must look for a vanishing eigenvalue, i.e., solve $\sum_j G_{ij} \Phi_j = 0$ which reads explicitly

$$\frac{m}{\tau^2}(\Phi_{i+1} - 2\Phi_i + \Phi_{i-1}) = -V''(q_i)\Phi_i. \tag{8.2.27}$$

This so-called Jacobi equation may be read as the Newtonian equation of motion (8.2.7) linearized around a classical path $q(t_i) = q_i$ from $q_0 = q'$ to $q_n = q$ in discretized time. Here, of course, we must look for a solution Φ_i that satisfies the boundary conditions $\Phi_0 = \Phi_n = 0$ and is normalizable as $\sum_i \Phi_i^2 = 1$. If such a solution exists, the "final" time $t_n = t$ is a t^c, and the point q is conjugate to q'.

A geometrical meaning of conjugate points is worth a look. It requires considering the family of classical paths $q_i(p') = q(p', t_i)$ with $i = 1, 2, \ldots$ originating
from one and the same initial $q_0 \equiv q'$ with different momenta p'. Two such paths
move apart as $q_i(p' + \delta p') - q_i(p') = \frac{\partial q_i(p')}{\partial p'} \delta p'$. The "response function"

$$J(p', t_i) = \frac{\partial q_i(p')}{\partial p'} \qquad (8.2.28)$$

measuring that divergence vanishes initially since the initial coordinate is fixed
as independent of p'. At the final time $t_n = t$, however, the response function
obeys

$$J(p', t) = \frac{\partial q(p')}{\partial p'} = -\left[\frac{\partial^2 S(q, q', t)}{\partial q \partial q'}\right]^{-1} \qquad (8.2.29)$$

where the last member is given by (8.2.18) since $p' = \frac{m}{\tau}(q_1 - q_0)$; it reveals
that the response function determines the preexponential factor in the Van Vleck
propagator (8.2.25). Each path of the family considered obeys the equation of
motion (8.2.7) which upon differentiation with respect to p' yields

$$\frac{m}{\tau^2}[J(p', t_{i+1}) - 2J(p', t_i) + J(p', t_{i-1})] = -V''(q_i)J(p', t_i).$$

The Jacobi equation (8.2.27) is met once more, and this time truly meant as
a linearized equation of motion. If one wants to determine $J(p', t_i)$ from here,
one needs a second initial condition that follows from the initial momentum as
$q_1 = q_0 + \frac{\tau}{m}p'$. Should the response function $J(p', t_i)$ vanish again at the final time
$t_n = t$, all member paths in the family would reunite in one and the same final
point $q(p', t_n) = q$; moreover, that final point must be conjugate to q' since, like
the eigenvector of the stability matrix G with vanishing eigenvalue the response
function $J(p', t_i)$ obeys the Jacobi equation and vanishes both initially and in
the end. To underscore the geometrical image of the family fanning out from q'
and reconverging in q^c, conjugate points are also called focal points.

One more time anticipating material to be presented in the next section, I
would like to mention yet another meaning of conjugate points. The propagator
$\langle q | e^{-iHt/\hbar} | q' \rangle$ sharply specifies the initial coordinate as q' but puts no restriction
on the initial momentum. From the classical point of view, a straight line L' (for
systems with a single degree of freedom) in phase space that runs parallel to the
momentum axis is thus determined. If every point on that line is dispatched along
the classical trajectory generated by the classical Hamiltonian function $H(q, p)$,
a time-dependent image L of the initial L' will arise. For very short times, L will
still be straight but will appear tilted against L' (see Fig. 8.2). At some finite
time t^c, when the potential energy has become effective, the image may and in
general will develop a first caustic above the q-axis, i.e., a point with vertical
slope, $\partial p/\partial q = \infty \Leftrightarrow \partial q/\partial p = 0$; but inasmuch as all points on L can still be
uniquely labelled by the initial momentum p' on L', these caustics can also be
characterized by $\partial q/\partial p' = 0$, i.e., by the conjugate-point condition. Therefore,

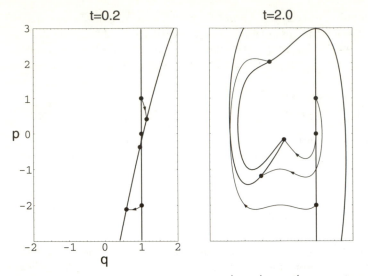

Fig. 8.2. Initial Lagrangian manifold L' at $q' = 1, t' = 0$ and its time-evolved images L at $t = 0.2$ (left) and $t = 2.0$ (right) for the one-dimensional double-well oscillator with the potential $V(q) = q^4 - q^2$. The directed thin lines depict orbits. The early-time L appears tilted clockwise against L' and still rather straight, as is typical for systems with Hamiltonians $H = p^2/2m + V(q)$. At the larger time $t = 2.0$, L has developed "whorls" and "tendrils" due to the anharmonicity of $V(q)$; moreover, four configuration-space caustics appear. This type of Lagrangian manifold is associated with the time-dependent propagator; it is transverse to the trajectories which carry it along as time evolves. Courtesy of Littlejohn [264]

a point in configuration space conjugate to the initial q' yields a caustic of L. Subsequent to the appearance of the first such caustic on L, the Van Vleck propagator can no longer consist of a single WKB branch in (8.2.15) but must temporarily comprise two such since there are two classical paths leading from the initial configuration-space point q' to the final q within the time span t (see Fig. 8.2). Eventually, when further caustics have appeared on L, more WKB branches arise in the propagator, as labelled by the index α and summed over in (8.2.25).

No essential difficulty is added for f degrees of freedom. Then the use of non-Cartesian coordinates and replacement of the Newtonian form of the equation of motion by the Lagrangian or Hamiltonian may be indicated. The fact is worth mentioning that conjugate points may and often do acquire multiplicities inasmuch as (at most f) eigenvalues λ may vanish simultaneously. Then, the Morse index counts the conjugate points with their multiplicities.

8.3 Trace Formulae and the Density of Levels à la Gutzwiller

Here, I shall derive the Gutzwiller type traces of quantum propagators. These are the principal ingredients for semiclassical approximations of quantum (quasi) energy spectra and and of measures of spectral fluctuations. For maps, the relevant propagator is the time-dependent one, $\langle q|U^n|q'\rangle$, while for autonomous flows, the tradition founded by Gutzwiller demands looking at the energy-dependent propagator alias resolvent, $(E - H)^{-1}$. The traces to be established will be semiclassical, i.e., valid only up to corrections of relative order $1/\hbar$ and under additional restrictions to be revealed as we progress. The starting point is the Van Vleck propagator which itself is a sum of contributions from classical orbits. The latter structure is inherited by the traces, and the contributing orbits are constrained to be periodic: Closure in configuration space is an immediate consequence of the traces being sums (or integrals) of diagonal elements $\langle q|U^n|q\rangle$ or $\langle q|(E-H)^{-1}|q\rangle$; then, periodicity in phase space results from a stationary-phase approximation in doing the integral which yields the condition $p = p'$.

8.3.1 Maps

As already explained in Chap. 4, the traces

$$
t_n = \int_{-\infty}^{+\infty} dx\, \langle x|U^n|x\rangle \tag{8.3.1}
$$

$$
\approx \sum_\alpha \int \frac{dx}{\sqrt{\mathrm{i}2\pi\hbar}} \sqrt{\left|\frac{\partial^2 S^{(n,\alpha)}}{\partial x \partial x'}\right|_{x=x'}}\; e^{\,\mathrm{i}\left\{S^{(n,\alpha)}(x,x)/\hbar - \nu^\alpha \pi/2\right\}}
$$

are the Fourier coefficients of the density of levels (see (4.13.1), reproduced as (8.3.8) below) and building blocks for the secular coefficients through Newton's formulae (see Sect. 4.14.3). Thus, their ($\hbar \to 0$)-approximants are the clues to a semiclassical discussion of quasi-energy spectra. Having given the semiclassical propagator (8.2.19), we just need to do the single x-integral in (8.3.1) and there once more employ the stationary-phase approximation. The stationary-phase condition, $dS^{(n,\alpha)}(q,q)/dq = [(\partial/\partial q + \partial/\partial q')S^{(n,\alpha)}(q,q')]_{q=q'} = p - p' = 0$, restricts the contributing paths to periodic orbits, for which the initial phase-space point q', p' and its nth classical iterate q, p coincide. The period n may be "primitive" or an integral multiple of a shorter primitive with period $n_0 = n/r$ and the number r of traversals a divisor of n. All n_0 distinct points along any one period-n orbit make the same (yes, the same, see below) additive contribution to the integral which may again be found from the Gaussian approximation $\exp\{\mathrm{i}S^{(n,\alpha)}(x,x)/\hbar\} \approx \exp\{(\mathrm{i}/\hbar)[S^{(n,\alpha)}(q,q) + \frac{1}{2}(S^{(n,\alpha)}(q,q))''(x - q)^2]\}$ and the Fresnel integral (8.2.12). So, we obtain Gutzwiller's trace formula

$$
t_n \approx \sum_\alpha n_0 \sqrt{\frac{1}{|(S^{(n,\alpha)}(q,q))''|}}\; \left|\frac{\partial^2 S^{(n,\alpha)}(q,q')}{\partial q \partial q'}\right|_{q=q'}\; e^{\,\mathrm{i}\left\{S^{(n,\alpha)}(q,q)/\hbar - \mu^{(n,\alpha)}\pi/2\right\}}
\tag{8.3.2}
$$

where the sum is over all period-n orbits, q is the coordinate of any of the n points on the αth such orbit, and $\mu^{(n,\alpha)}$ is the Maslov index[5]

$$\mu^{(n,\alpha)} = \nu^\alpha(q,q) + \frac{1}{2}\{1 - \mathrm{sign}((S^{(n,\alpha)}(q,q))'')\} \equiv \mu^\alpha_{\mathrm{prop}} + \mu^\alpha_{\mathrm{trace}}. \qquad (8.3.3)$$

A word is in order on the Morse index $\nu^\alpha(q,q) \equiv \mu^\alpha_{\mathrm{prop}}$ in the integrand in (8.3.1). I already pointed out in the previous subsection that $\nu^\alpha(q,q')$ depends on q,q' and the connecting path. Upon equating q and q' to the integration variable x, one must reckon with $\nu^\alpha(x,x)$ as an x-dependent quantity. On the other hand, $\nu^\alpha(x,x)$ is an integer. The n-step path under consideration will change continuously with x unless it gets lost as a classically allowed path; concurrent with continuous changes of x, its $\nu^\alpha(x,x)$ may jump from one integer value to another at some $\hat{x}$, but the asymptotic $\hbar \to 0$ contribution to the integral won't take notice unless $\hat{x}$ happened to be a stationary point of the action $S^{(n,\alpha)}(x,x)$, i.e., unless the n-step path were a period-n orbit in phase space rather than only a closed orbit in configuration space. There is no reason to expect a point $\hat{x}$ of jumping $\nu^\alpha(x,x)$ in coincidence with a point of stationary phase: Stationarity is a requirement for the *first* derivatives of the action while jumps of the Morse index take place when an eigenvalue of the matrix $G^{(n,\alpha)}_{ij}$ defining the *second* variation of the action around the path according to (8.2.11, 15) passes through zero.[6] Thus, the contribution of a point q on the αth period-n orbit involves the Morse index $\nu^\alpha(q,q)$ which characterizes a whole family of closed-in-configuration-space n-step paths near the period-n orbit in question. The question remains whether all n points on the period-n orbit have *one and the same* Morse index; they do, but I beg the reader's patience for the proof of that fact until further below.

An alternative form of the prefactor of the exponential in the semiclassical trace (8.3.2) explicitly displays the stability properties of the αth period-n orbit by involving its stability matrix (called monodromy matrix by many authors)

$$M = \left[\begin{pmatrix} \frac{\partial q}{\partial q'} \end{pmatrix}_{p'} \begin{pmatrix} \frac{\partial q}{\partial p'} \end{pmatrix}_{q'} \\ \begin{pmatrix} \frac{\partial p}{\partial q'} \end{pmatrix}_{p'} \begin{pmatrix} \frac{\partial p}{\partial p'} \end{pmatrix}_{q'} \right] = \begin{pmatrix} -\frac{S_{q',q'}}{S_{q,q'}} & -\frac{1}{S_{q,q'}} \\ \frac{(S_{q,q'})^2 - S_{q,q}S_{q',q'}}{S_{q,q'}} & -\frac{S_{q,q}}{S_{q,q'}} \end{pmatrix} \qquad (8.3.4)$$

which latter defines the n-step phase-space map as linearized about the initial point q',p'. The second of the above equations follows from the generating-function property (8.2.9) of the action; to save on war paint, subscripts on the action denote partial derivatives while the superscripts (n,α) have been and will from here on mostly remain dropped. Due to area conservation, $\det M = 1$, as is indeed clear from the last member in (8.3.4). Therefore, the two eigenvalues of M are reciprocal to one another; if they are also mutual complex conjugates, we confront a stable orbit; positive eigenvalues signal hyperbolic behavior

---

[5] Somewhat arbitrarily but consistently, I shall speak of Morse indices in propagators and Maslov indices in trace formulae.

[6] Should such nongeneric a disaster happen, there is a way out shown by Maslov: one changes to, say, the momentum representation; see next section.

and negative ones signal inverse hyperbolicity. According to (8.3.4), the trace of M is related to derivatives of the action such that the prefactor takes the form $\sqrt{\cdots} = 1/\sqrt{|\det(M-1)|} = 1/\sqrt{|2 - \mathrm{Tr}M|}$, whereupon we may write t_n as [241–243][7]

$$t_n \approx \sum \frac{n_0}{\sqrt{|2 - \mathrm{Tr}M|}} \, \mathrm{e}^{\mathrm{i}\{S(q,q)/\hbar - \mu\pi/2\}} \,. \tag{8.3.5}$$

It is quite remarkable that the trace of the n-step propagator is thus expressed in terms of canonically invariant properties of classical period-n orbits. To check on the invariance of $S^{(n,\alpha)}(q,q)$, first, think of a general periodically driven Hamiltonian system with $H(p, q, t + \tau) = H(p, q, t)$, and let a canonical transformation $p, q \to P, Q$ be achieved by the generating function $F(Q, q)$ according to $P = \partial F(Q, q)/\partial Q \equiv F_Q$ and $p = -\partial F(Q, q)/\partial q \equiv -F_q$. The action accumulated along a classical path from q' to q (equivalently, from Q' to Q) during a time span t may be calculated with either set of coordinates, and the most basic property of the generating function yields the relation $\int_{q'}^{q}[\tilde{p}d\tilde{q} - H(\tilde{q}, \tilde{p}, \tilde{t})d\tilde{t}] = \int_{Q'}^{Q}\{\tilde{P}d\tilde{Q} - H[q(\tilde{Q}, \tilde{P}), p(\tilde{Q}, \tilde{P}), \tilde{t}]d\tilde{t}\} + F(Q, q) - F(Q', q')$; but for a periodic path with period $n\tau$ from q to $q' = q$ and similarly $Q = Q'$, the difference $F(Q, q) - F(Q', q')$ vanishes such that the two actions in question turn out to be equal.

A little more labor is needed to verify the canonical invariance of the trace of the stability matrix. Starting from $(\frac{\partial P}{\partial P'})_{Q'} + (\frac{\partial Q}{\partial Q'})_{P'}$ one thinks of the new phase-space coordinates P, Q as functions of the old ones and uses the chain rule to write

$$
\begin{aligned}
\left(\frac{\partial Q}{\partial Q'}\right)_{P'} &= \left(\frac{\partial Q}{\partial q}\right)_p\left(\frac{\partial q}{\partial Q'}\right)_{P'} + \left(\frac{\partial Q}{\partial p}\right)_q\left(\frac{\partial p}{\partial Q'}\right)_{P'} \\
&= \left(\frac{\partial Q}{\partial q}\right)_p\left[\left(\frac{\partial q}{\partial q'}\right)_{p'}\left(\frac{\partial q'}{\partial Q'}\right)_{P'} + \left(\frac{\partial q}{\partial p'}\right)_{q'}\left(\frac{\partial p'}{\partial Q'}\right)_{P'}\right] \\
&\quad + \left(\frac{\partial Q}{\partial p}\right)_q\left[\left(\frac{\partial p}{\partial q'}\right)_{p'}\left(\frac{\partial q'}{\partial Q'}\right)_{P'} + \left(\frac{\partial p}{\partial p'}\right)_{q'}\left(\frac{\partial p'}{\partial Q'}\right)_{P'}\right]
\end{aligned}
\tag{8.3.6}
$$

and similarly for the second term in the trace. Upon employing the generating function $F(Q, q, t)$, we get $dP = F_{QQ}dQ + F_{Qq}dq$, $dp = -F_{qQ}dQ - F_{qq}dq$ and thus the Jacobian matrices (in analogy to (8.3.4))

$$
\frac{\partial(Q, P)}{\partial(q, p)} = \begin{pmatrix} -\frac{F_{qq}}{F_{qQ}} & -\frac{1}{F_{qQ}} \\ \frac{F_{qQ}F_{Qq} - F_{QQ}F_{qq}}{F_{qQ}} & -\frac{F_{QQ}}{F_{qQ}} \end{pmatrix}, \quad \frac{\partial(q, p)}{\partial(Q, P)} = \begin{pmatrix} -\frac{F_{QQ}}{F_{Qq}} & \frac{1}{F_{Qq}} \\ \frac{F_{QQ}F_{qq} - F_{qQ}F_{Qq}}{F_{Qq}} & -\frac{F_{qq}}{F_{Qq}} \end{pmatrix}.
$$

With these Jacobians inserted in (8.3.6) and the corresponding $\partial P/\partial P'$ and realizing that for a periodic point $q = q', p = p'$ as well as $Q = Q', P = P'$ we immediately conclude that

$$\left(\frac{\partial Q}{\partial Q'}\right)_{P'} + \left(\frac{\partial P}{\partial P'}\right)_{Q'} = \left(\frac{\partial q}{\partial q'}\right)_{p'} + \left(\frac{\partial p}{\partial p'}\right)_{q'}, \tag{8.3.7}$$

---

[7] Remarkably, the semiclassically approximate equality in (8.3.5) becomes a rigorous equality for the cat map, as was shown by *Keating* in Ref. [243].

i.e., the asserted canonical invariance of the trace of the stability matrix of a periodic orbit. The more difficult proof of the canonical invariance of the Maslov index will be discussed in Sect. 8.4.

At this point we can partially check, that in the trace formulae (8.3.2, 5), the n_0 distinct points along an orbit with primitive period n_0 contribute identically such that when summing over periodic orbits rather than periodic points we encounter the primitive period as a factor: According to (8.2.8) the function $S^{(n,\alpha)}(q_i, q_i)$ differs for the n_0 points q_i only in the irrelevant order of terms in the sum over single steps. The square root in the prefactor must also be the same for all points on the periodic orbit, simply since the stroboscopic time evolution may itself be seen as a canonical transformation and since the trace of the stability matrix is a canonical invariant, as shown right above. Similarly, once we have shown that the Maslov index is a canonical invariant, we can be sure of its independence of the periodic points along a periodic orbit.

Inspection of (8.2.8) also reveals that the contribution of a period-n orbit with shorter primitive period $n_0 = n/r$ may be rewritten as $S^{(n,\alpha)} = rS^{(n_0,\alpha)}$. As regards the prefactor for a repeated orbit of primitive period $n_0 = n/r$, we may invoke the meaning of M as the map linearized about the orbit to conclude the multiplicativity $M^{(n)} = (M^{(n_0)})^r$. Once again, the Maslov index is more obstinate than the other ingredients of the trace formula; its additivity for repeated traversals, $\mu^{(n,\alpha)} = r\mu^{(n_0,\alpha)} \equiv r\mu_\alpha$, which unfortunately holds only for unstable orbits, will be shown in Sect. 8.4. The Maslov index for stable orbits, as appear for systems with mixed phase spaces, need not be additive with respect to repeated traversals.

Now, the quasi-energy density is accessible through the Fourier transform

$$\varrho(\phi) = \frac{1}{2\pi N} \sum_{n=-\infty}^{\infty} t_n e^{in\phi} = \frac{1}{2\pi N}\left\{ N + 2\,\mathrm{Re} \sum_{n=1}^{\infty} t_n e^{in\phi} \right\}, \tag{8.3.8}$$

where a finite dimension N is assumed for the Hilbert space. I immediately proceed to a spectral average of the product of two densities,

$$
\begin{aligned}
\langle \varrho(\phi + \frac{2\pi}{N}e)\varrho(\phi)\rangle &= \int_0^{2\pi} \frac{d\phi}{2\pi} \varrho(\phi + \frac{2\pi}{N}e)\varrho(\phi) \\
&= \left(\frac{1}{2\pi N}\right)^2 \sum_{n=-\infty}^{\infty} |t_n|^2 \exp\{i\,n\frac{2\pi}{N}e\} \\
&= \left(\frac{1}{2\pi N}\right)^2 \left[N^2 + 2\sum_{n=1}^{\infty} |t_n|^2 \cos\left(n\frac{2\pi}{N}e\right)\right].
\end{aligned} \tag{8.3.9}
$$

This can be changed into the two-point cluster function in the usual way.[8]

It may be well to emphasize that a smooth two-point function results only after one more integration since the densities provide products of two delta functions. The form factor $|t_n|^2$ is such an integral and thus free of delta spikes (but

---

[8] Subtract and then divide by the product $(1/2\pi)^2$ of two mean densities, take out the delta function provided by the self-correlation terms, and change the overall sign.

still displays wild fluctuations in its n-dependence (see Sect. 4.16)). A cheap way of smoothing the above two-point function is to truncate the sum over n at some finite $n_{\max}$. Already in the quasi-energy density (8.3.8), that truncation will "regularize", roughly as $\delta(\phi - \phi_i) \to \sin[(n_{\max} + \frac{1}{2})(\phi - \phi_i)] / \sin[\frac{1}{2}(\phi - \phi_i)]$.

To fully define the spectrum within the semiclassical treatment we are pursuing we would need all periodic orbits with periods between 1 and $N/2$ since the traces t_n with larger n can be determined with the help of self-inversiveness, Newton's formulae, and the Hamilton Cayley theorem, as discussed in Sect. 4.14.3.

The most important information about the specifics of a given dynamic system with global classical chaos is expected to be encoded in short periodic orbits, since as the period n grows to infinity, typical orbits will follow the universal trend to ergodic exploration of the phase space. Such classical universality corresponds to the quantum universality of the spectral fluctuations on the (quasi)energy scale of the mean nearest neighbor spacing. The expectations just formulated constitute a strong motivation to seek semiclassical justification of the success of random-matrix theory. Now, the frame is set for a discussion of such efforts.

8.3.2 Flows

Gutzwiller suggests [237] proceeding toward the energy spectrum through the energy-dependent propagator defined by the one-sided Fourier transform

$$G(q, q', E) = \frac{1}{i\hbar} \int_0^\infty dt \, e^{iEt/\hbar} \langle q | e^{-iHt/\hbar} | q' \rangle = \langle q | \frac{1}{E - H} | q' \rangle. \qquad (8.3.10)$$

To ensure convergence, the energy variable E here must be endowed with a positive imaginary part. Indeed, the trace of the imaginary part of $G(q, q', E)$ yields the density of levels as $\mathrm{Im}E \to 0^+$. After inserting the Van Vleck propagator (8.2.25), it is natural again to invoke the semiclassical limit and evaluate the time integral by a stationary-phase approximation. The new stationarity condition is

$$E + \frac{\partial S^\alpha(q, q', t)}{\partial t} = E - H(q, q', t) = 0. \qquad (8.3.11)$$

Note that the Hamiltonian function here does not appear with its natural variables and therefore is time-dependent through $H(q, p) = H\big(q, p(q, q', t)\big) = H\big(q', p'(q, q', t)\big) = H(q, q', t)$. The phase of the exponential in our time integral is stationary for the set of times solving (8.3.11). A new selection of classical paths is so taken since besides q and q', it is now the energy E rather than the time span t that is prescribed. Just as we could have labelled the paths contributing to the time-dependent propagator by their energy, we could so employ their duration now. Actually, one usually does neither but rather puts some innocent looking label on all quantities characterizing a contributing path. Calling $t^\alpha(q, q', E)$ the roots of the new stationary-phase condition (8.3.11) and confining ourselves to the usual quadratic approximation in the expansion of the phase in powers of $t - t^\alpha$, we get

$$
\begin{aligned}
G_{\mathrm{osc}}(q, q', E) &= \frac{1}{\mathrm{i}\hbar} \sum_\alpha \sqrt{\left| \frac{\partial^2 S^\alpha(q, q', t^\alpha)}{\partial q \partial q'} \right|} \exp\left\{ \frac{\mathrm{i}}{\hbar}[Et^\alpha + S^\alpha(q, q', t^\alpha)] - \mathrm{i}\nu^\alpha \frac{\pi}{2} \right\} \\
&\quad \times \int_0^\infty \frac{dt}{\sqrt{\mathrm{i}2\pi\hbar}} \exp\left[\frac{\mathrm{i}}{2\hbar} \ddot{S}^\alpha(q, q', t^\alpha)(t - t^\alpha)^2 \right] \\
&= \frac{1}{\mathrm{i}\hbar} \sum_\alpha \sqrt{\left| \frac{1}{\ddot{S}^\alpha(q, q', t^\alpha)} \frac{\partial^2 S^\alpha(q, q', t^\alpha)}{\partial q \partial q'} \right|} \\
&\quad \times \exp\left\{ \frac{\mathrm{i}}{\hbar}[Et^\alpha + S^\alpha(q, q', t^\alpha)] - \mathrm{i}\kappa^\alpha \frac{\pi}{2} \right\}
\end{aligned}
\tag{8.3.12}
$$

where the double dot on $\ddot{S}^\alpha(q, q', t)$ requires two partial differentiations with respect to time at constant q and q'. The "osc" attached to $G_{\mathrm{osc}}(q, q', E)$ signals that the Gaussian approximation about the times t^α of stationary phase cannot do justice to very short paths where t^α is very close to the lower limit 0 of the time integral; their contributions will be added below under the name $\langle G \rangle$ such that eventually $G \approx \langle G \rangle + G_{\mathrm{osc}}$. The Morse index ν^α counts the number of conjugate points along the αth path up to time t^α, while for the successor κ^α in the last member of the foregoing chain, by appeal to the Fresnel integral (8.2.12),

$$
\kappa^\alpha = \nu^\alpha + \frac{1}{2}\left\{ 1 - \mathrm{sign}(\ddot{S}^\alpha(q, q', t^\alpha)) \right\}.
\tag{8.3.13}
$$

I shall come back to that index in the next section and refer to it as the Morse index of the energy-dependent propagator; we shall see that it is related to the number of turning points (points where the velocity $\dot{q}(t)$ changes sign) passed along the orbit. Now it is more urgent to point to another feature of the exponential, the appearance of the time-independent action

$$
S_0(q, q', E) = S(q, q', t) + Et
\tag{8.3.14}
$$

which may be seen as related to the original action by a Legendre transformation. It is natural to express the preexponential factor in the last member of (8.3.12) in terms of S_0 and its derivatives with respect to its "natural variables" as well. One obtains[9]

$$
\begin{aligned}
G_{\mathrm{osc}}(q, q', E) &= \frac{1}{\mathrm{i}\hbar} \sum_\alpha A^\alpha(q, q', E) \exp\left[\frac{\mathrm{i}}{\hbar} S_0^\alpha(q, q', E) - \mathrm{i}\kappa^\alpha \frac{\pi}{2} \right], \\
A^\alpha(q, q', E) &= \sqrt{\left| \det\begin{pmatrix} \frac{\partial^2 S_0^\alpha}{\partial q \partial q'} & \frac{\partial^2 S_0^\alpha}{\partial q \partial E} \\ \frac{\partial^2 S_0^\alpha}{\partial E \partial q'} & \frac{\partial^2 S_0^\alpha}{\partial E^2} \end{pmatrix} \right|}
\end{aligned}
\tag{8.3.15}
$$

by the usual prestidigitation of changing variables; here, the transformation $t^\alpha \to E$ according to a solution $t^\alpha(q, q', E)$ of the stationary-phase condition; to avoid

---

[9] The structures of (8.2.25) and (8.3.15) are similar; one might see the difference as the result of an extension of the phase space by inclusion of the pair E, t.

notational hardship, I wave good-bye to the index α, write S^0 for S_0^α, and denote partial derivatives of $S(q, q', t), S^0(q, q', E), t(q, q', E)$, and $E(q, q', t)$ by suffixes. Starting with the stationary-phase condition (8.3.11), for the first derivatives of the two action functions,

$$
\begin{aligned}
S_t &= -E \\
S_E^0 &= S_t t_E + t + E t_E = t \\
S_q^0 &= S_q + S_t t_q + E t_q = S_q = p, \quad S_{q'}^0 = S_{q'} = -p';
\end{aligned}
\tag{8.3.16}
$$

the first of the foregoing identities, together with (8.3.14), constitutes the Legendre transformation from the time-dependent to the energy-dependent action, inasmuch as it yields the time $t(q, q', E)$ taken by a classical phase-space trajectory from q' to q on the energy shell; the third line in (8.3.16) reveals that the energy-dependent action is the generating function for the classical transition on the energy shell, $p = \partial S^0(q, q', E)/\partial q$ and similarly $p' = -\partial S^0(q, q', E)/\partial q'$. Now, we fearlessly proceed to the second derivatives, starting with $S_{q,q'} = S_{q,q'}^0 + S_{q,E}^0 E_{q'}$; but here $E_{q'}$ can be expressed in terms of other second derivatives of S^0 since $dt = t_q dq + t_{q'} dq' + t_E dE$ at constant q and t yields $E_{q'} = -t_{q'}/t_E = -S_{Eq'}^0/S_{EE}^0$ such that indeed $S_{qq'}$ is expressed through derivatives of S^0 as

$$
S_{qq'} = S_{qq'}^0 - \frac{S_{qE}^0 S_{Eq'}^0}{S_{EE}^0}.
\tag{8.3.17}
$$

Finally, $S_{tt} = -E_t = -1/t_E = -1/S_{EE}^0$, whereupon the radicand under study,

$$
\frac{S_{qq'}}{S_{tt}} = S_{qE}^0 S_{Eq'}^0 - S_{EE}^0 S_{qq'}^0,
\tag{8.3.18}
$$

assumes the asserted determinantal form.

Inasmuch as dynamics with nonintegrable classical limits are to be included we must generalize to f degrees of freedom, and $f \geq 2$. The change is but little for $G(q, q' E)$: The determinant in the prefactor A becomes $(f+1) \times (f+1)$, with $\partial^2 S/\partial q \partial q'$ an $f \times f$ matrix bordered by an f-component column $\partial^2 S/\partial q \partial E$ to the right and a row $\partial^2 S/\partial E \partial q$ from below.[10] An overall factor $\sqrt{1/\mathrm{i}2\pi\hbar}^{f-1}$ arises, as is easy to check by going back to (8.2.3) where the free-particle propagator in f dimensions comes with a prefactor $\sqrt{m/\mathrm{i}2\pi\hbar\tau}^f$; of that, the part m/τ is eaten up for good by the f-dimensional generalization of (8.2.19) such that the fate of $\sqrt{1/\mathrm{i}2\pi\hbar}^f$ must be followed; the time integral in (8.3.12) takes away one of the f factors $1/\sqrt{\mathrm{i}2\pi\hbar}$; these remarks should also prepare the reader for the further changes in the prefactor toward the final trace formula. I shall not bother with arbitrary f but rather specialize to $f = 2$ from here on since this is enough for most applications known and also simplifies the work to be done.

---

[10] While this generalization is suggested by the very appearance of the prefactor in (8.3.15) and the previous footnote, and thus easy to guess, serious work is needed to actually verify the result.

Now, we must attack the trace

$$\int d^2x \, G_{\text{osc}}(x, x, E) = \frac{1}{i\hbar\sqrt{i2\pi\hbar}} \sum_{\alpha} \int d^2x \, A \, e^{i\{S^0/\hbar - \kappa\pi/2\}} \qquad (8.3.19)$$

by doing the twofold configuration-space integral, surely invoking the stationary-phase approximation. The stationary-phase condition, $(\frac{\partial}{\partial x} + \frac{\partial}{\partial x'})S^0(x, x')|_{x=x'} = p - p' = 0$, again restricts the contributing paths to periodic ones. But in contrast to the period-n orbits encountered in the previous subsection, which were a sequence of n points (q_i, p_i) with contributions to be summed over, now we encounter a continuous sequence of periodic points along each periodic orbit. The sum over their contributions takes the form of a single integral along the orbit, and that integral cannot be simplified by any Gaussian approximation of the integrand. It is only the integration transverse to a periodic orbit for which the stationary-phase approximation invites a meaningful Gaussian approximation. I shall specify the two configuration-space coordinates summarily denoted by x above as one along the orbit, $x_\parallel$, and a transverse one, $x_\perp$; the transverse coordinate is kin to the one and only one incurred for maps with $f = 1$ in the preceding subsection, cf (8.3.1). The two coordinates $x_\parallel, x_\perp$ are assumed orthogonal such that the corresponding axes necessarily change along the orbit and $d^2x = dx_\parallel dx_\perp$; moreover, the $x_\perp$-axis is taken as centered at the orbit, i.e., $x_\perp = 0$ and $\dot{x}_\perp = 0$ thereon. Such a choice helps with the twofold integral over $x_\parallel$ and $x_\perp$, first by giving a simple form to the 3×3 determinant in the prefactor A. To see this, we take various derivatives of the conservation law $H(q, p) = E$, $H(q', p') = E$, considering the canonical momenta as well as the action $S^0(q, q', E)$ as functions of q, q', and E; also observing $q = (x_\parallel, x_\perp)$, $p = (p_\parallel, p_\perp)$, and $\partial H/\partial p_\perp = \dot{x}_\perp = 0$ (and similarly for the primed quantities), we get

$$\left[\frac{\partial H(q, p)}{\partial E}\right]_q = 1 = \frac{\partial H(q, p)}{\partial p_\parallel}\frac{\partial p_\parallel}{\partial E} = \dot{x}_\parallel \frac{\partial p_\parallel}{\partial E} = \dot{x}_\parallel S^0_{x_\parallel E},$$

$$\left[\frac{\partial H(q', p')}{\partial E}\right]_{q'} = 1 = \dot{x}'_\parallel \frac{\partial p'_\parallel}{\partial E} = -\dot{x}'_\parallel S^0_{x'_\parallel E},$$

$$\left[\frac{\partial H(q', p')}{\partial x_\perp}\right]_{q', E} = 0 = \dot{x}'_\parallel \frac{\partial p'_\parallel}{\partial x_\perp} = -\dot{x}'_\parallel S^0_{x'_\parallel x_\perp}, \qquad (8.3.20)$$

$$\left[\frac{\partial H(q', p')}{\partial x_\parallel}\right]_{q', E} = 0 = \dot{x}'_\parallel \frac{\partial p'_\parallel}{\partial x_\parallel} = -\dot{x}'_\parallel S^0_{x_\parallel x'_\parallel},$$

$$\left[\frac{\partial H(q, p)}{\partial x'_\perp}\right]_{q, E} = 0 = \dot{x}_\parallel \frac{\partial p_\parallel}{\partial x'_\perp} = \dot{x}_\parallel S^0_{x_\parallel x'_\perp},$$

where on the left-hand sides the suffixes attached to the closing brackets display the variables to be held constant while taking derivatives.

Thus, the prefactor A from (8.3.15) becomes

$$A(q, q', E) = \sqrt{\left| \det \begin{pmatrix} 0 & 0 & 1/\dot{x}_\| \\ 0 & S^0_{x_\perp x'_\perp} & S^0_{x_\perp E} \\ -1/\dot{x}'_\| & S^0_{E x'_\perp} & S^0_{EE} \end{pmatrix} \right|} = \sqrt{\left| \frac{1}{\dot{x}_\| \dot{x}'_\|} \frac{\partial^2 S^0}{\partial x_\perp \partial x'_\perp} \right|},$$

(8.3.21)

whereupon the trace (8.3.19) appears in the form

$$\int d^2 x \, G_{\mathrm{osc}}(x, x, E) = \frac{1}{i\hbar} \sum_\alpha$$

(8.3.22)

$$\oint \frac{dx_\|}{|\dot{x}_\||} \int \frac{dx_\perp}{\sqrt{i 2\pi\hbar}} \sqrt{\left. \left| \frac{\partial^2 S^0(x_\perp, x_\|, x'_\perp, x_\|, E)}{\partial x_\perp \partial x'_\perp} \right| \right|_{x_\perp = x'_\perp}} \, \mathrm{e}^{\,i\{S^0(x, x, E)/\hbar - \kappa\pi/2\}}.$$

Integrating over $x_\perp$ with the usual quadratic approximation in the exponent, $S^0(x, x, E) = S(E) + [(\partial/\partial x_\perp + \partial/\partial x'_\perp)^2 S(x_\perp, x_\|, x'_\perp, x_\|, E)]_{x_\perp = x'_\perp = 0} \, x_\perp^2/2$, we get

$$\int \frac{dx_\perp}{\sqrt{i 2\pi\hbar}} \sqrt{\left. \left| \frac{\partial^2 S(x_\perp, x_\|, x'_\perp, x_\|, E)}{\partial x_\perp \partial x'_\perp} \right| \right|_{x_\perp = x'_\perp}} \, \mathrm{e}^{\,i S(x, x, E)/\hbar}$$

(8.3.23)

$$\approx \sqrt{\left. \left| \frac{\partial^2 S(x_\perp, x_\|, x'_\perp, x_\|, E)/\partial x_\perp \partial x'_\perp}{(\frac{\partial}{\partial x_\perp} + \frac{\partial}{\partial x'_\perp})^2 S(x_\perp, x_\|, x'_\perp, x_\|, E)} \right| \right|_{x_\perp = x'_\perp = 0}} \, \mathrm{e}^{\,i\{S(E)/\hbar - \mu_{\mathrm{trace}}\pi/2\}}$$

with the index ν according to the Fresnel integral (8.2.12),

$$\mu_{\mathrm{trace}} = \frac{1}{2}(1 - \mathrm{sign}\{[(\partial/\partial x_\perp + \partial/\partial x'_\perp)^2 S(x, x_\|, x', x_\|, E)]_{x = x' = 0}\}). \quad (8.3.24)$$

Note that I have somewhat sloppily written $S(E)$ for $S^0(0, x_\|, 0, x_\|, E)$ thus expressing the fact that the action is independent of the coordinate $x_\|$ along the periodic orbit and judging redundant the superscipt "0" once writing the energy as an argument of the action.

The index μ_{trace} resulting from the trace operation combines with the Maslov index κ from the energy-dependent propagator to the full Maslov index of the periodic orbit

$$\mu = \kappa + \mu_{\mathrm{trace}} \equiv \mu_{\mathrm{prop}} + \mu_{\mathrm{trace}}. \quad (8.3.25)$$

In further implementing the restriction to periodic orbits, it is well to realize that the transverse derivatives (w.r.t. $x_\perp$ and $x'_\perp$) of the action taken at $x_\perp = x'_\perp = 0$ are also independent of $x_\|$. If we take for granted, subject to later proof, that the full Maslov index μ is $x_\|$-independent as well, the final $x_\|$-integral $\oint dx_\|/\dot{x}_\| = T$ yields the primitive period of the orbit; it is the primitive period since the original configuration-space integral "sees" the orbit as a geometrical object without noticing repeated traversals. Thus, the trace we pursue becomes

$$\int d^2x \, G_{\text{osc}}(x, x, E) \approx \tag{8.3.26}$$

$$\frac{1}{i\hbar} \sum_\alpha T \sqrt{\left| \frac{\partial^2 S(x_\perp, x_\parallel, x'_\perp, x'_\parallel, E)/\partial x_\perp \partial x'_\perp}{(\frac{\partial}{\partial x_\perp} + \frac{\partial}{\partial x'_\perp})^2 S(x_\perp, x_\parallel, x'_\perp, x_\parallel, E)} \right|_{x_\perp = x'_\perp = 0}} \, e^{i\{S(E)/\hbar - \mu\pi/2\}} .$$

This trace formula looks remarkably similar to that obtained for maps with $f = 1$; see (8.3.2). In particular, the radicands in the prefactors are in complete correspondence such that by simply repeating the arguments of the previous subsection we express the present radicand in terms of the 2×2 stability matrix for a loop around the αth periodic orbit which should be envisaged as giving a map in a Poincaré section spanned by $x_\perp, p_\perp$. Thus, the final result for our trace, in analogy to (8.3.5), reads

$$\int d^2x \, G_{\text{osc}}(x, x, E) \approx \frac{1}{i\hbar} \sum_\alpha \frac{T}{\sqrt{|2 - \text{Tr}M|}} \, e^{iS(E)/\hbar} \equiv g_{\text{osc}}(E) \tag{8.3.27}$$

where for notational convenience the "action" $\mathcal{S}(E) \equiv S(E) - \hbar\mu\pi/2$ is defined to include the Maslov phase. Upon taking the imaginary part, we arrive at the oscillatory part of the density of levels[11]

$$\varrho_{\text{osc}}(E) \approx \frac{1}{\pi\hbar} \sum \frac{T}{\sqrt{|2 - \text{Tr}M|}} \cos(\mathcal{S}/\hbar). \tag{8.3.28}$$

It is appropriate to emphasize that orbits which are r-fold traversals of shorter primitive ones contribute with their primitive period T while the stability matrix M is the rth power of the one pertaining to the primitive orbit, just as for maps.

8.3.3 Weyl's Law

An integral of the form $\int_\tau^\infty dt \, A(t) e^{i\nu\Phi(t)}$ with large ν draws its leading contributions not only from the "points" t_α of stationary phase but also from near the boundary τ of the integration range. The pertinent asymptotic result is [247]

$$\int_\tau^\infty dt \, A(t) e^{i\nu\Phi(t)} \approx \frac{iA(\tau) e^{i\nu\Phi(\tau)}}{\nu\dot\Phi(\tau)} + \{\text{stationary-phase contributions}\}. \tag{8.3.29}$$

The one-sided Fourier transform defining the energy-dependent propagator in (8.3.10) is such an integral with $1/\hbar$ as ν and $\tau \to 0$. While the stationary-phase contributions were evaluated in the preceding subsection as $G_{\text{osc}}(E)$, now we

---

[11] Departing from the convention mostly adhered to in this book, here I do not restrict myself to Hilbert spaces with the finite dimension N and thus do not write a factor $1/N$ into the density of levels; the reader is always well advised to check from the context on the normalization of $\varrho(E)$.

turn to the boundary contribution $\langle G(E) \rangle$. Employing the Van Vleck form of the time-dependent propagator, we run into the phase $\Phi(t) = Et + S(q, q', t)$ and its derivative $\dot{\Phi}(t) = E - H(q, q', t)$. Thus, the boundary term reads

$$\int d^f q \, \langle G(q, q, E) \rangle \approx \lim_{\tau \to 0} \int d^f q d^f q' \, \frac{\delta(q - q')}{E - H(q, q', \tau)} \, \langle q | e^{-i\hat{H}\tau/\hbar} | q' \rangle \qquad (8.3.30)$$

where for brevity $\langle q | e^{-i\hat{H}\tau/\hbar} | q' \rangle$ stands for the Van Vleck propagator; the reader will recognize $\hat{H}$ as the Hamiltonian operator and distinguish it from the Hamiltonian function $H(q, q', \tau)$ appearing in the denominator. To see that the foregoing $(\tau \to 0)$-limit is well defined (rather than a bewildering $0 \times \infty$), we represent the delta function by a Fourier integral, $\delta(q) = (2\pi\hbar)^{-f} \int_{-\infty}^{\infty} d^f p \, e^{-ipq/\hbar}$, and think of the q'-integral as done by a stationary phase. The stationary-phase condition $\frac{\partial}{\partial q'}(S(q, q', \tau) + pq') = p - p'(q, q', \tau) = 0$ restores the natural variables q, p to the Hamiltonian function in the denominator, such that we may write

$$\int d^f q \, \langle G(q, q, E) \rangle \approx \lim_{\tau \to 0} \frac{1}{h^f} \int d^f q d^f q' d^f p \, \frac{e^{-ip(q-q')/\hbar}}{E - H(q, p)} \, \langle q | e^{-i\hat{H}\tau/\hbar} | q' \rangle .$$
$$(8.3.31)$$

Note that while taking advantage of the benefit $H(q, q', t) \to H(q, p)$ of the aforementioned stationary-phase approximation, I have otherwise still refrained from integrating over q'. Such momentary hesitation is not necessary but convenient since after first taking the limit $\tau \to 0$ to get $\langle q | e^{-i\hat{H}\tau/\hbar} | q' \rangle \to \delta(q - q')$ the, q'-integral becomes trivial. Thus, the desired result

$$\int d^f q \, \langle G(q, q, E) \rangle \approx \frac{1}{h^f} \int d^f q d^f p \, \frac{1}{E - H(q, p)} \qquad (8.3.32)$$

is reached. Providing the energy E with a vanishingly small imaginary part and using the familiar identity $\mathrm{Im}\{1/(x - i0^+)\} = \pi\delta(x)$, we arrive at Weyl's law,

$$\langle \varrho(E) \rangle \approx \frac{1}{h^f} \int d^f q d^f p \, \delta(E - H) , \qquad (8.3.33)$$

which expresses the nonoscillatory alias average density of states as the number of Planck cells contained in the classical energy shell. Following common usage, I refer to $\langle \varrho(E) \rangle$ also as the Thomas–Fermi density. Variants of and corrections to Weyl's law can be found in Refs. [248–251]. Such corrections are not only interesting in their own right but become indispensable ingredients for semiclassical determinations of spectra for systems with more than two freedoms [252].

8.3.4 Limits of Validity and Outlook

Most obvious is the limitation to the lowest order in $\hbar$, due to the neglect of higher orders when approximating sums and integrals by stationary phase. Readers interested in corrections of higher order in $\hbar$ may consult *Gaspard*'s work [253] as well as the treatments of three-dimensional billiards by *Primack* and *Smilansky* [252] and *Prosen* [254]. Interestingly, there are dynamical systems for which the trace formulae become rigorous rather than lowest order approximations. Prominent among these are the cat map [243], billiards on surfaces of constant negative curvature [255], and graphs [256].

Inasmuch as all periodic orbits are treated to make independent additive contributions to the trace, one implicitly assumes that the orbits are isolated. Strictly speaking, one is thus reduced to hyperbolic dynamics with only unstable orbits. In practice, systems with stable orbits and mixed phase spaces can be admitted, provided one stays clear of bifurcations. Near a bifurcation, the periodic orbits involved, about to disappear or just arisen, are nearly degenerate in action and must be treated as clusters; that necessity is signalled by the divergence of the prefactor A in the single-orbit contribution $Ae^{iS/\hbar}$, due to the appearance of eigenvalues unity in the stability matrix. Some more remarks on mixed-phase spaces will be presented in Sect. 8.12.

The greatest worry is caused by phase-space structures finer than a Planck cell, dragged into the trace formulae by orbits with long periods. Imagine a stroboscopic map for a system with a compact phase space with total volume Ω, a Hilbert space of finite dimension N, and a Planck cell of size $h^f \approx \Omega/N$. Since the number of periodic orbits with periods up to n grows roughly like $e^{\lambda n}$, a Planck cell, on average, begins to contain more than one periodic point once the period grows larger than the Ehrenfest time, $n_E = \lambda^{-1} \ln(\Omega h^{-f})$, and is crowded by exponentially many periodic points for longer periods. For maps with finite N, one escapes the worst consequences of that exponential proliferation by taking the periodic-orbit version of the t_n only for $1 \le n \le N/2$ and using unitarity, Newton's formulae, and the Hamilton–Cayley theorem to express all other traces in terms of the first $N/2$ traces, as explained in Sect. 4.14.2. In particular, the N eigenphases can be obtained once the first $N/2$ traces t_n are known.

The proliferation seems more serious for the trace formulae (8.3.27, 28) for flows since orbits of all periods, even infinitely long ones, are involved. Much relief was afforded by a series of papers by *Berry* and *Keating* that culminated [257–259] in a resummation, relating the contributions of long orbits to those of short ones in a certain functional equation to be presented in Sect. 8.9.

It is interesting to see that convergence may be enforced for the trace formula (8.3.27) by letting the energy E become complex, $\mathcal{E} = E + i\epsilon$ with [260],

$$\epsilon = \mathrm{Im}\,\mathcal{E} > \epsilon_c \equiv \hbar\lambda/2, \tag{8.3.34}$$

where λ is an average of the positive Lyapunov exponent that roughly gives the eigenvalues of the stability matrix M as $e^{\pm\lambda T}$. Indeed, assuming that all orbits unstable and recalling that exponential proliferation implies that the number of orbits with periods near T is roughly $e^{\lambda T}$, we conclude that the contributions of long orbits run away as $\{\#\,\text{of p.o.'s with periods near T}\}/\sqrt{\det M} \approx e^{\lambda T/2}$; but the addition of a classically small imaginary part to the energy (ϵ of order $\hbar$) provides an exponential attenuation $\exp \mathrm{Im}\,S(E + i\epsilon)/\hbar = \exp(-\epsilon T)$ sufficient to overwhelm the exponential runaway if $\epsilon > \epsilon_c$. Deplorably, an imaginary part $\mathrm{Im}\,\mathcal{E} = \mathcal{O}(\hbar)$ wipes out all structure in the density of levels on the scale of a mean level spacing since the latter is, according to Weyl's law, of the order $\hbar^f$, i.e., already smaller than the convergence-enforcing ϵ_c for the smallest dimension of interest, $f = 2$ for autonomous systems; for periodically driven single-freedom systems whose quasi-energy spacing is $\mathcal{O}(\hbar)$, complexifying the quasi-energies might be more helpful but has not, to my knowledge, been put to test yet.

An interesting attempt at unifying the treatment of maps and autonomous flows and at fighting the divergence due to long orbits was initiated by *Bogomolny* [261]. It is based on semiclassically quantizing the Poincaré map for autonomous flows. Similar in spirit and as fruitful is the scattering approach of *Smilansky* and coworkers [262]; these authors exploit a duality between bound states inside and scattering states outside billiards. Respectful reference to the original papers must suffice here, to keep the promise of steering a short course through semiclassical terrain.

As yet insufficient knowledge prevents expounding yet another promising strategy of avoiding divergences for flows in compact phase spaces which would describe flows through stroboscopic maps. One would choose some strobe period T and aim at the traces $t_n = \mathrm{Tr}\, \mathrm{e}^{-inTH/\hbar}$ with integer n up to half the Heisenberg time. Thus, the unimodular eigenvalues of $\mathrm{e}^{-iHT/\hbar}$ are accessible, and the periods of the classical orbits involved not exceeding half the Heisenberg time. The strobe period T would have to be chosen sufficiently small such that the energy levels yield phases $E_i T/\hbar$ fitting into a single (2π)-interval.

To efficiently determine spectra from semiclassical theory, we must learn how to make do with periodic orbits much shorter than half the Heisenberg time, possibly with periods only up to the Ehrenfest time. Interesting steps in that direction have been taken by *Vergini* [263].

8.4 Lagrangian Manifolds, Maslov Theory, and Maslov Indices

8.4.1 Lagrangian Manifolds

To bridge some of the gaps left in the proceeding two sections and to get a fuller understanding of semiclassical approximations, I pick up a new geometrical tool, Lagrangian manifolds [264]. These are f-dimensional submanifolds of a $(2f)$-dimensional phase space to whose definition I propose to gently lead the willing here. We shall eventually see that these hypersurfaces have "generating functions" that are just the actions appearing in semiclassical wave functions or propagators of the structure $A\mathrm{e}^{iS/\hbar}$. Therefore, Lagrangian manifolds provide the caustics at which semiclassical wavefunctions diverge. Above its q-space caustics, one may climb up or down a Lagrangian manifold while appearing to stay put in q-space.

A simple such manifold was encountered in Fig. 8.2 for $f = 1$ as the set of all points with a single fixed value of the coordinate and arbitrary momentum; it naturally arose as the manifold specified by the initial condition of the propagator, $\langle q | \mathrm{e}^{-iHt} | q' \rangle_{t=0} = \delta(q - q')$. Inasmuch as all points on that manifold project onto a single point in q-space, the latter point is a highly degenerate caustic. The image of that most elementary Lagrangian manifold under classical Hamiltonian evolution at some later time t also qualifies as Lagrangian.

For $f \geq 1$, the foregoing examples are immediately generalized to all of momentum space above a single point in configuration space and the images thereof under time evolution. These may be seen as f-component vector fields $p(q)$ defined on the f-dimensional configuration space, but not all vector fields qualify as momentum fields for classical Hamiltonian dynamics. Inasmuch as phase space allows for reparametrization by canonical transformations, we may think of the admissible momentum fields as gradients $p = \partial S/\partial q$ of some scalar generating function S, and such vector fields are curl-free, $\partial p_i/\partial q_j = \partial p_j/\partial q_i$. Curl-free vector fields are Lagrangian manifolds albeit not the most general ones. It has already become clear in the preceding section that the initial manifold specified by the propagator may and in general does develop caustics under time evolution, i.e., points in configuration space where some of the derivatives $\partial p_i/\partial q_j$ become infinite and in some neighborhood of which the momentum field is necessarily multivalued (see Fig. 8.2 again).

Therefore, the proper definition of Lagrangian manifolds avoids derivatives: *An f-dimensional manifold in phase space is called Lagrangian if in any of its points any two of its $(2f)$-component tangent vectors, $\delta z^1 = (\delta q^1, \delta p^1)$ and $\delta z^2 = (\delta q^2, \delta p^2)$, have an antisymmetric product $\omega(\delta z^1, \delta z^2) \equiv \delta p^1 \cdot \delta q^2 - \delta p^2 \cdot \delta q^1$ which vanishes,*

$$\omega(\delta z^1, \delta z^2) \equiv \delta p^1 \cdot \delta q^2 - \delta p^2 \cdot \delta q^1 = 0; \tag{8.4.1}$$

here the dot between two f-component vectors means the usual scalar product, like $\delta p^1 \cdot \delta q^2 = \sum_{i=1}^{f} \delta p_i^1 \delta q_i^2$. I leave to the reader as problem 8.8 to show that *the antisymmetric product $\omega(\delta z^1, \delta z^2)$ is invariant under canonical transformations, i.e., independent of the phase-space coordinates used to compute it.*

We can easily check that the definition (8.4.1) comprises all of the aforementioned examples. For $f = 1$, the antisymmetric product $\omega(\delta z^1, \delta z^2)$ measures the area of the parallelogram spanned by the two vectors in phase space; requiring that area to vanish means restricting the two two-component vectors in question so as to be parallel to one another; indeed, then, we realize that all curves $p(q)$ are Lagrangian manifolds for single-freedom systems since all tangent vectors in a given point are parallel. For $f > 1$, one may use the antisymmetric product $\omega(\delta z^1, \delta z^2)$ to define phase-space area in two-dimensional subspaces of the $(2f)$-dimensional phase space. Moreover, we can consider two tangent vectors such that δz^1 has only a single nonvanishing q-ish component, say δq_i, and likewise only a single nonvanishing p-ish one, δp_j, while, similarly, δz^2 has only δq_j and δp_i as nonvanishing entries. For the f-dimensional manifold in question to be Lagrangian, we must have $\delta p_j \delta q_j = \delta p_i \delta q_i$; now we may divide that equation by the nonvanishing product $\delta q_i \delta q_j$ and conclude that curl-free vector fields yield Lagrangian manifolds. Finally, we check that {all of p-space above a fixed point in q-space} (as distinguished by the initial condition of the propagator in the q-representation) as well as {all of q-space above a fixed point in p-space} (distinguished by the initial condition of the propagator in the p-representation) fit the definition (8.4.1). Then, the image of the latter manifold under time evolution is Lagrangian as well since time evolution is a canonical transformation.

It may be well to furnish examples for concrete dynamical systems. The simplest is the harmonic oscillator with $f = 1$ where the propagator in the coordinate representation distinguishes the straight line $q = q'$ in the phase plane as the initial manifold L'. The time-evolved image L arises through rotation about the origin by the angle ωt; once every period, at the times $n\pi/\omega$ with $n = 2, 4, 6, \ldots$, L coincides with L'. Clearly, that initial Lagrangian manifold is caustic and even highly degenerate. One more such caustic per period arises at $q = -q'$, at the times $n\pi/\omega$ with $n = 1, 3, 5, \ldots$. None of these caustics corresponds to turning points unless the initial momentum is chosen as $p' = 0$. The caustics at $q = \pm q'$ define mutual conjugate points in coordinate space since all trajectories fanning out of them refocus there again (and again). For a less trivial example, the reader is referred back to Fig. 8.1 which refers to a kicked top with a two-dimensional spherical phase space and depicts an initial Lagrangian manifold spanned by all of configuration space at fixed momentum.

A type of Lagrangian manifold not contained in the set of examples given above arises when dealing with the energy-dependent propagator (8.3.15); I shall denote such manifolds by L_E. Within the $(2f)$-dimensional phase space is the $(2f - 1)$-dimensional energy shell, wherein upon picking a point q' in coordinate space, we consider a sub-manifold of $f - 1$ dimensions as well as its time-evolved images for arbitrary positive times. In the course of time, these time-evolved images sweep out an f-dimensional manifold that we recognize as Lagrangian as follows. For the sake of simplicity, I spell out the reasoning for $f = 2$ for which Fig. 8.3 provides a visual aid; moreover, I assume a Hamiltonian of the form $H = p^2/2m + V(q)$. The manifold L_E to be revealed as Lagrangian originates by time evolution from a set of points along, say, the p_1-direction above the picked point q' in the $(q_1 q_2)$-plane; in the p_2-direction, $p_2 = \pm\sqrt{2m(E - V(q)) - p_1^2}$ determined by the energy through $H = E$; the latter also restricts the set of admitted values of p_1 to the interval $-\sqrt{2m(E - V(q))} \leq p_1 \leq +\sqrt{2m(E - V(q))}$. The initial one-dimensional manifold so characterized sweeps out L_E as every initial point is dispatched along the trajectory generated by the Hamiltonian H. Within the three-dimensional $q_1 q_2 p_1$-space illustrated in Fig. 8.3, L_E appears as a sheet which starts out flat but may eventually develop ripples and thus caus-

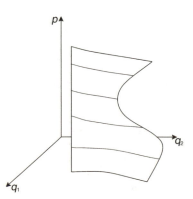

Fig. 8.3. Lagrangian manifold L_E associated with the energy-dependent propagator for $f = 2$, depicted as a surface in the three-dimensional energy shell. L_E is swept out by the time-evolved images of the initial line of fixed q_1, q_2; it starts out flat but in general tends to develop ripples and thus caustics. Note that the trajectories lie within L_E. Courtesy of Creagh, Robbins, and Littlejohn [266]

tics. By complementing the three coordinates with the uniquely determined p_2 attached to each point on the sheet, one gets L_E as a two-dimensional submanifold of the four-dimensional phase space.

Two independent tangent vectors offer themselves naturally at each point on L_E: One along the trajectory through the point, $\delta z^1 = \{\dot{q}(t; q', p_1'), \dot{p}(t; q', p_1')\}\delta t$, and the other, $\delta z^2 = \{(\partial q(t; q', p_1')/\partial p_1'), \partial p(t; q', p_1')/\partial p_1')\}\delta p_1'$, pointing toward a neighboring trajectory. Once more, by expressing the fourth components of these vectors in terms of the first three and invoking Hamilton's equations, we immediately check that their antisymmetric product vanishes, $\omega(\delta z^1, \delta z^2) = 0$, and thus reveals L_E as Lagrangian. To acquire familiarity with the technical background of this reasoning, the interested reader will want to write out the pertinent few lines of calculation.

The two tangent vectors just mentioned may be seen as elements of a texture covering the Lagrangian manifold L_E associated with the energy shell. In particular, the *trajectories running along L_E climb "vertically" in some momentum direction as they pass through a caustic of L_E. It follows that a caustic arising in the energy-dependent propagator corresponds to turning points of the trajectory involved* (see Fig. 8.4). *This is in contrast to the contributions of classical trajectories to the time-dependent Van Vleck propagator for which we have seen that caustics are related to points conjugate to the initial one, rather than to turning points.*

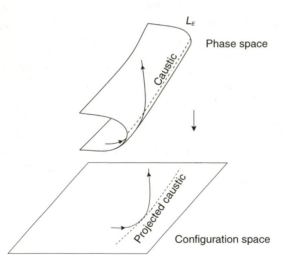

Fig. 8.4. While momentarily climbing vertically in L_E, a trajectory goes through a caustic that is associated with a turning point. Courtesy of Creagh, Robbins, and Littlejohn [266]

The statement that *the stable manifold L_s of any periodic orbit is Lagrangian* is useful for our subsequent discussion of Maslov indices.[12]. Imagine a point (q_0, p_0) on L_s and a tangent vector $\delta z = (\delta q, \delta p)$ attached to it. As this tangent vector changes into $\delta z(t)$ by transport along the trajectory through q_0, p_0, eventu-

---

[12] The stable and unstable manifolds of a periodic orbit are defined as the sets of points which the dynamics asymptotically carries toward the orbit as, respectively, $t \to +\infty$ and $t \to -\infty$; the reader is kindly asked to think for a moment about why L_s, as well as the unstable manifold L_u, are f-dimensional.

ally, by the definition of a stable manifold, it must become parallel to the flow vector $\delta z_{\text{p.o.}}(t) = (\partial H/\partial p, -\partial H/\partial q)_{\text{p.o.}}$ along the periodic orbit whose stable manifold is under consideration. Thus, the antisymmetric product $\omega(\delta z(t), \delta z_{\text{p.o.}}(t))$ vanishes for $t \to \infty$; but since it is also unchanged in time by Hamiltonian evolution, it must vanish at all times. So L_s is indeed Lagrangian. This statement will turn out to be useful for the discussion of Maslov indices in Subsect. (8.4.3) below.

While the notion of a Lagrangian manifold is defined without reference to a particular set of coordinates in phase space, their caustics are very much coordinate-dependent phenomena, as is clear already for $f = 1$ from Fig. 8.5. There we see a "configuration-space" caustic, i.e., a point with diverging $\partial p/\partial q$; the canonical transformation $q \to P, p \to -Q$ turns the curve $p(q)$ into one $P(Q)$ for which the previous caustic becomes a point with $\partial P/\partial Q = 0$ such that no configuration-space caustic appears; instead, we incur what should be called a "momentum-space" caustic inasmuch as the point in question comes with $\partial Q/\partial P = \infty$. Conversely, we could say that in the original coordinates the coordinate-space caustic is not a momentum-space caustic there since $\partial q/\partial p = 0$.

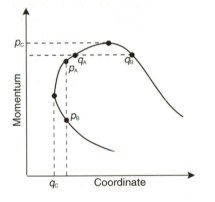

Fig. 8.5. A configuration-space caustic (at q_c) looks level from momentum space, as does a momentum-space caustic (at p_c) from configuration space. Near a caustic a Lagrangian manifold is double-valued.

For $f > 1$, a little more care is indicated in explaining caustics. One may uniquely label points on a Lagrangian manifold L by f suitable coordinates $\xi_1, \xi_2, \ldots, \xi_f$ such that q and p become functions $q(\xi)$ and $p_\parallel$) on L. Wherever the $q(\xi)$ are locally invertible functions, $\xi = \xi(q)$, and we can deal with the momenta as a vector field $p[\xi(q)]$; such invertibility requires that the Jacobian $\det(\partial q/\partial \xi)$ not vanish. Conversely, (configuration-space) caustics arise where $\det(\partial q/\partial \xi) = 0$ since in such points one finds tangent vectors $\delta z = (\delta q, \delta p)$ to L such that $\delta q = (\partial q/\partial \xi)\delta \xi = 0$, even though $\delta \xi \neq 0$. For every independent such $\delta \xi$, the Lagrangian manifold L appears perpendicular to coordinate space, unless it happens that $\delta \xi$ is also annihilated by $\partial p/\partial \xi$. The number of such independent $\delta \xi$ is called the order of the caustic; the most common value of that order is 1, but orders up to f may arise, and the latter extreme case indeed arises for the L distinguished by the initial form of the propagator $\langle q|e^{-iHt/\hbar}|q'\rangle$. Again, caustics are phenomena depending on the coordinates used; if one encounters a coordinate-space caustic, one can canonically transform, as $q_i \to P_i, p_i \to -Q_i$, for one or more of the f components so as to ban the caustic from coordinate space to momentum space. When dealing with the semiclassical approximation of a propagator, such a change corresponds to switching from the q_i-representation to the p_i-representation.

In a region devoid of configuration-space caustics, a Lagrangian manifold L is characterized by a curl-free momentum field $p(q)$ which in turn can be regarded as the gradient $\partial S/\partial q$ of a scalar field S. Taking up *Littlejohn*'s charming manner of speaking [264], I shall call S the generating function of L; this name reminds us of the other, already familiar role S plays in semiclassical games, that of the generating function of classical time evolution as a canonical transformation. Indeed, in the short-time version of the Van Vleck propagator (8.2.25), the action $S(q, q', t)$ is a generating function in both meanings. As the generating function of L, we may get S from a path-independent integral in configuration space

$$S(q) = \int^q dx\, p(x)\,. \tag{8.4.2}$$

The starting point for that integral is arbitrary in accordance with the fact that the generating function of L is determined only up to an additive constant. Even within the semiclassical form $Ae^{iS/\hbar}$ of a wave function such as the short-time version of the Van Vleck propagator, the additive constant in question is of no interest inasmuch as a constant overall phase factor is an unobservable attribute of a wave function.

However, if L has caustics such as shown in Figs. 8.5 and 8.6, one may divide it into caustic-free regions separated by the caustics, here each region possesses a different generating function $S^\alpha(q)$ and momentum field $p^\alpha(q) = \partial S^\alpha(q)/\partial q$. In other words, the momentum field is no longer single-valued when caustics are around. Then, semiclassical wave functions take the form of sums over several branches, $\psi = \sum_\alpha A^\alpha e^{iS^\alpha/\hbar}$. From a classical point of view, one might be content for each S^α to have an undetermined additive constant. For a wave function only, an overall phase factor is acceptable as undetermined, however; relative phases between the various $A^\alpha e^{iS^\alpha/\hbar}$ determine interference between them. Therefore, we can leave open only one additive constant for one of the S^α's and must uniquely determine all the other S^α's relative to an arbitrarily picked "first" one. With this remark, we are back to Maslov indices.

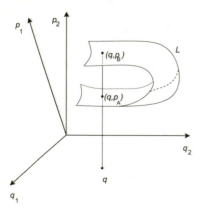

Fig. 8.6. Two branches (sheets, rather) of a Lagrangian manifold which may be seen as divided by a caustic (a configuration-space caustic here, depicted as the dashed line). Courtesy of Littlejohn [264]

8.4.2 Elements of Maslov Theory

In Chap. 5.3, I already mentioned Maslov's idea to temporarily switch to the momentum representation when a configuration-space caustic is encountered. It is now about time at least to sketch the implementation of that idea.

Let us first recall the time-dependent Van Vleck propagator (8.2.25) which in general consists of several "branches", i.e., terms of the structure $A^\alpha e^{i(S^\alpha/\hbar - \nu^\alpha \pi/2)}$. Each of these corresponds to a classical trajectory leading from some point on the initial Lagrangian manifold L' at q' to a point with coordinate q on the final manifold (See Fig. 8.7). The various such trajectories have different initial and final momenta, $p'^\alpha = -\partial S^\alpha/\partial q'$ and $p^\alpha = \partial S^\alpha/\partial q$, respectively. But as explained above, for Hamiltonians of the structure $H = T + V$ and sufficiently small times t, only a single term of the indicated structure arises, corresponding to a Lagrangian manifold L still without configuration-space caustics.

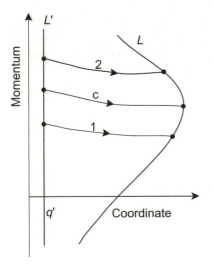

Fig. 8.7. Trajectory c has reached a caustic at the final time t, while that incident is imminent for orbit 1 and has already passed for orbit 2. Note again that the Lagrangian manifold associated with the time-dependent propagator is transverse to the trajectories. Courtesy of Littlejohn [264]

Now assuming t so large that L has developed caustics, I consider a region in configuration space around one of them. If dealing with $f = 1$, it suffices to imagine that there are only two branches of the curve $p(q)$ making up L "above" the q-interval in question; these branches are joined together at the phase-space point whose projection onto configuration space is the caustic. As pointed out in the previous subsection, there is no momentum-space caustic around then. For $f > 1$, I analogously consider a part of L with a single configuration-space caustic and no momentum-space caustic. Then, it is advisable to follow Maslov into the momentum representation where Schrödinger's equation reads

$$i\hbar\dot{\tilde{\psi}}(p,t) = \left[\frac{p^2}{2m} + V\left(i\hbar\frac{\partial}{\partial p}\right)\right]\tilde{\psi}(p,t). \tag{8.4.3}$$

A semiclassical solution may here be sought through the WKB ansatz

$$\tilde{\psi}(p,t) = \tilde{A}(p,t)e^{i\tilde{S}(p,t)/\hbar}. \tag{8.4.4}$$

Inasmuch as this is a single-branch function due to the assumed absence of momentum-space caustics, no problem with Maslov or Morse phases arises.

In full analogy to the usual procedure in the q-representation (see Problem 8.2), we find that to leading order in $\hbar$ the phase $\tilde{S}(p,t)/\hbar$ obeys the Hamilton–Jacobi equation

$$\dot{\tilde{S}}(p,t) + \left[\frac{p^2}{2m} + V\left(-\frac{\partial \tilde{S}(p,t)}{\partial p}\right)\right] = 0\,. \tag{8.4.5}$$

and thus may be interpreted as a momentum-space action. The next-to-leading order yields a continuity equation of the form

$$\frac{\partial}{\partial t}|\tilde{A}(p,t)|^2 - \frac{\partial}{\partial p}\{|\tilde{A}(p,t)|^2 V'[-\tilde{S}(p,t)]\} = 0\,; \tag{8.4.6}$$

here, the force $-V'$ provides the velocity in momentum space that makes $-|\tilde{A}(p,t)|^2 V'(-\tilde{S}(p,t))$ a probability current density in momentum space.

Coordinate-space and momentum-space wave functions are of course related by a Fourier transform. In our present semiclassical context, that Fourier transform is to be evaluated by stationary phase, in keeping with the leading order in $\hbar$ to which we are already committed. Therefore, I may write

$$\psi(q,t) = \int \frac{d^f p}{(2\pi\hbar)^{f/2}}\, e^{ipq/\hbar}\, \tilde{\psi}(p,t) \tag{8.4.7}$$

with the single-branch momentum-space wave function (8.4.4).

Momentarily setting $f = 1$, we arrive at the stationary-phase condition

$$q = -\partial \tilde{S}/\partial p \tag{8.4.8}$$

which for the envisaged situation has two solutions $p^\alpha(q)$ with $\alpha = 1, 2$. Thus, the single-branch $\tilde{\psi}(p)$ gives rise to a two-branch $\psi(q)$ which with the help of the Fresnel integral (8.2.12) we find as

$$
\begin{aligned}
\psi(q) &= \sum_{\alpha=1}^{2} A^\alpha(q) e^{iS^\alpha(q) - i\nu^\alpha \pi/2} \\
S^\alpha(q) &= p^\alpha(q)q + \tilde{S}(p^\alpha(q)) \\
A^\alpha(q) &= \tilde{A}(p^\alpha(q))\sqrt{\left|\frac{d^2\tilde{S}(p)}{dp^2}\right|^{-1}_{p=p^\alpha}} = \tilde{A}(p^\alpha(q))\sqrt{\left|\frac{dp^\alpha(q)}{dq}\right|} \\
\nu^\alpha &= \frac{1}{2}\left[1 - \text{sign}\left(\frac{d^2\tilde{S}(p)}{dp^2}\right)\right] = \frac{1}{2}\left[1 + \text{sign}\left(\frac{dp^\alpha(q)}{dq}\right)\right].
\end{aligned}
\tag{8.4.9}
$$

The two branches of the coordinate-space wave function differ not only by having their own action and prefactor each but also by the relative Maslov alias Morse phase $\pi/2$.[13]

---

[13] In this subsection, it would seem overly pedantic to insist on the name Morse phase or index for wave functions or propagators, reserving the name Maslov index for what appears in traces of propagators; as a compromise I speak here of Morse alias Maslov phases for multibranch wave functions or propagators.

A word and one more inspection of Fig. 8.7 are in order about the classical trajectories associated with the two branches of the wave function $\psi(q,t)$ in (8.4.9); each goes from a point on the initial Lagrangian manifold L' to a point on the image L of L' at time t. One of them, that labelled "1", has at some previous moment passed the caustic shown, while the trajectory labelled "2" has not yet gone through the "cliff"; the critical moment t_c has arrived for the trajectory designated "c", when the final point q is conjugate to the initial q'. If, as shown in the figure, the slope dp/dq of the Lagrangian manifold at the caustic changes from negative to positive (while passing through infinity), the Maslov alias Morse index of branch "1" is larger by one than that of branch "2", according to the above expression for ν^{α}. This is in accord with our previous interpretation of the Morse index: The slope dp/dq of L should be written as the partial derivative $\partial p(q, q')/\partial q$ when the initial-value problem pertaining to the propagator is at issue; but when $\partial p/\partial q$ changes from negative to positive so does $\partial q/\partial p$ and also, since points on L may still be uniquely labelled by the initial momentum p', the derivative $\partial q/\partial p'$; we had seen that $\partial q/\partial p' = 0$ is the conjugate-point condition and that the Morse index increases by one when a conjugate point is passed.

A similar picture arises for the energy dependent-propagator. The only difference is that the relevant Lagrangian manifold L_E has the trajectories lying within itself (rather than piercing, as is the case for the time-dependent propagator and its L). Even more intuitively, then, the trajectory which has climbed through the vertical cliff in L_E with the slope changing from negative to positive yields a Maslov/Morse index larger by one than that of the trajectory not yet through the cliff.

When attempting to solve the initial-value problem for $\psi(q,t)$ by a WKB ansatz in the coordinate representation, one finds a single-branch solution as long as the Lagrangian manifold L originating from the initial L' is free of caustics. The single-branch $\psi(q)$ diverges at precisely that moment when a caustic of L arrives at the configuration-space point q. A little later, a WKB form of $\psi(q,t)$ arises again but now as a two-branch function. That sequence of events might leave a spectator puzzled who is borné to the coordinate representation. Maslov's excursion into momentum space opens a caustic-free perspective and thus avoids the catastrophic but only momentary breakdown of WKB. Conversely, when in the momentum representation a momentum-space caustic threatens failure of WKB, one may take refuge in coordinate space (representation). At any rate, Maslov's reasoning explains quite naturally why the breakdown of the coordinate-space Van Vleck propagator at a coordinate-space caustic is only a momentary catastrophe. Incidentally, the divergence of the time-dependent propagator upon arrival of a caustic at q may but does not necessarily imply breakdown of the semiclassical approximation. The harmonic oscillator treated in Problem 8.6 provides an example: The divergence of the semiclassical Van Vleck propagator takes the form of a delta function, $G(q, q', t_c) = \delta(q+q')$ for times $t_c = (2n+1)T/2$, $n = 1, 2, \ldots$ and $G(q, q', t_c) = \delta(q-q')$ for times $t_c = nT$, $n = 1, 2, \ldots$, i.e., whenever a point conjugate to the initial q' is reached; but this behavior faithfully reproduces that of the exact propagator. Conjugacy means simultaneous reunification at q of a one-parameter family of orbits which originated from the common q' with different initial momenta and that family of orbits interferes constructively in building the "critical" propagator $G(q, q', t_c) = \langle q|e^{-iHt_c/\hbar}|q'\rangle$.

The occurrence of multibranch WKB functions may be understood as due to several effective caustics. Maslov's procedure then naturally assigns to each branch its own action branch $S^\alpha(q)$ and Maslov alias Morse phase $-\nu^\alpha\pi/2$, leaving open only a single physically irrelevant overall phase factor. It is well to keep in mind that each branch is provided by a separate classical trajectory and that the Maslov index equals the net number of passages of the trajectory through vertical cliffs in the relevant Lagrangian manifold where the slope dp/dq changes from negative to positive; "net" means that a passage in the inverse direction decreases the index by one.

To liven up my words a bit, I shall henceforth speak of clockwise rotation of the tangent to L or L_E at the caustic when the slope dp/dq changes from negative to positive and of anticlockwise rotation when the slope becomes negative. Thus, the net number of clockwise passages of that tangent through the momentum axis along a single traversal of a periodic orbit gives the Morse alias Maslov index in the contribution of that orbit to the propagator.

The above ($f{=}1$) result (8.4.9) can be generalized to more degrees of freedom [264, 265]. The essence of the change is to replace $|dp^\alpha(q)/dq|$ by $|\det dp^\alpha(q)/dq|$ in the amplitude $A^\alpha(q)$. If only a single eigenvalue of the matrix $\partial^2 S(p)/\partial p\partial p$ vanishes at the coordinate space caustic, one may imagine axes chosen such that in the above expression for the Maslov index $dp(q)/dq \to \partial p_l(q)/\partial q_m$ where the subscripts indicate the components of p and q with respect to which the coordinate-space caustic appears with $\partial p_l(q)/\partial q_m = \infty$.

8.4.3 Maslov Indices of Unstable Periodic Orbits as Winding Numbers

We can finally proceed to scrutinizing an unstable periodic orbit for its Maslov index μ, staying with autonomous systems of two degrees of freedom. The presentation will closely follow *Creagh, Robbins,* and *Littlejohn* [266] but *Mather's* pioneering contribution [267] should at least be acknowledged. The generalization to arbitrary f can be found in *Robbins'* paper [268]. Intuition will be furthered by focusing on the three-dimensional energy shell. Therein the stable (and the unstable) manifold of the periodic orbit is a two-dimensional surface. Now imagine a surface of section Σ, also two-dimensional, transverse to the orbit, and let that Σ move around the periodic orbit. As in deriving the trace formula in Sect. 8.3.2, I shall specify the location of Σ by the coordinate $x_\parallel$ along the orbit and parametrize $\Sigma(x_\parallel)$ by the transverse coordinate $x_\perp$ and its conjugate momentum $p_\perp$.

The stable manifold of the periodic orbit intersects $\Sigma(x_\parallel)$ in a line $\sigma(x_\parallel)$ through the origin where the orbit pierces through $\Sigma(x_\parallel)$; near the origin, $\sigma(x_\parallel)$ will appear straight. As $\Sigma(x_\parallel)$ is pushed along the orbit, the stable manifold $\sigma(x_\parallel)$ will rotate, possibly clockwise at times and then anticlockwise, as may happen. But $\sigma(x_\parallel)$ must return to itself when $\Sigma(x_\parallel)$ has gone around the periodic orbit. The net number of clockwise windings of $\sigma(x_\parallel)$ about the periodic orbit during a single round trip of $\Sigma(x_\parallel)$ is obviously an intrinsic property of that orbit, independent of the phase-space coordinates employed.

We shall see presently that twice the net number of clockwise windings of the stable manifold for one traversal of the periodic orbit in the sense of growing time *is* the Maslov index μ appearing in the trace formula (8.3.32). With that fact established, the constancy of μ along the periodic orbit (i.e., the independence of $x_\parallel$) is obvious, as is the additivity with respect to repeated traversals: If μ refers to a primitive orbit, its r-fold repetition has the Maslov index $r\mu$. Thus, the trace formula (8.3.31) for the density of levels may be rewritten by explicitly accounting for repetitions of primitive orbits,

$$\varrho_{\mathrm{osc}}(E) = \frac{1}{\pi\hbar} \sum_{r=1}^{\infty} \sum_{\mathrm{prim.orb.}} \frac{T}{\sqrt{|2 - \mathrm{Tr}M^r|}} \cos r(S/\hbar - \mu\pi/2), \qquad (8.4.10)$$

where T, M, S, and μ all refer to primitive orbits and complete "hyperbolicity" is assumed, i.e., absence of stable orbits.

To work, then! We shall have to deal with mapping the "initial" surface of section $\Sigma(x'_\parallel)$ to the "running" one, $\Sigma(x_\parallel)$, as linearized about the origin (i.e., about the orbit in question),

$$\begin{pmatrix} \delta x_\perp \\ \delta p_\perp \end{pmatrix} = \begin{bmatrix} \left(\frac{\partial x_\perp}{\partial x'_\perp}\right)_{p'_\perp} & \left(\frac{\partial x_\perp}{\partial p'_\perp}\right)_{x'_\perp} \\ \left(\frac{\partial p_\perp}{\partial x'_\perp}\right)_{p'_\perp} & \left(\frac{\partial p_\perp}{\partial p'_\perp}\right)_{x'_\perp} \end{bmatrix} \begin{pmatrix} \delta x'_\perp \\ \delta p'_\perp \end{pmatrix} = \begin{pmatrix} a & b \\ c & d \end{pmatrix} \begin{pmatrix} \delta x'_\perp \\ \delta p'_\perp \end{pmatrix} = M(x_\parallel, x'_\parallel) \begin{pmatrix} \delta x'_\perp \\ \delta p'_\perp \end{pmatrix}.$$

The special case of the matrix $M(x_\parallel, x'_\parallel)$ pertaining to a full traversal of the periodic orbit is the stability matrix M entering the trace formula (8.3.32); kin to the present $M(x_\parallel, x'_\parallel)$ is that found for area-preserving maps in (8.3.4); indeed, the close analogy of area preserving stroboscopic maps to mappings between surfaces of section for Hamiltonian flows permeates all of this chapter.

Now instead of immediately concentrating on the (section with $\Sigma(x_\parallel)$ of the) stable manifold $\sigma(x_\parallel)$, it is convenient first to look at the section $\lambda(x_\parallel)$ of $\Sigma(x_\parallel)$ with the Lagrangian manifold L_E associated with the energy-dependent propagator. The reader will recall that L_E is the two-dimensional sheet swept out by the set of all trajectories emanating from the set of all points within the energy shell projecting to one fixed initial point in configuration space. The line $\lambda(x_\parallel)$ arises as the image under the mapping $M(x_\parallel, x'_\parallel)$ of the momentum axis in $\Sigma(x'_\parallel)$ or rather that interval of the momentum axis accommodated in the energy shell. Like $\sigma(x_\parallel)$, the line $\lambda(x_\parallel)$ goes through the origin of the axes used in $\Sigma(x_\parallel)$; close to that origin $\lambda(x_\parallel)$ looks straight and can thus be characterized by the vector (b, d) in $\Sigma(x_\parallel)$; that vector even provides an orientation to λ when attached to the origin like a clock hand. I shall refer to the part of λ extending from the origin along the vector b, d as the half line λ_+.

Now, recall from (8.3.30) that the full Maslov index of a periodic orbit, $\mu = \mu_{\mathrm{prop}} + \mu_{\mathrm{trace}}$, gets the contribution μ_{prop} from the energy-dependent propagator while the trace operation furnishes μ_{trace}. The first of these has already been given a geometric interpretation in Sect. 8.4.2. The consideration there was worded for a two-dimensional phase space and thus fits our present section $\Sigma(x_\parallel)$, and the slope of the Lagrangian manifold here is given by d/b. The net number of clockwise rotations of the vector (b, d) through the momentum axis while one goes around the periodic orbit gives μ_{prop}. Only a subtle point regarding the initial Lagrangian manifold, i.e., the momentum axis (rather, the part thereof fitting in the energy shell), needs extra comment. That initial $\lambda(x_\parallel')$ is as a whole one caustic and one that is not actually climbed through. Whether or not the index μ_{prop} gets a contribution here depends on the sign of $\partial p_\perp/\partial x_\perp$ at infinitesimally small positive times, $t = 0^+$; if positive, no contribution to the index arises, whereas for negative slope, the index suffers a decrement of 1; this simply follows from our discussions of the Van Vleck propagator. In fact, we already saw that Hamiltonians of the structure $H = T + V$ always give rise to minimal (rather than only extremal) actions for small times, such that the initial caustic is immaterial for μ_{prop} there.

To proceed toward μ_{trace}, we must determine where the vector (b, d) ends up after one round trip through the periodic orbit. Since we are concerned with an unstable orbit, we can narrow down the final direction with the help of the stable and unstable manifolds whose tangent vectors at the origin are given by the two real eigenvectors e_s and e_u of the 2×2 matrix M pertaining to the completed round trip; the respective eigenvalues will be called $1/\tau$ and τ with $|\tau| > 1$. As shown in Fig. 8.8, these directions divide $\Sigma(x_\parallel)$ into four quadrants, labelled clockwise H, I, J, and K. H is distinguished by containing the upper momentum axis, so the positive-p_x part of the initial $\lambda(x_\parallel')$ lies therein. The final vector (b, d) must lie in that same quadrant if the orbit is hyperbolic ($\tau > 2$); conversely, for hyperbolicity with reflection ($\tau < -2$), the vector ends up in the opposite quadrant, J.

The index μ_{trace} depends on where the vector (b, d) and thus the half line λ_+ end up within the quadrants H or J. The definition (8.3.29) implies that $\mu_{\text{trace}} = 0$ if the quantity

$$
\begin{aligned}
w &\equiv \{[(\partial/\partial x_\perp + \partial/\partial x_\perp')^2 S(x_\perp, x_\parallel, x_\perp', x_\parallel, E)]_{x_\perp = x_\perp' = 0}\} \\
&= \frac{a + d - 2}{b} = \frac{\text{Tr}M - 2}{b} = \frac{\tau + 1/\tau - 2}{b}
\end{aligned} \tag{8.4.11}
$$

is positive while $\mu_{\text{trace}} = 1$ if $w < 0$. But since $|\tau + 1/\tau| > 2$, the signs of $\tau + 1/\tau - 2, \tau + 1/\tau$, and τ all coincide. It follows that the numerator of w is positive when M is hyperbolic and negative if M is hyperbolic with reflection. The denominator b, on the other hand, is positive or negative when the vector (b, d) ends up, respectively, to the right or the left of the momentum axis. Therefore, it is indicated to subdivide the quadrants H and J into the sectors H_+, H_-, J_+, and J_-, as shown in Fig. 8.8. The negative subscripts denote sectors counterclockwise to the momentum axis while the sectors clockwise of it have positive subscripts. It follows that $w > 0$ if the vector (b, d) ends up in either H_+ or J_+ since in that case, τ and b have the same sign; conversely, $w < 0$ if the said vector ends up in either H_- or J_-, τ and b differs in sign then. What finally matters is that

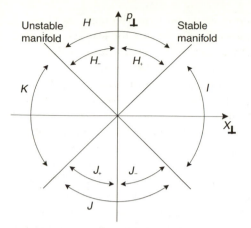

Fig. 8.8. The surface of section Σ transverse to a periodic orbit is divided into four sectors H, I, J, and K by the (sections with Σ of the) stable and unstable manifolds which look locally straight. At some arbitrarily chosen initial moment, the section λ of the Lagrangian manifold with Σ starts out as the momentum axis; its positive part, the half line λ_+, rotates as the section $\Sigma(x_\parallel)$ is carried around the periodic orbit and ends up, after one full traversal, either in sector H (hyperbolic case) or in sector J (inverse hyperbolic case). If λ_+ ends up in H_+ or J_+, $\mu_{\text{trace}} = 0$ while arrival in H_- or J_- implies $\mu_{\text{trace}} = 1$. Courtesy of Creagh, Robbins, and Littlejohn [266]

$\mu_{\text{trace}} = 0$ if the half line λ_+ ends up running through the sectors H_+ and J_+ while $\mu_{\text{trace}} = 1$ if it ends up running through H_- and J_-.

With both μ_{prop} and μ_{trace} geometrically interpreted, now we can tackle their sum, the full Maslov index. To that we extend the definitions of the quadrants H, I, J, and K to the intermediate surfaces of section $\Sigma(x_\parallel)$. We do that by introducing the vectors $e_s^*(x_\parallel)$ and $e_u^*(x_\parallel)$ such that their components along the x-axis and the p_x-axis in $\Sigma(x_\parallel)$ remain the same as those of e_s and e_u in the initial surface of section $\Sigma(x_\parallel')$. Note that these starred vectors are *not* the ones specifying the stable and unstable manifolds in $\Sigma(x_\parallel)$ through the eigenvectors $e_{s,u}^*$ of $M(x_\parallel, x_\parallel')$, i.e., $e_{s,u}^*(x_\parallel) \neq e_{s,u}(x_\parallel)$. Rather, the starred vectors carry the quadrants H, I, J, and K along in rigid connection with the x- and p_x-axes, as we go through the sequence of sections $\Sigma(x_\parallel)$ around the orbit. During that round trip, we may define the current value of μ_{prop} as the net number of clockwise passages of the vector $(b, d)(x_\parallel)$ through the momentum axis along the part of the voyage done. A little more creativity is called for in defining intermediate values of μ_{trace} and thus the sum μ since the connection with the trace must be severed. It proves useful to define $\mu_{\text{trace}} = 0$ if $\lambda(x_\parallel)$ and thus the vector $(b, d)(x_\parallel)$ is clockwise of the momentum axis, i.e., in the sectors H_+ or J_+ and $\mu_{\text{trace}} = 1$ if $\lambda(x_\parallel)$ is anticlockwise of it in the sectors H_- or J_-. There is no need to define μ_{trace} for the quadrants I and K since, as we have seen above, λ cannot end up therein.

Now the full Maslov index μ behaves quite simply as we go through the orbit. When $\lambda(x_\parallel)$ sweeps through the quadrants H and J, no change of μ occurs, since whenever $\lambda(x_\parallel)$ rotates through the momentum axis, the changes of μ_{prop} and μ_{trace} cancel in their sum. But μ does increase by one for every completed clockwise passage of the vector $(b, d)(x_\parallel)$ through either quadrant I or K.

Finally, I hurry to formulate the result just obtained in terms of the stable manifold $\sigma(x_\parallel)$. This is necessary since in contrast to the Lagrangian manifold $\lambda(x_\parallel)$, the stable manifold $\sigma(x_\parallel)$ is obliged to return to itself after one round trip through the periodic orbit. Incidentally, if $\lambda(x'_\parallel)$ does not coincide with $\sigma(x'_\parallel)$ initially, it cannot do so later either since the map $M(x_\parallel, x'_\parallel)$ is area-preserving and cannot bring about coincidence of originally distinct directions. This means that $\lambda(x_\parallel)$ shoves along $\sigma(x_\parallel)$, though not with a constant angle in between. But both lines must have the same net number of clockwise passages through the quadrants I and K. When $\sigma(x_\parallel)$ returns to itself after one traversal of the orbit it must have undergone an integer number of "half-rotations" in the surfaces of section $\Sigma(x_\parallel)$; half-rotation means a rotation by π about the origin. During a half-rotation $\sigma(x_\parallel)$ sweeps through I and K. The number of half-rotations that occurred during one round trip through the orbit is the full Maslov index; we might also say, that μ is twice the number of full 2π-rotations if we keep in mind that the number of full rotations is half-integer if the orbit is hyperbolic with reflection.

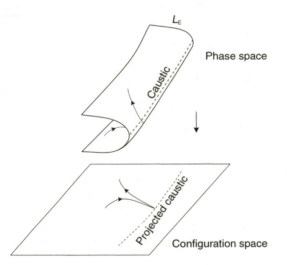

Fig. 8.9. An orbit momentarily coming to rest at a caustic. In this nongeneric case, the coordinates $x_\parallel, p_\parallel$ are momentarily ill defined. Then, the projection of the orbit onto configuration space displays a cusp. Courtesy of Creagh, Robbins, and Littlejohn [266]

Now the previous claim that the Maslov index is a winding number are shown true. In addition, we conclude that μ is even for hyperbolic orbits and odd for hyperbolicity with reflection. It may be well to remark that odd need not, but may be taken to mean either 1 or 3, and even either 0 or 2, since in the trace formula, μ appears only in a phase factor $e^{-i\mu\pi/2}$.

Before concluding, a sin must be confessed, committed in employing the specific coordinates $x_\parallel, p_\parallel, x_\perp, p_\perp$. Helpful as they indeed are, they may be ill defined at certain moments during the round trip along the periodic orbit. Such instants occur if when arriving at a caustic, the orbit momentarily comes to rest (see

Fig. 8.9), such that (not only $\dot{x}_\perp = 0$ which, by definition, is always the case but also) $\dot{x}_\parallel = 0$. In coordinate space, the orbit then traces out a cusp. In the tip of the cusp, the coordinates $x_\parallel, x_\perp$ are indeed ill defined since both axes become inverted there. Our final result for μ holds true, even when such cusps arise as one may check by temporarily switching to the momentum representation à la Maslov where the caustic is absent entirely.

While the considerations of the present section establish the canonical invariance of all ingredients of the Gutzwiller trace formula (8.4.10) for Hamiltonian flows of autonomous systems with two degrees of freedom, the question is still open for the case of maps with $f = 1$. To settle that question, I proceed to show the equivalence, both classical and quantum mechanical, of autonomous dynamics with $f = 2$ and periodically driven ones with $f = 1$.

8.5 Single-Freedom Periodic Driving as Two-Freedom Autonomy

Without much comment, I have placed skip between area-preserving maps with $f = 1$ in parallel with autonomous Hamiltonian dynamics with $f = 2$. The task of clarifying the relationship between the two cases is indeed overdue. While not all area-preserving maps can be seen as stroboscopic descriptions of Hamiltonian continuous-time dynamics with periodic driving, such stroboscopic maps at least are equivalent to certain autonomous flows with one more degree of freedom and that equivalence will be established now. Only then will the Maslov index in the trace formula (8.3.5) for maps rightfully enjoy the privilege of confirmed canonical invariance and additivity with respect to repeated traversals, at least for the class of stroboscopic maps just characterized. I confine myself to the simplest case, single-freedom maps versus two-freedom flows, leaving the easy generalization to the interested reader as Problem 8.9. Alternative embeddings of single-freedom maps in two-freedom flows can be found in [269, 270].

Starting with classical dynamics, I consider a Hamiltonian function that explicitly depends on the time t in a periodic manner,

$$H(q, p, t) = H(q, p, t + T). \tag{8.5.1}$$

The idea is to extend the two-dimensional phase space to four dimensions by treating the time t as a new coordinate, $Q = t$, and associating a conjugate momentum P with Q. That momentum is added to the original Hamiltonian, such that the Hamiltonian function in the extended phase space becomes

$$\mathcal{H}(q, p, Q, P) = H(q, p, Q) + P. \tag{8.5.2}$$

The added P serves to generate continuous translation with respect to Q. Calling the new time parameter τ we get Hamilton's equations in the extended phase space as

$$\frac{dQ}{d\tau} = \frac{\partial \mathcal{H}}{\partial P} = 1 \implies Q = \tau = t$$
$$\frac{dP}{d\tau} = -\frac{\partial \mathcal{H}}{\partial Q}$$
$$\frac{dq}{d\tau} = \frac{\partial \mathcal{H}}{\partial p} = \frac{\partial H}{\partial p}$$
$$\frac{dp}{d\tau} = \frac{\partial \mathcal{H}}{\partial q} = \frac{\partial H}{\partial q}. \tag{8.5.3}$$

The first of these relates Q to τ and t in the desired manner and the last two are just the original Hamilton equations; the second is of no consequence for the original driven dynamics.

In the three-dimensional energy shell $\mathcal{H} = \mathcal{E}$ the new momentum is uniquely expressed in terms of q, p, and Q as $P = \mathcal{E} - H(q, p, Q)$ such that the energy shell may be thought of as spanned by the latter three variables. As Q grows uniformly in time, the extended-phase space trajectories trace out curves all of which go through every plane Q=const in the energy shell. In particular, we may look at the equally spaced planes $Q = nT$ with $n = 0, 1, \ldots$ through each of which a trajectory pierces in a single point. However, due to the periodicity of the Hamiltonian $\mathcal{H}(q, p, Q, P)$ in, Q the coordinate Q becomes effective only modulo T (somewhat like an angular coordinate becoming effective only modulo 2π). The mapping of a point of the nth such plane to a point in the $(n + 1)$st is therefore nothing but the stroboscopic period-to-period map in the original phase space. Incidentally, that map could also be seen as a Poincaré map for the extended system, with the welcome peculiarity that all trajectories take precisely the same time T between subsequent visits to the Poincaré surface of section.

On to quantum mechanics! Even though quantization of the extended dynamics is neither necessary nor particularly useful for the main purpose of this section, it is sufficiently nice to warrant a few lines. (But I shall not bestow labels on equations, to discourage gossip about a slightly frivolous escapade.) It is natural to provide the extended system with a Hamiltonian operator $\hat{\mathcal{H}}$ identical in appearance to the classical Hamiltonian function $\mathcal{H}$. In the coordinate representation, the new momentum becomes the differential operator $P = \frac{\hbar}{i} \frac{\partial}{\partial Q}$. Closing the eye on difficulties of interpretation, I proceed to a wave function $\Psi(q, Q, \tau)$ and write the Schrödinger equation

$$i\hbar \frac{\partial}{\partial \tau} \Psi(q, Q, \tau) = \left(\hat{H}(q, p, Q) + \frac{\hbar}{i} \frac{\partial}{\partial Q} \right) \Psi(q, Q, \tau).$$

The ansatz

$$\Psi(q, Q, \tau) = e^{-i\hat{P}\tau/\hbar} \tilde{\Psi}(q, Q, \tau)$$

purges that equation of the new momentum operator $\hat{P}$, at the expense of sneaking a time dependence into the old Hamiltonian according to $e^{i\hat{P}\tau/\hbar} Q e^{-i\hat{P}\tau/\hbar} = Q + \tau$. The resulting Schrödinger equation

$$i\hbar \frac{\partial}{\partial \tau} \tilde{\Psi}(q, Q, \tau) = \hat{H}(q, p, Q + \tau)\tilde{\Psi}(q, Q, \tau)$$

can be solved formally by the familiar time-ordered product as $\tilde{\Psi}(q, Q, \tau) = \left(\exp\left[-i \int_Q^{Q+\tau} d\tau' \hat{H}(q, p, \tau')/\hbar\right]\right)_+ \Psi(q, Q, 0)$. But since the unitary translation operator changes a function $f(Q)$ into $\exp(-iP\tau/\hbar) f(Q) = f(Q - \tau)$, the solution of the extended Schrödinger equation can be written as $\Psi(q, Q, \tau) = \left\{ \exp\left[-i \int_{Q-\tau}^Q d\tau' \hat{H}(q, p, \tau')/\hbar\right]\right\}_+ \Psi(q, Q-\tau, 0)$. Still ruthless about interpretability as a probability amplitude, I choose the only "natural" initial wave function $\Psi(q, Q, 0) = \psi(q, 0)\delta(Q)$ so as to get the time-dependent wave function

$$\Psi(q, Q, \tau) = \delta(Q - \tau) \left\{ \exp\left[-i \int_0^\tau d\tau' \hat{H}(q, p, \tau')/\hbar\right]\right\}_+ \psi(q, 0).$$

Herein the new coordinate Q remains sharp due to the linearity of the extended Hamiltonian in the new momentum P, while the part $\psi(q, \tau)$ is just the ordinary wave function of the original system whose time-evolution operator becomes the familiar Floquet operator of the stroboscopic quantum map when the time τ is set equal to the period T of the driving,

$$F = \left\{ \exp\left[-i \int_0^T d\tau' \hat{H}(q, p, \tau')/\hbar\right]\right\}_+,$$

$$F^n = \left\{ \exp\left[-i \int_0^{nT} d\tau' \hat{H}(q, p, \tau')/\hbar\right]\right\}_+, \tag{8.5.4}$$

$$\psi[q, (n+1)T] = F\psi(q, nT).$$

So the escapade, now terminated, has done nothing but bring back to sight the familiar quantum mechanics of the original system, to which I confine the remainder of this section.

To complete the discussion of Maslov indices $\mu = \mu_{\mathrm{prop}} + \mu_{\mathrm{trace}}$ for stroboscopic maps, we begin by looking back at the trace formula (8.3.5) for $(f = 1)$ maps and appreciating the identity of appearance with that of $(f = 2)$ flows, (8.3.32). The similarity was brought about by focusing, when treating flows, on two-dimensional subspaces of the four-dimensional phase space. The reduction from four to two dimensions proceeded in two steps: First, we crept into the energy shell by going over to the energy-dependent propagator and then therein we looked at sections transverse to each periodic orbit. That same reduction was employed when revealing the Maslov index $\mu = \mu_{\mathrm{prop}} + \mu_{\mathrm{trace}}$ as a winding number in the previous section.

All of our discussions of the Maslov index for $(f = 2)$ flows carry over to stroboscopic $(f=1)$ maps resulting from periodic driving. Here, the phase space is two-dimensional to begin with. Now, it is indeed helpful to imagine that two-dimensional phase space embedded into the extended four-dimensional one. Then, the classical stroboscopic map appears, as already mentioned above, as a Poincaré map. The geometric visualization of the embedding is even much simplified: The energy shell looks the same everywhere, $P = \mathcal{E} - H(q, p, Q)$; moreover, rather than having to construct a separate section for each periodic orbit, we get one and the same sequence of sections for all orbits in the equidistant planes $Q = nT$. A continuous-time orbit in the extended phase space reveals itself of period n for the stroboscopic map by its periodic sequence of piercings through these planes. The stable manifold of a continuous-time trajectory cuts each of the planes in a curve; in continuous time, it continuously winds around the trajectory; the number of its windings around a periodic orbit along a single traversal thereof gives the Maslov index $\mu = \mu_{\mathrm{prop}} + \mu_{\mathrm{trace}}$; the reasoning is the same as in 8.4.3. The reader must recall only that the time-dependent semiclassical propagator which must be discussed for maps has its μ_{prop} increasing discontinuously whenever the underlying orbit passes through a configuration-space point conjugate to the initial one; but these conjugate points are just the caustics.

A word of regret should be devoted to area-preserving maps which cannot be understood as stroboscopic versions of Hamiltonian continuous-time dynamics. A familiar example is the baker's map. No general statement can be made about their semiclassical treatment, Maslov indices, etc., since each must be quantized separately in some specific way, without a Hamiltonian function guiding the canonical quantization and the semiclassical approximation thereto.

8.6 The Sum Rule of Hannay and Ozorio de Almeida

Classical ergodicity has a remarkable consequence for the semiclassical limit of the form factor, which partly explains why fully chaotic dynamics conforms to the predictions of random-matrix theory in its quantum aspects.

Ergodic behavior implies that almost all trajectories visit everywhere in phase space and eventually cover it with uniform density, within the restrictions imposed by conservation laws.

The simplest case is that of an area-preserving map without conservation laws in a compact phase space of volume Ω. Denoting a phase-space point by $x \equiv (q, p) = (q_1, \ldots, q_f, p_1, \ldots, p_f)$ and the nth iterate of some initial point x_0 under the map by $x_n(x_0)$, we may write the assumption of ergodicity as

$$\lim_{N \to \infty} \frac{1}{N} \sum_{n=1}^{N} \delta\big(x - x_n(x_0)\big) = \frac{1}{\Omega}. \qquad (8.6.1)$$

Not contained in the set of almost all allowable initial points x_0 are periodic points of the map. For if x_0 is recurrent with period n the orbit visits nowhere except its n member points incessantly.

However, very long periodic orbits should not have preferred phase-space regions either such that the restriction "almost all" really bars only points on short periodic orbits. It is also appropriate to exclude stable periodic points, either by restricting the consideration to completely hyperbolic dynamics or, a bit more liberally, at least by requiring the classical phase space to have a negligible fraction of its volume covered by islands of regular motion. Then, one may reformulate the ergodic hypothesis so as to employ only long and unstable periodic orbits,

$$\lim_{n\to\infty} \frac{1}{\Delta n} \sum_{n'=n-\Delta n/2}^{n-\Delta n} \delta\big(x_0 - x_{n'}(x_0)\big) = \lim_{n\to\infty} \big\langle \delta\big(x_0 - x_{n'}(x_0)\big) \big\rangle = \frac{1}{\Omega}, \quad (8.6.2)$$

where the brackets $\langle \dots \rangle$ denote the time average over the time window Δn. We may and shall in fact drop the limit $n \to \infty$, keeping in mind that the orbits should be long in the sense $1 \ll \Delta n \ll n$. The distribution $\delta\big(x_0 - x_n(x_0)\big)$ is a sum of delta functions, one for each of the distinct points on each period-n orbit,

$$\delta\big(x_0 - x_n(x_0)\big) = \sum_\alpha^{(\text{period } n)} \sum_{i=1}^{n_0} \frac{1}{\big|\det\big(\frac{\partial x_n}{\partial x_0} - 1\big)\big|} \delta\big(x_0 - x_i^\alpha\big). \quad (8.6.3)$$

The distinct points are $n_0 = n$ in number if the orbit is primitive and $n_0 = n/r$ if it is the r-fold repetition of an orbit with primitive period n_0. Here again, we use the $2f \times 2f$ stability matrix $M = \partial x_n/\partial x_0$ of the periodic orbit; it is one and the same matrix for all points on the αth orbit but differs in general for different orbits. Inserting the expression (8.6.3) for the delta function in the ergodic hypothesis (8.6.2) and integrating over phase space, we get the HOdA sum rule

$$1 = \Big\langle \sum_\alpha \frac{n_0}{\big|\det\big(\frac{\partial x_n}{\partial x_0} - 1\big)\big|} \Big\rangle \approx n \Big\langle \sum_\alpha \frac{1}{\big|\det(M - 1)\big|} \Big\rangle. \quad (8.6.4)$$

To arrive at the approximate equality in the foregoing chain, I have, with appeal to the assumed largeness of n, neglected the contributions from repetitions of shorter orbits with primitive periods $n_0 = n/r$. For the justification, I revert to the discussion of the Van Vleck propagator in Sect. 8.2, again employing the notions of an average (largest) Lyapunov exponent and exponential proliferation. The first of these allows us to estimate roughly the determinants as $e^{n\lambda}$, for all period-n orbits, primitive or not; the second allows us to estimate the number of contributing orbits as $e^{n\lambda}$ for primitive orbits and $e^{n\lambda/r}$ for repeated shorter ones; clearly, then, repeated shorter orbits are outnumbered exponentially for large n. Finally, n could be taken out of the average due to the assumed inequality $1 \ll \Delta n \ll n$. For further reference, I write out the sum rule once more,

$$\frac{1}{\Delta n} \sum_\alpha^{(\text{periods} \in [n, n+\Delta n])} \frac{1}{\big|\det(M - 1)\big|} = \frac{1}{n} \quad \text{for} \quad 1 \ll \Delta n \ll n. \quad (8.6.5)$$

For flows (autonomous dynamics in continuous time), the sum rule looks rather the same,

$$\frac{1}{\Delta T} \sum_\alpha^{(\text{periods} \in [T, T+\Delta T])} \frac{1}{\big|\det(M - 1)\big|} = \frac{1}{T} \quad \text{for} \quad \Delta T \ll T, \quad (8.6.6)$$

but the stability matrix M is now $2(f-1) \times 2(f-1)$ and refers to motion transverse to the periodic orbit. As for maps, ergodicity is at the basis here but for the derivation, which requires some daring acrobatics on delta functions erected along continuous orbits, the interested reader is referred to the original article of Hannay and Ozorio de Almeida [271].

8.7 The "Diagonal" Approximation for the Short-Time Form Factor

As already announced at the beginning of the previous section the, HOdA sum rule proves important for a first step toward a semiclassical understanding of the reason that fully chaotic dynamic systems obey many predictions of random-matrix theory. Taking up an argument pioneered by *Berry* [272], I shall outline here the pertinent discussion of the form factor, starting with maps and finally proceeding to flows.

8.7.1 Maps

The HOdA sum rule (8.6.5) is in fact reminiscent of the form factor $|t_n|^2$ if we recall the trace formula (8.3.5),

$$t_n \approx \sum_p \sum_r \frac{n/r}{\sqrt{|\det\left(M_p^r - 1\right)|}} e^{ir(S_p - \mu_p \pi/2)} , \qquad (8.7.1)$$

where p labels primitive orbits and r repetitions; needless to say, an r-fold repetition is possible only when n/r is an integer.

At this point, I must pick up a subtle issue about which there was no reason to bother previously. There may be distinct orbits sharing the same action, Maslov index, and stability matrix. Their number, called action multiplicity g_p, is much talked about at present; typically, it takes the value 1 for systems without time reversal-invariance while with such symmetry one expects $g_p = 2$ except for self-retracing orbits; indeed, time reversal-symmetry requires every orbit to have a partner which, if distinct, still shares the properties mentioned (see Problem 8.9). It is important to be aware of such multiplicities since the g_p orbits of a multiplet interfere constructively in the trace formula, while different actions necessarily entail more or less destructive interference.

Upon neglecting repetitions and taking the squared modulus, we get the form factor as a double sum over primitive period-n orbits, which I now choose to write as a double sum over the action multiplets just mentioned,

$$|t_n|^2 \approx n^2 \sum_{[p],[p']} \frac{g_p g_{p'} \exp i[(S_p - S_{p'}) - (\mu_p - \mu_{p'})\pi/2]}{\sqrt{|\det(M_p - 1)\det(M_{p'} - 1)|}} ; \qquad (8.7.2)$$

the square brackets around the labels for primitive orbits in the summation are reminders that only differing multiplets of orbits are to be included; the multiplicities are accounted for in explicit factors. The diagonal part of that double sum, i.e., the terms with $p = p'$, resembles the HOdA sum rule (8.6.5), once we smooth it by averaging over the time window Δn. That average then yields

$$\langle |t_n|^2_{\text{diag}} \rangle \approx g|n| \tag{8.7.3}$$

where I have taken the freedom to invoke the symmetry of the form factor under $n \leftrightarrow -n$ and to replace the multiplicity g_p by its average g. Superficial quickness might suggest the squared multiplicity in (8.7.3), but of course the HOdA sum rule (8.6.5) also acquires a factor $1/g$ on the r.h.s. when the summation on the l.h.s. is chosen to go over multiplets of orbits.

Thus, a very nice surprise presents itself: For systems without time-reversal invariance, which have $g = 1$, the foregoing local time average of the diagonal approximation coincides exactly with the CUE prediction (4.13.14) for times n up to the Heisenberg time N. Dynamics with time reversal invariance have $g = 2$, and for these the present semiclassical result reproduces the leading term of the small-n expansions of the COE expression (4.13.28, 29), as well as of the CSE expression (4.13.35, 36). This is immediately obvious for the orthogonal universality class, where both theories yield $2|n|$. As regards dynamics from the symplectic universality class we must remember that the t_n for the symplectic ensemble were defined in Chap. 4 as half the actual traces, $t_n^{\text{CSE}} = \frac{1}{2} \text{Tr} F^n$ while we work here with the full traces, $t_n = \text{Tr} F^n$. Accounting for the factor $g(1/2)^2 = 1/2$ the present semiclassical treatment of dynamics belonging to the symplectic universality class does indeed reproduce the random-matrix behavior $(\frac{1}{2})^2 \langle |\text{Tr} F^n|^2 \rangle = \overline{|t_n^{\text{CSE}}|^2} = \frac{1}{2}|n|$ for small $|n|$.

Needless to say, the failure of (8.7.3) to conform to the prediction of random-matrix theory for large times n cannot be blamed on the HOdA sum rule (8.6.4); for large times, those approaching or even exceeding the Heisenberg time N the off-diagonal terms in the double sum (8.7.6) can no longer be neglected.

To discuss the limit of validity of the diagonal approximation for the form factor from within periodic-orbit theory, I should first recall that by invoking the HOdA sum rule we have excluded very short times, i.e. restricted ourselves to $1 \ll \Delta n \ll n$; that is not a great worry since very short periodic orbits have nothing much universal to them, classically or quantum mechanically. We are more concerned about the limiting time, up to which the foregoing result is valid. For the sake of concreteness, I consider a map in a $(2f)$-dimensional compact phase space of total volume Ω. It is human to present first an optimistic estimate n_H for a limiting time. Assigning a Planck cell of size h^f with a phase-space point to account roughly for quantum uncertainty, one could say that a period-n orbit covers the area nh^f. Once that area equals Ω, for $n_H h^f = \Omega$, the orbit in question can certainly not escape partially constructive interference with other periodic orbits visiting the same Planck cells. Inasmuch as the diagonal approximation obviously discards any constructive interference between orbits with different actions, it must have lost validity for times of the order n_H. But $n_H = \Omega/h^f$ is just the Heisenberg time, the time needed to resolve the discreteness of the quasi-energy spectrum, here related to the dimension N of the Hilbert space through Weyl's law. Such optimistic expectation in the validity of the diagonal approximation is further backed by an interesting argument by *Fishman* and*Keating* about billiards pierced by magnetic flux lines [273].

A pessimistic estimate n_E for the limiting time is the so-called "Ehrenfest time" which is also based on associating a Planck cell with a phase-space point. Such an originally spherical cell will become deformed by the classical map, that is stretched in the unstable directions and squeezed in the stable directions. For an order-of-magnitude argument, we may account for the stretching in any unstable direction with an average Lyapunov exponent λ in the exponential stretching factor $e^{\lambda n}$. To some degree, such stretching and the corresponding squeezing $e^{-\lambda n}$ in the stable directions will be reflected in the quantum evolution as well, but one must be prepared for a loss of mutual faithfulness of the quantum and classical evolution on a time scale given by $(\Omega/h^f)^{\frac{1}{2f}} \approx e^{\lambda n_E}$; this is because we may see quantum fluctuations as a perturbation of classical dynamics and the stretching factor $e^{\lambda n_E}$ as the error blowup typical of chaotic dynamics. At any rate, quantum interference between different periodic orbits may become effective on the Ehrenfest time scale so estimated. We should note that the Ehrenfest time, is proportional to $\ln \hbar$, and is much shorter than the Heisenberg time, $n_H \propto 1/h^f$, and this is why n_E is a pessimistic and n_H an optimistic estimate for the range of validity of the diagonal approximation.

If we boldly extend the linear time dependence of the form factor down to $\tau = n/N = 0$, somewhat illegitimately from within periodic-orbit theory, so as to secure a cusp $|\tau|$ at $\tau = 0$, we recover, for all universality classes corresponding to global classical chaos, the leading nonoscillatory part of the two-point cluster functions of random-matrix theory. The oscillatory contributions are inaccessible to the present diagonal approximation since they reflect singularities of the form factor and its derivatives away from $\tau = 0$. From this point of view, the most important shortcoming of the diagonal approximation is that it says nothing about the locations or the character of such further singularities. Readers who skipped Chap. 4 may need at least to study Problem 4.13 to appreciate the foregoing remarks.

8.7.2 Flows

As for maps, the form factor is defined as the Fourier transform of the two-point correlator of the density of levels. Thus, we must recall the semiclassical level density $\varrho(E) = \langle \varrho(E) \rangle + \varrho_{\mathrm{osc}}(E)$ with the smooth part $\langle \varrho(E) \rangle$ given by Weyl's law (8.3.33) and the oscillatory part given by Gutzwiller's trace formula (8.3.33). It may be well to write out the latter once more, this time with repetitions visible,

$$\varrho_{\mathrm{osc}}(E) = \frac{1}{\pi\hbar} \sum_p \sum_{r=1}^{\infty} \frac{T_p}{\sqrt{|\det M_p^r - 1|}} \cos r(S_p/\hbar - \mu_p\pi/2). \qquad (8.7.4)$$

It is customary to define a normalized connected correlator as

$$y(e) = \frac{1}{\langle \varrho(E) \rangle^2} \langle \varrho(E + e/\langle \varrho(E) \rangle) \varrho(E) \rangle - 1. \qquad (8.7.5)$$

The angular brackets denote a local spectral average over an energy interval ΔE large enough to contain many levels but sufficiently small such that the smooth part $\langle \varrho(E) \rangle$ of the density does not noticeably vary within. If we imagine the energy offset $e/\langle \varrho(E) \rangle$–which is measured in units of the local mean spacing– positive, we need not worry about the "self-correlation" term $\delta(e)$ such that the above is the cluster function $-Y_2(e)$ as, e.g., in (4.9.14); I denote it by $y(e)$ to emphasize the kinship as well as the difference (minus sign, local energy average instead of ensemble average, self-correlation term); the index 2 for two points is dispensable since I don't intend to talk about correlations of higher than second order here. The local energy average cancels the terms linear in $\varrho_{\text{osc}}(E)$ such that our semiclassical two-point cluster function reads

$$
\begin{aligned}
y(e) &= \frac{1}{\langle \varrho(E) \rangle^2} \langle \varrho_{\text{osc}}(E + e/\langle \varrho(E) \rangle) \varrho_{\text{osc}}(E) \rangle \\
&= \frac{1}{\pi^2 \hbar^2 \langle \varrho \rangle^2} \sum_{p,p'} \sum_{r,r'} \frac{T_p T_{p'}}{\sqrt{\left| \det (M_p^r - 1)(M_{p'}^{r'} - 1) \right|}} \\
&\quad \times \langle \cos r(S_p/\hbar - \mu_p \pi/2) \cos r'(S_{p'}/\hbar - \mu_{p'} \pi/2) \rangle .
\end{aligned}
\tag{8.7.6}
$$

So as not to overburden the eye, I have left out the energy arguments in the periodic-orbit summands; moreover, the energy average has been focused on the product of the cosines since the prefactors of the trace formula are expected to be smooth on the scale of oscillation of the cosines, essentially since the actions are referred to Planck's constant there. The product of cosines can be expressed as a sum of two cosines, with the sum and the difference of the previous arguments as the new arguments; then, the energy average eliminates the more rapidly oscillating cosine of the sum such that we are left with

$$
\begin{aligned}
y(e) &= \frac{2}{T_H^2} \sum_{p,p'} \sum_{r,r'} \frac{T_p T_{p'}}{\sqrt{\left| \det (M_p^r - 1)(M_{p'}^{r'} - 1) \right|}} \\
&\quad \times \langle \cos \left\{ r \left[S_p(E + e2\pi\hbar/T_H)/\hbar - \mu_p(E + e2\pi\hbar/T_H)\pi/2 \right] \right. \\
&\quad \left. - r' \left[S_{p'}(E)/\hbar - \mu_{p'}(E)\pi/2 \right] \right\} \rangle ;
\end{aligned}
\tag{8.7.7}
$$

I have expressed the mean density here in terms of the Heisenberg time as

$$
T_H = 2\pi\hbar \langle \varrho \rangle .
\tag{8.7.8}
$$

At this point, we may indulge in the expectation, motivated by experience with random-matrix theory or specific dynamical systems, that the cluster function $y(e)$ will decay to zero on an e-scale of order unity, i.e., on an E-scale given by not very many mean level spacings; on such a microscopic scale, the classical action cannot change by more than what the first nontrivial term in the Taylor expansion suggests, $S_p(E+e2\pi\hbar/T_H) - S_p(E) \approx e2\pi\hbar T_p/T_H$ since $\partial S_p(E)/\partial E = T_p$. Similar reasoning relates the Maslov indices $\mu_p(E + e2\pi\hbar/T_H)$ and $\mu_p(E)$, but these two cannot help being equal in the limit of a classically small offset since the Maslov index cannot change continuously; a given periodic orbit will be deformed a little under a small change of its energy but will not change a topological property such as its winding number.

Finally, a separation of the double sum over periodic orbits in a diagonal and an off-diagonal term suggests itself, $y(e) = y_{\text{diag}}(e) + y_{\text{off}}(e)$, and the latter is like the above (8.7.7) except for $\sum_{p,p',r,r'} \to \sum_{p \neq p', r \neq r'}$ and

$$y_{\text{diag}}(e) = \frac{2}{T_H^2} \sum_{[p]} \sum_{r=1}^{\infty} \frac{g_p^2 T_p^2}{\left|\det (M_p^r - 1)\right|} \cos r\left(e2\pi T_p/T_H\right). \tag{8.7.9}$$

As previously for maps, the multiplicity g_p has been sneaked in here and the periodic-orbit sum correspondingly restricted to a sum over multiplets of orbits. The reader is invited to employ the HOdA sum rule to evaluate the periodic-orbit sum in (8.7.9); see Problem 8.10; the result, $y_{\text{diag}} = -g/(2\pi^2 e^2)$, will also be arrived at in the following subsection.

Here, we go to the Fourier transform as

$$\tilde{y}(\tau) = \int_{-\infty}^{+\infty} de \, y(e) \, e^{\mathrm{i}2\pi e\tau} = K(\tau) - \delta(\tau) \tag{8.7.10}$$

and get the diagonal approximation to the form factor for $\tau > 0$,

$$K_{\text{diag}}(\tau) = \frac{1}{T_H^2} \sum_{[p]} \sum_{r=1}^{\infty} \frac{g_p^2 T_p^2}{\left|\det (M_p^r - 1)\right|} \delta\left(\tau - \frac{rT_p}{T_H}\right). \tag{8.7.11}$$

We eliminate the delta spikes by a local time average and neglect repetitions,

$$K_{\text{diag}}(\tau) = \frac{1}{T_H^2} \frac{1}{\Delta\tau} \sum_{[\tau \leq T_p/T_H \leq \tau + \Delta\tau]} \frac{g_p^2 T_p^2}{\left|\det (M_p - 1)\right|} ; \tag{8.7.12}$$

finally invoking the HOdA sum rule (8.7.6) and recalling that the form factor is even in τ, we find hat

$$K_{\text{diag}}(\tau) = g|\tau|, \tag{8.7.13}$$

in coincidence with the result (8.7.3) for maps. Note that the sum rule has again done away with one of the two multiplicity factors. As for maps, the result applies to all three universality classes associated with classically ergodic behavior.

The limits of validity discussed at the end of the previous subsection remain in force as well. But to emphasize those limits it may be appropriate to reformulate them in a somewhat more refined manner, suggested by (8.7.8) and Weyl's law, $T_H \propto \hbar^{1-f}$. The universal behavior $K(\tau) \approx K_{\text{diag}}(\tau) = g|\tau|$ prevails for $\hbar \to 0$ with τ fixed; conversely, if $\hbar$ is held fixed as τ goes to zero, one enters the non-universal system specific regime dominated by short orbits. Similarly, the validity of $y(e) \approx y_{\text{diag}}(e) = -g/(2\pi^2 e^2)$ requires $\hbar \to 0$ at fixed e.

8.7.3 A Classical Zeta Function

The diagonal approximation to the two-point function and the form factor has a certain classical flavor to it, inasmuch as quantum interferences between contributions of orbits with differing actions are taken to be fully destructive. It is worthwhile–and a good investment toward extensions–to unravel the classical background of the diagonal approximation, following *Cvitanović* and *Eckhardt* [274], *Bogomolny* and *Keating* [275], as well as *Agam, Altshuler,* and *Andreev* [276]. We shall see that the propagator of the classical phase-space density possesses an expansion in terms of periodic orbits that is closely related to the semiclassical diagonal approximations mentioned. Sharpening the demand on classical chaoticity, I shall assume that the propagator in question (which is often referred to as the Frobenius–Perron operator) has a discrete spectrum; a zero eigenvalue corresponds to the stationary probability density and all other eigenvalues, whose real parts signal decay, are bounded away from zero. The existence of the zero eigenvalue can be ascertained with the help of the HOdA sum rule, consistent with the idea that long periodic orbits are associated with equilibrium. Conversely, eigenvalues that neighbor zero manifest the non-universal behavior displayed by short periodic orbits.

Considering a hyperbolic Hamiltonian flow in a four-dimensional phase space[14] I temporarily adopt the shorthand X for a point in the three-dimensional energy shell and write $f(X_0, t)$ for the energy-shell trajectory, i.e. the point reached at the running time t from the initial point X_0 at time zero. Thus, an initial energy-shell density $P(X, 0)$ evolves into

$$P(X, t) = \int dX_0 \, \delta\big(X - f(X_0, t)\big) P(X_0, 0) \equiv \int dX_0 \, P(X, t | X_0) P(X_0, 0).$$

(8.7.14)

The kernel $P(X, t | X_0) = \delta\big(X - f(X_0, t)\big)$ may be seen as a classical propagator. For the sake of concreteness, I would like to imagine that the propagator has a discrete spectrum of "generalized eigenvalues" or resonances, $e^{-\gamma_0 t}, e^{-\gamma_1 t}, \ldots$ with $\mathrm{Re}\,\gamma_i > 0$ and multiplicities m_i, without worrying about the conditions under which that might be the case ("mixing" is required but, please, retreat to a quiet place with, e.g., *Gaspard*'s textbook [277] on nonlinear dynamics for illumination on that, as well as on a more careful treatment of the material to follow). At any rate, a "trace" for the propagator may be defined as

$$\int dX \, P(X, t | X) = \sum_i m_i \, e^{-\gamma_i t}.$$

(8.7.15)

An obvious interpretation of that trace is, that is an integrated return probability. The trace may be expressed in terms of properties of periodic orbits.[15] To pick

---

[14] Ref. [274] treats arbitrary f and is not even restricted to Hamiltonian dynamics.

[15] This is a good point to gratefully remember *Henri Poincaré* who, a century ago, didn't speak of but knew about "chaos" and saw periodic orbits as the skeleton on which the tissue of nonintegrable motion hangs in phase space.

up the contribution of the r-fold traversal of the pth primitive orbit, we need only restrict the integration range to a narrow toroidal tube around that orbit. The result is quickly obtained by parametrizing[16] the tube of integration by one coordinate along the orbit, $X_{||}$, and two transverse coordinates summarily denoted by $X_\perp$. Since the three-dimensional delta function in (8.7.15) factors analogously, the integral over $X_\perp$ gives

$$\int dX_\perp \, \delta_\perp \big(X - f(X, rT_p) \big) = \frac{1}{|\det(M_p^r - 1)|} \tag{8.7.16}$$

where once more we use the stability matrix M_p of the orbit, raised to the rth power for the r-fold traversal. For the integral along the orbit we conveniently tie up $X_{||}$ with the time as $X_{||} = Vt$, and the velocity $V = |\dot{f}|$ is uniquely given in every point in phase space. Then, $\delta_{||}\big(X - f(X, t) \big) = \sum_{r=1}^{\infty} V^{-1}\delta(t - rT_p)$, and $\oint dX_{||}\delta_{||}\big(X - f(X, t) \big) = T_p \sum_{r=1}^{\infty} \delta(t - rT_p)$. Thus, the trace of the classical propagator reads

$$\begin{aligned}
\int dX \, P(X, t|X) &= \sum_p T_p \sum_{r=1}^{\infty} \frac{\delta(t - rT_p)}{|\det(M_p^r - 1)|} \\
&= \frac{i}{2\pi} \int_{-\infty}^{\infty} dk \, e^{ikt} \frac{\partial}{\partial k} \sum_p \sum_{r=1}^{\infty} \frac{e^{-ikrT_p}}{r|\det(M_p^r - 1)|} .
\end{aligned} \tag{8.7.17}$$

The shorthand

$$Z(s) = \exp\left[-\sum_p \sum_{r=1}^{\infty} \frac{e^{srT_p}}{r|\det(M_p^r - 1)|} \right] \tag{8.7.18}$$

offers itself now; it allows us to write the trace under study as

$$\int dX \, P(X, t|X) = \frac{i}{2\pi} \int_{-i\infty}^{+i\infty} ds \, e^{-st} \frac{Z'(s)}{Z(s)} = \sum_i m_i \, e^{-\gamma_i t} . \tag{8.7.19}$$

But the foregoing identity means that the "resonances" γ_i can be determined from the zeros of the function $Z(s)$, just as if the latter were something like a secular determinant for the classical propagator; then, tradition demands that we call $Z(s)$ a zeta function. To facilitate comparison with the quantum zeta function to be met in Sects. 8.9 and 8.10, I pause for an interlude and reexpress the classical zeta function as a slightly different product over periodic orbits. To that end, we must recall that for Hamiltonian flows of two-freedom systems, the stability matrix has two eigenvalues that are mutually inverse; since we are restricted to unstable systems with no stable orbits, these eigenvalues are real and may be denoted as Λ_p and $1/\Lambda_p$, where $|\Lambda_p| > 1$. The inverse-determinant weight of the

---

[16] The reader will remember our use of similar coordinates in Sect. 8.4.3 for a similar purpose.

pth orbit may be expanded as

$$
\begin{aligned}
|\det(M_p^r - 1)|^{-1} &= |(1 - \Lambda_p^r)(1 - \Lambda_p^{-r})|^{-1} = |\Lambda_p|^{-r}(1 - \Lambda_p^{-r})^{-2} \\
&= \sum_{j=0}^{\infty}\sum_{k=0}^{\infty} |\Lambda_p|^{-r}\Lambda_p^{-(j+k)r} .
\end{aligned}
\tag{8.7.20}
$$

Once this expansion is inserted in the zeta function (8.7.18), the sum over repetitions of the pth orbit yields a logarithm,

$$
\ln Z(s) = \sum_p \sum_{j=0}^{\infty}\sum_{k=0}^{\infty} \ln\left(1 - \frac{e^{sT_p}}{|\Lambda_p|\Lambda_p^{j+k}}\right) .
\tag{8.7.21}
$$

Choosing $l \equiv k + j$ and k as summation variables, we can do the sum over k as well and upon reexponentiating, get the desired product form of the zeta function

$$
Z(s) = \prod_p \prod_{l=0}^{\infty}\left(1 - \frac{e^{sT_p}}{|\Lambda_p|\Lambda_p^l}\right)^{l+1} = \prod_{[p]}\prod_{l=0}^{\infty}\left(1 - \frac{e^{sT_p}}{|\Lambda_p|\Lambda_p^l}\right)^{(l+1)g_p} .
\tag{8.7.22}
$$

In the last member, as in the previous two subsections, I have grouped the periodic orbits into action multiplets, a precaution advisable in view of the intended comparison with similarly structured quantum objects.

Now, it is easy to see that for the unstable systems under consideration, the phase-space density equilibrates on the energy shell, consistent with the assumed property of mixing. This follows from the fact that the zeta function has a simple zero at $s = 0$ and thus the propagator has a unique unit eigenvalue, corresponding to a stationary eigenfunction. Since the small-s behavior must be tied up with long orbits, we may, looking at $\ln Z(s)$ as given by (8.7.22), expand $\ln(1 - e^{sT_p}|\Lambda_p|^{-1}\Lambda_p^{-l}) \approx -e^{sT_p}|\Lambda_p|^{-1}\Lambda_p^{-l}$ and then keep only the leading term of the l-sum to get

$$
\ln Z(s) \xrightarrow{s\to 0} -\sum_p \frac{e^{sT_p}}{|\Lambda_p|} \approx -\int_{T_0}^{\infty} dT \frac{e^{sT}}{T} \approx \ln(sT_0) ;
\tag{8.7.23}
$$

here I have invoked the HOdA sum rule in the form (8.6.6) to recast the sum over long orbits into an integral, with T_0 some reference time. We may imagine that T_0 is the period upward of which the HOdA sum rule begins to apply and thus $T_0 \ll T_H$; the precise value does not matter since T_0 will not show up in final results. At any rate, the claim $Z(s) \propto s$, as $s \to 0$, is borne out.

Finally, I come to the goal of this section and relate the diagonal part (8.7.9) of the two-point correlator to the classical zeta function [275, 278]. Comparing the periodic-orbit sums in (8.7.9) and (8.7.18), it is clear that such a relationship must exist; a convenient form is based on the definition

$$
Z_g(s) = \prod_{[p]}\prod_{l=0}^{\infty}\left(1 - \frac{e^{sT_p}}{|\Lambda_p|\Lambda_p^l}\right)^{(l+1)g_p^2}
\tag{8.7.24}
$$

which differs from the classical zeta function in the power of the multiplicity g_p. That is an intuitive difference to arise since the trace of the classical propagator involves a single sum over periodic orbits to begin with, whereas the quantum two-point correlator comes as a pair of sums each of which involves action multiplets of orbits. Inasmuch as dynamical systems typical of the orthogonal or unitary universality class have, respectively, $g_p = 2$ and $g_p = 1$, at least for almost all periodic orbits, we may replace $g_p \to g$ and then find $Z_g(s)$ as the gth power of the classical zeta function $Z(s)$,

$$Z_g(s) = Z(s)^g . \tag{8.7.25}$$

There are strongly chaotic systems for which the multiplicities g_p are not constant but vary appreciably from orbit to orbit [243, 279]. For these there is no simple relation between $Z_g(s)$ and the classical zeta function, but neither do these systems have spectral statistics anywhere near universal.

We still need to check that the diagonal part of the two-point correlator can be found from $Z_g(s)$ and thus, if $g_p = g$, from the classical zeta function. To that end, we simply take the second derivative of $\ln Z_g(s)$ w.r.t. s,

$$
\begin{aligned}
\left(\ln Z_g(s)\right)'' &= \sum_{[p]} g_p^2 \sum_{l=0}^{\infty} (l+1) \frac{d^2}{ds^2} \ln\left(1 - \frac{e^{sT_p}}{|\Lambda_p|\Lambda_p^l}\right) \\
&= -\sum_{[p]} g_p^2 \sum_{l=0}^{\infty} (l+1) \frac{T_p^2 |\Lambda_p|\Lambda_p^l e^{-sT_p}}{(|\Lambda_p|\Lambda_p^l e^{-sT_p} - 1)^2} .
\end{aligned}
$$

This is already closer to y_{diag} than it might appear; adding on the "correction"

$$a(s) = \sum_{[p]} g_p^2 \sum_{l=0}^{\infty} \frac{(l+1)T_p^2}{(|\Lambda_p|\Lambda_p^l e^{-sT_p} - 1)^2} ,$$

which by our usual estimate for Λ_p and regard of exponential proliferation is easily seen as convergent for $s = 0$ and thus indeed a small correction to the divergent main term,

$$
\begin{aligned}
\left[\ln Z_g(s)\right]'' + a(s) &= -\sum_{[p]} g_p^2 \sum_{l=0}^{\infty} \frac{(l+1)\,T_p^2}{|\Lambda_p|\Lambda_p^l e^{-sT_p} - 1} \\
&= -\sum_{[p]} g_p^2 T_p^2 \sum_{l=0}^{\infty} (l+1) \frac{e^{sT_p}|\Lambda_p|^{-1}\Lambda_p^{-l}}{1 - e^{sT_p}|\Lambda_p|^{-1}\Lambda_p^{-l}} \\
&= -\sum_{[p]} g_p^2 T_p^2 \sum_{k=1}^{\infty} \frac{e^{ksT_p}}{|\Lambda_p|^k} \sum_{l=0}^{\infty} \frac{l+1}{\Lambda_p^{kl}} \\
&= -\sum_{[p]} g_p^2 T_p^2 \sum_{k=1}^{\infty} \frac{e^{ksT_p}}{|\Lambda_p|^k(1 - \Lambda_p^{-k})^2} \\
&= -\sum_{[p]} g_p^2 T_p^2 \sum_{k=1}^{\infty} \frac{e^{ksT_p}}{|\det(M_p^k - 1)|} .
\end{aligned}
\tag{8.7.26}
$$

Each step in the foregoing chain of equations should be familiar from previous encounters, except $\sum_{l=0}^{\infty}(l+1)\Lambda_p^{-kl} = (1 - \Lambda_p^{-k})^{-2}$; the reader will appreciate the convergence of all intervening geometric series. The enjoyable conclusion

$$
\begin{aligned}
y_{\text{diag}}(e) &= -\frac{2}{T_H^2}\text{Re}\left[\frac{d^2}{ds^2}\ln Z_g(s) + a(s)\right]_{s=\text{i}2\pi e/T_H} \\
&= -\frac{2g}{T_H^2}\text{Re}\left[\frac{d^2}{ds^2}\ln Z(s) + a(s)\right]_{s=\text{i}2\pi e/T_H}
\end{aligned}
\tag{8.7.27}
$$

does fill the promise of relating the diagonal part of the two-point correlator to classical dynamics.

In the semiclassical limit $T_H \propto \hbar^{1-f} \to \infty$, the correction $a(s)$ may be dropped, and the zeta function is required near $s = 0$ where we have seen that it is $Z(s) \to T_0 s$. Since $(\ln T_0 s)'' = -1/s^2$, we conclude that

$$
y_{\text{diag}}(e) = -\frac{g}{2\pi^2 e^2}\ .
\tag{8.7.28}
$$

This remarkable result gives, with $g = 1$, the (negative of the) nonoscillatory part of the cluster function of the CUE/GUE, $Y_{\text{CUE}}(e) = (1 - \cos 2\pi e)/(2\pi^2 e^2)$, and thus leaves the oscillatory remainder to be expected from the off-diagonal part of the double sum over periodic orbits. For systems with time-reversal invariance and where the time-reversal operation squares to unity, we must use $g = 2$ and thus unfortunately do not find that the above exhausts the nonoscillatory part of y_{COE} in full; only the first term of the large-e expansion is obtained; see (4.13.30). With proper account of the normalization employed for the level density of matrices from the CSE in Chap. 4.13 (only one level from each Kramers' doublet is admitted in the density) the above semiclassical result also agrees with the random-matrix result, $(1/2)^2 2y_{\text{diag}}(e) \hat{=} - Y_{\text{CSE}}(e)$, as far as the leading nonoscillatory term $1/(2\pi e)^2$ is concerned; see (4.13.37). These conclusions were in fact already reached at the end of Sect. 8.7.1.

8.8 The Form Factor at Long Times

Having just revealed that the diagonal approximation to the form factor is valid for times at least up to the Ehrenfest time, possibly even some finite fraction of the Heisenberg time, we should begin trying to understand why most globally chaotic systems have a form factor approaching the random-matrix plateau $\langle|t_n|^2\rangle/N \to 1$ or $\tilde{y}(\tau) \to 1$ for times exceeding the Heisenberg time, $n/N \to \infty$ or $\tau \to \infty$. Readers who have studied the chapter on random-matrix theory will remember that an intuitive explanation of the asymptotic saturation of the form factor was given at the end of Sect. 4.13.3: For maps, the form factor $|t_n|^2 = \sum_{i,j=1}^{N}e^{-\text{i}(\phi_i-\phi_j)}$ is a sum of N^2 unimodular terms the overwhelming majority of them bear no noticeable mutual correlations in the limit $1 \ll N \ll n$. The asymptotic absence of correlations is due to two facts: (1) The phases $n\phi_i$ become effective only modulo 2π, and (2) even the rigidity of the spectrum (which favors the spacing $2\pi/N$ is washed out for $n > N$. The central limit theorem thus suggests the saturation in question, irrespective of whether there is time reversal. The reasoning is the same for flows since their form factor can be written in the analogous form $\tilde{y}(\tau) = \sum_{i,j} e^{-i(E_i - E_j)T_H \tau/\hbar}$ (see Problem 8.11).

Due to its quantum character the foregoing reasoning applies to dynamical systems as well as to random matrices from any of the Gaussian or circular ensembles of Chap. 4.

The same result follows from an essentially equivalent even though seemingly quite different semiclassical argument by *Berry* [272]. Even though limitations of space forbid presenting Berry's derivation here, I want to draw the reader's attention to it because it involves a certain bootstrap idea which may be seen as a precursor to and thus an appropriate preparation for the more recent developments which will be the topics of subsequent sections.

8.9 Bootstrap of Short and Long Orbits Through Unitarity

I have pointed out repeatedly, first in Chap. 4.14.2 and last in Sect. 8.3.4 of this chapter, that the spectrum of a unitary $N \times N$ matrix F is uniquely determined by the first $N/2$ traces $t_n = \mathrm{Tr}F^n$ if the dimension N is even; for N odd, the first $(N+1)/2$ traces are needed; in what follows, I shall summarily refer to $N/2$ and $(N+1)/2$ as half the Heisenberg time. Given that the nth trace is semiclassically expressed in terms of properties of classical period-n orbits, we see that no orbit of period larger than half the Heisenberg time is relevant for any spectral characteristic. We also conclude that unitarity must link properties of orbits with periods exceeding that dividing time to properties of shorter periodic orbits. The link is provided by the so-called self-inversiveness (4.14.26),

$$a_{N-n} = a_N a_n^*, \tag{8.9.1}$$

of the secular polynomial

$$\det(F - \lambda) = \sum_{n=0}^{N} (-\lambda)^n a_{N-n} \equiv P(\lambda) \tag{8.9.2}$$

and Newton's relationships (4.14.25) between the secular coefficients a_n and the t_n,

$$a_n = \frac{1}{n!} \begin{vmatrix} t_1 & 1 & 0 & \cdots & 0 & 0 \\ t_2 & t_1 & 2 & \cdots & 0 & 0 \\ \cdots & \cdots & \cdots & \cdots & \cdots & \cdots \\ t_{n-1} & t_{n-2} & t_{n-3} & \cdots & t_1 & n-1 \\ t_n & t_{n-1} & t_{n-2} & \cdots & t_2 & t_1 \end{vmatrix}, \tag{8.9.3}$$

once the trace formula (8.3.5) is used for the t_n.

What appears, to be an immediate consequence of well known facts of algebra for maps, is not easily transcribed to flows and their periodic-orbit sum (8.3.33) for the density of levels. Not even the names "functional equation" or "Riemann–Siegel look-alike" commonly used for the flow variant of self-inversiveness [257–259] reveal the algebraic background. In fact, Berry and Keating were rather more inspired by number theory and the famous Riemann-Siegel formula for Riemann's zeta function in their suggestion of a look-alike.

To prepare the reader for the ideas of Berry and Keating, it is helpful to stick with maps and algebra for a few more moments. The secular polynomial (8.9.2) is obviously not real for unimodular $\lambda = e^{-i\phi}$, i.e., for a real phase ϕ. But we easily find a function $B(\phi)$ without zeros for real ϕ that turns $P(e^{-i\phi})$ into a real "zeta function" as

$$\zeta(\phi) = B(\phi)P(e^{-i\phi}),$$

$$B(\phi) = \exp\left[\frac{i}{2}\sum_{i=1}^{N}(\phi + \pi + \phi_i)\right] = (-\lambda^*)^{\frac{N}{2}}(a_N^*)^{\frac{1}{2}}. \qquad (8.9.4)$$

The $B(\phi)$ so chosen does make ζ real for real ϕ,

$$\zeta(\phi) = \zeta(\phi)^* = \zeta(-\phi) \qquad (8.9.5)$$

and that latter identity goes by the name of functional equation; by inserting the secular polynomial we find it equivalent to the self-inversiveness (8.9.1). Needless to say, the real zeros of $\zeta(\phi)$ are just the N quasi-energies ϕ_i; when $\zeta(\phi)$ is semiclassically approximated, $\zeta(\phi) \approx \zeta_{s.cl.}(\phi)$, by employing the Gutzwiller traces (8.3.5) for n up to half the Heisenberg time and invoking self-inversiveness alias the functional equation, one gets a semiclassically approximated spectrum. (The flow analogue of $\zeta_{s.cl.}(\phi) = 0$ has been termed "semiclassical quantization" in Refs. [257–259]). As already discussed in Chap. 4.14.2, not all roots $\lambda_i = e^{-i\phi_i}$ of a self-inversive polynomial need be unimodular, i.e., not all of the ϕ_i need come out real; the reality of $\zeta_{s.cl.}(\phi)$ guarantees that the roots to be either real or come in complex conjugate pairs. Should such a disaster happen, the blame would have to go to the semiclassical approximation for the first $N/2$ or $(N+1)/2$ traces.

Readers with the strong nerves of physicists are now invited to adventure in mathematically unsafe territory, starting with the definition of a zeta function for a Hamiltonian $\hat{H}$

$$\zeta(\mathcal{E}) = \det\{A(\mathcal{E}, \hat{H})(\mathcal{E} - \hat{H})\} = \prod_j\{A(\mathcal{E}, E_j)(\mathcal{E} - E_j)\}$$

$$= \det A(\mathcal{E}, \hat{H})\exp\operatorname{Tr}\ln(\mathcal{E} - \hat{H}). \qquad (8.9.6)$$

To provide a notational distinction between definitely real and possibly complex energies, in this section I shall denote the former by E and the latter by $\mathcal{E}$. The function A is assumed to have no real zeros, to be real for real E, and to ensure convergence should N not be finite; it is otherwise arbitrary; elementary examples for such "regularizers" are found in [257] and Problems 8.12, 8.13. Then, the zeta function has the eigenvalues E_j as the only real zeros. The logarithm appearing above can be expressed as the integral of the resolvent trace

$$g(\mathcal{E}) = \operatorname{Tr}\frac{1}{\mathcal{E} - \hat{H}} \qquad (8.9.7)$$

which yields the level density for $\mathcal{E} = E + i0^+$ as $\varrho(E) = -\frac{1}{\pi} \operatorname{Im} g(E + i0^+)$ and thus the level staircase as

$$\mathcal{N}(E) = -\frac{1}{\pi} \int_0^E dE' \operatorname{Im} g(E' + i0^+) ; \tag{8.9.8}$$

for convenience I imagine that the spectrum of $\hat{H}$ begins immediately to the right of $E = 0$. Now, I replace, as announced, the logarithm,

$$\operatorname{Tr} \ln(\mathcal{E} - \hat{H}) - \operatorname{Tr} \ln(-\hat{H}) = \int_0^{\mathcal{E}} d\mathcal{E}' g(\mathcal{E}') \tag{8.9.9}$$

$$= \int_0^{\mathcal{E}} d\mathcal{E}' g_{\mathrm{osc}}(\mathcal{E}') + \int_0^{\mathcal{E}} d\mathcal{E}' \langle g(\mathcal{E}') \rangle ,$$

splitting the resolvent trace into an oscillatory and a smoothed part; the latter is defined as usual by a local spectral average. The smoothed part is again split into two additive pieces such that $\int_0^{\mathcal{E}} d\mathcal{E}' \langle g(\mathcal{E}') \rangle = -i\pi \langle \mathcal{N}(\mathcal{E}) \rangle + \{ \int_0^{\mathcal{E}} d\mathcal{E}' \langle g(\mathcal{E}') \rangle + i\pi \langle \mathcal{N}(\mathcal{E}) \rangle \}$; as indicated by a somewhat cavalier notation, the first of these is defined so as to become $-i\pi \langle \mathcal{N}(E) \rangle$ for $\mathcal{E} = E + i0^+$ while the second becomes real in that limit; the local spectral average $\langle \mathcal{N}(E) \rangle$ is the Weyl staircase, i.e., the number of Planck cells contained in the part of phase space with energy up to E,

$$\langle \mathcal{N}(E) \rangle = (2\pi\hbar)^{-f} \int d^f q d^f p \, \Theta[E - H(q, p)] . \tag{8.9.10}$$

Thus, the zeta function assumes the form

$$\zeta(\mathcal{E}) = B(\mathcal{E}) \exp \left\{ -i\pi \langle \mathcal{N}(\mathcal{E}) \rangle + \int_0^{\mathcal{E}} d\mathcal{E}' \{ g(\mathcal{E}') - \langle g(\mathcal{E}') \rangle \} \right\} ,$$

$$B(\mathcal{E}) = \det(-A\hat{H}) \exp \left\{ \int_0^{\mathcal{E}} d\mathcal{E}' \langle g(\mathcal{E}') \rangle + i\pi \langle \mathcal{N}(\mathcal{E}) \rangle \right\} . \tag{8.9.11}$$

The function $B(\mathcal{E})$ becomes real and non-zero for real $\mathcal{E} = E$. To this point, we have departed from the safe path indicated in (8.9.4) only for maps by writing the secular polynomial $\det(E - \hat{H})$ in the exponential form in the last member of (8.9.6) to prepare to bring in the resolvent trace for the oscillatory part of which we have the periodic-orbit sum (8.3.32).

Since we actually need the integral of g_{osc}, we should realize that the periodic-orbit sum (8.3.32),

$$g_{\mathrm{osc}}(\mathcal{E}) = \frac{1}{i\hbar} \sum_p \sum_{r=1}^{\infty} T_p \frac{e^{irS_p(\mathcal{E})/\hbar}}{\sqrt{|2 - \operatorname{Tr} M_p^r|}} , \tag{8.9.12}$$

is itself nearly a total derivative since due to $\partial S_p / \partial \mathcal{E} = T_p(\mathcal{E})$,

$$\frac{\partial}{\partial \mathcal{E}} \frac{e^{irS_p(\mathcal{E})/\hbar}}{\sqrt{|2 - \operatorname{Tr} M_p^r|}} = \frac{i}{\hbar} r T_p \frac{e^{irS_p(\mathcal{E})/\hbar}}{\sqrt{|2 - \operatorname{Tr} M_p^r|}} + \ldots ; \tag{8.9.13}$$

the dots stand for a term involving the derivative of the denominator which is smaller by one order in $\hbar$ than the first term and may be dropped. Upon inserting all of this periodic-orbit stuff into our zeta function from (8.9.11), we get the periodic-orbit approximation

$$\zeta(\mathcal{E}) \approx B(\mathcal{E}) \exp\left(-i\pi\langle\mathcal{N}(\mathcal{E})\rangle\right) \prod_p \exp\left\{-\sum_{r=1}^{\infty} \frac{e^{ir\mathcal{S}_p(\mathcal{E})/\hbar}}{r\sqrt{|2 - \operatorname{Tr} M_p^r|}}\right\}; \qquad (8.9.14)$$

from this point on I shall somewhat frivolously write "=" instead of "$\approx$", trusting that the reader will not take periodic-orbit approximations for more than the leading-order-in-$\hbar$ asymptotics that they are meant to provide. Safe ground would and eventually will be left when proceeding to real energy, $\mathcal{E} \to E$, in fact even as soon as $\operatorname{Im}\mathcal{E}$ becomes smaller than the convergence-enforcing limit $\epsilon_c = \hbar\lambda/2$ established at the end of Sect. 8.3.4. For the moment, we can still afford to stay clear of that difficulty and stick with $\operatorname{Im}\mathcal{E} > \epsilon_c$.

It is helpful to write the infinite product over periodic orbits and their repetitions in (8.9.14) as a series. To that end, we may first do the sum over repetitions, as we have already done in Sect. 8.7.3. Employing the eigenvalues[17] $\exp(\pm\lambda_p T_p)$ of the stability matrix M_p to express the determinant $\det(M_p^r - 1) = |2 - \operatorname{Tr} M_p^r|$ as

$$|\det(M_p^r - 1)| = e^{r\lambda_p T_p}\left(1 - e^{-r\lambda_p T_p}\right)^2 = \left(e^{r\lambda_p T_p/2} - e^{-r\lambda_p T_p/2}\right)^2, \quad (8.9.15)$$

we can write the stability prefactor as the geometric series

$$\frac{1}{\sqrt{|\det(M_p^r - 1)|}} = e^{-r\lambda_p T_p/2} \sum_{k=0}^{\infty} e^{-kr\lambda_p T_p}, \qquad (8.9.16)$$

whereupon the sum over repetitions yields $\ln\left(1 - \exp\left[i\mathcal{S}_p/\hbar - (k + \frac{1}{2})\lambda_p T_p\right]\right) = -\sum_{r=1}^{\infty} \frac{1}{r} \exp r[i\mathcal{S}_p/\hbar - (k + \frac{1}{2})\lambda_p T_p]$. In the zeta function we have thus bargained one product against another,

$$\zeta(\mathcal{E}) = B(\mathcal{E}) e^{-i\pi\langle\mathcal{N}(\mathcal{E})\rangle} \prod_p \prod_{k=0}^{\infty}\left\{1 - \exp\left[i\mathcal{S}_p/\hbar - (k + \tfrac{1}{2})\lambda_p T_p\right]\right\}. \quad (8.9.17)$$

A sum is to come through Euler's identity

$$\prod_{k=0}^{\infty}(1 - ax^k) = 1 + \sum_{r=1}^{\infty} \frac{(-a)^r x^{r(r-3)/4}}{(x^{-1/2} - x^{1/2})(x^{-1} - x)\dots(x^{-r/2} - x^{r/2})} \qquad (8.9.18)$$

and reads

$$\zeta(\mathcal{E}) = B(\mathcal{E}) e^{-i\pi\langle\mathcal{N}(\mathcal{E})\rangle} \qquad (8.9.19)$$

$$\times \prod_p \left\{1 + \sum_{r=1}^{\infty}(-1)^r \frac{\exp\left[ir\mathcal{S}_p/\hbar - \frac{1}{4}r(r-1)\lambda_p T_p\right]}{\prod_{j=1}^{r}|\det(M_p^j - 1)|^{1/2}}\right\}.$$

---

[17] Assuming an unstable system without stable orbits, here I do the sum for a locally hyperbolic map, reserving until later remarks on maps with local inverse hyperbolicity.

The phase factors suggest interpreting the summation variable r as a new repetition number. Upon expanding the product over primitive orbits, we finally arrive at the desired series

$$\zeta(\mathcal{E}) = B(\mathcal{E}) e^{-i\pi\langle\mathcal{N}(\mathcal{E})\rangle} \sum_{n=0}^{\infty} C_n(\mathcal{E}) e^{i\tilde{S}_n(\mathcal{E})/\hbar} . \tag{8.9.20}$$

Now, the summation is over the so-called pseudo-orbits in which each primitive orbit is repeated r_p times (possibly $r_p = 0$) such that n is really a multiple summation variable,

$$n = \{r_p\}, \quad r_p = 0, 1, 2, \ldots . \tag{8.9.21}$$

A pseudo-orbit has as its action and pseudoperiod the pertinent sums over the contributing primitive orbits,

$$\tilde{S}_n = \sum_p r_p S_p, \qquad \tilde{T}_n = \frac{\partial \tilde{S}_n}{\partial \mathcal{E}} = \sum_p r_p T_p . \tag{8.9.22}$$

The pseudo-orbit series (8.9.20) can be thought ordered by increasing pseudo-period $\tilde{T}_n$. The weight $C_n(\mathcal{E})$ of the nth pseudo-orbit is

$$C_n(\mathcal{E}) = \prod_p \left\{ (-1)^{r_p} \frac{\exp\left[-\frac{1}{4} r_p (r_p - 1)\lambda_p T_p\right]}{\prod_{j=1}^{r_p} |\det(M_p^j - 1)|^{1/2}} \right\} . \tag{8.9.23}$$

I leave to the reader to show that primitive orbits that are hyperbolic with reflection show up like hyperbolic ones in the pseudo-orbit sum (8.9.20), except for the replacement $(-1)^{r_p} \rightarrow (-1)^{\text{Int}[(r_p+1)/2]}$ in the weight $C_n(\mathcal{E})$.

Incidentally, multiple repetitions of a primitive orbit within a pseudo-orbit have rapidly decreasing weights as r_p grows, due to the factor

$$\frac{\exp\left[-\frac{1}{4} r_p (r_p - 1)\lambda_p T_p\right]}{\prod_{j=1}^{r_p} |\det(M_p^j - 1)|^{1/2}} \longrightarrow \exp\{-\frac{1}{2} r_p^2 \lambda_p T_p\} . \tag{8.9.24}$$

Thus, the most important long pseudo-orbits are those composed of singly traversed primitive orbits.

Mustering a lot of courage, now I follow Berry and Keating to real energy in the pseudo-orbit sum for the zeta function. Convergence is surely lost; reality, while required by the definition (8.9.6), is no longer manifest and definitely destroyed by any finite truncation of the infinite sum. To make the best of the seemingly desperate situation we may impose reality once more,

$$e^{-i\pi\langle\mathcal{N}(E)\rangle} \sum_{n=0}^{\infty} C_n(E) e^{\frac{i}{\hbar}\tilde{S}_n(E)} = e^{i\pi\langle\mathcal{N}(E)\rangle} \sum_{n=0}^{\infty} C_n(E) e^{-\frac{i}{\hbar}\tilde{S}_n(E)} , \tag{8.9.25}$$

in the hope of thus forbidding the divergence to play the most evil of games with us. But no amount of conjuration will retrieve convergence. Not even refuge to analytic continuation to complex energy is possible any longer since the l.h.s. would require $\mathrm{Im}\mathcal{E} > \hbar\lambda/2$ and the r.h.s. $\mathrm{Im}\mathcal{E} < -\hbar\lambda/2$, leaving no overlap of the domains of convergence. Keating modestly calls the foregoing reality condition the "formal functional equation"[258], before cunningly drawing the most useful consequences anticipated in the title of this section.

A Fourier transform $\int_{e-\alpha(\hbar)}^{e+\alpha(\hbar)} dE \exp\left(\mathrm{i}\tau E/\hbar\right)(\dots)$ with respect to a classically small but semiclassically large energy interval,

$$\alpha(\hbar) \to 0, \qquad \alpha(\hbar)/\hbar \to \infty \qquad \text{for} \quad \hbar \to 0, \tag{8.9.26}$$

is now done on the functional equation. Neither the Weyl staircase $\langle\mathcal{N}(E)\rangle$ nor the action $\mathcal{S}_n$ vary much over the small integration range and may thus both be expanded around e to first order such that the l.h.s. of (8.9.25) yields

$$\mathrm{e}^{\mathrm{i}[-\pi\langle\mathcal{N}(e)\rangle+\tau e/\hbar]} \int_{-\alpha(\hbar)}^{+\alpha(\hbar)} d\eta\, \mathrm{e}^{\mathrm{i}[-\eta\pi\langle\varrho(e)\rangle+\eta\tau/\hbar]} \sum_{n=0}^{\infty} C_n(e)\, \mathrm{e}^{\mathrm{i}(\tilde{S}_n+\eta\tilde{T}_n)/\hbar} \tag{8.9.27}$$

or, after rescaling the integration variable as $\eta/\hbar \to \eta$ and boldly interchanging the order of integration and summation,

$$\hbar\, \mathrm{e}^{\mathrm{i}[-\pi\langle\mathcal{N}(e)\rangle+\tau e/\hbar]} \sum_{n=0}^{\infty} C_n(e)\, \mathrm{e}^{\mathrm{i}\tilde{S}_n/\hbar} \int_{-\alpha(\hbar)/\hbar}^{+\alpha(\hbar)/\hbar} d\eta\, \mathrm{e}^{\mathrm{i}\eta[-\pi\hbar\langle\varrho(e)\rangle+\tau+\tilde{T}_n]}. \tag{8.9.28}$$

But now the integration range is large, and the η-integral gives a fattened delta function $2\pi\delta_{\hbar/\alpha(\hbar)}[-\pi\hbar\langle\varrho(e)\rangle+\tau+\tilde{T}_n]$ of width $\hbar/\alpha(\hbar)$; that width tends to zero as $\hbar \to 0$.

Exactly the same procedure brings the r.h.s. of the functional equation into to a form identical to (8.9.29) save for sign changes in front of $\langle\mathcal{N}\rangle, \langle\varrho\rangle, \tilde{S}_n$, and $\tilde{T}_n$. Hence, the functional equation reappears as

$$\mathrm{e}^{-\mathrm{i}\pi\langle\mathcal{N}(e)\rangle} \sum_{n=0}^{\infty} C_n(e)\, \mathrm{e}^{\mathrm{i}\tilde{S}_n(e)/\hbar}\, \delta_{\hbar/\alpha(\hbar)}\big(-\pi\hbar\langle\varrho(e)\rangle+\tau+\tilde{T}_n(e)\big) \tag{8.9.29}$$

$$\approx \mathrm{e}^{\mathrm{i}\pi\langle\mathcal{N}(e)\rangle} \sum_{n=0}^{\infty} C_n(e)\, \mathrm{e}^{-\mathrm{i}\tilde{S}_n(e)/\hbar}\, \delta_{\hbar/\alpha(\hbar)}\big(\pi\hbar\langle\varrho(e)\rangle+\tau-\tilde{T}_n(e)\big).$$

As a final step we integrate on both sides over τ from 0 to ∞ and recognize $\pi\hbar\langle\varrho(e)\rangle = T_H(e)/2$ as half the Heisenberg time:

$$\sum_{\tilde{T}_n \le \frac{T_H}{2}} C_n \exp\mathrm{i}(\tilde{S}_n/\hbar - \pi\langle\mathcal{N}\rangle) \approx \sum_{\tilde{T}_n > \frac{T_H}{2}} C_n \exp\mathrm{i}(-\tilde{S}_n/\hbar + \pi\langle\mathcal{N}\rangle). \tag{8.9.30}$$

We have arrived at a sum rule connecting the manifestly finite contribution from pseudo-orbits with pseudo-periods below half the Heisenberg time to the ill-defined contribution of the infinitely many longer pseudo-orbits; the two contributions are mutual complex conjugates. If we import that sum rule into the zeta function (8.9.20), we get a finite semiclassical approximation, the celebrated Riemann–Siegel look-alike

$$\zeta_{\text{s.cl.}}(E) = 2B(E) \sum_{\tilde{T}_n(E) \leq T_H(E)/2} C_n(E) \cos\left[\tilde{S}_n(E)/\hbar - \pi\langle\mathcal{N}(E)\rangle\right]. \quad (8.9.31)$$

What a relief indeed, after all the chagrin about lost convergence! The zeros of this finite semiclassical zeta function should and have been checked [280, 281]; they give semiclassically accurate energy eigenvalues up to the energy E which one is free to choose, provided that the periodic orbits with periods up to $T_H(E)/2$ are available. It is immaterial now whether the Hilbert space is finite- or infinite-dimensional. We should not forget the spiritual backbone of the bootstrap, i.e., the self-inversiveness (8.9.1) disguised as a functional equation (8.9.5) or in the semiclassical implementation with pseudo-orbits, (8.9.25).

As already mentioned, the Riemann–Siegel look-alike (8.9.31) was first conjectured as an extension of the famous Riemann–Siegel formula for the nontrivial zeros of Riemann's zeta function. The latter formula may be derived by arguments that are formally analogous to those employed here, but in fact are mathematically sound for the Riemann case. More recent work that limitations of space forbid discussing here, has achieved much more soundness for the dynamic zeta functions of chaotic systems, basically by analytic continuation toward complex $\hbar$ rather than complex energy [259].

8.10 Spectral Statistics Beyond the Diagonal Approximation

Now, we are equipped to appreciate the semiclassical understanding of the reason that generic chaotic dynamics have the spectral fluctuations predicted by random-matrix theory, reached by this time of writing. To unveil it, I invoke the now familiar splitting of periodic-orbit sums into a short-orbit part and a long-orbit remainder. Setting the dividing period at some value T^*, I propose to look at the short-orbit part of the level staircase,

$$\mathcal{N}_{T^*}(E) = \langle\mathcal{N}(E)\rangle + \mathcal{N}_{T^*}^{\text{osc}}(E) \quad (8.10.1)$$

$$\mathcal{N}_{T^*}^{\text{osc}}(E) = \sum_p^{T_p \leq T^*} \sum_{r=1}^{\infty} \frac{\sin(r\mathcal{S}_p/\hbar)}{\pi r \sqrt{|\det(M_p^r - 1)|}},$$

and its derivative (see (8.9.13)), the truncated density of levels $\varrho_{T^*}(E)$. If the dividing time T^* is chosen as of the order of the Heisenberg time, $T^* \approx T_H(E)$, the truncated density should show sufficiently pronounced peaks to allow for good estimates of the energy levels E_n up to E. Equivalently, one may determine semiclassical estimates $E_n(T^*)$ from the truncated staircase as [280, 282]

$$\mathcal{N}_{T^*}\left[E_n(T^*)\right] = n + \tfrac{1}{2} \quad (8.10.2)$$

since indeed the staircase is defined to mount one unit step for every successive level. The foregoing semiclassical quantization is a variant of that developed in the previous section, $\zeta_{\text{s.cl.}}(E) = 0$.

The task of the long orbits barred from $\varrho_{T^*}(E)$ may be seen as twofold: (1) They ought to fine-tune the levels to their exact values, $E_n(T^*) \rightarrow E_n$; one has foregone such rigor once embarked on any semiclassical approximation and should not shed tears about the loss now; (2) they should sharpen the peaks of $\varrho_{T^*}(E)$ to delta functions. That second role, however, can be played by an impostor density, $D_{T^*}(E) = \sum_n \delta[E - E_n(T^*)]$. I hurry to more respectfully call the latter a "bootstrapped" density because its Fourier transform has structure for all times, not just for times up to the cutoff T^*, and good structure indeed: As discovered by *Bogomolny* and *Keating* [275], $D_{T^*}(E)$ helps a great deal in calculating the two-point correlator of the level density beyond the diagonal approximation, without requiring orbits whose periods exceed T^*. I proceed to explain the theory of Bogomolny and Keating which, incidentally, was inspired by an important seminal paper by *Andreev* and *Altshuler* [283].

We start by rewriting $D_{T^*}(E)$ as

$$
D_{T^*}(E) \;=\; \varrho_{T^*}(E) \sum_n \delta\!\left(\mathcal{N}_{T^*}(E) - n - \tfrac{1}{2}\right)
$$

$$
=\; \varrho_{T^*}(E) \sum_{k=-\infty}^{\infty} (-1)^k \exp\!\left[\mathrm{i}2\pi k \mathcal{N}_{T^*}(E)\right] \tag{8.10.3}
$$

and use this as the density in the connected two-point correlator (8.7.5) to get

$$
y(e) \;=\; \frac{1}{\langle \varrho \rangle^2} \Big\langle \varrho_{T^*}(E)\varrho_{T^*}(E + e/\langle \varrho \rangle) \sum_{k_1,k_2} (-1)^{k_1 - k_2}
$$

$$
\times \exp\Big\{\mathrm{i}2\pi\big[k_1 \mathcal{N}_{T^*}(E) - k_2 \mathcal{N}_{T^*}(E + e/\langle \varrho \rangle)\big]\Big\}\Big\rangle - 1 . \tag{8.10.4}
$$

The term with $k_1 = k_2 = 0$ in the double sum, $\langle \varrho \rangle^{-2}\langle \varrho_{T^*}(E)\varrho_{T^*}(E+e/\langle \varrho \rangle)\rangle - 1$, deserves special attention. It differs from the full correlator by admitting only periodic orbits with periods up to T^*. Since we have stipulated that $T^* \approx T_H$, we may, remembering the optimistic estimate of the range of validity of the diagonal approximation from 8.7, identify the ($k_1 = k_2 = 0$) term with just the diagonal approximation $y_{\text{diag}}(e)$; recalling also the connection (8.7.27) of $y_{\text{diag}}(e)$ with the classical zeta function we have, as already set forth in (8.7.28),

$$
y_{\text{diag}}(e) \;=\; \langle \varrho \rangle^{-2}\langle \varrho_{T^*}(E)\varrho_{T^*}(E + e/\langle \varrho \rangle)\rangle - 1
$$

$$
\approx\; -\frac{2g}{T_H^2}\,\mathrm{Re}\left[\frac{d^2}{ds^2}\ln Z(s)\right]_{s=\mathrm{i}2\pi e/T_H} \tag{8.10.5}
$$

$$
=\; -\frac{g}{2(\pi e)^2} .
$$

Note that I have (1) not bothered to include the correction term $a(s)$ which is irrelevant for the goal of this section, unveiling random-matrix behavior and (2) assumed that the action multiplicities g_p are representable by the single number g.

Now on to the remainder of the double sum in (8.10.4) which can't help being the complement to $y_{\text{diag}}(e)$ in $y(e)$, i.e., the elusive "off-diagonal" part $y_{\text{off}}(e)$. An important simplification arises here from the observation that only the diagonal terms with $k_1 = k_2$ can contribute, since the local energy average annuls the exponential $\exp[\mathrm{i}2\pi(k_1 - k_2)\mathcal{N}_{T_*}(E)]$ for $k_1 \neq k_2$. It follows that

$$y_{\text{off}}(e) = \frac{1}{\langle\varrho\rangle^2}\Big\langle \varrho_{T^*}(E)\varrho_{T^*}(E + e/\langle\varrho\rangle)$$
$$\times \sum_{k\neq 0} \exp\Big\{\mathrm{i}2\pi k\big[\mathcal{N}_{T^*}(E) - \mathcal{N}_{T^*}(E + e/\langle\varrho\rangle)\big]\Big\}\Big\rangle. \qquad (8.10.6)$$

But the feast of simplifications is not over yet: To leading order in $\hbar$, we may identify ϱ_{T^*} with $\langle\varrho\rangle$ and then twice cancel these two quantities, thus obtaining

$$y_{\text{off}}(e) = \Big\langle \sum_{k\neq 0} \exp\Big\{\mathrm{i}2\pi k\big[\mathcal{N}_{T^*}(E) - \mathcal{N}_{T^*}(E + e/\langle\varrho\rangle)\big]\Big\}\Big\rangle. \qquad (8.10.7)$$

When we split the staircase into mean and oscillatory parts, the difference of the two mean parts yields $\langle\mathcal{N}_{T^*}(E)\rangle - \langle\mathcal{N}_{T^*}(E + e/\langle\varrho\rangle)\rangle \to -e$ such that

$$y_{\text{off}}(e) = \sum_{k\neq 0} \mathrm{e}^{-\mathrm{i}2\pi ke}\Big\langle \exp\Big\{\mathrm{i}2\pi k\big[\mathcal{N}_{T^*}^{\text{osc}}(E) - \mathcal{N}_{T^*}^{\text{osc}}(E + e/\langle\varrho\rangle)\big]\Big\}\Big\rangle. \qquad (8.10.8)$$

A sweet temptation arises at this point: If the periodic-orbit contributions to the truncated staircase $\mathcal{N}_{T^*}^{\text{osc}}(E)$ were statistically independent, their large number would entitle us to invoke the central limit theorem and ascribe Gaussian statistics to the truncated staircase; then, the averaged exponential could be replaced as $\langle\exp[\dots]\rangle \to \exp\frac{1}{2}\langle[\dots]^2\rangle$. Momentarily suppressing all scruples about validity, I succumb and assume such independence. The fruit is

$$y_{\text{off}}(e) = \sum_{k\neq 0} \mathrm{e}^{-\mathrm{i}2\pi ke}\exp\Big\{-2\pi^2 k^2\Big\langle\big[\mathcal{N}_{T^*}^{\text{osc}}(E) - \mathcal{N}_{T^*}^{\text{osc}}(E + e/\langle\varrho\rangle)\big]^2\Big\rangle\Big\}$$
$$= \sum_{k\neq 0} \mathrm{e}^{-\mathrm{i}2\pi ke}\left[\frac{\exp 4\pi^2\langle\mathcal{N}_{T^*}^{\text{osc}}(E + e/\langle\varrho\rangle)\mathcal{N}_{T^*}^{\text{osc}}(E)\rangle}{\exp 4\pi^2\langle\mathcal{N}_{T^*}^{\text{osc}}(E)^2\rangle}\right]^{k^2}. \qquad (8.10.9)$$

The two-point correlator of the truncated staircase turns out to be the principal ingredient in $y_{\text{off}}(e)$ and that is a remarkable bootstrap indeed: Due to $T^* \approx T_H$, the truncated-staircase correlator can be represented by the diagonal approximation in terms of periodic orbits,

$$\langle\mathcal{N}_{T^*}^{\text{osc}}(E + e/\langle\varrho\rangle)\mathcal{N}_{T^*}^{\text{osc}}(E)\rangle \qquad (8.10.10)$$
$$= \sum_{[p]}^{T_p \leq T^*} \sum_{r=1}^{\infty} g_p^2 \left\langle\frac{\sin(r\mathcal{S}_p(E + e/\langle\varrho\rangle)/\hbar)\sin(r\mathcal{S}_p(E)/\hbar)}{\pi^2 r^2|\det(M_p^r - 1)|}\right\rangle$$
$$= \sum_{[p]}^{T_p \leq T^*} \sum_{r=1}^{\infty} \frac{g_p^2\cos(re2\pi T_p/T_H)}{2\pi^2 r^2|\det(M_p^r - 1)|}.$$

The few lines of calculation left out in between the members of the foregoing chain are the same as those leading from (8.7.6) to (8.7.9).

Would the repetition number here enter with the factor $1/r$ rather than $1/r^2$, we could easily and rigorously relate the exponentiated version of the truncated-staircase correlator required in (8.10.9),

$$\exp 4\pi^2 \langle \mathcal{N}_{T^*}^{\mathrm{osc}}(E + e/\langle \varrho \rangle)\mathcal{N}_{T^*}^{\mathrm{osc}}(E) \rangle = \left| \exp \sum_{[p]}^{T_p \leq T^*} \sum_{r=1}^{\infty} \frac{g_p^2 e^{(ire2\pi T_p/T_H)}}{r^2 |\det(M_p^r - 1)|} \right|^2 ,$$

(8.10.11)

with the classical zeta function $Z(s)$ defined in (8.7.18). But repetitions cannot get in our way of checking for universality in spectral fluctuations. Both in classical and quantum mechanics, universal behavior can be brought about only by long orbits that explore "all" of phase space; among the orbits with a given long period, the primitive ones exponentially outnumber and outweigh the repeated shorter ones, as is easily seen from our usual account of exponential proliferation and an average Lyapunov exponent. With repetitions neglected and the restriction to systems with $g_p = g$,

$$\exp 4\pi^2 \langle \mathcal{N}_{T^*}^{\mathrm{osc}}(E + e/\langle \varrho \rangle)\mathcal{N}_{T^*}^{\mathrm{osc}}(E) \rangle = |Z_{T^*}(i2\pi e/T_H)|^{-2g} .$$

(8.10.12)

As signalled by the subscript T^*, the function $Z_{T^*}(s)$ differs from the classical zeta function by the cutoff in the periodic-orbit sum at the time T^*; whether we keep repetitions in Z_{T^*} is a matter of taste since their contributions are negligible anyway. Thus, the off-diagonal part of the two-point correlator becomes

$$y_{\mathrm{off}}(e) = \sum_{k \neq 0} e^{-i2\pi ke} \left| \frac{Z_{T^*}(i2\pi e/T_H)}{Z_{T^*}(0)} \right|^{-2gk^2} .$$

(8.10.13)

The critical reader may wonder about the appearance of $Z_{T^*}(0)$ since I had argued in Sect. 8.7.3, that the classical zeta function $Z(s)$ vanishes with vanishing argument. The cutoff T^* is essential here; it must prevent $Z_{T^*}(s)$ from vanishing for $s \to 0$ since it is precisely the excluded orbits with periods up to infinity that ergodically explore all of the classical phase space and bring about the stationary behavior of the phase-space density reflected in the existence of the simple zero of $Z(s)$ at $s = 0$. The following calculation paralleling (8.7.23) indeed shows that

$$Z_{T^*}(0) = \exp\left[-\sum_{p}^{T \leq T^*} \frac{1}{|\det(M_p - 1)|} \right] = \exp\left(-\int_{T_0}^{T^*} \frac{dt}{t} \right) = \frac{T_0}{T^*} .$$

(8.10.14)

Similarly evaluating $|Z_{T^*}(ix)|$ with real x,

$$Z_{T^*}(ix) = \exp\left(-\int_{T_0}^{T^*} dt \frac{\cos xt}{t} \right) = \exp\left[\mathrm{Ci}\,(xT_0) - \mathrm{Ci}\,(xT^*) \right] ,$$

(8.10.15)

where the cosine integral $\text{Ci}\,(x) = -\int_x^\infty dt\,\frac{\cos t}{t}$ with $x \geq 0$ can be written as [284]

$$\text{Ci}\,(x) - \gamma - \ln x = \int_0^x dt\,\frac{\cos t - 1}{t} = \sum_{n=1}^\infty \frac{(-1)^n x^{2n}}{2n(2n)!}\ ; \tag{8.10.16}$$

γ denotes Euler's constant. Clearly, $Z_{T^*}(\mathrm{i}x) \to Z_{T^*}(0)$ as $x \to 0$. On the other hand, if we intend to use $Z_{T^*}(\mathrm{i}x)$ only for $x > 0$, we may let $T^* \approx T_H \to \infty$ to obtain $Z_{T^*}(\mathrm{i}x) \to Z(\mathrm{i}x) = \exp\big(\text{Ci}\,(xT_0)\big)$ and the familiar $Z(0) = 0$; we conclude that the two limits $x \to 0$ and $T^* \to \infty$ do not commute for $Z_{T^*}(\mathrm{i}x)$.

With the foregoing remarks in mind we find that $Z_{T^*}(\mathrm{i}x)/Z_{T^*}(0) \approx \mathrm{e}^\gamma \mathrm{e} 2\pi T^*/T_H$ so that for e of order unity, the series (8.10.13) for $y_{\text{off}}(e)$ has rapidly decreasing contributions as $|k|$ grows and therefore can be well approximated by the terms with $k = \pm 1$. Thus, we find

$$y_{\text{off}}(e) = 2\cos(2\pi e)\left|\frac{Z_{T_H}(\mathrm{i}2\pi e/T_H)}{Z_{T_H}(0)}\right|^{-2g} = \frac{2\cos 2\pi e}{(2\pi e)^{2g}}\left(\frac{T_H}{T^*\mathrm{e}^\gamma}\right)^{2g}. \tag{8.10.17}$$

Combining this with $y_{\text{diag}}(e)$ from (8.7.28), we get the periodic-orbit theory prediction for the connected two-point function as

$$y_{\text{p.o.t.}}(e) = -\frac{g}{2\pi^2 e^2} + \frac{2\cos 2\pi e}{(2\pi e)^{2g}}\left(\frac{T_H}{T^*\mathrm{e}^\gamma}\right)^{2g}. \tag{8.10.18}$$

This is to be compared with the negative two-point cluster function $-Y(e)$ predicted by the various ensembles of random-matrix theory. We must set aside our pride and determine the cutoff time T^* by fitting the coefficient to that encountered in random-matrix theory.

For systems without time-reversal symmetry, we must take $g = 1$. After setting $T^* = T_H \mathrm{e}^{-\gamma} \approx 0.56T_H$, we can enjoy the full agreement $y(e) = -Y_{\text{CUE}}(e)$. The bootstrap (8.10.17) has in effect imparted knowledge of the jump of the first derivative of the form factor at the Heisenberg time to $y(e)$. The exact result is reached since there is no structure in the CUE form factor beyond the jumps of the first derivative at $\tau = 0$ and $|\tau| = 1$.

More amazing and truly non-trivial is the message of (8.10.18) about dynamics from the orthogonal universality class, even though the success is not complete. Simply setting $g = 2$ and $T^* = T_H\sqrt{2}\mathrm{e}^{-\gamma} \approx 0.73T_H$, we get the leading contribution $-(\pi e)^{-2}$ to the nonoscillatory term complemented by the leading contribution $(2\pi^4 e^4)^{-1}\cos 2\pi e$ to the oscillatory term of the COE result (4.13.30), as if the bootstrap had incorporated the jump of the third derivative of the COE form factor at the Heisenberg time $|\tau| = 1$.

A little less complete yet, but still quite satisfactory, is the yield for dynamics from the symplectic universality class. As already explained in Sect. 8.7, we must use $g = 2$ because of time-reversal invariance and throw in a factor $(1/2)^2$ to account for the normalization of the density used for the symplectic ensemble in Chap. 4; moreover, time was measured in units of half the Heisenberg time in Sect. 4.13.5 which fact obliges us to insert a factor 2 in the argument of the cosine in (8.10.18). The choice to be made for the cutoff time is the same as for the unitary universality class. Thus, the periodic-orbit result takes the form

$$y_{\text{p.o.t.}}^{\text{sympl}}(e) = -\frac{1}{(2\pi e)^2} + \frac{\cos 4\pi e}{2(2\pi e)^4} \tag{8.10.19}$$

which correctly reproduces the leading contributions to those terms in the random-matrix result (4.13.37) that are due to the singularities of the form factor at $\tau = 0$ and $\tau = \tau_H$. Even the leading contribution to the term oscillating with the frequency 2π is missed out on, which the CSE predicts as $(4|e|)^{-1}\cos 2\pi e$; this is to say that the logarithmic singularity of the form factor at $\tau = \tau_H/2$ escapes the present bootstrap.

The need to fit the cutoff time of course means that the weights of the oscillatory terms caught by bootstrapping are not a result of the theory; only the frequencies (and thus the locations of the singularities of the form factor) and the powers of e (and thus the nature of the singularity) are. There is, however, consolation in the fact that the cutoff time comes out nearer $\tau_H/2$ than τ_H. Thus, we may attribute some self-consistency to the assumption of statistical independence made before. For the nonoscillatory term, the weights do come out right without any fitting since the celebrated HOdA sum rule gives the size of the jump of the first derivative of the form factor.

There is no hiding the most painful failure of all semiclassical efforts to explain universality in spectral fluctuations: The small-e behavior and the different degrees of level repulsion keep mocking periodic orbits, unless one insists that $y_{\text{p.o.t.}}(e)$ is exact for dynamics without time-reversal symmetry and thus does imply quadratic level repulsion. What a challenge!

8.11 Singularity Skeleton of Form Factor and Cluster Function

When trying to measure up against the challenge just marveled at, it cannot hurt to learn from random-matrix theory what to look for. Therefore, I bring forth from hiding in Problem 4.13 a bit of distribution-theory lore on Fourier transforms: When the $2n$th derivative $\tilde{f}^{(2n)}(\tau)$ of a real even function $\tilde{f}(\tau)$ has delta-function singularities at $\tau = \pm\tau_0$, the Fourier transform $f(e) = \int_{-\infty}^{\infty} d\tau\, \tilde{f}(\tau)\cos 2\pi e\tau$ has a contribution $\propto (e)^{-2n}\cos 2\pi e\tau_0$,

$$\tilde{f}^{(2n)}(\tau) = a\delta(|\tau| - \tau_0) + \ldots \Leftrightarrow f(e) = \frac{(-1)^n 2a}{(2\pi e)^{2n}}\cos 2\pi e\tau_0 + \ldots\,.$$

The proof of that identity was the task in Problem 4.13; it is in essence repeated below. It follows that the jumps of the first derivative of the form factor at $\tau_0 = 0$ for all three ensembles of random matrices and at $\tau_0 = \tau_H$ for the CUE give rise to terms in the cluster function proportional to e^{-2} and $e^{-2}\cos 2\pi e\tau_H$, respectively. The jump of the third derivative of the form factor at $\tau = \tau_H$ for the COE and

CSE gives rise to $e^{-4}\cos 2\pi e\tau_H$. To that point, we had caught up to dynamical systems, basically with the HOdA sum rule and bootstrapping.

Terms proportional to $|e|^{-(2n+1)}\cos 2\pi e\tau_0$ in the cluster function would be due to logarithmic singularities of the form factor at $\tau = \tau_0$. To prove that statement in the spirit of distribution theory, we consider the Fourier integral

$$\tilde{f}(\tau) = \int_{-\infty}^{\infty} de \frac{\cos 2\pi e\tau_0}{|e|^{2n+1}} \cos 2\pi e\tau$$

$$= \int_{-\infty}^{\infty} de \frac{1}{2|e|^{2n+1}} \left[\cos 2\pi e(\tau - \tau_0) + \cos 2\pi e(\tau + \tau_0) \right] \qquad (8.11.1)$$

and take the $(2n+1)$st derivative w.r.t. τ,

$$\tilde{f}^{(2n+1)}(\tau) = (-1)^n (2\pi)^{2n+1} \int_0^{\infty} de \left[\sin 2\pi |e|(\tau - \tau_0) + \sin 2\pi |e|(\tau + \tau_0) \right]$$

$$= (-1)^n (2\pi)^{2n} \mathcal{P} \left(\frac{1}{\tau - \tau_0} + \frac{1}{\tau + \tau_0} \right), \qquad (8.11.2)$$

where $\mathcal{P}$ denotes the principal value. Therefore, the function $\tilde{f}(\tau)$ must be the $(2n)$th indefinite integral of $(-1)^n (2\pi)^{2n} \left(\ln|\tau - \tau_0| + \ln|\tau + \tau_0| \right)$. The simplest special case, $n = 0$, is needed to relate the logarithmic singularity of the form factor of the CSE at half the Heisenberg time to the corresponding oscillatory term in the cluster function; note that the coefficient $-1/4$ shows up both in front of the logarithm and its oscillatory correspondent. We had already concluded in the previous section that the semiclassical theory of dynamic systems must still catch these terms for the symplectic universality class, possibly by some improved bootstrap exercised on the HOdA-sum-rule behavior at small times.

But now we turn to the degree of level repulsion that can also be discussed with the help of the foregoing statements about delta-function and logarithmic singularities. Let me begin with the orthogonal case where random-matrix theory predicts linear repulsion, as reflected in the term $\propto |e|$ in the cluster function (4.13.30) at small $|e|$. We are encountering a jumping first derivative of the cluster function at the origin which corresponds to a nonoscillatory term $\propto 1/\tau^2$ in the form factor at large times; one indeed easily checks $K_{\text{COE}}(\tau) \to 1 - 1/(12\tau^2) \ldots$. So, to ascertain linear level repulsion for dynamic systems from the orthogonal class, we may look for a non-oscillatory $1/\tau^2$ term in the form factor.

Conversely, the absence of a $1/\tau^2$ term excludes linear level repulsion. We may go on to say that the absence of nonoscillatory terms $\propto 1/|\tau|^m$ with $m = 1, 2, \ldots$ up to some m_{max} means that the cluster function has a correspondingly regular behavior at the origin, $y(e) = 1 + \frac{1}{2}y''(0)e^2 + \frac{1}{4!}y^{(4)}(0)e^4 + \ldots$. In general, one would expect that $y''(0) \neq 0$, i.e., quadratic level repulsion; that is what happens in the CUE and for dynamic systems without time-reversal symmetry. We need to nail down the symmetry argument imposing $y''(0) = 0$ in the symplectic case.

In the absence of support for level repulsion from periodic-orbit theory, one may recall with relief that the statistical justification for the von Neumann–Wigner theorem in Sect. 3.4 has a semiclassical basis, inasmuch as a single "large" spectrum ($N \gg 1$) can provide a sufficient data basis from which to extract a self-averaging distribution of spacings. On the other hand, the degree of level repulsion is given by the codimension of level crossings and thus a quantum notion remote from the world of classical orbits.

In any case, it is profitable to reverse the above reasoning about Fourier transforms so as to draw conclusions for the decay of the form factor $K(\tau)$ for large times from the degree of level repulsion manifest in the cluster function $y(e)$. Linear repulsion, i.e., $y(e) \propto |e|$ for $e \to 0$, requires that the form factor have a long-time tail $K(\tau) - 1 \propto 1/\tau^2$. Quadratic repulsion, i.e., $y(e) \propto e^2$ without a logarithmic correction $\propto e^2 \ln|e|$, requires that $K(\tau) - 1$ decay at least as fast as $1/\tau^4$ for $\tau \to \infty$. Finally, quartic repulsion according to $y(e) \to e^4$ without a logarithmic correction $\propto e^4 \ln|e|$ obliges the form factor to decay as $1/\tau^6$ or faster.

We may still hope for a better understanding of bootstrapping so as to relate the large-τ behavior, just characterized, to the limit of small τ in the form factor and thus to get information beyond that provided by the HOdA sum rule.

8.12 Mixed Phase Space

Up to this point, we have mostly and tacitly taken for granted that the classical dynamics under consideration are hyperbolic, i.e., have no stable periodic orbits. Generic systems, however, are neither integrable nor fully chaotic in the sense mentioned but rather fill their "mixed" phase spaces with chaotic as well as stable regions. If a regular region is so small that it is not resolved by a Planck cell, it leaves but feeble signatures in quantum behavior. For differences to fully hyperbolic systems to become sizable, the islands of regular motion must contain a good fraction of the total number of Planck cells at all visited. A thorough semiclassical treatment of mixed phase spaces would fill a separate chapter if not a book. This exposition will be confined to a few remarks and references.

Chaotic regions are populated by hyperbolic periodic orbits that allow for semiclassical treatment by Gutzwiller's trace formula. Islands of regular motion, on the other hand, have central elliptic periodic orbits as their backbones; these are surrounded by chains of further elliptic and hyperbolic orbits. Under refined resolution, that phase space structure appears repeated to ever finer scales in a self-similar fashion: there is an infinite hierarchy of islands within islands [285]. Needless to say, quantum mechanics is immune to such excesses, and Planck's constant sets the limit of resolution.

Some spectral characteristics can, at least summarily, be understood by accounting for the separate chaotic and regular regions with their phase space volumes. According to a rather successful hypothesis by *Berry* and *Robnik* [286], for instance, the spacing distribution $P(S)$ for the spectrum of a system with $\omega = \Omega_{\rm c}/\Omega$ and $1 - \omega = \Omega_{\rm r}/\Omega$ as phase-space fractions is approximated by the correspondingly weighted sum of the Wigner distribution (4.4.2) and the exponential characteristic of regular spectra (see also Problem 6.19),

$$P(S) = \omega P_\beta(S) + (1 - \omega)e^{-S} . \qquad (8.12.1)$$

As a control parameter is varied to steer a dynamical system from regular toward increasingly chaotic behavior, one passes through bifurcations at which new periodic orbits are born. All periodic orbits, elliptic or hyperbolic, with periods up to half the Heisenberg time T_H are relevant for a semiclassical description of spectral properties. The number of times a single Planck cell (located in a chaos dominated region) is visited by periodic orbits with periods up to some value T grows larger than unity once T exceeds the Ehrenfest time T_E. At $T \approx T_H/2$, a typical Planck cell is crowded by exponentially many periodic orbits. We still do not really know how to deal with such a multitude of orbits which tends to give a quantum mechanically illegitimate structure to Planck cells. Worse yet, since all of these orbits must have arisen through bifurcations that it is clear that the cell in question must typically contain many orbits that have just arisen in bifurcations.

Fortunately, some progress has been achieved lately in dealing with phase-space structures like chains of islands around central orbits. Such local structures are classically generated by collective actions that can be classified by so-called normal forms typical for the bifurcation at which the chain structure is born from the central orbit. Normal forms themselves are characterized by their codimension, i.e., the number of controllable parameters needed for their location: Codimension-one bifurcations are seen when a single parameter is varied; if one gets close to some bifurcation by changing a single parameter but does not quite hit, one may zero in by fine-tuning a second parameter and then has located a case of codimension two. Codimension-one bifurcations were discussed by *Meyer* [287] and codimension-two ones more recently by *Schomerus* [288].

Phase-space structures generated by a normal-form action collectively contribute to the trace of the quantum propagator through terms of the form $Ae^{iS/\hbar}$ where S is the normal form in question and A is a suitable prefactor. Such collective contributions uniformly regularize the sum of single-orbit terms that diverge at the underlying bifurcation. Such uniform approximations were pioneered by *Ozorio de Almeida* and *Hannay* [289] and brought to recent fruition by *Tomsovic, Grinberg*, and *Ullmo* [290, 291] and *Schomerus* and *Sieber* [288, 292–295]. It is well to acknowledge that uniform approximations have a long history in optics and in catastrophe theory [296].

A most interesting phenomenon related to bifurcations is the semiclassical relevance of so-called ghost orbits. These are complex solutions of the nonlinear classical equations of motion which of course have no classical reality to them; they arise as saddle-point contributions to integral representations of quantum propagators, with equal formal right as the stationary-phase contributions of real periodic orbits, provided the original path of integration can legitimately be deformed so as to pass over the corresponding saddles; that condition always rules out ghost orbits whose complex actions have a positive real part such that $e^{iS/\hbar}$ would diverge for $\hbar \to 0$. Contributions from ghost orbits are usually suppressed exponentially in the latter limit except in some neighborhood of a bifurcation where the competition of the two "limits" $\hbar \to 0$ and $\mathrm{Im}\,S \to 0$ may render them important, sometimes even so far away from the bifurcation that the ghost can be accounted for with a separate term $Ae^{iS/\hbar}$ rather than including it in a collective normal-form term [297].

If Gutzwiller's trace formula is augmented by contributions from stable orbits, ghost orbits and clusters of orbits making up phase-space structures characteristic of bifurcations, reasonable semiclassical approximations for whole spectra become accessible, and the mean error for a level is a small single-digit percentage of the mean spacing [298].

The traces $t_n = \mathrm{Tr}\,F^n$ of powers of the Floquet operator of periodically driven systems and thus the form factor display anomalously large variations in their dependence on control parameters as a bifurcation is passed. The same is true for fluctuations of the level staircase. Every type of bifurcation produces a peaked contribution, and the maximum amplitude follows a power law with a bifurcation specific exponent for its $\hbar$-dependence [299–302].

Interesting consequences of the coexistence of regular and chaotic motion can be seen in the time evolution of the classical phase-space density, which obeys the Liouville equation in the Hamiltonian case. The pertinent time-evolution operator, commonly called the Frobenius–Perron operator is unitary w.r.t. the Hilbert space of functions one may define on phase space. Thus, the spectrum of the Frobenius-Perron operator lies on the unit circle in the complex plane. However, the resolvent of the Frobenius–Perron operator may have poles inside the unit circle on a second Riemann sheet; these so-called Pollicott–Ruelle resonances [277] have a solid physical meaning as the rates of probability loss from the phase space regions supporting the corresponding eigenfunctions. These resonances can rather simply be detected by looking on phase space with limited resolution. Such blurring automatically arises when the infinite unitary Frobenius-Perron operator is cut to finite size, say $N \times N$, in a basis whose functions are ordered by resolution [303, 304]. Of course, the $N \times N$ approximant of the Frobenius-Perron operator is non-unitary and has eigenvalues not exceeding unity in modulus. Those eigenvalues of the finite approximating matrix which are insensitive to variations in N once N is sufficiently large, turn out to be Pollicott-Ruelle resonances. The corresponding eigenfunctions are located on and immediately around elliptic periodic orbits for resonances approaching unity in modulus for large N; resonances smaller than unity in modulus, on the other hand, have eigenfunctions supported by the unstable manifolds of hyperbolic periodic orbits; at any rate, these eigenfunctions are strongly scarred [304]. Once a resonance is identified, it can be recovered through the so-called cycle expansion, i.e., a representation of the spectral determinant in the fashion explained in Sect. 8.9, where the periodic orbits included are those visible in the scars (and their repetitions). These classical findings have quantum analogues. Instead of the phase-space density, one must employ any of the so-called quasi-probability densities that are defined so as to have the phase space coordinates as independent variables and to represent the density operator; examples are the Wigner and Husimi or Glauber's Q-function [305]. The Liouville-von Neumann equation for the density operator may be written as an evolution equation for the quasi-probability used, say the Q-function.

Thus, a Q propagator arises whose classical analogue is the Frobenius–Perron operator. In contrast to the Frobenius–Perron operator, however, the Q-propagator may be a finite matrix whose dimension $N_q = \Omega/\hbar^f$ is given by Weyl's law as the number of Planck cells contained in the classically accessible phase-space volume Ω. The spectrum of the Q propagator must lie on the unit circle due to the unitarity time evolution of the density operator. It turns out that classical and quantum dynamics become indistinguishable if looked upon with a resolution much coarser than a Planck cell: $N \times N$ approximants of the Frobenius–Perron operator and the Q propagator with $N \ll N_q$ yield the same resonances and scarred eigenfunctions [306], which is a rather intuitive and satisfactory result indeed.

Of the host of observations of mixed phase spaces in real systems, at least some representative examples deserve mention. Conductivity measurements for two-dimensional antidot structures in semiconductors [307] have revealed classical [308] and semiclassical [309] manifestations of mixed phase spaces. Magneto-transport through chaotic quantum wells has recently attracted interest [310], as has the radiation pattern emerging from mesoscopic chaotic resonators [311]. The mass asymmetry in nuclear fission has found a semiclassical interpretation [312]. Atomic hydrogen in a strong magnetic field remains a prime challenge for atomic spectroscopy [313]. Finally, quasi-particle excitations of Bose–Einstein condensates trapped in anisotropic (even axially symmetric) parabolic potentials are nonintegrable if their energies are comparable to the chemical potential [314]; these latter phenomena make for a particularly nice enrichment of known chaotic waves, inasmuch as the underlying wave equation is the nonlinear Gross–Pitaevski equation.

8.13 Problems

8.1. Revive your mechanics education and recall: The action S for any path $q(t')$, classically allowed or not, between an initial point q' at time 0 and a final point q at time t is defined as the integral $S = \int_0^t dt' L[q(t'), \dot{q}(t')]$. The extremal value taken by that integral along a classically allowed path is a function $S(q, q', t)$ of the initial and final coordinates and the time span mostly called "action along the path"; it generates the canonical transformation from initial to final phase-space points as $p' = -\partial S(q, q', t)/\partial q'$, $p = \partial S(q, q', t)/\partial q$. With this in mind, show for yourself that $\partial S/\partial t = -H$ or $dS = p\,dq - p'\,dq' - H\,dt$. If the Hamiltonian function does not explicitly depend on t, one can proceed to the time-independent action $S_0(q, q', E) = \int_{q'}^q p\,dq$ by $S(q, q', t) = S_0(q, q', E) - Et$ which looks like a Legendre transform.

8.2. Now freshen up your quantum mechanics, and retrace your first steps into WKB terrain: Entering Schrödinger's equation $i\hbar\dot{\psi} = (p^2/2m + V(x))\psi$ with the ansatz $\psi = A(x, t)e^{iS(x,t)/\hbar}$ show that to leading order in $\hbar$ the phase $S/\hbar$ obeys the Hamilton–Jacobi equation $\dot{S} + H(x, \nabla S) = 0$ while the next-to-leading order yields the continuity equation $\dot{\varrho} + \operatorname{div} j = 0$ for the probability density $\varrho = |A|^2$ and the probability current density $j = \varrho \nabla S/m$.

8.3. Modify the Van Vleck propagator (8.2.19) and the trace formula (8.3.5) for f degrees of freedom.

8.4. The baker's map operates on the unit square as phase space in two steps, like a baker on incompressible dough: Compression to half height in the q-direction and double breadth in p is followed by stacking the right half of the resulting rectangle on top of the left. Show graphically that the number of intersections of the nth iterate of a line of constant q has 2^n intersections with any line of constant p.

8.5. Transcribing the reasoning of Sect. 8.3.2 from discrete to continuous time shows that free motion has minimal action. You will enjoy being led in this classical problem to the Schrödinger equation of a particle in a box.

8.6. Show that the harmonic oscillator has (1) the action $S(q, q', t) = \frac{m\omega}{2 \sin \omega t}[(q^2 + q')^2 \cos \omega t - 2qq']$, (2) a Van Vleck propagator coinciding with the exact one and shrinking to a delta function every half period, and (3) a conjugate point every half period. After having studied Sect. 8.4, draw out the fate of the initial Lagrangian manifold.

8.7. The Kepler ellipse can be parametrically represented by

$$r = a(1 - e \cos x_{\|})$$

$$t = \sqrt{\frac{ma^3}{GM}}(x_{\|} - e \sin x_{\|})$$

where the radial coordinate r is reckoned from a focus, a is the semimajor axis, e the excentricity, m the reduced and M the total mass, and G the gravitational constant. Using $a = -GmM/2E$ and $e = \sqrt{1 + 2EL^2/m^3M^2G^2}$ with E the energy and L the angular momentum, find the configuration-space caustics and check that $\partial^2 S/\partial t^2 = -\partial E/\partial t$ diverges thereon. Hint: Look at the turning points of the radial oscillation. Note that for $f > 1$, in contrast to oscillations of a single degree of freedom, the turning points of one coordinate are *not* characterized by vanishing kinetic energy.

8.8. Show that the symplectic form (8.4.1) is independent of the coordinates used to compute it. Hint: Use a generating function $F(Q, q)$ to transform canonically from q, p to Q, P.

8.9. Generalizing the reasoning of Sect. 8.5 show the equivalence of f-freedom stroboscopic maps with $(f + 1)$-freedom autonomous flows.

8.10. Use the HOdA sum rule to evaluate the diagonal approximation for the cluster function (8.7.9) to get $y_{\text{diag}}(e) = -g/(2\pi^2 e^2)$.

8.11. Use the definition $\varrho(E) = \sum_i \delta(E - E_i)$ of the level density of autonomous systems to write the form factor as a sum of unimodular terms $e^{-i(E_i - E_j)T_H \tau/\hbar}$.

Before embarking on the few lines of calculation, think about what expression for the Heisenberg time T_H you will come up with.

8.12. A particle in a one-dimensional box has the energy spectrum $E_j = j^2$ where for notational convenience $\hbar^2/2m$ is set equal to unity. Use the infinite-product representation of the sine,

$$\sin z = z \prod_{j=1}^{\infty} (1 - \frac{z^2}{\pi^2 j^2}) \tag{8.13.1}$$

to show that a convergent zeta function

$$\zeta(E) = \prod_{j=1}^{\infty} \{A_j(E - j^2)\} = \frac{\sin \pi \sqrt{E}}{\pi \sqrt{E}}$$

results from $A_j = -1/E_j$. Note that the energy levels are the zeros of $\zeta(E)$.

8.13. Consider the harmonic oscillator with the spectrum $E = j - 1/2$ for $j = 1, 2, \ldots$. Use Euler's infinite product for the Gamma function

$$\frac{1}{\Gamma(z)} = z e^{\gamma z} \prod_{j=1}^{\infty} \left[\left(1 + \frac{z}{j}\right) e^{-z/j} \right] \tag{8.13.2}$$

and the product (8.13.1) once more to show that the regularizer

$$A(E, E_j) = -\frac{1}{E_j + \frac{1}{2}} \exp \frac{E + \frac{1}{2}}{E_j + \frac{1}{2}}$$

yields the zeta function

$$\zeta(E) = \frac{1}{\pi} e^{\gamma(E + \frac{1}{2})} \Gamma(E + \frac{1}{2}) \sin \pi (E + \frac{1}{2})$$

where γ is Euler's constant. Check that $\zeta(E)$ has the eigenvalues as its zeros, i.e., that the additional zeros of the sine are cancelled by the poles of the Gamma function.

9. Dissipative Systems

9.1 Preliminaries

Regular classical trajectories of dissipative systems eventually end up on limit cycles or settle on fixed points. Chaotic trajectories, on the other hand, approach so-called strange attractors whose geometry is determined by Cantor sets and their fractal dimension. In analogy with the Hamiltonian case, the two classical possibilities of simple and strange attractors are washed out by quantum fluctuations. Nevertheless, genuinely quantum mechanical distinctions between regular and irregular motion can be identified. The main goal of this chapter is the development of one such distinction, based on the generalization of energy levels to complex quantities whose imaginary parts are related to damping. An important time scale separation arising for dissipative quantum systems will also be expounded: Coherences between macroscopically distinct states tend to decay much more rapidly than quantities with well-defined classical limits. An example relevant in the present context is the dissipative destruction of quantum localization for the kicked rotator.

9.2 Hamiltonian Embeddings and Generators of Damped Motion

Any dissipative system S may be looked upon as part of a larger Hamiltonian system. In many cases of practical interest, the Hamiltonian embedding involves weak coupling to a heat bath R whose thermal equilibrium is not noticeably perturbed by S. Strictly speaking, as long as S + R is finite in its number of degrees of freedom, the Hamiltonian nature entails quasi-periodic rather than truly irreversible temporal behavior. However, for practical purposes, quasi-periodicity and true irreversibility cannot be distinguished in the "effectively" dissipative systems S in question. The equilibration time(s) imposed on S by the coupling to R are typically large compared to the time scales τ_R of all intrabath processes probed by the coupling; in such situations, to which discussions will be confined, the dissipative motion of S acquires a certain Markovian character. Then, the density operator $\varrho(t)$ of S obeys a "master" equation of the form

$$\dot{\varrho}(t) = l\varrho(t) \tag{9.2.1}$$

with a suitable time-independent generator l of infinitesimal time translations. The dissipative motion described by (9.2.1) must, of course, preserve the Hermiticity, positivity, and normalization of ϱ. Now, I shall proceed to sketch how the master equation (9.2.1) for the density operator ϱ of S can be derived from the microscopic Hamiltonian dynamics [315, 316] of S + R.

The starting point is the Liouville–von Neumann equation for the density operator $W(t)$ of S + R,

$$\dot{W}(t) = -\frac{i}{\hbar} [H, W(t)] \equiv LW(t) . \tag{9.2.2}$$

Here H denotes the Hamiltonian which comprises "free" terms H_S and H_R for S and R, as well as an interaction part H_{SR},

$$H = H_S + H_R + H_{SR} . \tag{9.2.3}$$

A similar decomposition holds for the Liouvillian L. The formal integral of (9.2.2) reads

$$W(t) = e^{Lt} W(0) . \tag{9.2.4}$$

To slightly simplify the algebra to follow, we assume stationarity with respect to H_R and the absence of correlations between S and R for the initial density operator,

$$W(0) = \varrho(0)R , \quad [H_R, R] = 0 . \tag{9.2.5}$$

Inasmuch as the bath is macroscopic, it is reasonable to require that

$$R = Z_R^{-1} e^{-\beta H_R} , \quad \text{Tr}_R \{R\} = 1 . \tag{9.2.6}$$

The formal solution (9.2.4) of the initial-value problem (9.2.2, 9.2.5) suggests the definition of a formal time-evolution operator $U(t)$ for S,

$$\varrho(t) = U(t)\varrho(0) ,$$
$$U(t) = \text{Tr}_R \{e^{Lt} R\} . \tag{9.2.7}$$

Due to the coupling of S and R, this $U(t)$ is not unitary. Assuming that $U(t)$ possesses an inverse, we arrive at an equation of motion for $\varrho(t)$,

$$\dot{\varrho}(t) = l(t)\varrho(t) ,$$
$$l(t) = \dot{U}(t)U(t)^{-1} , \tag{9.2.8}$$

in which $l(t)$ may be interpreted as a generator of infinitesimal time translations. In general, $l(t)$ will be explicitly time-dependent but will approach a stationary limit

$$l = \lim_{t \to \infty} \dot{U}(t)U(t)^{-1} \tag{9.2.9}$$

on the time scale τ_R characteristic of the intrabath processes probed by the coupling. On the much larger time scales typical for S, the asymptotic operator l generates an effectively Markovian process.

Only in very special cases can l be constructed rigorously [317, 318]. A perturbative evaluation with respect to the interaction Hamiltonian H_{SR} is always feasible and in fact quite appropriate for weak coupling. Assuming for simplicity that

$$\mathrm{Tr}_R \{H_{SR} R\} = 0 , \tag{9.2.10}$$

one easily obtains the perturbation expansion of l as

$$l = L_S + \int_0^\infty dt \, \mathrm{Tr}_R \left\{ L_{SR} e^{(L_R+L_S)t} L_{SR} e^{-(L_R+L_S)t} R \right\} \tag{9.2.11}$$

where third- and higher order terms are neglected (Born approximation). The few step derivation of (9.2.11) uses the expansion of e^{Lt} in powers of L_{SR},

$$e^{Lt} = e^{(L_R+L_S)t} + \int_0^t dt' e^{(L_R+L_S)t'} L_{SR} e^{(L_R+L_S)(t-t')} + \cdots , \tag{9.2.12}$$

the identities $\mathrm{Tr}_R \{L_S(\dots)\} = L_S \, \mathrm{Tr}_R \{(\dots)\}$, $\mathrm{Tr}_R \{L_R(\dots)\} = 0$, and the assumed properties (9.2.5, 9.2.10) of R and H_{SR}.

Of special interest for the remainder of this chapter will be an application to spin relaxation. Therefore, we shall work out the generator l for that case. The spin or angular momentum is represented by the operators J_z, $J_\pm$ which obey

$$[J_+, J_-] = 2J_z , \quad [J_z, J_\pm] = \pm J_\pm . \tag{9.2.13}$$

The free spin dynamics may be generated by the Hamiltonian

$$H_S = \hbar\omega J_z \tag{9.2.14}$$

and for the spin bath interaction, we take

$$H_{SR} = \hbar \left(J_+ B + J_- B^\dagger \right) \tag{9.2.15}$$

where B and $B^\dagger$ are a pair of Hermitian-conjugate bath operators. We shall speak of the observables entering the interaction Hamiltonian as coupling agents. The Hamiltonian of the free bath need not be specified. By inserting H_{SR} from (9.2.14) in the second-order term of l given in (9.2.11), we obtain

$$l^{(2)}\varrho = \int_0^\infty dt \, \langle B(t)B^\dagger \rangle \, e^{i\omega t} [J_- \varrho, J_+]$$
$$+ \int_0^\infty dt \, \langle B^\dagger B(t) \rangle \, e^{i\omega t} [J_+, \varrho J_-] + \mathrm{H.c.} . \tag{9.2.16}$$

Interestingly, the c-number coefficients appearing here are the Laplace transforms of the bath correlation functions $\langle B(t)B^\dagger \rangle$ and $\langle B^\dagger B(t) \rangle$, taken at the frequency ω

of the free precession (apart from the imaginary unit i). The correlation functions refer to the reference state R of the bath which is assumed to be the thermal equilibrium state given in (9.2.6); their time dependence is generated by the free-bath Hamiltonian H_R. For the sake of simplicity, I have assumed, in writing (9.2.16), that $\langle B(t)B(0)\rangle = \langle B^\dagger(t)B^\dagger(0)\rangle = 0$.

The Laplace transforms of the bath correlation functions are related to their Fourier transforms by

$$\int_0^\infty dt\, e^{i\omega t} f(t) = \tfrac{1}{2}f(\omega) + \tfrac{i}{2\pi}\mathcal{P}\int_{-\infty}^{+\infty} d\nu \tfrac{f(\nu)}{\omega-\nu}\,,$$
$$f(\omega) = \int_{-\infty}^{+\infty} dt\, e^{i\omega t} f(t)\,, \tag{9.2.17}$$

where $f(t)$ stands for either $\langle B(t)B^\dagger\rangle$ or $\langle B^\dagger B(t)\rangle$ and $\mathcal{P}$ denotes the principal value. Furthermore, since one and the same bath Hamiltonian H_R generates the time dependence of $B(t)$ and determines the canonical equilibrium (9.2.6), the well-known fluctuation-dissipation theorem holds in the form [319]

$$\int_{-\infty}^{+\infty} dt\, e^{i\omega t}\langle B(t)B^\dagger\rangle = 2\kappa(\omega)\left[1 + n_{\mathrm{th}}(\omega)\right]$$
$$\int_{-\infty}^{+\infty} dt\, e^{i\omega t}\langle B^\dagger B(t)\rangle = 2\kappa(\omega)n_{\mathrm{th}}(\omega) \tag{9.2.18}$$

where

$$2\kappa(\omega) = \int_{-\infty}^{+\infty} dt\, e^{i\omega t}\,\langle [B(t),\, B^\dagger]\rangle \tag{9.2.19}$$

is a real Fourier-transformed bath response function and

$$n_{\mathrm{th}}(\omega) = \frac{1}{e^{\beta\hbar\omega} - 1} \tag{9.2.20}$$

is the average number of quanta in an oscillator of frequency ω at temperature $1/\beta$. Combining (9.2.11, 9.2.14–9.2.19) we obtain the master equation

$$\begin{aligned}
\dot{\varrho} = l\varrho \;=\;& -i\left[(\omega + \delta + \delta_{\mathrm{th}})J_z - \delta J_z^2, \varrho\right]\\
&+\kappa(1 + n_{\mathrm{th}})\left\{[J_-, \varrho J_+] + [J_-\varrho, J_+]\right\}\\
&+\kappa n_{\mathrm{th}}\left\{[J_+, \varrho J_-] + [J_+\varrho, J_-]\right\}
\end{aligned} \tag{9.2.21}$$

with the "frequency shifts"

$$\delta = \frac{1}{\pi}\mathcal{P}\int_{-\infty}^{+\infty} d\nu \frac{\kappa(\nu)}{\omega - \nu}\,, \quad \delta_{\mathrm{th}} = \frac{1}{\pi}\mathcal{P}\int_{-\infty}^{+\infty} d\nu \frac{n_{\mathrm{th}}(\nu)\kappa(\nu)}{\omega - \nu}\,. \tag{9.2.22}$$

This master equation was first proposed to describe superfluorescence in Ref. [320] and put to experimental test in that context in Ref. [321].

The first commutator in (9.2.21) describes a reversible motion according to a Hamiltonian $\hbar(\omega + \delta + \delta_{\mathrm{th}})J_z - \hbar\delta J_z^2$; the first term of this generates linear precession around the z-axis at a shifted frequency $\omega + \delta + \delta_{\mathrm{th}}$, whereas the second term yields a nonlinear precession around the z-axis. The remaining terms in the master equation are manifestly irreversible.

In view of the temperature dependence of n_{th}, the stationary solution $\bar{\varrho}$ of the master equation (9.2.21) must be separately stationary with respect to both the reversible and the irreversible terms. While the reversible terms admit any function of J_z, the irreversible ones single out

$$\bar{\varrho} = Z_{\text{S}}^{-1} e^{-\beta H_{\text{S}}} , \tag{9.2.23}$$

i.e., the unperturbed canonical operator with the temperature $\propto 1/\beta$ equalling that of the bath. The stationarity of this $\bar{\varrho}$ may be checked by a little calculation with the help of the identity

$$J_{\pm} e^{x J_z} = e^{\pm x} e^{-x J_z} J_{\pm} \tag{9.2.24}$$

but in fact follows from very general arguments [322]: Due to the structure of (9.2.21), $\bar{\varrho}$ must be independent of δ and κ, i.e. of zeroth order in H_{SR}. In the limit of vanishing coupling, however, the angular momentum and the bath become statistically independent and their total energy tends to the sum of the respective unperturbed parts. The only function of H_{S} or, equivalently, J_z meeting these requirements is the exponential (9.2.23).

The master equation obviously respects the conservation law for the squared angular momentum, as indeed it must since $[H_{\text{S}} + H_{\text{SR}} + H_{\text{R}}, \boldsymbol{J}^2] = 0$,

$$\boldsymbol{J}^2 = j(j+1) = \text{const} . \tag{9.2.25}$$

Once the quantum number j is fixed, the Hilbert space is restricted to $2j + 1$ dimensions, and the density operator is representable by a $(2j + 1) \times (2j + 1)$ matrix. A convenient basis is provided by the joint eigenvectors $|jm\rangle$ of $\boldsymbol{J}^2$ and J_z pertaining to the respective eigenvalues $j(j+1)$ and m. By employing

$$J_{\pm} |jm\rangle = \sqrt{(j \mp m)(j \pm m + 1)} \, |j, m \pm 1\rangle , \tag{9.2.26}$$

the master equation (9.2.21) is easily rewritten as a set of coupled differential equations for the matrix elements $\varrho_{mm'}(t)$.

Since no phase is distinguished for the polarizations $J_{\pm}$, there is a closed set of "rate equations" for the probabilities $\varrho_{mm}(t) \equiv \varrho_m(t)$,

$$\dot{\varrho}_m(t) = 2\kappa(1 + n_{\text{th}})[(j - m)(j + m + 1)\varrho_{m+1} - (j - m + 1)(j + m)\varrho_m]$$
$$+ 2\kappa n_{\text{th}} [(j - m + 1)(j + m)\varrho_{m-1} - (j - m)(j + m + 1)\varrho_m] . \tag{9.2.27}$$

These actually form a master equation of the Pauli type with energy-lowering transition rates $w(m+1 \to m) = 2\kappa(1+n_{\text{th}})(j-m)(j+m+1)$ and upward rates $w(m \to m + 1) = 2\kappa n_{\text{th}}(j - m)(j + m + 1)$. As revealed by the factor $1 + n_{\text{th}}$, downward transitions of the angular momentum are accompanied by spontaneous or induced emissions of quanta $\hbar\omega$ into the bath. Upward transitions, on the other hand, require the absorption of quanta $\hbar\omega$ from the bath and thus are proportional to n_{th}. Of course, neither absorption nor induced emission can take place when the heat bath temperature is too low, i.e., when $k_{\text{B}} T \ll \hbar\omega$. In this latter limit, the angular momentum keeps dissipating its energy, i.e., emitting quanta into the bath, until it settles into the ground state, $m = -j$.

It is as well to note in passing that the master equation (9.2.21) or (9.2.27) is nothing but a slightly fancy version of Fermi's golden rule. Indeed, the transition rates $w(m + 1 \to m)$ and $w(m \to m + 1)$ may be calculated directly with that rule; then, the constant $\kappa(\omega) \equiv \kappa$ appears as proportional to the number of bath modes capable, by resonance, of receiving quanta $\hbar\omega$ emitted by the precessing angular momentum.

I should also add a word about the limit of validity of the master equation (9.2.21). Due to the perturbative derivation, we are confined to the case of small coupling,

$$\kappa, \delta \ll \omega ,\tag{9.2.28}$$

in which the free precession is only weakly perturbed. Moreover, the damping constant κ, in energy units, must be smaller than the bath temperature,

$$\hbar\kappa \ll k_{\mathrm{B}}T .\tag{9.2.29}$$

To explain this latter restriction, it is useful to consider (9.2.20) and to replace the free-precession frequency ω by a complex variable z. The function $n_{\mathrm{th}}(z)$ has imaginary poles at $\beta\hbar z_\mu = 2\pi\mathrm{i}\mu$, $\mu = 0, \pm1, \pm2, \dots$. Then, one may reconstruct the correlation functions $\langle B(t)B^\dagger \rangle$ and $\langle B^\dagger B(t) \rangle$ by inverting the Fourier transforms (9.2.18). Closing the frequency integrals in the upper half of the z plane, as is necessary for $t > 0$, one encounters thermal transients $\mathrm{e}^{\mathrm{i}z_\mu t} = \mathrm{e}^{-|z_\mu|t}$; these transients must decay much faster than $\mathrm{e}^{-\kappa t}$ or else the assumed time-scale separation for the bath and the damped subsystem is violated, and the asymptotic generator l becomes meaningless.

A limit of special relevance for the remainder of this chapter is that of low temperatures, $n_{\mathrm{th}} \ll 1$, and large angular momentum, $j \gg 1$. In this quasi-classical low-temperature regime, classical trajectories become relevant as a reference behavior about which the quantum system displays fluctuations. Classical dynamics may be obtained from (9.2.21) by first extracting equations of motion for the mean values $\langle \boldsymbol{J}(t) \rangle$ and then factorizing as $\langle J_z J_\pm \rangle = \langle J_z \rangle \langle J_\pm \rangle$ etc. In the limit of negligible n_{th} and with the help of a spherical-coordinate representation,

$$\langle J_z \rangle = j \cos \Theta , \quad \langle J_\pm \rangle = j \sin \Theta \mathrm{e}^{\pm\mathrm{i}\phi} ,\tag{9.2.30}$$

classical dynamics turns out to be that of the overdamped pendulum,

$$\dot{\Theta} = \Gamma \sin \Theta , \quad \dot{\phi} = 0 ,$$
$$\Gamma = 2\kappa j .\tag{9.2.31}$$

Evidently, the classical limit must be taken as $j \to \infty$ with constant Γ. Then, the angle Θ relaxes toward the stable equilibrium $\Theta(\infty) = \pi$ according to

$$\tan\frac{\Theta(t)}{2} = \mathrm{e}^{\Gamma t}\tan\frac{\Theta(0)}{2} .\tag{9.2.32}$$

9.3 Time-Scale Separation for Probabilities and Coherences

Off-diagonal density matrix elements between states differing in energy by not too many quanta display lifetimes of the same order of magnitude, under the dissipative influence of heat baths, as diagonal elements. A typical example is provided by the damped angular momentum considered above, when the quantum number j is specified as $j = 1/2$. The damping rates for the off-diagonal element $\varrho_{1/2,-1/2}$ and the population probability $\varrho_{1/2,1/2}$ are easily found from (9.2.21), respectively, as

$$
\begin{aligned}
\gamma_\perp &= \kappa(1 + 2n_{\mathrm{th}}) \\
\gamma_\| &= 2\kappa(1 + 2n_{\mathrm{th}}) \,,
\end{aligned}
\tag{9.3.1}
$$

i.e., they are indeed of the same order of magnitude.

A different situation arises for large quantum numbers. Coherences between mesoscopically or even macroscopically distinct states usually have lifetimes much shorter than those of occupation probabilities. I propose to illustrate that phenomenon of "accelerated decoherence" for an angular momentum subject to the low-temperature version ($n_{\mathrm{th}} = 0$) of the damping process (9.2.21); it suffices to consider the dissipative part of that process whose generator Λ is defined by

$$
\dot{\varrho} = \Lambda\varrho = \kappa\big\{[J_-, \varrho J_+] + [J_-\varrho, J_+]\big\} \,;
\tag{9.3.2}
$$

the classical limit of that process was characterized above in (9.2.31, 32).

For reasons to be explained presently, it is convenient to start with a superposition of two coherent states (7.6.12),

$$
|\,\rangle = c|\Theta\phi\rangle + c'|\Theta'\phi'\rangle
\tag{9.3.3}
$$

where the complex amplitudes c and c' are normalized as $|c|^2 + |c'|^2 = 1$. In the limit of large j, the two component states may indeed be said to be macroscopically distinguishable; the superposition is often called a Schrödinger cat state, reminiscent of Schrödinger's bewilderment with the notorious absence of quantum interference with macroscopically distinct states from the classical world. In fact, the phenomenon of accelerated decoherence that we are about to reveal explains why such interferences are usually impossible to observe.

The rigorous solution $\exp\{\Lambda t\}|\,\rangle\langle\,|$ of the master equation (9.3.2) originating from the initial density operator $|\,\rangle\langle\,|$ for the cat state (9.3.3) can be constructed without much difficulty; see Sect. 9.14 and Refs. [331–334]. However, we reach our goal more rapidly by considering the piece

$$
\varrho_{\Theta\phi,\Theta'\phi'}(t) = \exp\{\Lambda t\}|\Theta\phi\rangle\langle\Theta'\phi'|
\tag{9.3.4}
$$

and estimating its lifetime from the initial time rate of change of its norm

$$
\mathcal{N}_{\Theta\phi,\Theta'\phi'}(t) = \mathrm{Tr}\varrho_{\Theta\phi,\Theta'\phi'}(t)^\dagger \varrho_{\Theta\phi,\Theta'\phi'}(t) \,.
\tag{9.3.5}
$$

By differentiating w.r.t. time, setting $t = 0$, inserting the generator Λ, and using (7.6.17), we immediately get

$$\dot{\mathcal{N}}_{\Theta\phi,\Theta'\phi'}(t) = -(2\kappa j)\{\sin^2\Theta + \sin^2\Theta' - \cos(\phi - \phi')\sin\Theta\sin\Theta' \\ + \tfrac{1}{2}[(1+\cos\Theta)^2 + (1+\cos\Theta')^2]\}. \tag{9.3.6}$$

Now, the announced time-scale separation may be read off: The probabilities described by $\Theta = \Theta', \phi = \phi'$ have the classical rate of change $\Gamma = 2\kappa j$; in contrast, for $\sin\Theta \neq \sin\Theta'$ and/or $\phi \neq \phi'$, i.e., coherences, the first square bracket in the foregoing rate does not vanish and carries the "acceleration factor" j. Inasmuch as j is large, we conclude that the cat state (9.3.3) rapidly (on the time scale $1/j\Gamma$) decoheres to a mixture,

$$\left(c|\Theta\phi\rangle + c'|\Theta'\phi'\rangle\right)\left(c\langle\Theta\phi| + c'\langle\Theta'\phi'|\right) \\ \longrightarrow |c|^2|\Theta\phi\rangle\langle\Theta\phi| + |c'|^2|\Theta'\phi'\rangle\langle\Theta'\phi'|, \tag{9.3.7}$$

while the weights of the component states begin to change only much later at times of the order $1/\Gamma$.

An exception of the general rule of accelerated decoherence is highly interesting: Coherences between coherent states which lie symmetrically to the "equator" ($\Theta' = \pi - \Theta$) on one and the same "great circle" ($\phi' = \phi$) are exempt from the rule, simply because $\sin\Theta = \sin(\pi - \Theta)$. These give rise to long-lived Schrödinger cat states (9.3.3) [334].

A deeper reason for the immunity of these exceptional cat states will appear once we have clarified the distinction of the coherent states for the process under consideration. To that end, we must come back to the coupling agents $J_\pm$ entering the interaction Hamiltonian (9.2.15). These happen to have coherent states as approximate eigenstates in the semiclassical limit of large j,

$$J_\pm|\Theta\phi\rangle \approx je^{\pm i\phi}\sin\Theta\,|\Theta\phi\rangle \qquad \text{if} \quad \tan^2\Theta \gg 1/j. \tag{9.3.8}$$

That property is worth checking since the notion of an approximate eigenvector is not met all too often. But indeed, it is easily seen, again with the help of (7.6.17), that the two vectors $J_\pm|\Theta\phi\rangle$ and $je^{\pm i\phi}\sin\Theta\,|\Theta\phi\rangle$ have norms with a relative difference of order $1/j$ and comprise an angle (in Hilbert space) in between themselves which is of order $1/\sqrt{j}$, provided that the coherent state in question does not appreciably overlap the polar ones at $\Theta = 0$ and $\Theta = \pi$. The approximate eigenvalues $e^{\pm i\phi}\sin\Theta$ are doubly degenerate, and the two pertinent approximate eigenstates have the same azimuthal angle ϕ whereas the polar angles are symmetric about $\pi/2$. The Schrödinger cat state with equatorial symmetry cannot be rapidly decohered by the damping mechanism in consideration since its two component states are precisely such a doublet.

To appreciate the foregoing reasoning better, we may think of the unitary motion generated by some Hamiltonian that has a degenerate eigenvalue E. An initial superposition of states from the degenerate subspace will subsequently acquire the overall prefactor $\exp(-iEt/\hbar)$ but will not undergo any internal change. Or, to stay with dissipative motions, we may consider an interactive Hamiltonian $H_{SR} = XB$ with Hermitian coupling agents X and B for the system S and the reservoir R, respectively. Then, the Fermi golden rule type arguments of Sect. 9.2 yield a master equation with the dissipative generator

$$\Lambda\varrho = (\kappa/x_0^2)\{[X, \varrho X] + [X\varrho, X]\} \tag{9.3.9}$$

where κ is a rate constant and x_0 is a constant of the same dimension as the coupling agent X. Obviously now, occupation probabilities of eigenstates $|x\rangle$ of the coupling agent X are not influenced at all by the damping: $\Lambda|x\rangle\langle x| = 0$. Coherences are affected, however, since $\Lambda|x\rangle\langle x'| = \kappa\big((x - x')/x_0\big)^2|x\rangle\langle x'|$, and the proportionality of the decoherence rate to $(x - x')^2/x_0^2$ signals accelerated decoherence. Acceleration would fail only for superpositions of eigenstates $|x, \alpha\rangle$ belonging to one degenerate eigenvalue x. In principle, such states could still be very different, even macroscopically, as suggested by the toy example $X = Y^2$ for which the eigenstates $|\pm y\rangle$ of Y form a doublet with common eigenvalue y^2 of X.

The phenomenon considered here is universal. The unobservability of superpositions of different pointer positions in measurement devices finds its natural explanation here and so does the reduction of superpositions to mixtures in microscopic objects coupled to macroscopic measurement or preparative apparatuses [323–326]. Examples of current experimental interest include superconducting quantum interference devices (SQUIDS), Bloch oscillations in Josephson junctions, and optical bistability. The common goal of such investigations is to realize the largest possible distinction, hopefully mesoscopic, between two quantum states that still gives rise to detectable coherences in spite of weak dissipation. The first actual observations of accelerated decoherence were reported for photons in microwave cavities in Ref. [327].

A negative implication of the phenomenon in question also deserves mention. The classical alternatives of regular and chaotic motion can be reflected, as shown in Chap. 7, in different appearances of the sequence of quantum recurrences in the time evolution of certain expectation values. Quantum recurrence events of whatever appearance, however, are constructive interference phenomena requiring preservation of coherences between successive events. Inasmuch as recurrence events involve coherences between classically distinct states, rather feeble damping suffices to destroy the phase relations necessary for substantial constructive interference.

9.4 Dissipative Death of Quantum Recurrences

We proceed to a quantitative analysis of recurrences in the presence of feeble damping [328–330]. Due to the finiteness of its Hilbert space, the kicked top will be a convenient example to work with again. The main conclusions, however, will be of quite general validity.

The clue to the treatment lies in the fact that very weak dissipation suffices to suppress recurrence events. Thus, the damping may be considered a small perturbation of the unitary part of the dynamics. Furthermore, under conditions of classical chaos ideas borrowed from random-matrix theory allow a full implementation of first-order perturbation theory.

The kicked top to be employed here must be described now by a dissipative map for the density operator $\varrho(t)$ after the tth kick,

$$\varrho(t+1) = e^{\Lambda}e^{L}\varrho(t) \tag{9.4.1}$$

where e^{L} represents the unitary motion discussed in previous chapters,

$$e^{L}\varrho = F\varrho F^{\dagger}$$
$$F = e^{-ipJ_z}e^{-i\lambda J_y^2/2j} . \tag{9.4.2}$$

For the sake of convenience, the axis of the nonlinear precession is now taken to be the y-axis and that of the linear precession to be the z-axis. The nonunitary factor e^{Λ} describes the dissipative part of the evolution. To keep the situation simple, we will assume low temperatures, $n_{\text{th}} \approx 0$, and neglect the frequency shift terms, $\delta = 0$. Then, the generator Λ is obtained from (9.2.21) as

$$\Lambda\varrho = \frac{\Gamma}{2j} \left([J_-, \varrho J_+] + [J_-\varrho, J_+] \right) \tag{9.4.3}$$

except that here the classical damping constant $\Gamma = 2j\kappa$ is used to facilitate comparison with the classical behavior (9.2.31, 9.2.32). The free-precession term in (9.2.21) is not included in Λ but is written instead as a separate factor in the unitary evolution operator e^{L}; no approximation is involved in that separation since $\Lambda[J_z, \varrho] = [J_z, \Lambda\varrho]$. Thus, the whole dissipative quantum map (9.4.1) may be said to allow linear precession around the z-axis with concurrent damping, whereas the nonlinear precession prevails during a separate phase of the driving period. Note that the time $t = 0, 1, 2, \ldots$ and the coupling constants Γ, λ, p are all taken as dimensionless and that $\hbar = 1$.

In the weak-damping limit, the generator of the map can be approximated as

$$e^{\Lambda}e^{L} = (1 + \Lambda + \ldots)e^{L} . \tag{9.4.4}$$

With the eigenstates of the unitary Floquet operator F denoted by $|\mu\rangle$, the dyadic eigenvectors of e^{L} read $|\mu\rangle\langle\nu|$ and the corresponding unimodular eigenvalues $\exp(-i\phi_\mu + i\phi_\nu)$. To first order in Λ, the perturbed eigenvalue, for $\mu \neq \nu$, is

$$\lambda_{\mu\nu} = e^{-i(\phi_\mu - \phi_\nu)}(1 - \Lambda_{\mu\nu}) \approx e^{-i(\phi_\mu - \phi_\nu) - \Lambda_{\mu\nu}} , \quad \mu \neq \nu , \tag{9.4.5}$$

where the real damping constant

$$\begin{aligned} \Lambda_{\mu\nu} &= -\text{Tr}\left\{ (|\mu\rangle\langle\nu|)^{\dagger} \Lambda|\mu\rangle\langle\nu| \right\} \\ &= \frac{\Gamma}{2j} \left(-2\langle\mu|J_-|\mu\rangle\langle\nu|J_+|\nu\rangle + \langle\nu|J_+J_-|\nu\rangle + \langle\mu|J_+J_-|\mu\rangle \right) \end{aligned} \tag{9.4.6}$$

is the $\mu\nu$ element of the dyad $-\Lambda|\mu\rangle\langle\nu|$; the latter is the infinitesimal increment of $|\mu\rangle\langle\nu|$ generated by Λ during the time interval dt, divided by dt. With Λ expressed in the $|j,m\rangle$ basis, the damping constant $\Lambda_{\mu\nu}$ takes the form

$$
\Lambda_{\mu\nu} = \sum_{m,n=-j}^{+j} \Big[a(m,n)\,|\langle\mu|j,m\rangle|^2\,|\langle\nu|j,n\rangle|^2
$$

$$
-b(m,n)\langle\mu|j,m\rangle\langle j,m+1|\mu\rangle\langle\nu|j,n+1\rangle\langle j,n|\nu\rangle \Big] , \tag{9.4.7}
$$

$$
a(m,n) = \frac{\Gamma}{2j}\,[(j+m)(j-m+1)+(j+n)(j-n+1)]
$$

$$
b(m,n) = \frac{\Gamma}{j}\,[(j-m)(j+m+1)(j-n)(j+n+1)]^{1/2} .
$$

Once the eigenvectors $|\mu\rangle$ of the Floquet operator F are known, the sums in (9.4.7) can be evaluated.

The classical dynamics implied by (9.4.1–9.4.3) in the limit $j \to \infty$ is chaotic on the overwhelming part of the sphere $(\boldsymbol{J}/j)^2 = 1$ for $\lambda \approx 10$, $p \approx \pi/2$, $\Gamma \approx 10^{-4}$, the case to be studied in the following. The strange attractor accommodating the chaotic trajectories covers the classical sphere almost uniformly. Of course, for dissipation to become manifest in the classical dynamics, a time of the order $1/\Gamma = 10^4$ must elapse, and the strangeness of the attractor is revealed only on even larger time scales.

In the situation of global classical chaos chosen here, the Floquet operator F may be considered a random matrix drawn from Dyson's circular orthogonal ensemble. Then, the second term in $\Lambda_{\mu\nu}$, as given by (9.4.7), is, for $j \gg 1$, unable to make a non-zero contribution to the sum over m,n since the individual eigenvector components are random in sign. The remaining term in $\Lambda_{\mu\nu}$ may be simplified as

$$
\Lambda_{\mu\nu} = \frac{\Gamma}{2j} \sum_{m=-j}^{+j} (j+m)(j-m+1)\left(|\langle\mu|j,m\rangle|^2 + |\langle\nu|jm\rangle|^2 \right) . \tag{9.4.8}
$$

The damping constants $\Lambda_{\mu\nu}$ now appear to be random numbers. Due to the sum over m, however, the relative root-mean-square deviation from the mean is $\sim 1/\sqrt{j}$ and to within this accuracy the $\Lambda_{\mu\nu}$ are in fact all equal to one another and to their ensemble mean (Problem 9.8). Therefore, they must also equal their arithmetic mean,

$$
\Lambda_{\mu\nu} = \frac{\Gamma}{j(2j+1)} \sum_{\mu=1}^{2j+1}\sum_{m}(j+m)(j-m+1)\,|\langle\mu|j,m\rangle|^2
$$

$$
= \frac{\Gamma}{j(2j+1)} \sum_{m}(j+m)(j-m+1)
$$

$$
= \tfrac{2}{3}\Gamma(j+1) \approx \tfrac{2}{3}\Gamma j , \quad \mu \neq \nu . \tag{9.4.9}
$$

In short, the random numbers $\Lambda_{\mu\nu}$ are self-averaging in the limit of large j.

The result just obtained is in fact quite remarkable. It is independent of the coupling constants λ and p and even comes with the claim of validity for all tops with global chaos in the classical limit which are weakly damped according to (9.4.3). Furthermore, since $\Lambda_{\mu\nu}$ turns out to be independent of μ and ν for $\mu \neq \nu$, all off-diagonal dyadic eigenvectors $|\mu\rangle\langle\nu|$ of e^L display the same attenuation. Finally, the proportionality of $\Lambda_{\mu\nu}$ to j reflects the the phenomenon of accelerated decoherence. The life time of the coherences $\Lambda_{\mu\nu}$ is $\sim 1/\Gamma j$ while classically the damping is noticeable only on a time scale larger by a factor j.

For perturbation theory to work well, the damping constants $\Lambda_{\mu\nu}$ must be small. As to how small, we may argue that our first-order estimate should not be outweighed by second-order corrections. In the latter, small denominators of order $1/j^2$ appear since $2j(2j+1)$ nonvanishing zero-order eigenvalues $\phi_\mu - \phi_\nu$ lie in the accessible (2π)-interval. Thus, we are led to require that

$$\Gamma \ll \frac{1}{j^3} . \tag{9.4.10}$$

A word about the diagonal dyads $|\mu\rangle\langle\mu|$ may be in order. They are all degenerate in the conservative limit, because they are eigenvectors of e^L with eigenvalue unity. The fate of that eigenvalue for weak damping must be determined by degenerate perturbation theory, i.e., by diagonalizing Λ in the $(2j+1)$-dimensional space spanned by the $|\mu\rangle\langle\mu|$, a task in fact easily accomplished.

The elements of the $(2j+1) \times (2j+1)$ matrix in question read

$$\begin{aligned}
\text{Tr} &\{|\mu\rangle\langle\mu|(\Lambda|\nu\rangle\langle\nu|)\} \\
&= -\frac{\Gamma}{j} \left\{ \delta_{\mu\nu}\langle\mu|J_+ J_-|\mu\rangle - |\langle\mu|J_-|\nu\rangle|^2 \right\} \\
&= -\frac{\Gamma}{j} \Bigg\{ \delta_{\mu\nu} \sum_m |\langle\mu|m\rangle|^2 (j+m)(j-m+1) \\
&\quad - \sum_{mm'} [(j+m)(j-m+1)(j+m')(j-m'+1)]^{1/2} \\
&\quad \times \langle\nu|j,m-1\rangle\langle j,m|\mu\rangle\langle\mu|j,m'\rangle\langle j,m'-1|\nu\rangle \Bigg\} . \tag{9.4.11}
\end{aligned}$$

They are quite similar in structure to the damping constants $\Lambda_{\mu\nu}$ given in (9.4.7). As was the case for the latter, random-matrix theory can be invoked in evaluating the various sums in (9.4.11), provided there is global chaos in the classical limit. Due to the sums over m and m', all matrix elements (9.4.11) are again self-averaging, i.e., have root-mean-square deviations from their ensemble mean that are smaller than the mean by a factor of order $1/\sqrt{j}$. In the ensemble mean, only the terms with $m = m'$ in the double sum contribute. For the off-diagonal elements, $\mu \neq \nu$, the probabilities $|\langle jm|\mu\rangle|^2$ and $|\langle j, m-1|\nu\rangle|^2$ may be considered independent, and therefore both ensembles average to $1/(2j+1) \approx 1/2j$. For the diagonal elements, $\mu = \nu$, the contribution of the second sum in (9.4.11) need not be considered further since it is smaller than the first by a factor of order $1/j$.

Thus, the matrix to be diagonalized reads

$$\langle \mu \,|(\Lambda|\nu\rangle\langle\nu|)|\,\mu\rangle = -\frac{2}{3}\Gamma j \left[\delta_{\mu\nu} - \frac{1-\delta_{\mu\nu}}{2j} \right] . \tag{9.4.12}$$

Its diagonal elements precisely equal the eigenvalues $-\Lambda_{\mu\nu}$ given in (9.4.9) while its off-diagonal elements are smaller by the factor $1/2j$. Interestingly, in spite of their relative smallness, the off-diagonal elements may not be neglected since probability conservation requires that

$$\sum_{\mu=1}^{2j+1} \langle \mu \,|(\Lambda|\nu\rangle\langle\nu|)|\,\mu\rangle = 0 ; \tag{9.4.13}$$

indeed, $\langle\mu|(\Lambda|\nu\rangle\langle\nu|)|\mu\rangle dt$ is the differential increment, due to the damping, in the probability of finding the state $|\mu\rangle$ a time span dt after the state $|\nu\rangle$ had probability one.

   The matrix (9.4.12) is easily diagonalized. It has the simple eigenvalue zero and the $2j$-fold eigenvalue $(-2\Gamma j/3)$. The corresponding eigenvalues of the map $e^{\Lambda}e^{L}$ are

$$\lambda = 1 \text{ simple },$$
$$\lambda = e^{-2\Gamma j/3} \text{ } 2j\text{-fold} . \tag{9.4.14}$$

Clearly, the simple eigenvalue unity pertains to the stationary solution of the quantum map (9.4.1). Special comments are in order concerning the equality of the $2j$-fold eigenvalue in (9.4.14) and the modulus of those considered previously in (9.4.5, 9.4.9). Even though this equality implies one and the same time scale for the relaxation of the coherences $\langle\mu|(\Lambda|\mu\rangle\langle\nu|)|\nu\rangle$ and the probabilities $\langle\mu|(\Lambda|\nu\rangle\langle\nu|)|\mu\rangle$, there is no contradiction with the time-scale separation discussed in Sect. 9.3; the latter refers to coherences $\langle j, m|\varrho|j, m'\rangle$ and probabilities $\langle j, m|\varrho|j, m\rangle$. The eigenstates of $\boldsymbol{J}^2$ and J_z form a natural basis when the damping under consideration is the only dynamics present, whereas the eigenstates $|\mu\rangle$ of $\boldsymbol{J}^2$ and the Floquet operator F are the natural basis vectors when the damping is weak and chaos prevails in the classical limit. The equality of all damping rates encountered here suggests an interpretation in which, under conditions of global classical chaos, each eigenstate $|\mu\rangle$ and thus also each probability $\langle\mu|(\Lambda|\nu\rangle\langle\nu|)|\mu\rangle$ comprises coherences between angular momentum eigenstates $|j, m\rangle$ and $|j, m'\rangle$ where $|m - m'|$ takes values up to the order j. Such an interpretation is indeed tantamount to the applicability of random-matrix theory, which assigns, in the ensemble mean, one and the same value $1/(2j+1) \approx 1/2j$ to all probabilities for finding $J_z = m$ in a state $|\mu\rangle$. This may be viewed as a quantum manifestation of the assumed globality of classical chaos: The strange attractor covers the classical sphere $(\boldsymbol{J}/j)^2 = 1$ uniformly.

   Global uniform support is evident in the stationary solution of the quantum map (9.4.1): Indeed, it is easily verified that the eigenvector of the matrix (9.4.12) pertaining to the simple vanishing eigenvalue has all components equal to one another. Therefore, the unique density operator invariant under the map (9.4.1) is the equipartition mixture of all $2j + 1$ Floquet eigenstates $|\mu\rangle$,

$$\bar{\varrho} = \frac{1}{2j + 1} \sum_{\mu=1}^{2j+1} |\mu\rangle\langle\mu| . \qquad (9.4.15)$$

These results for the eigenvalues of the map $e^\Lambda e^L$ enable strong statements about the time dependence of observables. For instance, all observables with a vanishing stationary mean display the universal attenuation law

$$\langle O(t)\rangle = e^{-2\Gamma jt/3} \langle O(t)\rangle_{\Gamma=0} . \qquad (9.4.16)$$

The angular momentum components J_i provide examples of such observables. For instance,

$$\begin{aligned}
\langle J_z(\infty)\rangle &= \frac{1}{2j + 1} \sum_{\mu=1}^{2j+1} \langle\mu|J_z|\mu\rangle \\
&= \frac{1}{2j + 1} \sum_{\mu=1}^{2j+1} \sum_{m=-j}^{+j} m\,|\langle m|\mu\rangle|^2 \\
&= \frac{1}{2j + 1} \sum_{m=-j}^{+j} m = 0 .
\end{aligned} \qquad (9.4.17)$$

Figure 9.1 compares the ratio $\langle J_z(t)\rangle/\langle J_z(t)\rangle_{\Gamma=0}$, calculated by numerically iterating the map (9.4.1), for $\Gamma = 1.25 \times 10^{-4}$ and $\Gamma = 0$ with the random-matrix prediction (9.4.16). The agreement is quite impressive even though the numerical work was done for j as small as 40.

An implication of the above concerns the fate of quantum recurrences under dissipation. Quasi-periodic reconstructions of mean values in the conservative limit must suffer the attenuation described by (9.4.16). The mean frequency of large fluctuations for $\Gamma = 0$ roughly equals the quasi-energy spacing $2\pi/(2j + 1)$. Therefore, the critical damping above which recurrences are noticeably suppressed can be estimated as $\pi/j \approx 2\Gamma j/3$, i.e.,

$$\Gamma_{\mathrm{crit}} \approx \frac{1}{j^2} . \qquad (9.4.18)$$

For $\Gamma < \Gamma_{\mathrm{crit}}$, the damping constants $2\Gamma j/3$ are smaller than even a typical nearest neighbor spacing $2\pi/(2j + 1)$ of conservative quasi-energies. Then, the quasi-periodicity of the conservative limit must still be resolvable, of course.

Figure 9.2 displays the fate of recurrences in $\langle J_z(t)\rangle$ for the map (9.4.1–9.4.3) under conditions of global classical chaos. The survival condition $\Gamma < \Gamma_{\mathrm{crit}}$ for recurrences is clearly visible. The amplitude of the quasi-periodic temporal fluctuations of $\langle J_z(t)\rangle$ displayed in Fig. 9.2 calls for comment. The ordinates in that

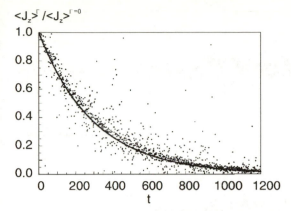

Fig. 9.1. Ratio of the time-dependent expectation value $\langle J_z(t)\rangle$ with and without damping for the kicked top (9.4.1, 2). Parameters were chosen as $j = 40$, $p = 1.7$, $\lambda = 10$. The coherent initial state was localized at $\langle J_x\rangle/j = 0.43$, $\langle J_y\rangle/j = 0$, $\langle J_z\rangle/j = 0.9$. The smooth curve represents the prediction (9.4.16) of random-matrix theory. The optimal fit for the decay rate to the numerical data is $0.61j\Gamma$ whereas random-matrix theory predicts $(2/3)j\Gamma$

figure are scaled in the expectation that the variance of the temporal fluctuations behaves as

$$\sigma = \frac{1}{j}\left[\overline{\left(\langle J_z\rangle - \overline{\langle J_z\rangle}\right)^2}\right]^{1/2} \sim \frac{1}{\sqrt{j}} \tag{9.4.19}$$

where the bar denotes a time average, $\overline{\langle J_z\rangle} = \frac{1}{T}\int_0^T dt\langle J_z(t)\rangle$ with $T \gtrsim j$. This behavior is indeed borne out in Fig. 9.2 and somewhat more explicitly in Fig. 9.3. The scaling (9.4.19) can be understood on the basis of random-matrix theory. Since the damping is taken care of by the factor $\exp(-2\Gamma jt/3)$ according to (9.4.16), the argument may be given for $\Gamma = 0$; it uses the scaling $J_z \sim j$ and the eigenvector statistics from Sect. 4.8. Since similar arguments have been expounded before, this particular one is left to the reader as Problem 9.11.

As a further check of random-matrix theory the damping constants $\Lambda_{\mu\nu}$ of (9.4.7) and the eigenvalues of the matrix (9.4.11) were evaluated numerically, using the numerically determined eigenstates $|\mu\rangle$ of the Floquet operator given in (9.4.2). The calculations were done for various pairs of p, λ, all corresponding to global classical chaos and for $j = 5, 10, 20, 40, 80$. The arithmetic mean of all $(2j+1)^2 - (2j+1)$ off-diagonal damping constants so determined is independent of p and λ and matches the random-matrix result to within less than $1/\sqrt{j}$. The variation with j of the relative root-mean-square deviations from the mean is consistent with a proportionality to $1/\sqrt{j}$; see Fig. 9.4.

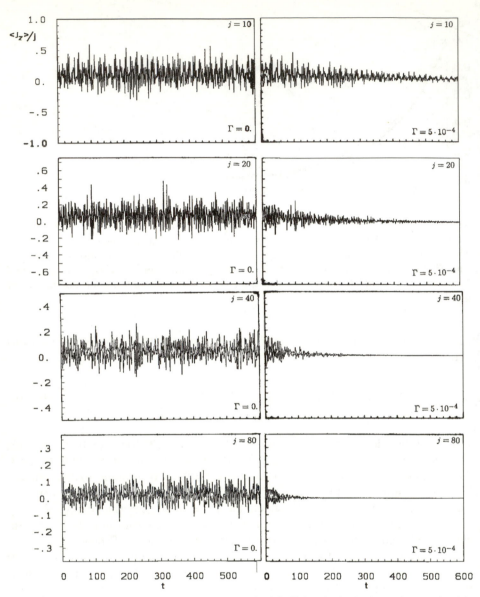

Fig. 9.2. Time-dependent expectation value $\langle J_z \rangle / \sqrt{j}$ for the kicked top (9.4.1, 2) without damping (*left column*) and with $\Gamma = 5 \times 10^{-4}$ for several values of j. Initial conditions and coupling constants p, λ are as in Fig. 9.1. The influence of j on the lifetime of recurrences is clearly visible. The mean $\langle J_z \rangle$ is referred to $\sqrt{j}$ in the expectation that the strength of the temporal fluctuations of $\langle J_z \rangle / j$ is of the order $1/\sqrt{j}$; see also Fig. 9.3

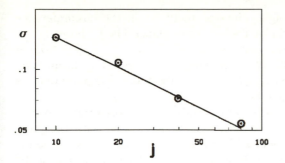

Fig. 9.3. Variance of the temporal fluctuations of $\langle J_z \rangle$ according to (9.4.19). Dynamics and parameters are as in Figs. 9.1, 9.2. The straight line represents the prediction of random-matrix theory, $\sigma \sim 1/\sqrt{j}$. The four points were obtained by numerically iterating the map (9.4.1, 9.4.2) and performing the time average indicated in the text; linear regression on the four points suggests an exponent -0.51

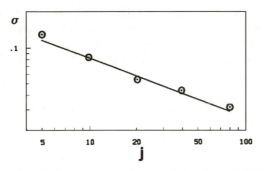

Fig. 9.4. Check on the equality of all "off-diagonal" damping constants $\Lambda_{\mu\nu}$ in the weak-damping limit $\Gamma j \ll 1$ of the kicked top (9.4.1, 9.4.2) with $p = 1.7$, $\lambda = 6$. The plot shows the variation with j of the root-mean-square deviation of the $\Lambda_{\mu\nu}$ from their arithmetic mean $\bar{\Lambda}_{\mu\nu}$, $\sigma = \left(\overline{[(\Lambda_{\mu\nu} - \bar{\Lambda}_{\mu\nu})^2]} \right)^{1/2} / \Gamma j$. The five points, based on $j = 5, 10, 20, 40, 80$, yield, by linear regression, the power law $\sigma \sim j^{-0.50}$ which is precisely the prediction of random-matrix theory (*straight line*)

9.5 Complex Energies and Quasi-Energies

Dissipative quantum maps of the form

$$\varrho(t+1) = M\varrho(t) \tag{9.5.1}$$

have generators M that are neither Hermitian nor unitary. Their eigenvalues $e^{-i\phi}$ are complex numbers that, for stable systems, are constrained to have moduli not exceeding unity,

$$|e^{-i\phi}| \leq 1 . \tag{9.5.2}$$

Indeed, an eigenvector of M with $\mathrm{Im}\{\phi\} > 0$ would grow indefinitely in weight with the number of iterations of the map.

In the conservative limit, all eigenvalues actually lie on the circumference of the unit circle around the origin of the complex plane. Their phases ϕ are differences of eigenphases of the unitary Floquet operator, $\phi_{\mu\nu} = \phi_\mu - \phi_\nu$. If the Hilbert space is N dimensional, N of the conservative $\phi_{\mu\nu}$ vanish, and the remainder falls into $N(N-1)/2$ pairs $\pm|\phi_\mu - \phi_\nu|$. Only N of these pairs refer to adjacent quasi-energy levels, i.e., give nearest neighbor spacings.

As the damping is increased, the eigenvalues $e^{-i\phi}$ tend to wander inward from the unit circle and eventually can no longer be associated with the pairs of conservative quasi-energies from which they originated; such an association is possible only within the range of applicability of a perturbative treatment of the damping, such as that given in the last section. In general, only one eigenvalue, that pertaining to the stationary solution of the map (9.5.1), is excepted from the inward migration and actually rests at unity. Figure 9.5 depicts the inward motion of the $e^{-i\phi}$ for a kicked top. We shall refer to the complex ϕ as complex or generalized quasi-energies.

Given that, for conservative maps in N-dimensional Hilbert spaces with $N \gg 1$, statistical analyses of the Floquet spectrum yield important characteristics and, especially, the possibility of distinguishing "regular" and "chaotic" dynamics, the question naturally arises whether statistical methods can be usefully applied to dissipative maps as well [336].

The density of eigenvalues is worthy of particular attention. In the conservative limit, the quasi-energies tend to be uniformly distributed along the circumference of the unit circle. A linear distribution of uniform density along a circle of smaller radius will still arise for weak damping, as long as first-order perturbation theory is still reliable and provided that random-matrix theory is applicable in zeroth order, i.e., provided chaos prevails in the classical limit. Indeed, as shown in the preceding section, the generalized quasi-energies $\phi_{\mu\nu}$ begin their inward journey with uniform radial "speed" $\partial\phi/\partial\Gamma$; the relative spread of speed is $\sim 1/\sqrt{j}$. However, for large damping, a two-dimensional surface distribution tends to arise in place of the one-dimensional line distribution. Numerical results like those displayed in Fig. 9.5 suggest isotropy but radial nonuniformity for that surface distribution.

To render density fluctuations in different parts of the spectrum meaningfully comparable, the coordinate mesh within the unit circle must in general be rescaled to achieve a uniform mean density of points throughout. This is similar in spirit to the unfolding of real energy spectra discussed in Sect. 4.7. Now, fluctuations may be characterized by, e.g., the distribution of nearest neighbor spacings; when the spacing between two points in the plane is defined as their Euclidean separation, a nearest neighbor can be found for each point and a spacing distribution established.

All of the above considerations require only slight modifications for damped systems under temporally constant external conditions. In such cases, the density operator obeys a master equation of the form

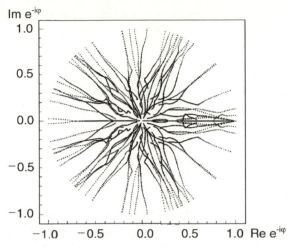

Fig. 9.5. Inward migration from the unit circle in the complex plane of the eigenvalues of the dissipative map (9.4.1, 9.6.1) for a kicked top with $j = 6$, $p = 2$, $\lambda_0 = 8$, $\lambda_1 = 10$ as the damping constant Γ is increased from 0 to 0.4 in steps of 0.005. A crossing of two eigenvalue "lines" in this picture does not in general correspond to a degeneracy since the crossing point may have different values of Γ on the two lines. The reflection symmetry about the real axis is a consequence of Hermiticity conservation for the density operator by the dissipative quantum map; see Sect. 8.9

$$\dot{\varrho}(t) = l\varrho(t) , \quad l = L + \Lambda , \tag{9.5.3}$$

where the generator l of infinitesimal time translations generally has a conservative part L and a damping part Λ. The eigenvalues of l, which are more analogous to the exponents $i\phi$ than to the eigenvalues $e^{-i\phi}$ of discrete-time maps, must have negative real parts for stable systems; distributed along the imaginary axis in the conservative limit, they tend to spread throughout the left half of the complex plane once the damping becomes strong.

In view of the conservative limit and the analogy to periodically driven systems, it is convenient to denote the eigenvalues of l by $-iE$ and to regard the E as complex or generalized energies. It should be kept in mind, though, that for vanishing damping, the eigenvalue in question becomes the difference of two eigenenergies of the Hamiltonian, $E_{\mu\nu} = E_\mu - E_\nu$, and the corresponding eigendyad of l becomes $|\mu\rangle\langle\nu|$ with $|\mu\rangle$ and $|\nu\rangle$ energy eigenstates.

9.6 Different Degrees of Level Repulsion for Regular and Chaotic Motion

Now, we consider strong damping outside the range of applicability of first-order perturbation theory. For greater flexibility, the kicked-top dynamics of Sect. 9.4

will be generalized slightly to allow two separate conservative kicks per period

$$F = e^{-ipJ_z} e^{-i\lambda_0 J_z^2/2j} e^{-i\lambda_1 J_y^2/2j} \ . \tag{9.6.1}$$

The damping generator Λ in the map (9.4.1) is kept unchanged, i.e., given by (9.4.3). Now, the map $e^\Lambda e^L$ must be diagonalized numerically. As in the zero-damping case, the angular-momentum basis $|jm\rangle$ proves convenient for that purpose, especially since the dissipative part e^Λ of the map can be given in closed form as the tetrad

$$(e^\Lambda)_{mn,pq} = \langle jm| \left(e^\Lambda |jp\rangle\langle jq| \right) |jn\rangle \ , \tag{9.6.2}$$

which represents the exact solution of the master equation $\dot{\varrho} = \Lambda\varrho$ at $t = 1$ originating from an arbitrary initial dyad $|jp\rangle\langle jq|$. For the explicit form of the tetrad, the reader is referred to [320, 331]; the method of construction will be presented in Sect. 9.14.

Figure 9.6 displays the eigenvalues of the map $e^\Lambda e^L$ which were obtained by numerical diagonalization for $j = 15$, $\Gamma = 0.07$. The coupling constants p, λ_0, λ_1 in the Floquet operator were chosen to yield either regular classical motion (Fig. 9.6a; $p = 2$, $\lambda_0 = 11.7$, $\lambda_1 = 0$) or predominantly chaotic classical motion (Fig. 9.6b; $p = 2$, $\lambda_0 = 11.7$, $\lambda_1 = 10$). Mere inspection of the two annular clouds suggests a lesser inhibition with respect to close proximity under conditions of classically regular motion (Fig. 9.6a).

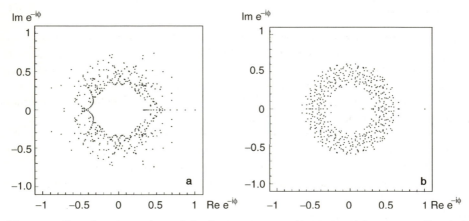

Fig. 9.6. Complex eigenvalues of the dissipative map (9.4.1, 9.6.1) for $j = 15$, $\Gamma = 0.07$, $p = 2$, $\lambda_0 = 11.7$ under conditions of (**a**) classically regular motion ($\lambda_1 = 0$) and (**b**) chaos ($\lambda_1 = 10$). Note again the reflection symmetry about the real axis. The inhibition to close proximity is clearly lower in the regular than in the chaotic case

To unfold the spectra, a local mean density was determined around each point as $\bar{\varrho} = n/\pi d_n^2$ where d_n is the distance to the nth nearest neighbor; after making sure that the precise value of n does not matter, the choice $n = 10$ was made as a compromise respecting both limits in $1 \ll n \ll (2j+1)^2$. Then, the distance to the first neighbor was rescaled as

$$S = d_1 \sqrt{\bar{\varrho}} \tag{9.6.3}$$

in analogy to (4.7.11). (Note that in (4.7.11), $\bar{\varrho}$ is normalized as a probability density and refers to a distribution of points along a line.) Figure 9.7 depicts the spacing staircases $I(S)$ for the two clouds of Fig. 9.6. Clearly, linear repulsion [i.e., a quadratic rise of $I(S)$ for small S] is obtained when the classical dynamics is regular, whereas cubic repulsion [$I(S) \propto S^4$] prevails under conditions of classical chaos.

The distinction between linear and cubic repulsion of the complex levels and its correlation with the classical distinction between predominantly regular motion and global chaos does not seem to be a peculiarity of the dynamics chosen. At any rate, for the model considered, linear repulsion prevails not only in the strictly integrable case $\lambda_1 = 0$ but also for all λ_1 in the range $0 \leq \lambda_1 \leq 0.2$ which corresponds to the predominance of regular trajectories on the classical sphere; cubic repulsion, on the other hand, arises not only for $\lambda_1 = 6$ but was also found for larger values of λ_1.

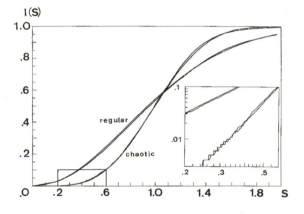

Fig. 9.7. Integrated spacing distributions $I(S)$ for two different universality classes. One curve in each of the two pairs refers to the map (9.4.1, 9.6.1) with $j = 10$, $\Gamma = 0.1$, $p = 2$, and 100 different values of λ_0 from $10 \leq \lambda_0 \leq 12$. The "regular" case has $\lambda_1 = 0$ and the "chaotic" one $\lambda_1 = 8$. The second line in the regular pair represents the Poissonian process in the complex plane according to (9.7.4); in the "chaotic" pair, the spacing staircase of general non-Hermitian matrices of dimension $2j + 1 = 21$ appears. The insert reveals linear and cubic repulsion

These results certainly suggest universality of the two degrees of level repulsion, but they cannot be considered conclusive proof thereof. An interesting corroboration and, in fact, extension comes from the recent observation that the damped periodically driven noisy rotator displays linear and cubic repulsion of the complex Floquet eigenvalues of its Fokker-Planck equation, depending on whether the noiseless deterministic limit is, respectively, regular or chaotic [337].

The reader will have noticed that Fig. 9.7 depicts a pair of curves for each of the "regular" and "chaotic" cases. One curve in a pair pertains to the map in consideration while the other represents a tentative interpretation which we shall discuss presently. In intuitive generalizations of the conservative cases, the interpretations involve a Poissonian random process in the plane (rather than along a line) for the regular limit and a Gaussian ensemble of random matrices restricted neither by unitarity nor by Hermiticity in the chaotic limit. Figure 9.7 suggests that both interpretations work quite well.

It would be nice to have an exactly solvable damped system with generic spectral fluctuations, i.e.; linear repulsion. A good candidate seems to be the map $e^{\Lambda}e^{L}$ with

$$F = e^{-ipJ_z}e^{-i\lambda J_z^2/2j}$$

$$\Lambda\varrho = \frac{\Gamma}{2j}\left([J_z, \varrho J_z] + [J_z\varrho, J_z]\right) + \frac{\gamma}{j^3}\left([J_z^2, \varrho J_z^2] + [J_z^2\varrho, J_z^2]\right) . \tag{9.6.4}$$

Its "eigen-dyads" $|j,m\rangle\langle j,m'|$ are built by the eigenvectors of $\boldsymbol{J}^2$ and J_z. The eigenvalues can be read off immediately as

$$\lambda_{mm'} = \exp\left\{-i\left[(m - m')p + (m^2 - m'^2)\frac{\lambda}{2j}\right]\right.$$
$$\left. -(m - m')^2\frac{\Gamma}{2j} - (m^2 - m'^2)^2\frac{\gamma}{j^3}\right\} . \tag{9.6.5}$$

A peculiarity of the dynamics (9.6.4) is that the damping generator does not permit transitions between the conservative Floquet eigenstates $|j, m\rangle$ but only phase relaxation in the coherences between such states. Consequently, the $(2j + 1)$-fold degeneracy of the conservative eigenvalue unity, common to all diagonal dyads $|jm\rangle\langle jm|$, is still present in the $\lambda_{mm'}$ for $\gamma, \Gamma > 0$. One might expect generic spacing fluctuations since both the real and the imaginary part of $\ln \lambda_{mm'}$ depend on the two quantum numbers $m \pm m'$. Thus, the whole spectrum

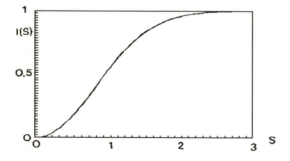

Fig. 9.8. Integrated spacing distribution $I(S)$ for the eigenvalues (9.6.5) of the classically integrable dissipative map (9.6.4) with $j = 30$, $p = 17.3$, $\lambda = 24.9$, $\Gamma = 0.012$, $\gamma = 0.011$. The staircase is hardly distinguishable from the prediction for the two dimensional Poissonian process, Sect. 8.7

may be looked upon as a superposition of many effectively independent ones. This expectation is indeed borne out by the spacing staircase of Fig. 9.8.

9.7 Poissonian Random Process in the Plane

Imagine that N points are thrown onto a circular disc of radius R, with uniform mean density and no correlation between throws. The mean area per point is $\pi R^2/N$, and the mean separation $\bar{s}$ between nearest neighbors is of the order $R/\sqrt{N}$. The distribution of nearest-neighbor spacings and the precise mean separation $\bar{s}$ are also readily determined.

The probability $P(s)ds$ of finding the nearest neighbor of a given point at a distance between s and $s + ds$ equals the probability that one of the $(N - 1)$ points is located in the circular ring of thickness ds around the given point and that all $(N - 2)$ remaining points lie beyond this:

$$P(s)ds = (N - 1)\frac{2\pi s\,ds}{\pi R^2}\left(1 - \frac{\pi s^2}{\pi R^2}\right)^{N-2}. \tag{9.7.1}$$

It is easy to check that this distribution is correctly normalized, $\int_0^R ds P(s) = 1$. The mean spacing reads

$$\bar{s} = \frac{R}{\sqrt{N}}\left[\sqrt{N}(N - 1)2\int_0^1 dt\, t^2(1 - t^2)^{N-2}\right]. \tag{9.7.2}$$

After rescaling the integration variable as $t = \tau/\sqrt{N}$, the square bracket in (9.7.2) is seen, as $N \to \infty$, to approach the limit

$$2\int_0^{\sqrt{N}} d\tau\, \tau^2\left(1 - \frac{\tau^2}{N}\right)^N \to 2\int_0^\infty d\tau\, \tau^2 e^{-\tau^2} = \frac{\sqrt{\pi}}{2}. \tag{9.7.3}$$

To find the asymptotic spacing distribution for large N, the spacing must be referred to its mean $\bar{s} = R\sqrt{\pi/4N}$. After introducing the appropriately rescaled variable $S = s/\bar{s}$ and performing the limit as in (9.7.3), the distribution (9.7.1) is turned into

$$P(S) = \tfrac{1}{2}\pi S e^{-\pi S^2/4}. \tag{9.7.4}$$

Amazingly, (9.7.4) gives Wigner's conjecture exactly for the spacing distribution in the GOE and COE discussed in Chap. 4. As was shown there, the Wigner distribution is rigorous for the GOE of 2×2 matrices but is only an approximation, albeit a rather good one, for $N \times N$ matrices with $N > 2$. In the present context, the Wigner distribution arises as a rigorous result in the limit $N \to \infty$. According to Figs. 9.7, 9.8, this distribution is reasonably faithful to the spacing distributions of the respective regular dynamics.

The question arises *why* the Poissonian random process in the plane should so accurately reproduce the spacing fluctuations of generic integrable systems with damping. To achieve an intuitive understanding, the reader may recall from Chap. 5 the most naive argument supporting the analogous statement for Hamiltonian dynamics: the spectrum of an integrable Hamiltonian with f degrees of freedom may be approximated by torus quantization. Its levels are labelled by f quantum numbers, and the spectrum may be thought of as consisting of many independent subspectra.

An analogous statement can be made about the spectrum of the map $e^{\Lambda} e^{L}$ with the conservative Floquet operator F from (9.6.1) and the damping generator Λ from (9.4.3), provided $\lambda_1 = 0$ in F to make the classical limit integrable. As is easily checked, this map has the symmetry

$$[J_z, e^{\Lambda} e^{L} \varrho] = e^{\Lambda} e^{L} [J_z, \varrho] \qquad (9.7.5)$$

for arbitrary ϱ. It follows that the map in question does not couple dyads $|j, m + k\rangle\langle j, m|$ with different values of k. Indeed, by using $|j, m+k\rangle\langle j, m|$ for ϱ in (9.7.5) and writing

$$e^{\Lambda} e^{L} |j, m + k\rangle\langle j, m| = \sum_{m'k'} C^{mk}_{m'k'} |j, m' + k'\rangle\langle j, m'| \,,$$

one obtains from (9.7.5) the identity $(k' - k)C^{m'k'}_{mk} = 0$, i.e., $C^{m'k'}_{mk} \propto \delta_{k'k}$. Consequently, k is a good quantum number for the map in question and the cloud of eigenvalues depicted in Fig. 9.6a may be segregated into many subclouds, one for each value of k.

When the map defined by (9.4.3) and (9.6.1) is turned into a classically nonintegrable map by allowing $\lambda_1 > 0$, the symmetry (9.7.5) is broken. Consequently, now the previously segregated ($\lambda_1 = 0$) subspectra begin to interact. It is still not counterintuitive that the spacing distribution remains Poissonian as long as λ_1 is small ($0 \le \lambda_1 \le 0.2$); experience with conservative level dynamics suggests that a large number of close encounters of levels must have taken place before strong level repulsion can arise throughout the spectrum.

9.8 Ginibre's Ensemble of Random Matrices

By dropping the requirement of Hermiticity, *Ginibre* [338] was led from the Gaussian unitary ensemble to a Gaussian ensemble of matrices with arbitrary complex eigenvalues z. The joint probability density for the N eigenvalues of such $N \times N$ matrices looks similar to (4.3.15) for the GOE,

$$P(z_1, \ldots, z_N) = \frac{\mathcal{N}^{-1}}{\pi^N} \prod_{i<j}^{1 \ldots N} |z_i - z_j|^2 \exp\left(-\sum_{i=1}^{N} |z_i|^2 \right) \,, \qquad (9.8.1)$$

but is to be normalized by integrating each z_i over the complex plane.

The goal of this section is to extract the distribution of nearest-neighbor spacings from (9.8.1) and to compare it with the numerically obtained spacing distribution of the classically chaotic top discussed in Sect. 9.6. The calculations to follow fall in several separate parts, each of which merits interest in its own right. Some of them are alternatives to *Mehta's* [339] and Ginibre's classic treatments. First, we shall determine the normalization factor in (9.8.1), then proceed to evaluating the reduced joint densities $P_j(z_1, z_2, \ldots, z_j)$ with $j = 1, 2, 3, \ldots$, and finally use the results obtained to find the spacing distribution $P(S)$.

9.8.1 Normalizing the Joint Density

The product of differences in (9.8.1) is related to a Vandermonde determinant,

$$\prod_{\substack{i<j}}^{1 \ldots N} (z_i - z_j) = \det\left(z_i^{k-1}\right) = \begin{vmatrix} 1 & z_1 & z_1^2 & \cdots & z_1^{N-1} \\ 1 & z_2 & z_2^2 & \cdots & z_2^{N-1} \\ \vdots & & & & \vdots \\ 1 & z_N & z_N^2 & \cdots & z_N^{N-1} \end{vmatrix}, \tag{9.8.2}$$

which was already encountered in Sect. 4.11. Then, the squared modulus of the product in (9.8.2) takes the form

$$\prod_{\substack{i<j}}^{1 \ldots N} |z_i - z_j|^2 = \det\left(\sum_{j=1}^{N} z_j^{i-1} z_j^{*\, k-1}\right). \tag{9.8.3}$$

It will be convenient for the following to represent again an $n \times n$ determinant as the Gaussian integral over anticommuting variables (4.12.1),

$$\det A = \int \left(\prod_{j}^{1 \ldots n} d\eta_j^* d\eta_j\right) \exp\left(-\sum_{ik}^{1 \ldots n} \eta_i^* A_{ik} \eta_k\right) \tag{9.8.4}$$

introduced in Sect. 4.12. We employ this identity to the determinant in (9.8.3) for which the exponential $\exp(-\eta^* A\eta)$ in (9.8.4) may be simplified slightly. Since pairs $\eta_i^* \eta_k$ commute among themselves, the general representation yields

$$\det\left(\sum_{j=1}^{N} z_j^{i-1} z_j^{*\, k-1}\right) =$$

$$\int \left(\prod_{j}^{1 \ldots N} d\eta_j^* d\eta_j\right) \prod_{j}^{1 \ldots N} \exp\left(-\sum_{ik}^{1 \ldots N} \eta_i^* \eta_k z_j^{i-1} z_j^{*\, k-1}\right). \tag{9.8.5}$$

Moreover, when each of the N exponentials on the r.h.s. is expanded in a Taylor series, second- and higher order terms vanish identically. Indeed,

$$\left(\sum_{ik} \eta_i^* \eta_k z^{i-1} z^{*\, k-1}\right)^2 = \sum_{ijkl} \eta_i^* \eta_j \eta_k^* \eta_l z^{i-1} z^{*\, j-1} z^{k-1} z^{*\, l-1} = 0 \tag{9.8.6}$$

since the product of the four Grassmann variables is antisymmetric in j and l and the product of the four ordinary numbers is symmetric. It follows that

$$\det\left(\sum_{j=1}^{N} z_j^{i-1} z_j^{*\,k-1}\right)$$

$$= \int\left(\prod_{j}^{1\dots N} d\eta_j^* d\eta_j\right) \prod_{j}^{1\dots N}\left(1 - \sum_{ik} \eta_i^* \eta_k z_j^{i-1} z_j^{*\,k-1}\right)$$

$$= \int\left(\prod_{j}^{1\dots N} d\eta_j^* d\eta_j\right) \prod_{j}^{1\dots N}\left(- \sum_{ik} \eta_i^* \eta_k z_j^{i-1} z_j^{*\,k-1}\right) \tag{9.8.7}$$

where the last step is based on the vanishing of any Grassmann integral over a constant integrand.

With the help of this integral representation of the product of differences in (9.8.1), the calculation of the normalization factor $\mathcal{N}$ becomes an easy matter. When the z integrals are done first, each of them yields a factor

$$\int d^2 z \frac{1}{\pi} e^{-|z|^2} \sum_{ik} \eta_i^* \eta_k z^{i-1} z^{*\,k-1} = \sum_i \eta_i^* \eta_i (i-1)! \tag{9.8.8}$$

and the normalization factor $\mathcal{N}$ is determined by

$$\mathcal{N} = \int\left(\prod_j d\eta_j^* d\eta_j\right)\left[-\sum_{i=1}^{N} \eta_i^* \eta_i (i-1)!\right]^N . \tag{9.8.9}$$

For non-zero contributions to arise, each of the N identical factors in the integrand must provide a different pair $\eta_i^* \eta_i$; there are $N!$ such contributions, and they are all numerically equal since the pairs $\eta_i^* \eta_i$, $\eta_j^* \eta_j$ commute,

$$\mathcal{N} = N!(-1)^N\left[\prod_{i=1}^{N}(i-1)!\right]\int\left(\prod_j d\eta_j^* d\eta_j \eta_j^* \eta_j\right) = \prod_{j=1}^{N} j! . \tag{9.8.10}$$

Thus, the correctly normalized joint probability density of all N eigenvalues reads

$$P(\{z\}) = \left(\prod_{k}^{1\dots N} \frac{e^{-|z_k|^2}}{\pi\, k!}\right) \prod_{i<j} |z_i - z_j|^2 . \tag{9.8.11}$$

9.8.2 The Density of Eigenvalues

The probability density for a single eigenvalue is obtained by integrating the joint density over all but one variable,

$$
\begin{aligned}
P_1(z_1) \;=\;& \int d^2 z_2 \ldots d^2 z_N \, P(z_1, \ldots, z_N) \\[2mm]
=\;& \left(\prod_{j}^{1\ldots N} j! \right)^{-1} \frac{e^{-|z_1|^2}}{\pi} \int \left(\prod_{k}^{2\ldots N} d^2 z_k \frac{e^{-|z_k|^2}}{\pi} \right) \\[2mm]
& \times \int \left(\prod_{l}^{1\ldots N} d\eta_l^* d\eta_l \right) \prod_{m}^{1\ldots N} \left(-\sum_{in} \eta_i^* \eta_n z_m^{i-1} z_m^{*\,n-1} \right) .
\end{aligned}
\tag{9.8.12}
$$

Here again we employ the representation (9.8.4, 9.8.7) for the product of differences in the joint distribution (9.8.1). After doing the $N-1$ integrals over the z_k, the density $P_1(z)$ reads

$$
\begin{aligned}
P_1(z) =\;& \left(\prod_{j}^{1\ldots N} j! \right)^{-1} \frac{e^{-|z|^2}}{\pi} \int \left(\prod_{j}^{1\ldots N} d\eta_j^* d\eta_j \right) \left(-\sum_{ik} \eta_i^* \eta_k z^{i-1} z^{*\,k-1} \right) \\[2mm]
& \times \left[-\sum_{l} \eta_l^* \eta_l (l-1)! \right]^{N-1} .
\end{aligned}
\tag{9.8.13}
$$

Since $N-1$ of the factors in the integrand involve pairs $\eta_l^* \eta_l$ with coinciding indices, the remaining factor must have this property as well for the $2n$-fold Grassmann integral to pick up a nonzero contribution,

$$
\begin{aligned}
P_1(z) =\;& \left(\prod_{j}^{1\ldots N} j! \right)^{-1} \frac{e^{-|z|^2}}{\pi} \int \left(\prod_{j}^{1\ldots N} d\eta_j^* d\eta_j \right) \left(-\sum_{i}^{1\ldots N} \eta_i^* \eta_i |z|^{2(i-1)} \right) \\[2mm]
& \times \left[-\sum_{l}^{1\ldots N} \eta_l^* \eta_l (l-1)! \right]^{N-1} .
\end{aligned}
\tag{9.8.14}
$$

For every fixed value of the summation index i in the first factor of the integrand, there are $(N-1)!$ ways of assigning indices to the remaining pairs $\eta^* \eta$ such that each of the naturals $1, 2, \ldots, N$ appears once as an index on a pair $\eta^* \eta$; thus all factorials in the normalization factor are cancelled except for $1/[(i-1)!N]$,

$$
P_1(z) = \frac{1}{N\pi} e^{-|z|^2} e_{N-1}(|z|^2) ,
\tag{9.8.15}
$$

where $e_n(x)$ denotes the nth order polynomial in x equalling the first $n+1$ terms of the Taylor expansion of e^x about the origin,

$$
e_n(x) = 1 + \frac{x}{1!} + \ldots \frac{x^n}{n!} .
\tag{9.8.16}
$$

It is immediately obvious from (9.8.15, 16) that $P_1(z)$ approaches the constant $1/N\pi$ in the limit $N \to \infty$ with z held fixed. Conversely, when N is kept constant and z grows indefinitely, $P_1(z)$ vanishes. The transition region between the two limiting cases lies at $|z|^2 \approx N$. To investigate the transition it is convenient to note that

$$\frac{\partial}{\partial x}e_n(x) = e_{n-1}(x) \tag{9.8.17}$$

and therefore,

$$e^{-x}e_n(x) = \int_x^\infty dy \frac{y^n}{n!}e^{-y} . \tag{9.8.18}$$

For large n, the integrand in (9.8.18) has a sharp maximum at $y = n$; its width is of the order $\sqrt{n}$. When x lies to the left of the maximum by several times the width, the integral deviates only slightly from unity; for x well to the right of the maximum, however, the integral is exponentially small since it collects only contributions from the wing of the integrand. For x near the maximum, the asymptotic large-n behavior reveals itself when the variable $\xi = (x - n)/\sqrt{2n}$ is introduced and the limit $n \to \infty$ taken with the help of Stirling's formula. Setting $f(\xi) = e_n(x)e^{-x}$, one finds $f'(\xi) = (1/\sqrt{\pi})e^{-\xi^2}$ and thus,

$$e_n(x)e^{-x} \approx \frac{1}{2}\operatorname{erfc}\left(\frac{x - n}{\sqrt{2n}}\right) . \tag{9.8.19}$$

Interestingly, this asymptotic formula still encompasses the correct limits 1 and 0 for x well below and well above n, respectively. Most importantly, since the relative width $\sim \sqrt{n}/n$ of the transition vanishes with $n \to \infty$, the function $e_n(x)e^{-x}$ behaves essentially like the unit step function $\Theta(n - x)$.

To sum up, as $N \to \infty$, the eigenvalues z tend to cover the circular disc of radius $\sqrt{N}$ around the origin of the complex plane with uniform probability density $1/\pi N$. Needless to say, by rescaling as $z \to z\sqrt{N}$, the circle in question is contracted to the unit circle that was encountered above for complex matrices representing dissipative quantum maps.

9.8.3 The Reduced Joint Densities

By integrating $P(z_1, \ldots , z_N)$ over $N - n$ variables, the joint probability density of n eigenvalues is obtained. Instead of (9.8.13) for $n = 1$, now,

$$P_n(z_1, \ldots , z_n) = \left(\prod_j^{1 \ldots N} j!\right)^{-1} \frac{e^{-|z_1|^2}}{\pi} \cdots \frac{e^{-|z_n|^2}}{\pi} \int \left(\prod_j^{1 \ldots N} d\eta_j^* d\eta_j\right)$$

$$\times \left[\prod_l^{1 \ldots n}\left(-\sum_{ik} \eta_i^* \eta_k z_l^{i-1} z_l^{*\,k-1}\right)\right]\left\{-\sum_m \eta_m^* \eta_m (m - 1)!\right\}^{N-n} . \tag{9.8.20}$$

As before, each of the naturals $1, 2, \ldots N$ must appear precisely once as the index of an η^* and an η. For a given set of n indices $\{i\}$ on the η^* and the identical set $\{k\}$ (apart from ordering) on the η in the square bracket, there are $(N - n)!$ possibilities of using up the remaining naturals for the $(N - n)$ curly brackets in (9.8.20). The latter then provide the factor $\prod_{j \notin \{i\}} (-\eta_j^* \eta_j)(j - 1)!$. With the $N - n$ pairs $\eta_j^* \eta_j$ integrated out, the reduced density becomes

$$P_n(z_1, \ldots, z_n) = \frac{(N - n)!}{N!} \frac{e^{-|z_1|^2}}{\pi} \cdots \frac{e^{-|z_n|^2}}{\pi} \left[(-1)^n \int \left(\prod_j^{1 \ldots n} d\eta_j^* d\eta_j \right) \right.$$

$$\times \sum_{\substack{\{i\}, \{k\}}}^{1 \ldots N} \eta_{i_1}^* \eta_{k_1} \cdots \eta_{i_n}^* \eta_{k_n} z_1^{i_1 - 1} z_1^{* k_1 - 1} \cdots$$

$$\left. \times z_n^{i_n - 1} z_n^{* k_n - 1} / (i_1 - 1)! \cdots (i_n - 1)! \right] . \qquad (9.8.21)$$

The equality of the sets $\{i\}$ and $\{k\}$ can be realized in $n!$ different ways, corresponding to the $n!$ permutations $\{\mathcal{P}i\}$ of the $\{i\}$. For each such permutation, the $2n$ integrals can be carried out to yield the factor $(-1)^n (-1)^{\mathcal{P}}$ where $(-1)^{\mathcal{P}}$ is the signature of the permutation. The square bracket in (9.8.21) becomes

$$[\,] = \sum_{\{i\}}^{1 \ldots N} \frac{z_1^{i_1 - 1}}{(i_1 - 1)!} \cdots \frac{z_n^{i_n - 1}}{(i_n - 1)!} \sum_{\mathcal{P}} (-1)^{\mathcal{P}} (z_1^*)^{-1 + \mathcal{P}i_1} \cdots (z_n^*)^{-1 + \mathcal{P}i_n} .$$

At this point, the restriction that all the i be different becomes superfluous since the sum over the permutations $\mathcal{P}$ vanishes if any of the i coincide. Instead of summing over the permutation in the exponents, one may sum over the permutations of indices on the z^*:

$$[\,] = \sum_{\mathcal{P}} (-1)^{\mathcal{P}} \sum_{\{i\}}^{1 \ldots N} \frac{(z_1 z_{\mathcal{P}1}^*)^{i_1 - 1}}{(i_1 - 1)!} \cdots \frac{(z_n z_{\mathcal{P}n}^*)^{i_n - 1}}{(i_n - 1)!}$$

$$= \sum_{\mathcal{P}} (-1)^{\mathcal{P}} e_{N-1}(z_1 z_{\mathcal{P}1}^*) e_{N-1}(z_2 z_{\mathcal{P}2}^*) \cdots e_{N-1}(z_n z_{\mathcal{P}n}^*)$$

$$= \det \left[e_{N-1}(z_i z_k^*) \right] . \qquad (9.8.22)$$

Thus, the reduced density reads

$$P_n(z_1, \ldots, z_n) = \frac{(N - n)!}{N!} \left(\prod_j^{1 \ldots n} \frac{e^{-|z_j|^2}}{\pi} \right) \det \left[e_{N-1}(z_i z_k^*) \right] . \qquad (9.8.23)$$

This includes (9.8.15) as the special case $n = 1$ and even implies a new representation of the full joint density (9.8.1) for $n = N$.

Similarly to the behavior of $P_1(z_1)$, the joint distribution $P_n(z_1 \ldots z_n)$ vanishes in the limit $N \to \infty$ if at least one of the variables z lies outside the circle of radius $\sqrt{N}$, $|z|^2 > N$. When all the z lie within that circle, on the other hand, the truncated exponentials approach the full ones such that

$$P_n(z_1 \ldots z_n) = \pi^{-N} \frac{(N-n)!}{N!} \det\left(e^{(z_i - z_k)z_k^*}\right) .$$

(9.8.24)

Obviously, homogeneity within the circle $|z_i|^2 < N$ is established since the asymptotic distribution (9.8.24) is invariant under a common shift, $z_i \to z_i + \delta$, of all variables.

9.8.4 The Spacing Distribution

It will be convenient to determine the spacing distribution $P(s)$ from the probability $H(s)$ that an eigenvalue has its nearest neighbor further away than s, i.e.,

$$P(s) = -\frac{d}{ds}H(s) .$$

(9.8.25)

We shall refer to $H(s)$ as the hole probability. The more familiar integrated spacing distribution $I(s)$, i.e., the probability that an eigenvalue has its nearest neighbor closer than s, is related to $H(s)$ by $I(s) = 1 - H(s)$. With $\langle \ldots \rangle$ denoting the average over all N eigenvalues, and the weight is the joint density (9.8.1), the hole distribution reads

$$H(s) = \langle \Theta\left(|z_2 - z_1| - s\right) \Theta\left(|z_3 - z_1| - s\right) \ldots \Theta\left(|z_N - z_1| - s\right)\rangle .$$

(9.8.26)

It is left to the reader in Problem 9.15 to consider the case $N = 2$ and to verify that

$$\begin{aligned} H_2(s) &= e_1\left(\tfrac{s^2}{2}\right) e^{-s^2/2} , \\ P_2(s) &= \tfrac{1}{2}s^3 e^{-s^2/2} \end{aligned}$$

(9.8.27)

or, with the mean spacing rescaled to unity,

$$P_2(S) = (3^4 \pi^2 2^{-7}) S^3 \exp\left[-(3^2 \pi^2 2^{-4})S^2\right] .$$

(9.8.28)

In the following, arbitrary values of N will be admitted initially, but eventually the limit $N \gg 1$ will be taken. To that end, we rewrite (9.8.26) as

$$\begin{aligned} H(s) &= \left\langle \prod_{j=2}^{N} \left[1 - \Theta\left(s^2 - |z_1 - z_j|^2\right)\right]\right\rangle \\ &= 1 + \sum_{\nu=1}^{N-1} \frac{(-1)^\nu}{\nu!} \sum_{n_1 \neq n_2 \neq \ldots \neq n_\nu}^{2 \ldots N} \\ &\quad \times \left\langle \Theta\left(s^2 - |z_1 - z_{n_1}|^2\right) \ldots \Theta\left(s^2 - |z_1 - z_{n_\nu}|^2\right)\right\rangle \end{aligned}$$

(9.8.29)

and invoke the joint density of $\nu + 1$ eigenvalues (9.8.23),

$$
H(s) - 1 = \sum_{\nu=1}^{N-1} \frac{(-1)^\nu}{\nu!N} \int_{|z_1|<\sqrt{N}} \frac{d^2 z_1}{\pi} e^{-|z_1|^2}
$$

$$
\times \int_{|z_2-z_1|<s} \frac{d^2 z_2}{\pi} e^{-|z_2|^2} \cdots \int_{|z_{\nu+1}-z_1|<s} \frac{d^2 z_{\nu+1}}{\pi} e^{-|z_{\nu+1}|^2}
$$

$$
\times \det\left[e_{N-1}(z_i z_k^*) \right] . \tag{9.8.30}
$$

The redundant limit $|z_1| < \sqrt{N}$ on the z_1 integral emphasizes the asymptotic $(N \to \infty)$ disappearance of the integrand for z_1 outside the circle of radius $\sqrt{N}$. With this in mind and noting (or strictly speaking, anticipating) that the nearest neighbor spacing is extremely likely to be much smaller than $\sqrt{N}$, all $e_{N-1}(z_i z_k^*)$ in the $(\nu+1) \times (\nu+1)$ determinant in (9.8.30) may be replaced by the exponentials $e^{z_i z_k^*}$. After shifting $z_2, z_3, \ldots z_{\nu+1}$ by z_1, the hole distribution takes the form

$$
H(s) - 1 = \sum_{\nu=1}^{N-1} \frac{(-1)^\nu}{\nu!N} \int_{|z_1|<\sqrt{N}} \frac{d^2 z_1}{\pi} e^{-|z_1|^2}
$$

$$
\times \int_{|z_2|<s} \frac{d^2 z_2}{\pi} e^{-|z_1+z_2|^2} \cdots \int_{|z_{\nu+1}|<s} \frac{d^2 z_{\nu+1}}{\pi} e^{-|z_{\nu+1}+z_1|^2}
$$

$$
\times \det\left(\exp\left\{ \left[z_i + (1-\delta_{i1})z_1 \right] \left[z_k^* + (1-\delta_{k1})z_1^* \right] \right\} \right) . \tag{9.8.31}
$$

Now, from the first column of the determinant, we extract the factor $e^{|z_1|^2}$, from the jth column with $j = 2, 3, \ldots, \nu+1$ the factor $e^{z_1(z_j^*+z_1^*)}$, and from the jth row the factor $e^{z_j z_1^*}$ and arrive at

$$
H(s) - 1 = \sum_{\nu=1}^{N-1} \frac{(-1)^\nu}{\nu!N} \int_{|z_1|<\sqrt{N}} \frac{d^2 z_1}{\pi} \left(\prod_{j=2}^{\nu+1} \int_{|z_j|<s} \frac{d^2 z_j}{\pi} e^{-|z_j|^2} \right)
$$

$$
\times \begin{vmatrix}
1 & 1 & 1 & \cdots & 1 \\
1 & e^{|z_2|^2} & e^{z_2 z_3^*} & \cdots & e^{z_2 z_{\nu+1}^*} \\
1 & e^{z_3 z_2^*} & e^{|z_3|^2} & \cdots & e^{z_3 z_{\nu+1}^*} \\
\vdots & \vdots & \vdots & & \vdots \\
1 & e^{z_{\nu+1} z_2^*} & e^{z_{\nu+1} z_3^*} & \cdots & e^{|z_{\nu+1}|^2}
\end{vmatrix} . \tag{9.8.32}
$$

The $(\nu+1) \times (\nu+1)$ determinant arising here may be written more concisely as $(\det e^{z_i z_k^*})_{z_1=0}$, where i and k run from 1 to $\nu+1$ and the prescription $z_1 = 0$ ensures that the elements in the first row and the first column are unity. At any rate, the integrand in (9.8.32) is independent of z_1 such that the z_1 integral yields the factor πN,

$$
H(s) - 1 = \sum_{\nu=1}^{N-1} \frac{(-1)^\nu}{\nu!} \left(\prod_{j=2}^{\nu+1} \int_{|z_j|<s} \frac{d^2 z_j}{\pi} e^{-|z_j|^2} \right) \left(\det e^{z_i z_k^*} \right)_{|z_1|=0} . \tag{9.8.33}
$$

Converting to polar coordinates for the remaining ν integration variables we first consider the ν-fold angular integral. To give it a more symmetrical appearance, it will be complemented with a dummy $(\nu+1)$th integral $\int_0^{2\pi} d\phi_1/2\pi$. With $n \equiv \nu + 1$, $d^n\phi$ for $d\phi_1 \ldots d\phi_n$, and the integration ranges $0 \le \phi_j \le 2\pi$ understood,

$$
\int \frac{d^n\phi}{(2\pi)^n} \left(\det e^{z_i z_k^*} \right)_{|z_1|=0}
$$

$$
= \left[\int \frac{d^n\phi}{(2\pi)^n} \sum_{\mathcal{P}} (-1)^{\mathcal{P}} \exp\left(z_1 z_{\mathcal{P}1}^* + z_2 z_{\mathcal{P}2}^* + \ldots z_n z_{\mathcal{P}n}^* \right) \right]_{|z_1|=0}
$$

$$
= \left[\sum_{\mu_1 \cdots \mu_n}^{0 \ldots \infty} \frac{1}{\mu_1! \cdots \mu_n!} \sum_{\mathcal{P}} (-1)^{\mathcal{P}} \right.
$$

$$
\left. \int \frac{d^n\phi}{(2\pi)^n} (z_1 z_{\mathcal{P}1}^*)^{\mu_1} \cdots (z_n z_{\mathcal{P}n}^*)^{\mu_n} \right]_{|z_1|=0} .
\tag{9.8.34}
$$

Instead of summing over the permutation of indices on the z^*, one may sum over the permutations of indices in the exponents of the z^*. Then, the above equation can be extended:

$$
\int \frac{d^n\phi}{(2\pi)^n} \left(\det e^{z_i z_k^*} \right)_{|z_1|=0}
$$

$$
= \left[\sum_{\{\mu\}} \frac{1}{\mu_1! \cdots \mu_n!} \sum_{\mathcal{P}} (-1)^{\mathcal{P}} \right.
$$

$$
\left. \times \int \frac{d\phi_1}{2\pi} z_1^{\mu_1} z_1^{*\,\mu_{\mathcal{P}1}} \cdots \int \frac{d\phi_n}{2\pi} z_n^{\mu_n} z_n^{*\,\mu_{\mathcal{P}n}} \right]_{|z_1|=0}
$$

$$
= \left[\sum_{\{\mu\}} \frac{|z_1|^{2\mu_1}}{\mu_1!} \cdots \frac{|z_n|^{2\mu_n}}{\mu_n!} \sum_{\mathcal{P}} (-1)^{\mathcal{P}} \delta_{\mu_1,\mu_{\mathcal{P}1}} \cdots \delta_{\mu_n,\mu_{\mathcal{P}n}} \right]_{|z_1|=0}
$$

$$
= \left[\sum_{\{\mu\}} \frac{|z_1|^{2\mu_1}}{\mu_1!} \cdots \frac{|z_n|^{2\mu_n}}{\mu_n!} \det \left(\delta_{\mu_i,\mu_k} \right) \right]_{|z_1|=0} .
\tag{9.8.35}
$$

The determinant $\det \left(\delta_{\mu_i,\mu_k} \right)$ is unity when all of the μ are different and is zero otherwise. Finally, setting $|z_1| = 0$, we obtain

$$
\int \frac{d^n\phi}{(2\pi)^n} \left(\det e^{z_i z_k^*} \right)_{|z_1|=0} = \sum_{\substack{\mu_2 \cdots \mu_n \\ (\neq)}}^{1,2,\ldots} \frac{|z_2|^{2\mu_2}}{\mu_2!} \cdots \frac{|z_n|^{2\mu_n}}{\mu_n!}
\tag{9.8.36}
$$

and thus for the hole distribution,

$$
H(s) = 1 + \sum_{\nu=1}^{N-1} \frac{(-1)^\nu}{\nu!} \sum_{\substack{\mu_2 \cdots \mu_{\nu+1} \\ (\neq)}}^{1,2,\ldots} \prod_{j}^{2 \ldots \nu+1} \int_0^{s^2} dx_j \frac{x_j^{\mu_j}}{\mu_j!} e^{-x_j} .
\tag{9.8.37}
$$

At this point, the limit $N \to \infty$ can be taken, whereupon $H(s)$ takes the form of an infinite product

$$
\begin{aligned}
H(s) &= \prod_{n=1}^{\infty} \left(1 - \int_{0}^{s^2} dx \, \frac{x^n}{n!} e^{-x} \right) \\
&= \prod_{n=1}^{\infty} \left(\int_{s^2}^{\infty} dx \, \frac{x^n}{n!} e^{-x} \right) \\
&= \prod_{n=1}^{\infty} \left(e_n(s^2) e^{-s^2} \right) .
\end{aligned}
\tag{9.8.38}
$$

Fortunately, this infinite product converges quite rapidly. Its Taylor expansion around $s = 0$ is immediately accessible,

$$
H(s) = 1 - \frac{s^4}{2} + \frac{s^6}{6} - \frac{s^8}{24} + \frac{11}{120} s^{10} - \frac{71}{720} s^{12} + \cdots .
\tag{9.8.39}
$$

Its asymptotic large-s behavior is most conveniently studied through the logarithm for which the infinite product turns into the sum

$$
\ln H(s) = \sum_{n=1}^{\infty} \ln \left[e_n(s^2) e^{-s^2} \right] .
\tag{9.8.40}
$$

It follows from the integral representation (9.8.18) that for $s^2 \gg 1$, the sum (9.8.40) draws important contributions only from the range $1 \le n \lesssim s^2$; the cutoff at $n = s^2$ is sharper, the larger s^2. On the other hand, the nth order polynomial $e_n(s^2)$ is dominated by its nth order monomial when $s^2 \gg 1$ such that $\ln H(s)$ becomes

$$
\ln H(s) \approx \sum_{n=1}^{s^2} \ln \left(\frac{s^{2n}}{n!} e^{-s^2} \right) .
\tag{9.8.41}
$$

Finally, by approximating the sum as an integral and using Stirling's formula, we obtain

$$
\ln H(s) \approx \int_{0}^{s^2} dn \ln \left[s^{2n} \frac{e^{-s^2}}{\Gamma(n+1)} \right] \approx -\frac{1}{4} s^4 .
\tag{9.8.42}
$$

This slightly cavalier argument can be corroborated with the help of the Euler–MacLaurin summation formula [340] which also yields the next largest term,

$$
\ln H(s) = -\tfrac{1}{4} s^4 - s^2 \left[\ln s + O(1) \right] .
\tag{9.8.43}
$$

It is quite remarkable that the fall-off of $H(s)$ at large spacings is more rapid than Gaussian, in contrast to the cases of Hermitian and unitary matrices treated in Chap. 4.

To compare with numerical data it is convenient to rescale the spacing variable s linearly so as to achieve unit mean spacing, $s = S\bar{s}$, where $\bar{s} = \int_{0}^{\infty} ds \, H(s) =$

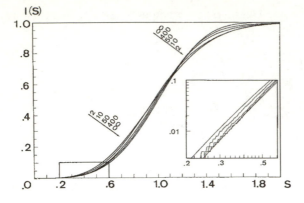

Fig. 9.9. Spacing staircase for general complex matrices of dimension 2 [see (9.8.27)], 10, 50, 400, ∞ [see (9.8.38)]. The insert reveals cubic repulsion

1.142929 .... . Figure 9.9 depicts the spacing staircase $I(S) = 1 - H(S)$ so obtained, together with that for $N = 2$ corresponding to (9.8.28). Also shown are numerically obtained spacing staircases for Ginibre's ensembles of random $N \times N$ matrices, with $N = 10, 50, 400$. The N dependence within the family of five staircases is rather more pronounced than for Hermitian and unitary matrices. This is due to a boundary effect. When eigenvalues closer to the confining circle than a mean spacing are omitted in numerically constructing the spacing staircases for random matrices, excellent agreement with the asymptotic ($N = \infty$) result is reached for $N > 20$.

The fine agreement of the asymptotic spacing staircase with that obtained by diagonalizing the dissipative quantum map for the classically chaotic top, shown in Fig. 9.7, arises without discarding any eigenvalues of the map. Boundary effects due to eigenvalues close to the inner or outer fringes of the annular cloud in Fig. 9.6b were suppressed by the unfolding procedure described in Sect. 9.6.

9.9 General Properties of Generators of Dissipative Dynamics

Generators of dissipative quantum dynamics are subject to restrictions that do not apply to members of Ginibre's matrix ensemble. Among these are stability and also conservation of Hermiticity, positivity, and normalization of the density matrix. Therefore, the faithfulness of quantum maps to Ginibre's ensemble with respect to the spacing distribution (and possibly other properties as well) needs explanation.

Probably of greatest interest is the conservation of Hermiticity,

$$(D\varrho)^\dagger = D\varrho \text{ for } \varrho = \varrho^\dagger , \tag{9.9.1}$$

since this entails an anti unitary symmetry of the generator D. For non-Hermitian matrices X, the identity (9.9.1) implies $(DX)^\dagger = DX^\dagger$ as is seen by decomposing as $X = X_1 + iX_2$ with $X_{1,2}$ Hermitian. With the help of the operator $\mathcal{A}$,

$$\mathcal{A}X = X^\dagger , \tag{9.9.2}$$

which turns any matrix into its Hermitian conjugate, the conservation of Hermiticity may be written as $\mathcal{A}DX = D\mathcal{A}X$, i.e.,

$$[\mathcal{A}, D] = 0 . \tag{9.9.3}$$

This property of the generator is quite appropriately called an antiunitary symmetry since $\mathcal{A}$ is antilinear by definition; moreover, with respect to the scalar product of matrices,

$$\langle\langle X|Y\rangle\rangle = \mathrm{Tr}\,\{X^\dagger Y\} , \tag{9.9.4}$$

which will be of importance in the remainder of this chapter, the operator $\mathcal{A}$ is in fact antiunitary, $\langle\langle \mathcal{A}X|\mathcal{A}Y\rangle\rangle = \langle\langle X|Y\rangle\rangle^* = \langle\langle Y|X\rangle\rangle$; finally, the definition (9.9.2) implies

$$\mathcal{A}^2 = 1 . \tag{9.9.5}$$

As was shown in Sect. 2.5, any matrix D with an antiunitary symmetry $\mathcal{A}$ obeying $\mathcal{A}^2 = 1$ can be given a real representation that can be constructed with the help of $\mathcal{A}$ from an arbitrary basis. It follows that it would be more appropriate to consider the generators D as members of the ensemble of real asymmetric matrices with Gaussian distributions for their elements rather than as members of Ginibre's ensemble of generically complex matrices.

Incidentally, the eigenvalues of generators (as of all real asymmetric matrices) are either real or come in complex-conjugate pairs. This is in fact obvious from (9.9.3) since with $DR = \lambda R$, one has $D\mathcal{A}R = \lambda^*\mathcal{A}R$. Of course, the annular clouds of eigenvalues of generators shown in Figs. 9.5 and 9.6 reflect the property in question: The clouds of eigenvalues are mirror symmetric about the real axis.

One would intuitively expect Ginibre's ensemble and the corresponding ensemble of real asymmetric matrices to have the same distributions of nearest-neighbor spacings in the limit of large dimension, provided that one disregards, in the case of real asymmetric matrices, eigenvalues closer to the real axis than a mean spacing, $|\mathrm{Im}\,\{z_i\}| \lesssim \bar{s}$. The latter exclusion shields the spacing distribution from the special correlations between real eigenvalues and also from those between eigenvalues that are complex conjugate to one another and simultaneously have a nearest-neighbor relationship. In fact, that exclusion was made in constructing the spacing staircases for the damped kicked top displayed in Fig. 9.7. It also underlies Fig. 9.10 which refers to the subensemble of Ginibre's ensemble constituted by real asymmetric matrices; each spacing staircase in Fig. 9.10 was obtained by numerically diagonalizing 100 such matrices; the expected cubic repulsion [i.e. $I(S) \sim S^4$] is clearly visible as is the agreement with the spacing staircases for Ginibre's ensemble shown in Fig. 9.9.

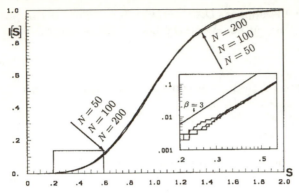

Fig. 9.10. Spacing staircase for general real matrices of dimension 50, 100, and 200. The insert reveals cubic repulsion; the straight line is a guide to the eye

Real matrices with Gaussian element densities have been studied by *Sommers* and *Lehmann* [341]. The long-standing problem of the statistics of their eigenvalues is now fully understood and in accord with the above qualitative reasoning. The technical difficulties involved prohibit a detailed presentation of the investigation of Sommers and Lehmann in this book. However, a simple proof of $I(S) \sim S^4$ for small S will be given in Sect. 9.11 below.

Having indicated why the restriction of Hermiticity conservation does not prevent generators of dissipative quantum maps (under conditions of fully developed classical chaos) from being faithful to Ginibre's ensemble, now we turn to the further restriction of stability. As already mentioned in Sect. 9.5, the eigenvalues z of generators D cannot exceed unity in modulus. After unfolding the spectrum to uniform mean density, the mean spacing between nearest neighbors is thus $\sim 1/\sqrt{N}$. Obviously then, the restriction in question does not distinguish generators D from members of Ginibre's ensemble except for a simple change of scale in the complex plane in which the eigenvalues are located.

Probability conservation,

$$\mathrm{Tr}\, D\varrho = \mathrm{Tr}\, \varrho = 1 \,,\tag{9.9.6}$$

cannot prevent generic spacing fluctuations either, as will be shown now. First, it is convenient to introduce a little new notation. Any basis of state vectors $|n\rangle$ with $n = 1, 2, \ldots, N$ provides a dyadic basis

$$|nm\rangle\rangle \equiv |n\rangle\langle m| \tag{9.9.7}$$

for representing observables X as

$$X = \sum_{nm} X_{nm}|nm\rangle\rangle \,. \tag{9.9.8}$$

With the help of the adjoint tetrad $\langle\langle kl|$, defined by

$$\langle\langle kl|ij\rangle\rangle = \mathrm{Tr}\left\{(|k\rangle\langle l|)^\dagger |i\rangle\langle j|\right\} = \delta_{ki}\delta_{lj} \,, \tag{9.9.9}$$

a generator D can be given the tetradic representation

$$D = \sum_{ijkl} |ij\rangle\rangle D_{ij,kl} \langle\langle kl| \tag{9.9.10}$$

which is consistent with (9.9.8) for both ϱ and $D\varrho$,

$$D\varrho = \sum_{ijkl} |i\rangle\langle j| D_{ij,kl} \varrho_{kl} . \tag{9.9.11}$$

Since D is not unitary with respect to the scalar product of observables (9.9.4), left and right eigen-matrices of D must be distinguished,

$$D|R_i\rangle\rangle = z_i|R_i\rangle\rangle \,\hat{=}\, DR_i = z_i R_i$$

$$\langle\langle L_i|D = z_i\langle\langle L_i| \Leftrightarrow D^\ddagger L_i = z_i^* L_i , \tag{9.9.12}$$

$$\langle\langle L_i|R_j\rangle\rangle = \delta_{ij} \text{ for } z_i \neq z_j .$$

Here $D^\ddagger$ is the adjoint of the generator D with respect to the scalar product of observables, $\langle\langle X|DY\rangle\rangle = \langle\langle D^\ddagger X|Y\rangle\rangle$.

There is one consequence of probability conservation for the spectrum of D worth mentioning,[1]

$$\begin{aligned} \operatorname{Tr} D\varrho &= \langle\langle \mathbf{1}|D\varrho\rangle\rangle = \langle\langle D^\ddagger \mathbf{1}|\varrho\rangle\rangle \\ &= \operatorname{Tr}\varrho = \langle\langle \mathbf{1}|\varrho\rangle\rangle \end{aligned} \tag{9.9.13}$$

which means that D has the left eigen-matrix $L_0 \equiv \mathbf{1}$ with eigenvalue unity,

$$\langle\langle \mathbf{1}|D = \langle\langle \mathbf{1}| \Leftrightarrow D^\ddagger|\mathbf{1}\rangle\rangle = |\mathbf{1}\rangle\rangle . \tag{9.9.14}$$

Therefore, a right eigen-matrix $R_0 \equiv \bar{\varrho}$ with eigenvalue unity, $D\bar{\varrho} = \bar{\varrho}$ must exist, too. In other words, probability conservation entails the existence of a stationary solution of the map. (Incidentally, this does not imply that all initial density matrices relax toward $\bar{\varrho}$ upon repeated iteration of the map.) It also follows, by the way, that all right eigen-matrices of D with eigenvalues different from unity are traceless,

$$\langle\langle L_0|R_i\rangle\rangle = \operatorname{Tr}\{R_i\} = 0 \text{ for } z_i \neq z_0 = 1 .$$

In a representation like (9.9.10, 11) probability conservation is expressed as

$$\sum_i D_{ii,kl} = \delta_{kl} . \tag{9.9.15}$$

This property is certainly not generic for Ginibre's ensemble of $N^2 \times N^2$ matrices. However, imagine a Ginibre ensemble of $(N^2 - 1) \times (N^2 - 1)$ matrices $m_{ij,kl}$. For any m the "row index" ij as well as the "column index" kl runs over $N^2 - 1$ values.

---

[1] For the sake of clarity, the unit matrix is denoted by $\mathbf{1}$ here.

Then, each m can be complemented to an $N^2 \times N^2$ matrix M by adding a "first row" with index $ij = 00$ and a "first column" with index $kl = 00$ as

$$\begin{pmatrix} M_{00,00} & M_{00,kl} \\ M_{ij,00} & m_{ij,kl} \end{pmatrix} = \begin{pmatrix} 1 & \delta_{kl} - \sum_i m_{ii,kl} \\ 0 & m_{ij,kl} \end{pmatrix} \tag{9.9.16}$$

which obeys (9.9.15). It is easy to see from the secular equation $\det(M - z\mathbf{1}) = 0$ that each matrix (9.9.16) has one eigenvalue unity while the $N^2 - 1$ remaining eigenvalues coincide with those of m. Thus, it is clear that, in the limit $N^2 \gg 1$, the ensemble of matrices (9.9.16) has the same spacing distribution for its eigenvalues as Ginibre's ensemble of matrices m. Similarly, probability conservation cannot influence the spacing distribution for generators of dissipative quantum maps.

It would be most appropriate to conclude this section by demonstrating the irrelevance of positivity conservation,

$$\varrho \geq 0 \Rightarrow D\varrho \geq 0 \tag{9.9.17}$$

for the spacing distribution. Unfortunately, however, I have not been able to find such a proof.

As a final remark I would like to add that all of the above considerations and conclusions carry over to generators l of infinitesimal time translations; probability conservation, for instance, takes the form $\mathrm{Tr}\, l\varrho = 0$ in place of (9.9.6).

9.10 Level Dynamics of Dissipative Systems

Our discussion of level dynamics in Chap. 6 was confined to Hermitian and unitary matrices. There are several good reasons to look for a generalization to the parametric motion of the complex eigenvalues of matrices not subject to either of these restrictions. For one, the generators of dissipative motion in master equations fall into that more general class, and one would certainly like to avail oneself of a systematic perturbation theory for them. The question also arises whether Ginibre's ensemble of random matrices can be interpreted as an equilibrium distribution of a generalized Pechukas–Yukawa gas, analogous to what we have seen for the Wigner–Dyson ensembles in Chap. 6. A fictitious gas with integrable classical Hamiltonian motion rendering such service has indeed been found [342, 343] and will be briefly described in the sequel. Before entering the detailed discussion, it is well to realize that the eigenvalues of an arbitrary matrix

$$X(\lambda) = X_0 \cos \lambda + Y_0 \sin \lambda \tag{9.10.1}$$

move about in a complex plane as the real control parameter λ is changed. Anticipating an interpretation of the eigenvalues as particle coordinates and of λ as a time, we must expect a two-dimensional rather than a one-dimensional Pechukas–Yukawa gas. A further complication is that an arbitrary matrix X cannot be diagonalized by a unitary transformation in contrast to a Hermitian or a unitary matrix. The best a unitary transformation can achieve is a triangular form where all elements below or above the principal diagonal are equal to zero; in either case the diagonal elements are the eigenvalues. Level dynamics must provide a larger set of differential equations than for Hermitian and unitary matrices since now the off-diagonal elements of the triangular matrix met have no counterparts in these previously treated simpler cases.

The treatment of general complex matrices requires a more powerful mathematical technique which it is best to explain for real symmetric matrices. The reader will appreciate the novel view of familiar material and then be able to follow the subsequent generalization without difficulty.

9.10.1 Hamiltonian Dynamics of Real Symmetric Matrices

Upon differentiating $X(\lambda)$ as given by (9.10.1) w. r. t. λ, we get the coupled evolution equations

$$\dot{X} = Y \qquad \dot{Y} = -X. \qquad (9.10.2)$$

If X and Y were single variables each rather than real symmetric matrices, these equations could be read as Hamilton's equations for a harmonic oscillator, where X is the coordinate and Y is the canonically conjugate momentum. In fact, such Hamiltonian interpretation remains valid for the matrix. The $N(N+1)/2$ independent elements X_{ij} with $i \leq j$ serve as coordinates and the Y_{ij} with $i \leq j$ as momenta; the Poisson brackets read

$$\begin{aligned}
\{Y_{ii}, X_{kk}\} &= \delta_{ik} \\
\{Y_{ij}, X_{kl}\} &= \frac{1}{2}\delta_{ik}\delta_{jl} \qquad \text{for} \qquad i < j, \ k < l, \\
\{Y_{ij}, Y_{kl}\} &= 0 = \{X_{ij}, X_{kl}\}.
\end{aligned} \qquad (9.10.3)$$

The evolution equations (9.10.2) result from the Hamiltonian

$$\mathcal{H} = \frac{1}{2}\sum_{i=1}^{N}(Y_{ii}^2 + X_{ii}^2) + \sum_{i<j}(Y_{ij}^2 + X_{ij}^2) = \frac{1}{2}\text{Tr}\,(Y^2 + X^2) \qquad (9.10.4)$$

through

$$\dot{X}_{ij} = \{\mathcal{H}, X_{ij}\}, \qquad \dot{Y}_{ij} = \{\mathcal{H}, Y_{ij}\}. \qquad (9.10.5)$$

The matrix dynamics thus introduced remains unchanged under the orthogonal transformations

$$X \to X' = O^{-1}XO, \qquad Y \to Y' = O^{-1}YO \qquad \text{with} \qquad O\tilde{O} = 1. \quad (9.10.6)$$

The invariance of the Hamiltonian function $\mathcal{H}$ is indeed obvious. A one-line calculation shows that the Poisson brackets do not change either, provided that the orthogonal matrix O is independent of X and Y. But we shall need to drop that latter restriction and admit nonlinear coordinate changes; the orthogonality of O

still makes the transformation canonical,[2] but it is *not* advisable to try checking this by direct calculation (unless you can draw on unlimited supplies of paper and time). Readers proficient in differential forms [344] will be able to verify the invariance of the Poisson brackets in a few lines. Perhaps most enjoyable is the following elementary proof: Letting ξ and η be real N-component vectors, we build the quadratic forms $x = \sum_{ij} \xi_i X_{ij} \xi_j$ and $y = \sum_{ij} \eta_i Y_{ij} \eta_j$ and find, with the help of (9.10.3), that their Poisson brackets read $\{y, x\} = (\sum_i \eta_i \xi_i)^2 = (\eta, \xi)^2$; these are equivalent to (9.10.3) since special choices for the vectors ξ, η reproduce the latter brackets. Using the same vectors to build $x' = \sum_{ij} \xi_i (\tilde{O}XO)_{ij} \xi_j$ and similarly y', we find $\{y', x'\} = (O\eta, O\xi)^2 = (\eta, \xi)^2 = \{y, x\}$ and conclude, irrespectively of whether or not O depends on X and Y, the invariance of the Poisson brackets $\{y, x\}$ under orthogonal transformations. Special choices of the vectors ξ, η immediately yield the same Poisson brackets for X' and Y' as for X and Y.

The symmetry so established comes with a constant of the motion,

$$\mu(X, Y) = [Y, X]. \tag{9.10.7}$$

The constancy of the commutator $[Y, X]$ is easily verified using (9.10.1), $[Y, X] = [Y_0, X_0](\cos^2 \lambda + \sin^2 \lambda) = [Y_0, X_0]$. The relation of the conservation law to the symmetry under (9.10.6) follows from Noether's theorem: With $a = -\tilde{a}$ a real antisymmetric matrix and ϵ an infinitesimal number, we can write an orthogonal matrix infinitesimally close to the unit matrix as $O = e^{\epsilon a}$. For a Hamiltonian function $\mathcal{H}$ to be invariant under such a transformation, it must satisfy

$$\mathrm{Tr} \left(\frac{\partial \mathcal{H}}{\partial X} [a, X] + \frac{\partial \mathcal{H}}{\partial Y} [a, Y] \right) = 0 \tag{9.10.8}$$

since $\delta X = \epsilon[a, X]$ and $\delta Y = \epsilon[a, Y]$ are the infinitesimal changes of X and Y brought about by the transformation O. A generating function $\mathcal{G}$ producing such changes as $\delta X = \epsilon\{\mathcal{G}, X\} = \epsilon \partial \mathcal{G}/\partial Y$, $\delta Y = \epsilon\{\mathcal{G}, Y\} = -\epsilon \partial \mathcal{G}/\partial X$ is easily found by integration,

$$\mathcal{G} = \mathrm{Tr}\, a[Y, X]. \tag{9.10.9}$$

The invariance of the Hamiltonian function $\mathcal{H}$ under the symmetry generated by $\mathcal{G}$ implies $\{\mathcal{H}, \mathcal{G}\} = 0$ and this in turn means conservation of $\mathcal{G}$. By a special choice for the matrix a, say $a_{ij} = -a_{ji} = \delta_{ik}\delta_{jl}$, we find $\mathcal{G}$ as $\mu_{kl} = [Y, X]_{kl}$, but this reasoning holds for all pairs of indices, and therefore, the whole matrix $\mu = [Y, X]$ is conserved.

We can exploit the invariance under orthogonal transformations to choose the special transformation that diagonalizes X. Subjecting the matrices Y and μ to

---

[2] Calling orthogonal transformations "canonical" in Chap. 2.5 may have appeared a somewhat arbitrary terminology; it is in line with established classical parlance here.

that same transformation,

$$
\begin{aligned}
X &\rightarrow Q \equiv \tilde{O}XO = \mathrm{diag}(q_1, q_2, \ldots q_N), \\
Y &\rightarrow P \equiv \tilde{O}YO, \\
\mu &\rightarrow l \equiv \tilde{O}\mu O = [P, Q].
\end{aligned}
\tag{9.10.10}
$$

From the point of view of classical Hamiltonian mechanics, we have thus performed a canonical transformation from the $N(N + 1)$ original variables to as many new ones which we choose as the N eigenvalues q_m of X, the N diagonal elements

$$
p_m \equiv P_{mm} = (\tilde{O}YO)_{mm},
\tag{9.10.11}
$$

of $\tilde{O}YO$, the $N(N-1)/2$ independent elements $l_{mn} = (\tilde{O}\mu O)_{mn}$ of the antisymmetric matrix $l = \tilde{O}\mu O$, and the $N(N-1)/2$ real parameters in the orthogonal matrix O. Our foregoing proof of the invariance of the Poisson brackets of X and Y under canonical transformations now carries over to the diagonal elements as $\{p_m, q_n\} = \delta_{mn}$ such that the p, q constitute N canonical pairs.

It remains to construct all other Poisson brackets for the set of new variables p, q, l, O. The mathematically respectable way of doing this uses exterior differential forms [342, 343]. Fortunately, for readers not willing to resort to such modernism there is a pretty extension of the elementary argument already invoked above to verify the invariance of the Poisson brackets of the matrices X, Y under orthogonal transformations. Employing four real N-component vectors α, β, ξ, η, we consider matrix elements of l like $(\alpha, l\beta) = (O\alpha, \mu O\beta)$. The Poisson brackets (9.10.3) of X and Y immediately yield

$$
2\{(\xi, l\eta), (\alpha, l\beta)\} = (\xi, \alpha)(\eta, l\beta) + (\xi, \beta)(\alpha, l\eta) + (\eta, \alpha)(\beta, l\xi) + (\eta, \beta)(\xi, l\alpha)
$$

and special choices for our vectors then give the brackets for the matrix elements,

$$
\{l_{mn}, l_{ij}\} = \frac{1}{2}\left(\delta_{in}l_{mj} + \delta_{im}l_{jn} + \delta_{jn}l_{im} + \delta_{jm}l_{ni}\right),
\tag{9.10.12}
$$

already used but not derived in Chap. 6; see (6.2.23). The reader will have no difficulty in extending this kind of reasoning and verifying $\{p_m, l_{ij}\} = 0 = \{q_m, l_{ij}\}$.

To complete the present fully algebraic derivation of level dynamics for real symmetric matrices of the form (9.10.1), we just need to exploit the invariance of the Hamiltonian function $\mathcal{H}$ under arbitrary orthogonal transformations (9.10.6). With the help of the definitions (9.10.10, 11), we express $\mathcal{H}$ in terms of the new variables q, p, l and find

$$
\mathcal{H} = \frac{1}{2}\sum_m (p_m^2 + q_m^2) + \frac{1}{2}\sum_{m \neq n} \frac{l_{mn}^2}{(q_m - q_n)^2},
\tag{9.10.13}
$$

our old friend (6.10.13). We see that we need not bother finding the Poisson brackets involving the diagonalizing orthogonal matrix O since both the Poisson algebra and the Hamiltonian flow are closed for the q, p, l.

It is quite remarkable that the originally $N(N+1)$-dimensional Hamiltonian flow for the matrices X, Y has a closed subdynamics of $2N + N(N-1)/2$ dimensions which is still Hamiltonian. Clearly, we are facing a consequence of the symmetry (9.10.6) that is somewhat reminiscent of and indeed deeply related to the well-known Hamiltonian behavior of the radial coordinate in the Kepler problem.

Needless to say, the algebraic method of finding the Poisson brackets of the angular momenta l_{mn} can be extended to the unitary and symplectic symmetry classes without difficulty; see Problem 9.19.

9.10.2 Hamiltonian Dynamics of General Complex Matrices

The extension of the foregoing subsection to general $N \times N$ matrices of the form (9.10.1) is quite straightforward. The flow (9.10.2) still holds but takes place in a $2N^2$-dimensional complex phase space. Of course, we can write that flow as a real one by considering the pair (X, Y) of complex matrices as a quadruple of real ones through $X = X^{(1)} + iX^{(2)}, Y = Y^{(1)} + iY^{(2)}$. Then, the phase space is a $4N^2$-dimensional real manifold.

To reveal the Hamiltonian character of the flow under consideration, we introduce Poisson brackets to let the elements of X and Y form canonically conjugate pairs,

$$\{Y_{ij}^{(a)}, X_{kl}^{(b)}\} = \delta_{ab}\delta_{ik}\delta_{jl}, \qquad \{Y_{ij}^{(a)}, Y_{kl}^{(b)}\} = 0 = \{X_{ij}^{(a)}, X_{kl}^{(b)}\}. \qquad (9.10.14)$$

The Hamiltonian function

$$\mathcal{H} = \frac{1}{2}\operatorname{Tr} Y^\dagger Y + \frac{1}{2}\operatorname{Tr} X^\dagger X \qquad (9.10.15)$$

is real and nonnegative and yields (9.10.1) as Hamilton's equations. A convenient shortcut for these and most of the following calculation consists in working with complex matrix elements throughout and using the Poisson brackets

$$\{Y_{ij}, X_{kl}^*\} = 2\delta_{ik}\delta_{jl}, \qquad (9.10.16)$$
$$\{Y, X\} = \{Y, Y\} = \{Y, Y^*\} = \{X, X\} = \{X, X^*\} = 0.$$

In fact, the reasoning from here on parallels that of the previous subsection and therefore need not be written out again in much detail.

The Hamiltonian matrix dynamics thus introduced is invariant under the unitary transformations

$$X \to U^\dagger X U, \quad Y \to U^\dagger Y U, \quad U^\dagger U = 1. \qquad (9.10.17)$$

That symmetry implies the conservation of the anti-Hermitian matrix

$$\mu = \tfrac{1}{2}\left([Y, X^\dagger] + [Y^\dagger, X]\right) \qquad (9.10.18)$$

and can be exploited to introduce new phase-space coordinates to include the eigenvalues of X as a subset and thus exhibit level dynamics. Still denoting the tridiagonalizing matrix by U, we write

$$
\begin{aligned}
X &\to Q \equiv U^\dagger X U, \qquad Q_{mn} = 0 \quad \text{for} \quad m > n, \\
Y &\to P \equiv U^\dagger Y U, \\
\mu &\to l \equiv U^\dagger \mu U = \tfrac{1}{2}\left([P, Q^\dagger] + [P^\dagger, Q]\right).
\end{aligned}
\tag{9.10.19}
$$

The matrix Q is of upper triangular form and its diagonal elements $Q_{mm} \equiv q_m$ are the eigenvalues of X.

It is important to realize that the unitary matrix U is not uniquely determined by requiring that it tridiagonalize X; given any such U, a manifold of others is obtained as UD with D an arbitrary diagonal unitary matrix. To uniquely specify the triangularizing matrix, we may therefore impose N suitable constraints, so as to be left with $N^2 - N$ real parameters in U to achieve upper triangular form for $Q = U^\dagger X U$. For instance, we could employ the $N(N-1)$ nondiagonal Hermitian generators T_a of the group $U(N)$ and uniquely parametrize $U = \exp[\mathrm{i} \sum_{a=1}^{N(N-1)} x_a T_a]$ in terms of the $N(N-1)$ real coefficients x_a.

Now, we proceed to reparametrize the manifold of pairs of $N \times N$ matrices Y, X. As new phase-space coordinates, we introduce (1) the elements of Q that make for $N(N+1)$ real variables $\mathrm{Re}\, Q_{mn}$ and $\mathrm{Im}\, Q_{mn}$ with $m \leq n$, (2) similarly $N(N+1)$ real variables $\mathrm{Re}\, P_{mn}$ and $\mathrm{Im}\, P_{mn}$ with $m \leq n$, (3) $N(N-1)$ real variables $\mathrm{Re}\, l_{mn}$ and $\mathrm{Im}\, l_{mn}$ with $m < n$ (recall that $l_{mn} = -l_{nm}^*$), and the aforementioned $N(N-1)$ real variables parametrizing the triangularizing matrix U. As in the original set of coordinates, we encounter $4N^2$ variables. Their Poisson brackets can be found along the lines traced out above. Of course, for the elements of Q and those in the upper triangle of P, the brackets are those given in (9.10.14) since $X \to Q, Y \to P$ is again a canonical transformation. The l_{mn}, on the other hand, obey the Poisson brackets (6.2.25) of generators of the Lie algebra $u(N)$.

The transformation to the new set is generically nonsingular. To fully implement it, we still need to eliminate the P_{mn} with $m > n$ in favor of the l_{mn} with the help of the definitions (9.10.19). Writing out l_{mn} with $m < n$, we get

$$
\begin{aligned}
2l_{mn} &= \sum_{k \geq n} P_{mk} Q_{nk}^* - \sum_{k \leq m} P_{kn} Q_{km}^* + \sum_{k \leq n} P_{km}^* Q_{kn} - \sum_{k \geq m} P_{nk}^* Q_{mk} \\
&= \sum_{k \geq n} \left(P_{mk} Q_{nk}^* - P_{nk}^* Q_{mk}\right) + \sum_{k \leq m} \left(P_{km}^* Q_{kn} - P_{kn} Q_{km}^*\right) \\
&\quad + \sum_{n \geq k > m} P_{km}^* Q_{kn} - \sum_{n > k \geq m} P_{nk}^* Q_{mk}.
\end{aligned}
\tag{9.10.20}
$$

These equations may be solved for the P_{mn}^* with $m > n$ in terms of the l_{ij} with $i < j$ and the P_{ij} with $i \leq j$.

It is obviously difficult to express the Hamiltonian function $\mathcal{H}$ explicitly in terms of the new dynamic variables. In particular, there are many-body interactions rather than only the pair interactions we met in the level dynamics of Hermitian and unitary matrices. However c omplicated $\mathcal{H}$ may look for arbitrary complex matrices, it does enjoy the property of being integrable. Integrability may be seen as due to the fact that solving level dynamics is equivalent to diagonalizing the finite matrix X. Rather than trying to establish an explicit form, we content ourselves with naming the variables on which it depends: These are Q, the upper triangular part $P^{(\mathrm{up})}$ of P composed of the P_{mn} with $m \leq n$, and the off-diagonal part $l^{(\mathrm{off})}$ of the matrix l. We shall write the Hamiltonian function as $\mathcal{H}(Q, P^{(\mathrm{up})}, l^{(\mathrm{off})})$. We must imagine it to be determined from (9.10.15) or

$$\mathcal{H}(Q, P^{(\mathrm{up})}, l^{(\mathrm{off})}) = \frac{1}{2}\mathrm{Tr}\, P^\dagger P + \frac{1}{2}\mathrm{Tr}\, Q^\dagger Q \tag{9.10.21}$$

wherein the elements P_{mn} with $m > n$ must be expressed in terms of the elements of $P^{(\mathrm{up})}$, $l^{(\mathrm{off})}$.

We shall presently need the Jacobian J for the transformation from the variables $(\mathrm{Re}\, P_{mn}, \mathrm{Im}\, P_{mn}, m > n)$ to the variables $(\mathrm{Re}\, l_{mn}, \mathrm{Im}\, l_{mn}, m \neq n)$ or, what is equivalent up to a constant factor, for the transformation from $(P_{mn}, P_{mn}^*, m > n)$ to $(l_{mn}, l_{mn}^*, m > n)$. To construct J, we rewrite (9.10.20) as

$$\begin{aligned}
2l_{mn} &= \sum_{n \geq k > m} P_{km}^* Q_{kn} - \sum_{n > k \geq m} P_{nk}^* Q_{mk} + \ldots \\
&= (Q_{nn} - Q_{mm})P_{nm}^* + \sum_{n > k > m} \left(P_{km}^* Q_{kn} - P_{nk}^* Q_{mk}\right) + \ldots \tag{9.10.22}
\end{aligned}$$

where the dots refer to terms independent of P_{ij}, P_{ij}^* with $i > j$ and thus incapable of contributing to the Jacobian under construction. Obviously, then, the l_{mn} with $m < n$ do not depend on the P_{ij} with $i > j$ but only on their conjugates P_{ij}^*. Hence, the modulus $|J|$ of J equals the squared modulus of the Jacobian J_1 of the transformation (9.10.22). Now, for $n = m + 1$, there is only a single term on the r. h. s. of (9.10.22) that contributes the factor $Q_{m+1,m+1} - Q_{mm}$ to J_1. We can expand J_1 w. r. t. the rows containing only one element of that form. In the reduced determinant, again there are rows containing only one element with $n = m + 2$. Having picked up the factor $Q_{m+2,m+2} - Q_{mm}$ we continue the expansion and end up with $J_1 = \prod_{i<j}(Q_{ii} - Q_{jj})$ and thus

$$J = \prod_{i > j} |Q_{ii} - Q_{jj}|^2 \,. \tag{9.10.23}$$

We should appreciate that the Jacobian J depends only on the eigenvalues $Q_{ii} = q_i$ of the matrix X.

9.10.3 Ginibre's Ensemble

Somewhat frivolously disregarding all constants of the motion of our level dynamics other than the Hamiltonian function $\mathcal{H}$ itself, we write the canonical phase-space distribution for the many-body system as

$$P(Q, P^{(\mathrm{up})}, l^{(\mathrm{off})}) \propto \exp[-\mathcal{H}(Q, P^{(\mathrm{up})}, l^{(\mathrm{off})})] \tag{9.10.24}$$

$$\propto \exp\left(-\frac{1}{2}\sum_m |q_m|^2\right) \exp\left(-\frac{1}{2}\sum_{m<n} |Q_{mn}|^2\right) \exp\left(-\frac{1}{2}\mathrm{Tr}\, P^\dagger P\right).$$

Even though level dynamics does not explore all of the energy surface $\mathcal{H}(Q, P^{(\mathrm{up})}, l^{(\mathrm{off})}) = \mathcal{E}$, our experience with Hermitian and unitary matrices suggests trying to make do without the other constants of the motion. Of greatest interest is the reduced distribution of the eigenvalues $q_m = Q_{mm}$. This is accessible by integrating out all other variables,

$$P(q) \propto \int P(Q, P^{(\mathrm{up})}, l^{(\mathrm{off})}) \prod_{i<j} d^2 Q_{ij} d^2 P_{ij} d^2 l_{ij} \prod_i d^2 P_{ii} \, .$$

The integrals over the Q_{ij} are Gaussian and just contribute a constant factor. The remaining integrals are also reduced to Gaussian ones over $\exp\{-\frac{1}{2}\mathrm{Tr}P^\dagger P\}$; the Jacobian for that transformation was provided with foresight in (9.10.23), and it yields

$$P(q) \propto \exp\left(-\tfrac{1}{2}\sum_i |q_i|^2\right) \prod_{i>j} |q_i - q_j|^2 \, . \tag{9.10.25}$$

But this is indeed the distribution (9.8.1) known from Ginibre's ensemble.

As we have seen in the preceding section, Ginibre's ensemble admits matrices which could not qualify as models of generators of dissipative dynamics. Nevertheless, the predictions for the spectral fluctuations are often quite well respected in dynamical systems.

9.11 Resistance of Cubic Repulsion to Antiunitary Symmetries

Conservation of Hermiticity of the density matrix was revealed in Sect. 9.9 as an antiunitary symmetry $\mathcal{A}$ of the generator D of dissipative quantum maps

$$[\mathcal{A}, D] = 0 \, , \quad \mathcal{A}^2 = +1 \, . \tag{9.11.1}$$

Thus, the eigenvalues z_i of D are either real or come in complex-conjugate pairs. I propose to show that cubic repulsion is effective in close encounters of two complex eigenvalues z_1 and z_2 with $z_1 \neq z_2^*$. The argument [345] involves nearly degenerate perturbation theory and runs partially parallel to the investigations presented in Chap. 4 for Hermitian and unitary matrices.

Imagine a close encounter of z_i and z_j as a parameter in D is varied. For a particular value of the parameter in question, the corresponding eigenvectors[3] of D may be assumed known. By diagonalizing D in that two-dimensional subspace, one can follow the fate of the two eigenvalues throughout the encounter. The approximate behavior found for the two eigenvalues will be faithful to their exact behavior, provided that the spacing $|z_1 - z_2|$ is smaller than the distance between either z_1 or z_2 and any other eigenvalue of D.

---

[3] For the purpose of this section, the tetradic nature of D is irrelevant; to simplify words and formulas, I shall speak of eigenvectors rather than eigen-matrices and write $|\,\rangle$ rather than $|\,\rangle\rangle$ for the basis kets.

The difference between the two eigenvalues is

$$D_+ - D_- = \left[(D_{11} - D_{22})^2 + 4D_{12}D_{21}\right]^{1/2} .$$ (9.11.2)

When the spectrum of D displays many close encounters, the density of spacings becomes accessible,

$$P(s) = \langle \delta(s - |D_+ - D_-|) \rangle ,$$ (9.11.3)

where the average $\langle \dots \rangle$ is over all encounters or, formally speaking, over the matrix elements D_{ij} with a suitable distribution $W(\{D_{ij}\})$. Inasmuch as we are interested only in the behavior of $P(s)$ for $s \to 0$, it suffices to specify the distribution $W(\{D_{ij}\})$ for $D_{ij} \to 0$. The crucial requirement ensuring cubic repulsion will turn out to be that W remains finite ($\neq 0, \infty$), as all or some of the four D_{ij} approach zero. For the sake of simplicity we shall assume independent distributions for all matrix elements. It is important to realize that the D_{ij} will in general be complex; this is not in conflict with the statement that D can, due to the symmetry (9.11.1), be given a real representation since the special representation underlying the present perturbative argument is based on eigenvectors of D for a particular value of the parameter controlling the close encounter; eigenvectors of D pertaining to complex eigenvalues are complex. Indeed, using a basis in which D is real, if $DR = zR$, then $DR^* = z^*R^*$; by assumption, $z \neq z^*$, and thus $R \neq R^*$. At any rate, the average in (9.11.3) takes the form

$$\langle (\dots) \rangle = \int \prod_{ij=11,12,21,22} d^2 D_{ij} W_{ij}(D_{ij})(\dots) .$$ (9.11.4)

In contrast to the situations encountered for Hermitian and unitary matrices in Chap. 4, the discriminant in (9.11.2) is *not* a sum of nonnegative terms. To find the limiting form of $P(s)$ for $s \to 0$, the integrations in (9.11.4) must actually be carried out explicitly. That elementary task involves the following four complex integrals:

$$\int d^2x d^2y W(x)W(y)\delta^2(z - xy) \propto \ln |z|$$
$$\int d^2x W(x)\delta^2(z - x^2) \propto \tfrac{1}{|z|}$$
$$\int d^2x d^2y |x|^{-1} \ln |y|\delta^2(z - x - y) \propto \text{const} \neq 0$$
$$\int d^2x W(x)\delta (s - |\sqrt{x}|) \propto s^3 .$$ (9.11.5)

These auxiliary integrals are all meant in the limit of small s and small $|z|$, and the element distributions $W(x)$ are assumed to obey $W(x) \to \text{const}(\neq 0, \infty)$ for $x \to 0$. It is not difficult to establish from (9.11.2–9.11.5) that

$$P(s) \propto s^3 \quad \text{for} \quad s \to 0 .$$ (9.11.6)

A variant of the antiunitary symmetry (9.11.1),

$$AD = D^\dagger A , \quad A^2 = 1 ,$$ (9.11.7)

deserves attention, even though no physical realization is known at present. The reader will notice the formal analogy of unitary Floquet operators with time-reversal covariance. As a consequence of (9.11.7), the right and left eigenvectors of D pertaining to the same eigenvalue z are related by $L = \mathcal{A}R$ since $DR = zR$ implies $D^\dagger|\mathcal{A}R\rangle = z^*|\mathcal{A}R\rangle$ and its adjoint $\langle \mathcal{A}R|D = z\langle \mathcal{A}R|$. Thus, the 2×2 matrix of nearly degenerate perturbation theory takes the form

$$\begin{pmatrix} D_{11} & D_{12} \\ D_{21} & D_{22} \end{pmatrix} = \begin{pmatrix} \langle \mathcal{A}1|D|1\rangle & \langle \mathcal{A}1|D|2\rangle \\ \langle \mathcal{A}2|D|1\rangle & \langle \mathcal{A}2|D|2\rangle \end{pmatrix} . \tag{9.11.8}$$

Moreover, this matrix is complex symmetric since

$$\langle \mathcal{A}2|D|1\rangle = \langle 2|D^\dagger|\mathcal{A}1\rangle^* = \langle \mathcal{A}1|D|2\rangle , \tag{9.11.9}$$

where we have invoked $\mathcal{A}^2 = 1$ and the antiunitarity of $\mathcal{A}$. Taking $D_{21} = D_{12}$ in (9.11.2) and using (9.11.5) together with the further auxiliary integral

$$\int d^2x \frac{1}{|z|} \frac{1}{|z-x|} \propto -\ln|z| , \tag{9.11.10}$$

one immediately finds that

$$P(s) \propto -s^3 \ln s \quad \text{for} \quad s \to 0 . \tag{9.11.11}$$

Because of the weakness of the logarithmic singularity, (9.11.11) may still be referred to as cubic repulsion. In fact, the variant (9.11.7) of the antiunitary symmetry (9.11.1) is interesting mainly because it further illustrates the robustness of cubic level repulsion for matrices not restricted by either Hermiticity or unitarity.

One can show similarly that cubic repulsion between complex eigenvalues also prevails in the presence of antiunitary symmetries squaring to minus unity [346].

9.12 Irrelevance of Microreversibility to Cubic Repulsion

Having ascertained that matrices unrestricted by Hermiticity or unitarity tend to display cubic level repulsion irrespective of whether they are real asymmetric, complex symmetric, or general complex, now, I embark upon a final attempt to probe the universality of that cubic degree. The idea to be followed has as its goal the Hamiltonian embedding S+R of the dissipative system discussed in Sect. 9.2. The Hamiltonian H of S+R may or may not have an antiunitary symmetry, and the corresponding (generalized) time-reversal operator T may obey $T^2 = +1$ or $T^2 = -1$. Then, the (real!) energies or quasi-energies in general repel each other with the appropriate degree, as discussed in Chap. 3.

At issue now is the fate of this distinction between the three universality classes of unitary and Hermitian matrices when a damping is switched on, i.e., when the unitarity or Hermiticity is destroyed. The matter is quickly settled for dampings so weak that the damping terms $\text{Re}\{\ln z\}$ of the eigenvalues z of generators D are smaller than the mean nearestneighbor spacings of the zero-damping quasi-energies.[4] In that case, the $\text{Im}\{\ln z\}$ can be associated with differences of zero-damping quasi-energies, $\text{Im}\{\ln z\} = -\phi_\mu + \phi_\nu$, and the original quasi-energies ϕ_μ can be identified. As shown in Sect. 9.4, the $\Lambda_{\mu\nu} = \text{Re}\{\ln z\}$ are, under conditions of global classical chaos, independent of μ and ν and may therefore be interpreted as twice the width of any "level" ϕ_μ. Then, the zero-damping degree of level repulsion must still prevail.

The strong-damping regime [345, 346] remains to be discussed. It is well known [315, 347] that time-reversal invariance for the Hamiltonian embedding S + R, at least for temporally homogeneous systems, entails detailed balance for the reduced dynamics of S. Now, we shall proceed to establish detailed balance for the generators D of dissipative quantum maps on the basis of time-reversal invariance (in this context often referred to as to microreversibility) for the Floquet operator of S + R. It will become clear that the presence or absence of microreversibility, i.e., of detailed balance, is of no consequence for the degree of repulsion of the eigenvalues z of D.

The starting point is the unitary stroboscopic map for the density operator W of S + R,

$$W(t+1) = FW(t)F^\dagger , \tag{9.12.1}$$

which is assumed to have a physically relevant stationary solution $\bar{W} = F\bar{W}F^\dagger$. Time-reversal invariance amounts to

$$TFT^{-1} = F^\dagger , \quad [T, \bar{W}] = 0 . \tag{9.12.2}$$

Following conventional arguments [348, 349] we consider the stationary correlation function of two, not necessarily Hermitian observables A and B

$$\langle A(t)B \rangle = \text{Tr}\left\{ (F^\dagger)^t A F^t B \bar{W} \right\} \tag{9.12.3}$$

for which the symmetry (9.12.2) implies the identity

$$\langle A(t)B \rangle = \langle \hat{B}(t)\hat{A} \rangle , \tag{9.12.4}$$
$$\hat{A} = TA^\dagger T^{-1} \equiv \tau A .$$

It is to be noted that the operation τ on observables is linear rather than anti-linear; moreover, it is an operation that obviously squares to unity, $\tau^2 = 1$. To prove the identity (9.12.4), first we employ the antiunitarity of T to secure $(TXT^{-1})^\dagger = TX^\dagger T^{-1}$ and $\text{Tr}\{TX^\dagger T^{-1}\} = \text{Tr}\{X\}$ for any observable X. It

---

[4] The subsequent discussion refers to discrete-time maps but would proceed in complete analogy for generators of infinitesimal time translations of autonomous systems.

follows that

$$
\begin{aligned}
\langle A(t)B\rangle &= \mathrm{Tr}\left\{T\bar{W}B^{\dagger}F^{\dagger t}A^{\dagger}F^{t}T^{-1}\right\} \\
&= \mathrm{Tr}\left\{TB^{\dagger}F^{\dagger t}A^{\dagger}F^{t}T^{-1}\bar{W}\right\} \\
&= \mathrm{Tr}\left\{\hat{B}TF^{\dagger t}T^{-1}\hat{A}TF^{t}T^{-1}\bar{W}\right\} \\
&= \mathrm{Tr}\left\{\hat{B}F^{t}\hat{A}F^{\dagger t}\bar{W}\right\} \\
&= \mathrm{Tr}\left\{F^{\dagger t}\hat{B}F^{t}\hat{A}\bar{W}\right\} = \langle\hat{B}(t)\hat{A}\rangle \ .
\end{aligned}
\tag{9.12.5}
$$

The correlation functions in (9.12.4) will be needed when A and B are observables of S alone, i.e., behave like unity with respect to the heat bath R. Then, both correlation functions in (9.12.4) can be calculated using the subdynamics of S, which, in the situation of interest, is described by a dissipative map for the reduced density operator $\varrho(t) = \mathrm{Tr}_{\mathrm{R}}\{W(t)\}$ of S,

$$
\varrho(t+1) = D\varrho(t) \ .
\tag{9.12.6}
$$

With the help of the generator D and its adjoint $D^{\ddagger}$, the left-hand correlation function in (9.12.4) may be rewritten as

$$
\begin{aligned}
\langle A(t)B\rangle &= \mathrm{Tr}_{\mathrm{S}}\left\{A\,D^{t}(B\bar{\varrho})\right\} \\
&= \mathrm{Tr}_{\mathrm{S}}\left\{B\bar{\varrho}(D^{\ddagger})^{t}A\right\} \ .
\end{aligned}
\tag{9.12.7}
$$

To express the right-hand correlation function in similarly, (9.12.4) a reduced version T_{S} of the antiunitary operator T must first be defined. For S + R, the operator T can be written as the product of the complex conjugation K and two unitary matrices U_{S} and U_{R} which refer to S and R, respectively:

$$
T = U_{\mathrm{R}}U_{\mathrm{S}}K \equiv U_{\mathrm{R}}T_{\mathrm{S}}
\tag{9.12.8}
$$

Since A is assumed to act like unity with respect to R, the observable $\hat{A}$ defined in (9.12.4) takes the form

$$
\begin{aligned}
\hat{A} &= \tau A = U_{\mathrm{R}}U_{\mathrm{S}}KA^{\dagger}KU_{\mathrm{S}}^{\dagger}U_{\mathrm{R}}^{\dagger} \\
&= U_{\mathrm{S}}KA^{\dagger}KU_{\mathrm{S}}^{\dagger} = T_{\mathrm{S}}A^{\dagger}T_{\mathrm{S}}^{-1} \equiv \tau_{\mathrm{S}}A \ ,
\end{aligned}
\tag{9.12.9}
$$

which indeed involves the reduced time-reversal operation T_{S}. Of course, T_{S} is again antiunitary, and the associated operation τ_{S} is linear. A corollary to this property, to be employed presently for arbitrary observables X and Y of S, is

$$
\mathrm{Tr}_{\mathrm{S}}\{\hat{X}Y\} = \mathrm{Tr}_{\mathrm{S}}\{X\hat{Y}\} \Leftrightarrow \langle\langle\hat{X}|Y\rangle\rangle = \langle\langle X|\hat{Y}\rangle\rangle \ .
\tag{9.12.10}
$$

Thus equipped, we rewrite the right-hand correlation function in (9.12.4) without further reference to the reservoir R,

$$
\begin{aligned}
\langle\hat{B}(t)\hat{A}\rangle &= \mathrm{Tr}_{\mathrm{S}}\{\hat{B}D^{t}\hat{A}\bar{\varrho}\} = \mathrm{Tr}_{\mathrm{S}}\left\{B\tau_{\mathrm{S}}(D^{t}\hat{A}\bar{\varrho})\right\} \\
&= \mathrm{Tr}_{\mathrm{S}}\left\{B\tau_{\mathrm{S}}\left(D^{t}\tau_{\mathrm{S}}(\varrho A)\right)\right\} \ .
\end{aligned}
\tag{9.12.11}
$$

Since A and B are arbitrary observables of S, the equality of $\langle A(t)B\rangle$ and $\langle \hat{B}(t)\hat{A}\rangle$ implies, comparing (9.12.7) and (9.12.11), the operator identity

$$\bar{\varrho}D^{\ddagger} = \tau_{\mathrm{S}}D\tau_{\mathrm{S}}\bar{\varrho} \ . \tag{9.12.12}$$

The relation (9.12.12) is identical in appearance to the condition of detailed balance for the generator l of infinitesimal time translations for temporally homogeneous situations [348, 349],

$$\bar{\varrho}l^{\ddagger} = \tau_{\mathrm{S}}l\tau_{\mathrm{S}}\bar{\varrho} \ . \tag{9.12.13}$$

In fact, (9.12.12) may be looked upon as a special case of (9.12.13): writing $D = \exp{(lt)}$ and letting t vary continuously, differentiating with respect to t and finally setting $t = 0$, one obtains (9.12.13) from (9.12.12).

Other specializations bring (9.12.12 or 9.12.13) into a more familiar form. For instance, if there is a representation in which T equals the complex conjugation, the identity (9.12.12) reads

$$\sum_m D_{ij\,mk}\bar{\varrho}_{ml} = \sum_m D^*_{lk\,mi}\bar{\varrho}_{jm} = \sum_m D_{kl\,im}\bar{\varrho}_{jm} \ ; \tag{9.12.14}$$

to obtain the right-most expression in (9.12.14) the Hermiticity conservation (9.9.1) must be invoked. Moreover, if ϱ remains diagonal in that representation under iterations of the map D, i.e., if $D_{ij\,kk} = D_{ii\,kk}\delta_{ij}$ with $D_{ii\,kk}$ as the transition rate from state k to state i, the well-known condition for detailed balance in Pauli-type rate equations results,

$$D_{ii\,kk}\varrho_{kk} = D_{kk\,ii}\varrho_{ii} \ . \tag{9.12.15}$$

Hermiticity conservation and detailed balance imply that any right-hand eigen-matrix R of D is accompanied by a left-hand eigen-matrix L pertaining to the same eigenvalue,

$$L = c(\tau\bar{\varrho})^{-1}R^{\dagger} \tag{9.12.16}$$

where a real normalization factor c is such that $\langle\langle L|R\rangle\rangle = 1$. To simplify the notation, the index S on τ has been dropped, and it is assumed that the stationary density matrix has an inverse. The proof of (9.12.16) is easily accomplished with the help of the definitions in (9.9.12).

The 2×2 matrix describing a close encounter of two eigenvalues of D may be written as

$$D_{ij} = \langle\langle L_i|DR_j\rangle\rangle \ , \tag{9.12.17}$$

where the R_i and L_i are two pairs of left- and right-hand eigen-matrices of a generator $D^{(0)}$ infinitesimally close to D with respect to some control parameter of the dissipative dynamics. We assume that detailed balance holds for both D and $D^{(0)}$; it is important to realize, though, that the condition (9.12.12) involves the, generally different, stationary eigen-matrices $\bar{\varrho}$ and $\bar{\varrho}^{(0)}$ of D and $D^{(0)}$. The diagonal elements D_{11} and D_{22} in (9.12.17) cannot be expected to be related to one another by detailed balance, since even in the zero-damping limit they are left independent by time-reversal invariance. Less obvious is the fact that the symmetry $D_{12} = D_{21}$, imposed by time-reversal invariance in the zero-damping limit (see below), breaks down for finite damping. But indeed, $D_{12} = \langle\langle L_1 | D R_2 \rangle\rangle$ may be rewritten by using (9.12.16) to express L_1 and R_2 in terms of $\bar{\varrho}^{(0)}$ and, respectively, R_1 and L_2,

$$
\begin{aligned}
\langle\langle L_1 | D | R_2 \rangle\rangle &= \left\langle\left\langle \bar{\varrho}^{(0)-1} \tau R_1^\dagger \Big| D(\tau \bar{\varrho}^{(0)} L_2)^\dagger \right\rangle\right\rangle \frac{c_1}{c_2} \\
&= \left\langle\left\langle D\tau \bar{\varrho}^{(0)} L_2 \Big| (\bar{\varrho}^{(0)-1} \tau R_1^\dagger)^\dagger \right\rangle\right\rangle \frac{c_1}{c_2} \\
&= \left\langle\left\langle \tau \bar{\varrho}^{(0)} L_2 \Big| D^\ddagger \tau \bar{\varrho}^{(0)-1} R_1 \right\rangle\right\rangle \frac{c_1}{c_2}
\end{aligned}
\tag{9.12.18}
$$

where we have used $(\bar{\varrho}^{(0)-1} \tau R_1^\dagger)^\dagger = (\tau R_1^\dagger)^\dagger \bar{\varrho}^{(0)-1} = (\tau R_1) \bar{\varrho}^{(0)-1} = \tau(\bar{\varrho}^{(0)-1} R_1)$ together with the definition of the adjoint generator $D^\ddagger$. After invoking detailed balance according to (9.12.12), the chain of equations (9.12.18) continues as

$$
\begin{aligned}
\langle\langle L_1 | D | R_2 \rangle\rangle &= \left\langle\left\langle \tau \bar{\varrho}^{(0)} L_2 \Big| \bar{\varrho}^{-1} \tau D \tau \bar{\varrho} \tau \bar{\varrho}^{(0)-1} R_1 \right\rangle\right\rangle \frac{c_1}{c_2} \\
&= \left\langle\left\langle \tau \bar{\varrho}^{(0)} L_2 \Big| \bar{\varrho}^{-1} \tau D \bar{\varrho}^{(0)-1} R_1 \bar{\varrho} \right\rangle\right\rangle \frac{c_1}{c_2} .
\end{aligned}
\tag{9.12.18a}
$$

With the help of (9.12.10), the operation τ in the above bra $\langle\langle \dots |$ can be shifted to the ket $| \dots \rangle\rangle$,

$$
\begin{aligned}
\langle\langle L_1 | D | R_2 \rangle\rangle &= \left\langle\left\langle \bar{\varrho}^{(0)} L_2 \Big| \tau \bar{\varrho}^{-1} \tau D \bar{\varrho}^{(0)-1} R_1 \bar{\varrho} \right\rangle\right\rangle \frac{c_1}{c_2} \\
&= \left\langle\left\langle \bar{\varrho}^{(0)} L_2 \Big| (D \bar{\varrho}^{(0)-1} R_1 \bar{\varrho}) \bar{\varrho}^{-1} \right\rangle\right\rangle \frac{c_1}{c_2} \\
&= \left\langle\left\langle \bar{\varrho}^{(0)} L_2 \bar{\varrho}^{-1} \Big| D \bar{\varrho}^{(0)-1} R_1 \bar{\varrho} \right\rangle\right\rangle \frac{c_1}{c_2} .
\end{aligned}
\tag{9.12.18b}
$$

The final identity for D_{12} reads

$$
\langle\langle L_1 | D R_2 \rangle\rangle = \left\langle\left\langle \bar{\varrho}^{(0)} L_2 \bar{\varrho}^{-1} \Big| D \bar{\varrho}^{(0)-1} R_1 \bar{\varrho} \right\rangle\right\rangle \frac{c_1}{c_2} .
\tag{9.12.19}
$$

It may be well to pause at this point and look back at Sect. 2.12 or 3.2 where it was shown that time-reversal covariant Floquet operators F are generically represented by symmetric unitary matrices. The above derivation of (9.12.19) generalizes these previous arguments to dissipative dynamics. Specializing (9.12.19) back to the zero-damping case will in fact be instructive. Then, the generators $D^{(0)}$ and D become unitary with respect to the scalar product (9.9.4) of observables, and the eigen-matrices of $D^{(0)}$ are the dyads $|\mu\nu\rangle = |\mu\rangle\langle\nu|$ formed by the eigenvectors $|\mu\rangle$ of the Floquet operator $F^{(0)}$, $D^{(0)}|\mu\nu\rangle = F^{(0)}|\mu\rangle\langle\nu|F^{(0)\dagger} = e^{-i(\phi_\mu - \phi_\nu)}|\mu\nu\rangle$. Left and right eigen-matrices of $D^{(0)}$ must then coincide. This indeed follows from (9.12.16): Let $R = |\mu\rangle\langle\nu|$ and $T^2 = 1$, for which case the proof of $R = L$ is easiest. Then, there is no loss of generality in taking $T|\mu\rangle = |\mu\rangle$ and thus, from (9.12.16), $L = c\bar{\varrho}^{(0)-1}|\mu\rangle\langle\nu|$. The stationary eigen-matrix $\bar{\varrho}^{(0)}$ of $D^{(0)}$ cannot be unique in the nondissipative case; its most general form is that of a mixture of diagonal dyads, $\bar{\varrho}^{(0)} = \sum_\lambda \varrho_\lambda^{(0)} |\lambda\rangle\langle\lambda|$; the probabilities $\varrho_\lambda^{(0)}$ may be thought of as fixed by letting a finite damping go to zero. At any rate, $L = (c/\bar{\varrho}_\lambda^{(0)})|\mu\rangle\langle\nu|$ and by normalization, $\langle\langle L|R\rangle\rangle = 1$, we arrive at the equality $L = R = |\mu\rangle\langle\nu|$. Now fully equipped to specialize (9.12.19), we choose $R_1 = L_1 = |\mu\nu\rangle\rangle$ and $R_2 = L_2 = |\alpha\beta\rangle\rangle$. Then, the left-hand member of (9.12.19) reads $D_{12} = \langle\langle L_1|DR_2\rangle\rangle = \mathrm{Tr}\left\{|\nu\rangle\langle\mu|F|\alpha\rangle\langle\beta|F^\dagger\right\} = F_{\mu\alpha}F^*_{\nu\beta}$. In the right-hand member, on the other hand, the factor c_1/c_2 cancels since $\bar{\varrho}^{(0)}L_2 = \varrho_\alpha^{(0)}L_2 = c_2 L_2$ and similarly $\bar{\varrho}^{(0)-1}R_1 = c_1 R_1$. Furthermore, since $\bar{\varrho} = F\bar{\varrho}F^\dagger$ the right-hand member takes the form $\langle\langle L_2|D|R_1\rangle\rangle = D_{21} = F_{\alpha\mu}F^*_{\beta\nu}$. Thus, the identity (9.12.19) reduces to the symmetry $D_{12} = D_{21}$ which is equivalent to the previously established symmetry $F_{\alpha\mu} = F_{\mu\alpha}$ of time-reversal covariant Floquet matrices.

Now, we return to the discussion of finite damping. Detailed balance in the form (9.12.19) implies no relation between the two matrix elements D_{12} and D_{21} alone. Rather, if $\bar{\varrho}^{(0)-1}R_1\bar{\varrho}$ is expanded in terms of the complete set of right eigen-matrices of $D^{(0)}$ and, similarly, $\bar{\varrho}^{(0)}L_2\bar{\varrho}^{-1}$ in terms of the corresponding left eigen-matrices, it becomes clear that (9.12.19) relates D_{12} linearly to all of the other matrix elements of D; such a "global" relationship does not, of course, restrict the 2×2 matrix (9.12.17) describing the near miss of two eigenvalues of D. One must face the inescapable conclusion that for finite damping, the repulsion of two eigenvalues is insensitive to whether or not D obeys detailed balance, i.e., to whether or not the embedding of the dissipative system into a larger Hamiltonian one is time-reversal invariant. Thus, cubic level repulsion appears as generic for strongly dissipative dynamics with a globally chaotic classical limit.

9.13 Dissipation of Quantum Localization

9.13.1 Zaslavsky's Map

A simple dissipative generalization of the classical kicked rotator has been suggested by *Zaslavsky* [350, 351] as[5]

$$p_{t+1} = d\,p_t + \lambda \sin \Theta_{t+1}$$
$$\Theta_{t+1} = (\Theta_t + p_t)\,\mathrm{mod}\,(2\pi) \tag{9.13.1}$$

with a fractional momentum survival d per cycle,

$$0 \le d \le 1. \tag{9.13.2}$$

The special case $d = 1$ corresponds to the standard map (7.3.15) while the limit $d = 0$, i.e., complete momentum loss in one period, yields the one-dimensional

---

[5] Throughout Sect. 9.13 the classical unit of action I/τ will be set equal to unity. Consequently, the symbol $\hbar$ will denote Planck's constant in units I/τ.

circle map

$$\Theta_{t+1} = (\Theta_t + \lambda \sin \Theta_t) \bmod (2\pi)$$
$$p_t = \lambda \sin \Theta_t .$$

(9.13.3)

Formally, the fractional momentum loss d is the Jacobian of the map (9.13.1) and thus measures the contraction of the phase-space volume per iteration. The more familiar notion of a damping rate may be associated with either $-\ln d$ or $1 - d$, quantities that practically coincide for weak damping. Interestingly, damping breaks the periodicity in p of the standard map.

Of interest here will be the range of the two parameters λ and d for which the map (9.13.1) generates a strange attractor. That attractor (as well as simple ones in other parameter ranges) must be confined to the strip

$$|p| \leq p_{\max} = \frac{\lambda}{1 - d} ,$$

(9.13.4)

since for $|p| > p_{\max}$ the momentum p is reduced in modulus by the map for all values of the angle Θ.

A crude picture of the strange attractor for strong damping may be derived from the limit (9.13.3). The strange attractor arising for small values of d resembles the infinite-damping ($d = 0$) limit $p = \lambda \sin \Theta$ if viewed with bad resolution; upon closer inspection a narrow band of a self-similar set of branches unfolds (Fig. 9.11). On the other hand, for damping not too strong, the strange attractor can be described by the random-phase approximation already used in the Hamiltonian limit. In place of (7.5.2), the discrete-time Langevin equation

$$p_{t+1} = d\, p_t + \xi_t$$

(9.13.5)

applies now with the noise force ξ_t, which has zero mean and a second moment free of memory,

$$\langle \xi_t \rangle = 0 , \quad \langle \xi_t \xi_{t'} \rangle = D\delta_{tt'} .$$

(9.13.6)

The higher moments of ξ_t are the subject of Problem 7.2. There is no difficulty in establishing the solution of the linear Langevin equation,

$$p_t = d^t p_0 + \sum_{i=0}^{t-1} d^{t-1-i} \xi_i ,$$

(9.13.7)

which yields, if p_0 is not correlated with the noise,

$$\langle p_t \rangle = d^t \langle p_0 \rangle \to 0 \text{ for } t \to \infty$$

$$\langle p_t^2 \rangle = d^{2t} \langle p_0^2 \rangle + (1 - d^{2t})\langle p_\infty^2 \rangle$$

(9.13.8)

$$\langle p_\infty^2 \rangle = \frac{D}{1 - d^2} .$$

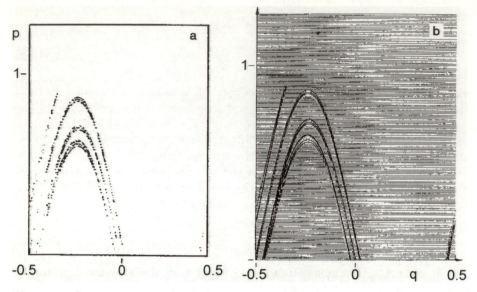

Fig. 9.11. Strange attractor for Zaslavsky's map (9.13.1) with $d = 0.3$, $\lambda = -5$. (Read q, p as $\Theta/2\pi$, $p/2\pi$). Part (**a**) shows the invariant manifold that forms the support of the stationary phase-space density. The latter is represented by contour lines in part (**b**). Courtesy of Dittrich and Graham [352]

For vanishing initial momentum, $p_0 = 0$, and weak dissipation, $d^2 \approx 1 + 2 \ln d$, the mean-squared momentum begins to grow diffusively, $\langle p_t^2 \rangle = Dt$. Eventually, however, that growth levels off, and the stationary value $\langle p_\infty^2 \rangle \approx -D/2 \ln d$ is approached.

Even though the higher moments of the noise are far from Gaussian, the momentum does tend to Gaussian behavior as t grows, provided that the damping is weak. Indeed, according to (9.13.7), p_∞ is the sum of many (of the order of $-1/\ln d$ in number) effectively independent contributions and thus, by the central limit theorem, it acquires Gaussian statistics as $-1/\ln d \to \infty$. Therefore, the stationary probability density of the momentum reads

$$P(p) = \left(\frac{1 - d^2}{2\pi D} \right)^{1/2} e^{-(1-d^2)p^2/2D} \,. \tag{9.13.9}$$

Consistency with (9.13.4) requires, with $D = \lambda^2/2$,

$$1 - d \ll 2(1 + d) \approx 4 \,, \tag{9.13.10}$$

i.e., indeed small damping. Problem 9.24 is concerned with corrections to the Gaussian behavior.

As a last preparation to the quantum treatment, it is worth pointing out how Zaslavsky's map may arise from a continuous-time description. Consider the equations of motion

$$\dot{p} = -\kappa p + \lambda \sin \Theta \sum_{n=-\infty}^{+\infty} \delta(t-n)$$

$$\dot{\Theta} = p \,.$$

(9.13.11)

These allow damping and free rotation to go on concurrently and continuously while the potential $\lambda \cos \Theta$ is switched on impulsively and periodically. Integrating from immediately after one kick to immediately after the next yields

$$p' = e^{-\kappa}p + \lambda \sin \Theta'$$

$$\Theta' = \Theta + \frac{1-e^{-\kappa}}{\kappa}p \,.$$

(9.13.12)

After setting

$$d = e^{-\kappa}, \quad p \to \frac{\ln d}{d-1}p, \quad \lambda \to \frac{\ln d}{d-1}\lambda \,,$$

(9.13.13)

this indeed becomes Zaslavsky's map. Note, however, that in the limit of feeble damping, the rescaling of p and λ becomes trivial since $(\ln d)/(d-1) = 1 + \mathrm{O}(\kappa^2)$.

9.13.2 The Quantum Mechanical Damped Rotator

In the manner introduced in Sect. 9.2, the rotator must be coupled to a heat bath to achieve damping. For the resulting master equation to imply Zaslavsky's map in the classical limit, a particular coupling H_{SR} is required. Following *Dittrich* and *Graham* [352] and staying close to Sect. 9.2 in notation, we choose

$$H_{SR} = \hbar(I_+B + I_-B^\dagger) \,, \quad H_S = \tfrac{1}{2}p^2$$

(9.13.14)

with bath operators B, $B^\dagger$ as in Sect. 9.2 and

$$I_- = I_+^\dagger = \sum_{n=0}^{\infty} \sqrt{n+1}\,(|n\rangle\langle n+1| + |-n\rangle\langle -n-1|) \,.$$

(9.13.15)

The state vectors employed here are the (angular) momentum eigenstates (7.3.19), $p|\pm n\rangle = \pm n\hbar|\pm n\rangle$. It is quite appropriate to call I_- a lowering operator: While $|\pm n\rangle$ with $n > 0$ are eigenstates of the kinetic energy $p^2/2$ with the common eigenvalue $\hbar^2 n^2/2$, the states $I_-|\pm n\rangle = \sqrt{n}|\pm n \mp 1\rangle$ have the smaller kinetic energy $\hbar^2(n-1)^2/2$. Similarly, I_+ raises the kinetic energy. The state of zero angular momentum is an exception since $I_-|0\rangle = \langle 0|I_+ = 0$.

Needless to say, the coupling (9.13.14, 9.13.15) is not intended to give a realistic microscopic description. The operators $I_\pm$ do not even lend themselves to classical interpretation. However, among the many models implying Zaslavsky's map in the classical limit, the one chosen here offers technical convenience; see Problem 9.26.

The Born and Markov approximations discussed in Sect. 9.2, together with suitable assumptions for the bath response function $\kappa(\omega)$ defined in (9.2.19), yield the following generator of infinitesimal time translations Λ for the density operator ϱ of the rotator:

$$\Lambda\varrho = -\mathrm{i}[p^2/2\hbar, \varrho] + \tfrac{1}{2}|\ln d| \{([I_-, \varrho I_+] + [I_-\varrho, I_+])\} \ . \tag{9.13.16}$$

Here the damping constant is denoted by $|\ln d|$ and the bath is assumed to be so cold as to be incapable of raising the energy of the free rotator ($n_{\text{th}} = 0$, see (9.2.21)). In the angular momentum representation, the time rate of change of ϱ reads

$$\begin{aligned}
\langle m|\Lambda\varrho|n\rangle \ &= \ -\mathrm{i}\frac{\hbar}{2}(m^2 - n^2)\langle m|\varrho|n\rangle \\
&+ |\ln d| \left(\Theta_{m \cdot n} \sqrt{(|m| + 1)(|n| + 1)} \left\langle m + \frac{m}{|m|}\middle|\varrho\middle|n + \frac{n}{|n|} \right\rangle \right. \\
&\left. - \tfrac{1}{2}(|m| + |n|)\langle m|\varrho|n\rangle\right)
\end{aligned} \tag{9.13.17}$$

where the unit step function is defined as

$$\Theta_m = \begin{cases} 1 & \text{for } m \geq 0 \\ 0 & \text{for } m < 0 \end{cases} . \tag{9.13.18}$$

Note that the master equation $\dot{\varrho} = \Lambda\varrho$ with the above generator Λ assigns a separate set of evolution equations to all matrix elements $\langle m|\varrho|m + \nu\rangle$ with a fixed integer ν. One such set arises for $\nu = 0$, i.e., for the diagonal elements, and that particular set has the structure of Pauli-type rate equations. The evolution of the mean momentum is easily extracted from these Pauli equations as $\langle \dot{p} \rangle = \langle \rangle \ln d$ or integrated over one unit of time, $\langle p_{t+1} \rangle = d\langle p_t \rangle$. The damping mechanism considered here thus indeed implies the desired classical behavior of the momentum. The diagonal elements $\langle m|\varrho|m\rangle$ also govern the stationary regime reached for $t \to \infty$. It is in fact immediately obvious from (9.13.16) and $I_-|0\rangle = 0$ that the stationary density operator is the projector $|0\rangle\langle 0|$ corresponding to the sharp value zero of the angular momentum.

The off-diagonal elements $\langle m|\varrho|m+\nu\rangle$, $\nu \neq 0$, all decay to zero as $t \to \infty$. The rates of decay are in accord with the discussion in Sect. 9.2, i.e., they increase with $|\nu|$. This is most easily seen for m and n differing in sign, in which case (9.2.17) implies the decay rate $|\ln d||m-n|/2$. For m and $n = m+\nu$ both positive (for the negative sign a similar statement holds) and large (such that $\langle m + 1|\varrho|n + 1\rangle = \langle m|\varrho|n\rangle + \mathcal{O}(\hbar)$; see Sect. 9.3), the decay rate is $|\ln d|(\sqrt{m} - \sqrt{n})^2/2$.

The dynamics of the angle Θ is determined by the off-diagonal elements of ϱ. To take proper account of the effective period 2π of Θ it is advantageous to discuss the angle dynamics in terms of $\mathrm{e}^{\mathrm{i}\Theta}$. By employing the Θ representation of the angular momentum eigenstates[6]

$$\langle \Theta|m\rangle = \frac{1}{\sqrt{2\pi}}\mathrm{e}^{\mathrm{i}m\Theta} \ , \tag{9.13.19}$$

---

[6] The reader should not be confused by the double meaning of Θ: While denoting the independent variable of wave-functions in (9.13.19), the symbol Θ otherwise refers to the angle coordinate of the rotator as a quantum operator.

the matrix elements and the expectation value of $e^{i\Theta}$ are easily found as

$$\langle m|e^{i\Theta}|n\rangle = \delta_{m,n+1} \tag{9.13.20}$$

$$\langle e^{i\Theta(t)}\rangle = \sum_m \langle m|\varrho(t)|m+1\rangle . \tag{9.13.21}$$

The subset of (9.13.17) where $n - m = \nu = 1$ thus applies. In the classical limit, large values of $|m|$ will be most important (formally, $m \sim 1/\hbar$). Assuming positive momentum at time t (the opposite sign is treated similarly), the subset in question reads

$$\begin{aligned}
\langle m|\dot{\varrho}|m+1\rangle &= \; i\hbar\left(m + \tfrac{1}{2}\right)\langle m|\varrho|m+1\rangle \\
&+ |\ln d| \left[\sqrt{(m+1)(m+2)}\langle m+1|\varrho|m+2\rangle \right. \\
&\left. - \left(m + \tfrac{1}{2}\right)\langle m|\varrho|m+1\rangle\right] .
\end{aligned} \tag{9.13.22}$$

Within corrections of relative weight $1/m \sim \hbar$, the simplification $\langle m|\dot{\varrho}|m+1\rangle = \mathrm{Im}\{\langle m|\varrho|m+1\rangle\}$ is valid and shows that the damping does not directly affect the angle Θ in the classical limit. Rather, in that limit $m = p = \langle p\rangle$ may be considered sharp whence the mean $\langle e^{i\Theta}\rangle$ obeys $\partial_t\langle e^{i\Theta}\rangle = \langle p\rangle\langle e^{i\Theta}\rangle$ due to (9.13.21, 9.13.22). Together with the momentum evolution $\partial_t\langle p\rangle = |\ln d|\langle p\rangle$ already established above, a closed set of classical equations for $\langle p\rangle$ and $\langle e^{i\Theta}\rangle$ results. Integration over one unit of time indeed yields the part of Zaslavsky's map referring to free rotation and damping,

$$\begin{aligned}
e^{i\Theta_{t+1}} &= \exp\left[i\left(\Theta_t + \tfrac{1-d}{|\ln d|}p_t\right)\right] \\
p_{t+1} &= dp_t .
\end{aligned} \tag{9.13.23}$$

To obtain the full map (either classically or quantum mechanically), the nonlinear kick must be accounted for as taking place after the free rotation and the damping.

In the quantum mechanical case the complete map for the density operator is obtained in terms of e^Λ (i.e. the exponentiated version of the generator Λ of infinitesimal time translations, to be evaluated for a unit time step) and the unitary evolution operator for the nonlinear kick,

$$F = e^{-(\lambda/\hbar)\cos\Theta} , \tag{9.13.24}$$

as

$$\varrho(t+1) = F\left(e^\Lambda \varrho(t)\right) F^\dagger . \tag{9.13.25}$$

To run this dissipative quantum map on a computer both F and e^Λ must be specified in a particular representation, for example, the angular momentum representation. The corresponding matrix F is given in (7.3.25, 7.3.26) while the tetrad

$$(e^\Lambda)_{mn,m'n'} = \langle m|(e^\Lambda|m'\rangle\langle n'|)|n\rangle \tag{9.13.26}$$

implied by (9.13.17) will be constructed in Sect. 9.14.

9.13.3 Destruction of Localization

Inasmuch as quantum localization for the kicked rotator is an interference phe-
nomenon involving coherences between angular momentum eigenstates $|n\rangle$ over
a span $l \gg 1$, it is clear from the general discussion of Sect. 9.3 that localization
must decay on a time scale $\sim 1/(l|\ln d|)$. This decay time must be compared with
the break time $t^* \approx l$ (see (7.5.6)) at which the kinetic energy of the undamped
kicked rotator stops growing diffusively in favor of quasi-periodic fluctuations
around a mean level $\langle p^2 \rangle \approx Dt^* \approx \hbar^2 l^2$. Localization cannot show up at all
when the decay time is smaller than t^*, i.e., when the damping exceeds a critical
strength,

$$| \ln d_{\mathrm{crit}}| = \frac{1}{l^2} \, . \tag{9.13.27}$$

The reader will appreciate the close analogy of (9.13.27) with (9.4.18) but should
also be aware of a difference in the precise meaning of the respective damping
thresholds: The top referred to in Sect. 9.4 does not in general display quantum
localization.

The remaining discussion will be restricted to the perhaps most interesting
case where the damping is much weaker than the critical damping (9.13.27),
i.e., $t^* \ll 1/(l|\ln d|)$. Figure 9.12 displays numerical results obtained by *Dittrich*
and *Graham* [352] on the basis of the map (9.13.25). Actually, due to the assumed smallness of the damping, that map could be linearized with respect to
the damping constant $|\ln d|$. The plots in Fig. 9.12 depict the time dependence of
the quantum mean $\langle p_t^2 \rangle$ for various values of $|\ln d|$; the initial state is always that
of zero angular momentum. As expected on the basis of the foregoing qualitative
remarks, in the time interval $0 \le t < t^*$, the kinetic energy grows diffusively,
$\langle p_t^2 \rangle = Dt$, just as if neither damping nor quantum effects were present.

Round about the break time, $t \approx t^* \approx l \approx \lambda^2/2\hbar^2$, while the damping is still
ineffective, the now familiar breakaway of $\langle p_t^2 \rangle$ from classical diffusion takes place;
the latter gives way to the quasi-periodic temporal fluctuations characteristic of
quantum localization.

Damping becomes conspicuous in the plots of Fig. 9.12 at somewhat later
times when t approaches the order of magnitude $1/(l|\ln d|)$, i.e., when quantum
coherences extending over one localization length (in the angular momentum
basis) begin to dissipate. Then, not only coherences between the initially excited
Floquet eigenstates tend to suffer from relaxation but also coherences within each
Floquet eigenstate. At any rate, the coherence decay drives the quantum mean
$\langle p_t^2 \rangle$ toward classical behavior with a tendency that is stronger, the larger the
damping. This tendency reflects the dissipative death of quantum localization, as
does the progressive smoothing of temporal fluctuations in the evolution of $\langle p_t^2 \rangle$.

A perhaps surprising feature displayed by the graphs of $\langle p_t^2 \rangle$ is a new dif-
fusive regime on the time scale $1/(l|\ln d|)$, albeit one with a diffusion constant
smaller than the classical one. Paradoxical as it may look at first sight, the diffusion constant in this new regime grows with increasing damping. In fact, this
phenomenon must result from the disruption of coherences by the incoherent

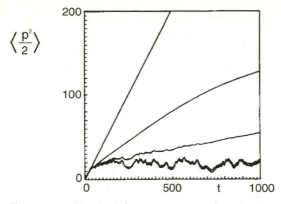

Fig. 9.12. Classical (*uppermost curve*) and quantum ($\hbar = 0.3/(\sqrt{5} - 1)$) mean kinetic energy of the kicked rotator with damping under conditions of classical chaos ($\lambda = 10$). The stronger the damping, the steeper the curves ($1 - d = 0.5 \times 10^{-6}$, 10^{-4}, 10^{-3}). In all cases, the damping is so weak that the classical curve is indistinguishable from the curve with no dissipation. Note the agreement of classical and quantum diffusion for times up to the break time t^* and the destruction of quantum localization by dissipation. Courtesy of Dittrich and Graham [352]

emission of quanta into the bath which constitutes the damping. Naively speaking, each such disruption cuts the wave packet into two pieces which may both resume diffusive spreading up to the width l. The incremental diffusion thus induced should be proportional to the rate of incoherent emission, i.e., $\propto |\ln d|$.

Finally, on the time scale $1/|\ln d|$ on which the damping makes itself felt even in the classical limit, the growth of $\langle p_t^2 \rangle$ comes to an end. Unfortunately, the numerical calculations could not be pushed far enough in time to check whether the stationary value $\langle p_\infty^2 \rangle$ coincides with the classical value given in (9.13.8). Nonetheless, a tendency toward the expected classical behavior is clearly discernible.

Both the intermediate incremental diffusion due to damping just described and the earlier signatures of dissipation near $t \gtrsim t^*$ are amenable to a simple semiquantitative treatment [352], to which we turn now. The Floquet eigenstates carrying the evolution of the wave packet, about l in number, must display linear repulsion of their quasi-energies roughly according to Wigner's distribution

$$P(s) = \tfrac{\pi}{2} \left(\tfrac{s}{\bar{s}^2}\right) \exp\left[-\tfrac{\pi}{4} \left(\tfrac{s}{\bar{s}}\right)^2\right]$$
$$\bar{s} = \tfrac{2\pi}{l} \, . \tag{9.13.28}$$

Since a quasi-energy spacing s is resolvable for times larger than $t_s = 2\pi/s$, the spacing distribution can be associated with a distribution of resolution times,

$$W(t_s) = \frac{2\pi^3}{\bar{s}^2 t_s^3} \exp\left[-\frac{\pi^3}{(\bar{s} t_s)^2}\right] \, , \tag{9.13.29}$$

whose integral, $\int_0^t dt_s W(t_s)$, gives the fraction of level spacings resolvable at time t. Only this fraction can have begun to contribute noticeably to quasi-periodic temporal fluctuations in $\langle p_t^2 \rangle$, i.e., has ceased to support the original diffusive growth. Conversely, as the quasi-periodic fluctuations are disrupted by incoherent transitions at a rate $l|\ln d|$, a fraction $(t - t_s)l|\ln d|$ of the spacings s ceases to support quantum quasi-periodicity and is thrown back to support diffusion. Thus, an incremental kinetic energy may be estimated as

$$\delta \langle p_t^2 \rangle = \hbar^2 l^2 \int_0^t dt_s (t - t_s) l |\ln d| W(t_s) . \tag{9.13.30}$$

The factor $\hbar^2 l^2$ in front of the integral indicates the angular momentum scale. The integral may be evaluated and yields, where $t^* = l$,

$$\delta \langle p_t^2 \rangle = \hbar^2 l^2 |\ln d| \left\{ \frac{t}{t^*} \exp \left[-\frac{\pi}{4} \left(\frac{t^*}{t} \right)^2 \right] - \frac{\pi}{2} \mathrm{erfc} \left(\frac{\sqrt{\pi} t^*}{2t} \right) \right\} . \tag{9.13.31}$$

For $t \gg t^*$, this indeed amounts to a diffusion,

$$\delta \langle p_t^2 \rangle \to \hbar l^2 |\ln d| \left(t - \frac{\pi}{2} t^* \right) \frac{1}{t^*} , \tag{9.13.32}$$

with a diffusion constant smaller than the classical one by a factor $|\ln d|$. This estimate is in reasonable agreement with the numerical results depicted in Fig. 9.12.

For times near the break time, $t \gtrsim t^*$, the result (9.13.31) may be expanded in powers of $(t - t^*)/t^*$. The first few terms of that expansion,

$$\begin{aligned} \delta \langle p_t^2 \rangle \approx \hbar^2 l^2 |\ln d| & \left[\left(1 - \frac{\pi}{2} \mathrm{erfc} \frac{\sqrt{\pi}}{2} \right) \right. \\ & \left. + \frac{t - t^*}{t^*} + \frac{\pi}{4} \left(\frac{t - t^*}{t^*} \right)^2 + \dots \right] , \end{aligned} \tag{9.13.33}$$

should describe the earliest manifestations of damping for $t \lesssim t^*$. Figure 9.13 indeed reveals a range of time with a quadratic increase of $\delta \langle p_t^2 \rangle$.

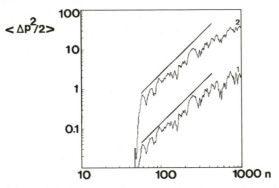

Fig. 9.13. Time evolution of the excess of the kinetic energy of the kicked rotator $(8.12, 24, 25, 17)$ over the undamped case $[\lambda = 10, \hbar = 0.3/(\sqrt{5} - 1); 1 - d = 5 \times 10^{-6}$ for plot 1, $1 - d = 10^{-4}$ for plot 2]. The logarithmic plot reveals a range of time with a quadratic increase. Courtesy of Dittrich and Graham [352]

9.14 Construction of Damping Propagators

Having played around quite a bit with the damping generators Λ for the top [see (9.2.21) or, for a simplified low-temperature version, (9.3.2 or 9.4.3)] and the rotator [see (9.13.16)], it would not seem right to close this chapter without indicating how the corresponding master equations $\dot{\varrho}(t) = \Lambda\varrho(t)$ can be solved. The general solutions $\varrho(t) = e^{\Lambda t}\varrho(0)$ were used to construct the complex quasi-energies of the map $e^{\Lambda}e^{L}$ for the kicked top in Sect. 9.6 and to generate the time dependence of the mean kinetic energy of the kicked rotator in Sect. 9.13. The method of solution is rather the same for the two cases. Readers not willing to be bothered by such technicalities will be excused.

9.14.1 Damped Spin

First, we turn to the top [320], write the master equation (9.3.2) on the basis formed by the eigenstates $|jm\rangle$ of $\boldsymbol{J}^2$ and J_z, and denote the elements of the density matrix as

$$\varrho_m^k \equiv \langle j, m+k|\varrho|j, m-k\rangle \quad \text{with} \tag{9.14.1}$$

$$k = 0, \pm\frac{1}{2}, \ldots, \pm j, \quad m = -j + |k|, \ldots, j - |k|.$$

These are probabilities for $k = 0$ and coherences of "skewness" $2k$ for $k \neq 0$. The master equation yields the coupled evolution equations

$$\frac{d}{d2\kappa t}\varrho_m^k = \sqrt{g_{m+k+1}g_{m-k+1}}\,\varrho_{m+1}^k - (g_m - k^2)\varrho_m^k, \tag{9.14.2}$$

$$g_m = j(j+1) - m(m-1) = (j+m)(j-m+1).$$

Two remarkable properties of the latter equations make for easy solvability. For one, not only the diagonal elements form a closed set; the coherences of fixed "skewness" k, i.e., the elements in the $(2k)$th "neben-diagonal" of the density matrix, form separate sets as well. Second, in each set we encounter only single-step downward couplings: $\dot{\varrho}_m^k(t)$ is a linear combination of only $\varrho_m^k(t)$ and $\varrho_{m+1}^k(t)$.

It is convenient to employ the propagator that establishes the linear relationship of the current-time density matrix to its initial version as

$$\varrho_m^k(t) = \sum_{n=m}^{j-|k|} D_{mn}^k(t)\varrho_n^k(0). \tag{9.14.3}$$

The propagator $D_{mn}^k(t)$ obviously obeys the evolution law (9.14.2) as well as does $\varrho_m^k(t)$, but its initial form is $D_{mn}^k(0) = \delta_{mn}$. For the Laplace transform,

$$\tilde{D}_{mn}^k(z) = \int_0^\infty dt\, e^{-zt} D_{mn}^k(t), \tag{9.14.4}$$

the evolution equation yields a separate set of single-step recursion relationships for each skewness $2k$ which is immediately solved as

$$\tilde{D}_{mn}^k(z) = \frac{1}{2\kappa\sqrt{g_{m-k}g_{m+k}}}\prod_{l=m}^{n}\frac{\sqrt{g_{l-k}g_{l+k}}}{\frac{z}{2\kappa} + g_l - k^2}. \tag{9.14.5}$$

Inverting the Laplace transform, we get

$$D_{mn}^k(t) = \sqrt{Q_{m-k}Q_{m+k}} \int_{b-i\infty}^{b+i\infty} \frac{dz/(2\kappa)}{2\pi i} e^{zt} \prod_{l=m}^{n} \frac{\sqrt{g_{l-k}g_{l+k}}}{\frac{z}{2\kappa}+g_l-k^2} , \quad (9.14.6)$$

$$Q_{mn} = \prod_{l=m+1}^{n} g_l = \frac{(j+n)!(j-m)!}{(j-n)!(j+m)!} \quad (9.14.7)$$

where the complex parameter b should lie to the left of all poles of the integrand in the complex z-plane and is otherwise arbitrary.

A family relationship between the principal diagonal ($k = 0$) and its neben-diagonals ($k \neq 0$), discovered in Ref. [331], becomes visible when the integration variable z in the inverse Laplace transformation is shifted to $z/(2\kappa) - k^2$,

$$D_{mn}^k(t) = D_{mn}^0(t) \sqrt{Q_{m-k,n-k}Q_{mm+k,n+k}/Q_{mn}^2} \exp\{2k^2\kappa t\} . \quad (9.14.8)$$

Therefore, it suffices to worry about the inverse Laplace transform for $k = 0$.

When doing that inverse Laplace transform, one should note that due to the parabolic behavior of g_m, the symmetry $g_m = g_{-m+1}$ gives rise to second-order poles if $n \geq 0$ and thus to contributions of the type $te^{-g_l t}$ as well as $e^{-g_l t}$; for $n < 0$, all poles are simple, and $D_{mn}^k(t)$ is a sum of decaying exponentials. For large j, i.e., in the semiclassical limit, the inverse Laplace transform is most conveniently evaluated using a saddle-point approximation [331–333, 335].

9.14.2 Damped Rotator

Here, we follow Ref. [352]. It is convenient to employ the angular momentum representation as in (9.13.17). Two types of matrix elements ϱ_{mn} must be distinguished, one with $m \cdot n < 0$ and the other with $m \cdot n \geq 0$. By inspecting (9.13.17) one can immediately verify that each matrix element ϱ_{mn} with $m \cdot n < 0$ is uncoupled from all other elements and decays as

$$\varrho_{mn}(t) = \exp\left[-i\tfrac{\hbar}{2}(m^2 - n^2)t - \tfrac{1}{2}(|m| + |n|)|\ln d|t\right] \varrho_{mn}(0) ,$$
$$\text{for } m \cdot n < 0 . \quad (9.14.9)$$

The elements with $m \cdot n \geq 0$, on the other hand, are coupled in separate sets, each of which has a fixed difference $m - n$; actually, two sets are associated with each fixed difference $m - n$, one with m and n both positive and the other with m and n both negative. These two sets have equivalent dynamics: Choosing two nonnegative integers l and $l + \nu$, we introduce a unified notation

$$\varrho(l,\nu,t) = \left\{ \begin{matrix} \langle l|\varrho(t)|l+\nu\rangle \\ \langle -l|\varrho(t)|-l-\nu\rangle \end{matrix} \right\} \quad l, l+\nu \geq 0 \quad (9.14.10)$$

for both cases and infer from (9.13.17) that

$$\dot{\varrho}(l,\nu) = \left[i\hbar\nu\left(l+\frac{\nu}{2}\right) - |\ln d|\left(l+\frac{\nu}{2}\right)\right]\varrho(l,\nu)$$
$$+ |\ln d|\sqrt{(l+1)(l+\nu+1)}\, \varrho(l+1,\nu) . \quad (9.14.11)$$

The symmetry between positive and negative angular momenta in (9.14.9, 9.14.10) is due to the fact that the low-temperature heat bath swallows positive amounts of kinetic energy $\hbar^2 |m^2 - (m-1)^2|/2$ in each absorption event.

Within each set of fixed ν, the coupling between matrix elements is directed down the l ladder: If initially

$$\varrho(l,\nu,0) \equiv \varrho(l,\nu,0|l_0) = \delta_{ll_0} , \tag{9.14.12}$$

then for $t > 0$ only, the $\varrho(l,\nu,t|l_0)$ with $0 \le l \le l_0$ develop nonvanishing values. Denoting the temporal Laplace transform of $\varrho(l,\nu,t|l_0)$ by $\tilde{\varrho}(l,\nu,z|l_0)$, we obtain the recursion relation

$$\tilde{\varrho}(l_0,\nu,z|l_0) = \frac{1}{z + \kappa(l_0 + \nu/2)} ,$$

$$\tilde{\varrho}(l,\nu,z|l_0) = \frac{|\ln d| \sqrt{(l+1)(l+\nu+1)}}{z + \kappa(l + \nu/2)} \tilde{\varrho}(l+1,\nu,z|l_0) , \tag{9.14.13}$$

$$\kappa \equiv |\ln d| - i\nu .$$

This has the solution

$$\tilde{\varrho}(l,\nu,z|l_0) = |\ln d|^{l_0-l} \sqrt{\frac{l_0!(l_0+\nu)!}{l!(l+\nu)!}} \prod_{m}^{l \ldots l_0} \frac{1}{z + \kappa(2m+\nu)} . \tag{9.14.14}$$

Upon inverting the Laplace transform, one encounters an $(l_0 - l)$-fold convolution of the form

$$\varrho(l,\nu,t|l_0) \sim e^{-\kappa(l+\nu/2)t} \int_0^t dt_1 \int_0^{t_1} dt_2 \ldots \int^{t_{l_0-l-1}} dt_{l_0-l}$$

$$\times e^{-\kappa(t_1+t_2+\ldots+t_{l_0-l})} . \tag{9.14.15}$$

Due to the symmetry of the integrand, the integral is easy to perform and yields

$$\varrho(l,\nu,t|l_0) = \left(\frac{|\ln d|}{\kappa} \right)^{l_0-l} \sqrt{\frac{l_0!(l_0+\nu)!}{l!(l+\nu)!}} \frac{1}{(l_0-l)!} d(t)^{l+\nu/2} [1 - d(t)]^{l_0-l} ,$$

$$d(t) = e^{-\kappa t} . \tag{9.14.16}$$

The subset with $\nu = 0$ is made up of the diagonal elements $\langle l|\varrho(t)|l \rangle$ or $\langle -l|\varrho(t)| - l \rangle$ originating from $\langle \pm l|\varrho(0)| \pm l \rangle = \delta_{ll_0}$. These diagonal elements constitute the binomial distribution

$$\varrho(l,0,t|l_0) = \binom{l_0}{l} d^{tl} (1 - d^t)^{l_0-l} . \tag{9.14.17}$$

On the other hand, the degree of off-diagonality is measured by ν. In accord with the results of Sect. 9.3, the $\varrho(l,\nu,t|l_0)$ suffer faster decay and more rapid oscillations as ν becomes larger.

9.15 Can the Kicked Top Be Realized?

A bit afraid of the dissipation of strength the reader may have suffered over a long chapter on dissipation, I hurry to serve a hopefully reconstructive treat. I already mentioned, in passing, previous experiments on microwave super-radiance. These could be modified to realize a periodically driven large spin alias Bloch vector resembling the kicked top [353].

Let us consider N identical two-level atoms in interaction with a single mode of the electromagnetic field in a cavity and allow for a detuning $\Delta = \omega_A - \omega_F$ between the atomic and mode frequencies. The pertinent Hamiltonian reads

$$H = \hbar\omega_F b^\dagger b + \hbar\omega_A J_z + \hbar g \left(b^\dagger J_- + b J_+\right) \tag{9.15.1}$$

where the coupling constant g, a single-atom-in-vacuo Rabi frequency, is proportional to the dipole matrix element for the atomic transition in question. The operators b and $b^\dagger$, respectively, annihilate and create photons and obey $[b, b^\dagger] = 1$ while the familiar angular momentum operators describe the degree of atomic excitation (J_z) and polarization ($J_\pm$) and obey $[J_+, J_-] = 2J_z$ and $[J_z, J_\pm] = \pm J_\pm$. However high the quality of the cavity, the mode will suffer losses if only due to radiation toward the outside world. Such losses assign to the mode of the empty cavity the behavior of a damped harmonic oscillator described by the master equation (9.16.1) which the reader will have derived in Problem 9.4. Assuming for simplicity a temperature sufficiently low for the number of thermal photons n_{th} to be negligible and setting $\omega_A = \omega + \delta$, we have the following master equation for the density operator W of mode and atoms:

$$\dot{W} = -\frac{i}{\hbar}[H, W] + \kappa \left\{[b, W b^\dagger] + [bW, b^\dagger]\right\}. \tag{9.15.2}$$

To reduce the foregoing dynamics to that of a spin or Bloch vector we need to assume that the detuning Δ or the damping constant κ or both are large compared to the Rabi frequency $\Omega = g\sqrt{N}$ with which the dipolar coupling in the Hamiltonian tends to endow the oscillatory exchange between atoms and mode. In that case, we can adiabatically eliminate the field mode from the master equation (9.15.2). By applying the elimination technique of Sect. 9.2 once more, we find that the reduced density operator $\rho = \text{Tr}\, W$ of the atoms obeys the master equation [354]

$$\dot{\rho} = -i\omega_A[J_z, \rho] - i\frac{g^2\Delta}{\kappa^2 + \Delta^2}[J_z^2 - J_z, \rho]$$

$$+ \frac{g^2\kappa}{\kappa^2 + \Delta^2}\left\{[J_-, \rho J_+] + [J_- \rho, J_+]\right\}. \tag{9.15.3}$$

Clearly, the terms proportional to g^2 represent the influence of the field mode on the atoms; notational differences apart, they constitute the (zero-temperature version of the) super-radiance master equation (9.2.21). Specializing to the large-detuning limit $\Delta \geq \kappa$, we can drop the dissipative terms and arrive at a conservative motion generated by the Hamiltonian $H = \frac{\hbar g^2 \Delta}{\kappa^2 + \Delta^2}(J_z^2 - J_z)$; the approximate disappearance of J_z against J_z^2 is valid in the interesting case of a large number of atoms $N = 2j$. That Hamiltonian would, of course, have an integrable classical limit due to the conservation of $\boldsymbol{J}^2$. Integrability is destroyed by switching on an external electric field with a periodic time dependence of period T, $f(t) = f(t+T)$ which is linearly polarized along, say, the x-direction. Then, after a suitable renaming of control parameters, the Hamiltonian becomes

$$H = \hbar \omega J_z + \frac{\hbar \lambda}{2j+1} J_z^2 + \hbar f(t) J_x \, . \tag{9.15.4}$$

This is the dynamics we have chosen to refer to as the periodically driven top; for $f(t)$ a train of delta kicks we have the kicked top. Experimenters with their own ideas about what is convenient will prefer monochromatic driving to kicking.

For another type of realization of the kicked top, one can consider small magnetic crystallites with an easy plane of magnetization, exposed to a magnetic field $B(t)$. Choosing the magnetic field along the x-axis and modulated in time by a periodic train of delta functions, and letting the plane of easy magnetization be distinguished by a crystal field $\sim J_z^2$, one is led to the Floquet operator (7.6.2). *Waldner*'s experiments [355] are of precisely that type, but unfortunately the spin quantum numbers realized up to now are too large to make any quantum effect observable. Finally, Josephson junctions may offer possibilities of realizing top dynamics [356].

9.16 Problems

9.1. For $j = 1/2$ determine the time-dependent solution of the master equation (9.2.21).

9.2. Let the bath be a collection of harmonic oscillators with the free Hamiltonian

$$H_R = \sum_i \hbar \omega_i b_i^+ b_i \, ,$$

and assume that the angular momentum couples to the bath variable

$$B = g \sum_i b_i$$

and its conjugate as

$$H_{SR} = \hbar (J_+ B + J_- B^\dagger) \, .$$

Show that Fermi's golden rule yields the transition rates $W(m \to m+1)$ and $W(m+1 \to m)$.

9.3. Let an angular momentum be coupled to a heat bath as $H_{SR} = \hbar J_z (B + B^\dagger)$. Show that the master equation in the weak-coupling limit is

$$
\begin{aligned}
\dot{\varrho} &= -\mathrm{i}[\omega J_z + \delta J_z^2, \varrho] \\
&\quad + \kappa(1 + 2n_{\mathrm{th}}) \left([J_z, \varrho J_z] + [J_z \varrho, J_z] \right) \, .
\end{aligned}
$$

Discuss the lifetimes of populations and coherences in the J_z representation.

9.4. Show that the master equation for the damped harmonic oscillator reads

$$\dot{\varrho}(t) = -i(\omega + \delta)[b^\dagger b, \varrho(t)] \qquad (9.16.1)$$

$$+\kappa(1 + n_{\text{th}})\left([b, \varrho b^\dagger] + [b\varrho, b^\dagger]\right) + \kappa n_{\text{th}}\left([b^\dagger, \varrho b] + [b^\dagger \varrho, b]\right) .$$

Give the stationary solution and the expectation values $\langle b(t) \rangle$ and $\langle b^\dagger(t)b(t) \rangle$. In the low-temperature limit ($n_{\text{th}} = 0$) also give the probability of finding m quanta in the oscillator at time $t > 0$, provided there were m_0 initially. Recall from your elementary quantum mechanics that the m-quantum state is defined by $b^\dagger b|m\rangle = m|m\rangle$ with $m = 0, 1, 2 \ldots$ and $b|m\rangle = \sqrt{m}|m-1\rangle$.

9.5. Show that the master equation (9.3.2) for the damped spin generally gives no time-scale separation for probabilities and coherences in the J_z basis. Use the initial rate of change of the norm of the temporal successor of $|j, m\rangle\langle j, m'|$. Note that one does not have a right to expect accelerated decoherence on that basis since J_- is not among the coupling agents!

9.6. In the low-temperature limit, rewrite the master equation (9.2.21) as an equation of motion for the weight function $P(\Theta, \phi, t)$ in the representation of $\varrho(t)$ by a diagonal mixture of coherent angular momentum states,

$$\varrho(t) = \int \sin \Theta \, d\Theta \, d\phi \, P(\Theta, \phi, t)|\Theta, \phi\rangle\langle\Theta, \phi| .$$

See the definition (7.6.12). Consult Ref. [357, 358] for help.

9.7. Return to Problem 9.4 for the damped harmonic oscillator. Now use *Glauber*'s coherent states [359] $|\beta\rangle$, defined by $b|\beta\rangle = \beta|\beta\rangle$ or $|\beta\rangle = \exp(-|\beta|^2/2) \times e^{\beta b^\dagger}|0\rangle$, to show that coherences between different such states decay faster than probabilities $\langle\beta|\varrho|\beta\rangle$. Consider the norm of $e^{\Lambda t}|\beta\rangle\langle\beta|$. For the first report of accelerated decoherence for the damped harmonic oscillator, you may want to check Ref. ([360]).

9.8. Use the distribution $P(y)$ from (4.8.13) to show that the root-mean-square deviation from the mean of the $\Lambda_{\mu\nu}$ given in (9.4.8) is proportional to $\sqrt{j}$ and that the relative deviation is proportional to $1/\sqrt{j}$.

9.9. Calculate the damping constants $\Lambda_{\mu\nu}$ analogous to (9.4.9) and the eigenvalues corresponding to (9.4.11) for the damping generator of Problem 8.3.

9.10. Show that $\langle J_{x,y}(\infty) \rangle = 0$ under conditions of global classical chaos. Proceed similarly to (9.4.17).

9.11. Calculate $\sqrt{(\langle J_z \rangle - \langle \bar{J}_z \rangle)^2}/j$ for $\Gamma = 0$ and the initial state chosen from the classically chaotic regime. ($\overline{\cdots}$) means averaging over time.

9.12. Find the nearest neighbor spacing distribution for the d-dimensional Poissonian process, $P_d(s) = \alpha d s^{d-1} e^{-\alpha s^d}$, and determine the parameter α.

9.13. Find the kth nearest neighbor spacing distribution for the d-dimensional Poissonian process,

$$P_d^k(s) = \frac{\alpha^k d}{(k-1)!} s^{(k-1)d} e^{-\alpha s^d} .$$

9.14. Normalize the joint distribution of the eigenvalues for Dyson's circular ensembles using Grassmann integration.

9.15. Establish the spacing distribution (9.8.27) for Ginibre's ensemble of 2×2 matrices with the help of (9.8.11) and (9.8.26).

9.16. Argue that a finite-size approximation to the hole distribution is obtained as

$$H(s) = \prod_{n=1}^{N-1} \left[e_n(s^2) e^{-s^2} \right] .$$

9.17. Establish the correction $-s^2 \ln s$ in (9.8.43) with the help of the Euler-McLaurin summation formula.

9.18. Show that (9.8.1) is equivalent to

$$D_{ijkl} = D_{jilk}^* .$$

9.19. Using the algebraic method of Sect. 9.10, show that the Poisson brackets of the "angular momenta" l_{mn} for the unitary universality class obey the Poisson brackets

$$\{l_{mn}, l_{ij}\} \quad = \quad l_{mj}\delta_{ni} - l_{in}\delta_{mj} .$$

With a bit of spare time, the symplectic case can be treated similarly; its Poisson brackets can be found in Ref. [361].

9.20. Recall how for $[M, \mathcal{A}] = 0$ with $\mathcal{A}$ antiunitary and $\mathcal{A}^2 = 1$, a representation can be constructed whose basis vectors are $\mathcal{A}$ invariant and which makes M a real matrix.

9.21. Show that the real eigenvalues of the generator D repel linearly.

9.22. Show that $P(S) = S^3/(8\lambda^4)K_0(S^2/4\lambda^2)$ for complex symmetric 2×2 matrices with a Gaussian element distribution $W(x) = (\lambda/\pi) \exp(-\lambda|x|^2)$, $\lambda = \lambda^*$, where K_0 is the modified Bessel function of order zero.

9.23. Calculate the normalization coefficient c in (9.12.16).

9.24. Using the moments $\langle \xi^{2n} \rangle = (\lambda/2)^{2n} \binom{2n}{n}$ determined in Problem 7.4, determine the deviation from Gaussian behavior of $P(p)$, as given by (9.13.9).

9.25. Specify the frequency dependence of the bath response function $\kappa(\omega)$ yielding the master equation (9.13.16); also give the condition on the bath temperatures implicit in (9.13.16).

9.26. As an alternative to the damping mechanism treated in the text, use $I_+ = I_- = \cos\Theta$ instead of (9.13.15). Specialize again to low temperatures, and employ the Born and Markov approximation in constructing the generator Λ. Which choice of $\kappa(\omega)$ secures $\langle\dot{p}\rangle = \langle p\rangle \ln d$?

10. Superanalysis for Random-Matrix Theory

10.1 Preliminaries

Gossip has it that the "supersymmetry technique" is difficult to learn. But quite to the contrary, representing determinants as Gaussian integrals over anticommuting alias Grassmann variables makes for great simplifications in computing averages over the underlying matrices, as we have seen in Chap. 4 and 9. Readers starting to read this book at this chapter will catch up soon and appreciate the power of Grassmann integrals as an immensely useful tool for dealing with various kinds of random matrices.

Superanalysis goes a few steps further. The first is to represent quotients of determinants as "mixed" Gaussian integrals, composed of ordinary ones for denominators and Grassmannian ones for numerators. The Gaussian integrand then has a sum of two quadratic forms in its exponent, one bilinear in the commuting integration variables and the second bilinear in the Grassmann integration variables. A second generalization suggests itself and turns out to be tremendously fruitful: The quotient of ordinary determinants is elevated to the superdeterminant of a supermatrix with both commuting and anticommuting elements. A mixed Gaussian integral equalling the superdeterminant then employs the supermatrix in question to determine the mixed quadratic form appearing in the Gaussian integrand.

Supersymmetry finally comes into play when after an average over a suitable ensemble of random matrices and certain further intermediate steps of calculation, we encounter non-Gaussian integrals over commuting and anticommuting variables that can naturally be thought of as elements of a supermatrix, such that the integrand is symmetric under a class of transformations of that supermatrix and the transformations themselves are represented by supermatrices.

The aim of this section is to provide a self-contained introduction to the supersymmetry technique and thus to subdue the gossip mentioned above.

We shall start with a homeopathic dose and derive Wigner's semicircle law for the GUE. No superhocus-pocus beyond trading a determinant for a Grassmann integral will be exercised. The subsequent introduction of supervectors, supermatrices, and superdeterminants (in short, superalgebra) and finally of differentiation and integration will appear as a welcome compaction of notation. Nontrivial supersymmetry first arises when we proceed to the two-point cluster function of the GUE and there encounter the celebrated nonlinear supermatrix sigma model. The power of the method will become manifest as we finally treat non-Gaussian ensembles, as well as ensembles of banded and sparse matrices.

The latter applications reflect the recent unification of the fields of disordered systems and quantum chaos. That development was of course pioneered by the discovery of the deep relationship between Anderson localization in disordered electronic systems and quantum localization in the kicked rotator which we have already discussed in Chap. 7. Superanalysis appears to strengthen such links.

In keeping with the introductory purpose of this chapter I shall confine the discussion to systems without time reversal-invariance. Readers who have understood the superanalytic method for that simplest case will not encounter serious difficulties when going to the original literature on systems with anti-unitary symmetries.

10.2 Semicircle Law for the Gaussian Unitary Ensemble

10.2.1 The Green Function and Its Average

The density of levels, $\varrho(E) = N^{-1} \sum_{i=1}^{N} \delta(E - E_i)$, can be obtained from the energy-dependent propagator alias Green function

$$G(\mathcal{E}) = \frac{1}{N} \mathrm{Tr} \frac{1}{\mathcal{E} - H} = \frac{1}{N} \sum_{i=1}^{N} \frac{1}{\mathcal{E} - E_i} \tag{10.2.1}$$

by letting the complex energy variable $\mathcal{E}$ approach the real axis from below, $\mathcal{E} \to E - i0^+ \equiv E^-$, and taking the imaginary part,

$$\varrho(E) = \frac{1}{\pi} \mathrm{Im}\, G(E^-); \tag{10.2.2}$$

the well-known identity $\frac{1}{E^-} = \frac{\mathcal{P}}{E} + i\pi\delta(E)$ is at work here.

In view of the intended GUE average by superintegrals, we must extract $G(\mathcal{E})$ from a quotient of two determinants. The generating function

$$Z(\mathcal{E}, j) = \frac{\det(\mathcal{E} - H)}{\det(\mathcal{E} - H - j)} = \frac{\det i(\mathcal{E} - H)}{\det i(\mathcal{E} - H - j)} \tag{10.2.3}$$

helps since it yields

$$G(\mathcal{E}) = \frac{1}{N} \left. \frac{\partial Z(\mathcal{E}, j)}{\partial j} \right|_{j=0}. \tag{10.2.4}$$

The reason for sneaking in the factors i in the last member of (10.2.3) will be given momentarily. One proves the identity (10.2.4) most simply by writing out $Z(\mathcal{E}, j)$ in the eigenbasis of H. To familiarize the reader with the machinery to be used abundantly later, it is well to also consider an alternative proof through the following steps:

$$\frac{\partial}{\partial j}\frac{1}{\det\left(\mathcal{E}-H-j\right)}\bigg|_0 = \frac{\partial}{\partial j}\exp\{-\operatorname{Tr}\ln\left(\mathcal{E}-H-j\right)\}\bigg|_0$$

$$= \frac{\partial}{\partial j}\exp\{-\operatorname{Tr}\ln\left(\mathcal{E}-H\right)+j\operatorname{Tr}\frac{1}{\mathcal{E}-H}+\mathcal{O}(j^2)\}\bigg|_0$$

$$= \frac{1}{\det\left(\mathcal{E}-H\right)}\frac{\partial}{\partial j}\exp\{j\operatorname{Tr}\frac{1}{\mathcal{E}-H}+\mathcal{O}(j^2)\}\bigg|_0$$

$$= \frac{1}{\det\left(\mathcal{E}-H\right)}\operatorname{Tr}\frac{1}{\mathcal{E}-H}. \tag{10.2.5}$$

Now, I represent the determinant in the numerator of the quotient in (10.2.3) by a Gaussian Grassmann integral à la (4.12.1) and the inverse determinant by a Gaussian integral with ordinary commuting integration variables as in (4.12.21),

$$Z(E^-,j) = \int\left(\prod_{k=1}^{N}\frac{d^2 z_k}{\pi}\,d\eta_k^* d\eta_k\right) \tag{10.2.6}$$

$$\times \exp\left\{i\sum_{m,n=1}^{N}\left(z_m^*\big(H_{mn}+(j-E^-)\delta_{mn}\big)z_n+\eta_m^*(H_{mn}-E^-\delta_{mn})\eta_n\right)\right\}.$$

The reader will recall that the kth ordinary double integral goes over the whole complex z_k plane while the Grassmann integrals need no limits of integration. A combination of both types of integrals as encountered above for the first time is called a superintegral. At any rate, it can be seen that the previously sneaked in factors i ensure convergence of the ordinary Gaussian integral in (10.2.6) since the real part of $-iE^-$ is negative.

All is well prepared for the GUE average

$$\overline{Z(E^-,j)} = \int\left(\prod_{k=1}^{N}\frac{d^2 z_k}{\pi}\,d\eta_k^* d\eta_k\right)\exp\left\{-i\sum_m\left[(E^--j)z_m^* z_m+E^-\eta_m^*\eta_m\right]\right\}$$

$$\times \exp\left[i\sum_{m,n}\overline{H_{mn}(z_m^* z_n+\eta_m^*\eta_n)}\right] \tag{10.2.7}$$

now that the random matrix elements H_{mn} appear linearly in the exponential to be averaged. It is worth noting that the coefficient of H_{mn} in that exponential is a "commuting variable" because it is bilinear in the Grassmannian variables. Moreover, the sum $\sum_{m,n}H_{mn}(z_m^* z_n+\eta_m^*\eta_n)$ in the exponent is a real quantity, provided we define complex conjugation of the Grassmannian generators η_k and η_k^* introduced in (4.12.2) by

$$(\eta_k)^* = \eta_k^*, \quad (\eta_k^*)^* = -\eta_k. \tag{10.2.8}$$

With this definition of complex conjugation extended to the whole Grassmann, algebra we secure the reality of $\eta_k^*\eta_k$ since $(\eta_k^*\eta_k)^* = (\eta_k^*)^*\eta_k^* = -\eta_k\eta_k^* = \eta_k^*\eta_k$ and indeed $(\sum H_{mn}\eta_m^*\eta_n)^* = \sum H_{mn}^*(-\eta_m\eta_n^*) = \sum H_{nm}\eta_n^*\eta_m = \sum H_{mn}\eta_m^*\eta_n$.

10.2.2 The GUE Average

We recall from Chap. 4 that the diagonal matrix elements H_{mm} and the pairs $H_{mn}, H_{nm} = H_{mn}^*$ for $1 \leq n \leq m \leq N$ are all statistically independent Gaussian random variables with zero means; the variances may be written as

$$\overline{H_{mm}^2} = \overline{|H_{mn}|^2} = \frac{\lambda^2}{N} \tag{10.2.9}$$

where λ is a real number setting the unit of energy.

The linear combination $\sum_{m,n} H_{mn} M_{nm}$ with any fixed nonrandom Hermitian matrix M is a real Gaussian random variable with the property

$$\overline{e^{i \sum H_{mn} M_{nm}}} = e^{-\frac{1}{2}\overline{(\sum H_{mn} M_{nm})^2}} = e^{-\frac{\lambda^2}{2N} \sum |M_{mn}|^2} . \tag{10.2.10}$$

Applying this to the average in (10.2.7) with $M_{mn} = z_m^* z_n + \eta_m^* \eta_n$, we obtain

$$\overline{\exp\left[i \sum_{m,n} H_{mn}(z_m^* z_n + \eta_m^* \eta_n)\right]}$$

$$= \exp\left[-\frac{\lambda^2}{2N} \sum_{m,n} (z_m^* z_n + \eta_m^* \eta_n)(z_n^* z_m + \eta_n^* \eta_m)\right] \tag{10.2.11}$$

$$= \exp\left\{-\frac{\lambda^2}{2N}\left[\left(\sum_m |z_m|^2\right)^2 - \left(\sum_m \eta_m^* \eta_m\right)^2 + 2\sum_m z_m \eta_m^* \sum_n z_n^* \eta_n\right]\right\} .$$

The GUE average is done, and with remarkable ease at that, as I may be permitted to say. The expense incurred is the remaining $4N$-fold superintegral in (10.2.7) to whose evaluation we now turn.

10.2.3 Doing the Superintegral

The most serious obstacle seems to be the non-Gaussian character of the average (10.2.11) with respect to the integration variables $\{z_m, z_m^*, \eta_m, \eta_m^*\}$. But that impediment can be removed by the so-called Hubbard–Stratonovich transformation of which we need to invoke three variants,

$$e^{-\frac{\lambda^2}{2N}\left(\sum |z_m|^2\right)^2} = \sqrt{\frac{N}{\lambda^2}} \int_{-\infty}^{\infty} \frac{da}{\sqrt{2\pi}} e^{-\frac{N}{2\lambda^2}a^2 + ia \sum |z_m|^2} ,$$

$$e^{\frac{\lambda^2}{2N}\left(\sum \eta_m^* \eta_m\right)^2} = \sqrt{\frac{N}{\lambda^2}} \int_{-\infty}^{\infty} \frac{db}{\sqrt{2\pi}} e^{-\frac{N}{2\lambda^2}b^2 - b \sum \eta_m^* \eta_m} , \tag{10.2.12}$$

$$e^{-\frac{\lambda^2}{N} \sum z_m \eta_m^* z_n^* \eta_n} = \frac{\lambda^2}{N} \int d\sigma^* d\sigma\, e^{-\frac{N}{\lambda^2}\sigma^* \sigma + i\sigma^* \sum z_m^* \eta_m + i \sum z_n \eta_n^* \sigma} .$$

The first two of these identities are elementary Gaussian integrals; in particular, the appearance of Grassmannians is not a problem since the bilinear form $\eta_m^* \eta_m$ commutes with everything. It will not hurt to explain the third a little. To that end, we first convince ourselves of the Grassmann character of $\Sigma \equiv \sum z_m^* \eta_m$ by checking that its square vanishes: Indeed, $\Sigma^2 = \sum_{mn} z_m^* z_n^* \eta_m \eta_n = 0$ since $z_m^* z_n^*$ is symmetric and $\eta_m \eta_n$ antisymmetric in the summation indices. The left-hand side of the third of the above identities may thus be written as $\exp[-\frac{\lambda^2}{N} \Sigma^* \Sigma] = 1 - \frac{\lambda^2}{N} \Sigma^* \Sigma = 1 + \frac{\lambda^2}{N} \Sigma \Sigma^*$; but to that very form the right-hand side is also easily brought by first expanding the integrand and then doing the integrals, r.h.s. $= \frac{\lambda^2}{N} \int d\sigma^* d\sigma \left[-\frac{N}{\lambda^2} \sigma^* \sigma - \frac{1}{2} (\sigma^* \Sigma + \Sigma^* \sigma)^2 \right] = -\frac{\lambda^2}{N} \int d\sigma^* d\sigma \left[\sigma^* \sigma (\frac{N}{\lambda^2} + 2\frac{1}{2} \Sigma \Sigma^*) \right] = 1 + \frac{\lambda^2}{N} \Sigma \Sigma^*$. Multiplying the three identities in (10.2.12), we get the GUE average (10.2.11) replaced by a fourfold superintegral

$$\exp\left[i \sum_{m,n} H_{mn} (z_m^* z_n + \eta_m^* \eta_n) \right] = \int \frac{da\,db}{2\pi} d\sigma^* d\sigma \, e^{-\frac{N}{2\lambda^2}(a^2 + b^2 + 2\sigma^* \sigma)}$$

$$\times \exp \sum_m \left[ia|z_m|^2 - b\eta_m^* \eta_m + i\sigma^* z_m^* \eta_m + iz_m \eta_m^* \sigma \right]. \tag{10.2.13}$$

Upon feeding this into the generating function (10.2.7), we may enjoy that the (4N)-fold integral over the $\{z_k, z_k^*, \eta_k, \eta_k^*\}$ has become Gaussian. Moreover, all of the N fourfold integrals over $z_k, z_k^*, \eta_k, \eta_k^*$ with $k = 1, \dots, N$ are equal such that their product equals the Nth power of one of them,

$$\overline{Z(E^-, j)} = \int \frac{da\,db}{2\pi} d\sigma^* d\sigma \, e^{-\frac{N}{2\lambda^2}(a^2 + b^2 + 2\sigma^* \sigma)} \tag{10.2.14}$$

$$\times \left(\int \frac{d^2 z}{\pi} d\eta^* d\eta \, \exp\left\{ -i \left[(E^- - j - a)|z|^2 + (E^- - ib)\,\eta^* \eta - \sigma^* z^* \eta - z\eta^* \sigma \right] \right\} \right)^N.$$

To integrate over η^* and η we proceed in the by now familiar fashion and expand the integrand,

$$\int d\eta^* d\eta \, \exp\left\{ -i \left[(\mathcal{E} - ib)\,\eta^* \eta - \sigma^* z^* \eta - z\eta^* \sigma \right] \right\} \tag{10.2.15}$$

$$= \int d\eta^* d\eta \, \left[-i(\mathcal{E} - ib)\,\eta^* \eta - \frac{1}{2} z^* z \left(\sigma^* \eta \eta^* \sigma + \eta^* \sigma \sigma^* \eta \right) \right]$$

$$= \int d\eta^* d\eta \, \eta^* \eta \left[-i(\mathcal{E} - ib) + z^* z \sigma^* \sigma \right]$$

$$= i(\mathcal{E} - ib) - z^* z \sigma^* \sigma.$$

For the remaining double integral over z to exist, we specify the complex energy as slightly below the real energy axis, $\mathcal{E} = E^-$, and benefit from the foresight of sneaking the factors i into the last member of (10.2.3); the integral yields

$$\int \frac{d^2 z}{\pi} [i(E^- - ib) - z^* z \sigma^* \sigma] \exp[-i(E^- - j - a)|z|^2]$$

$$= \frac{E^- - ib}{E^- - j - a} + \frac{\sigma^* \sigma}{(E^- - j - a)^2}, \tag{10.2.16}$$

whereupon the generating function takes the form

$$\overline{Z(E^-, j)} \;=\; \int \frac{da\,db}{2\pi} d\sigma^* d\sigma \, \exp\left[-\frac{N}{2\lambda^2}(a^2 + b^2 + 2\sigma^*\sigma) \right] \tag{10.2.17}$$

$$\times \left(\frac{E^- - \mathrm{i}\,b}{E^- - j - a} \right)^N \left[1 + \frac{\sigma^*\sigma}{(E^- - \mathrm{i}\,b)(E^- - j - a)} \right]^N .$$

Once in the mood of devouring integrals, we let those over σ and σ^* come next; as usual, we expand the integrand,

$$\int d\sigma^* d\sigma \, \exp\left(-\frac{N}{\lambda^2}\sigma^*\sigma \right) \left[1 + \frac{\sigma^*\sigma}{(E^- - \mathrm{i}\,b)(E^- - j - a)} \right]^N$$

$$= \int d\sigma^* d\sigma \left(1 - \frac{N}{\lambda^2}\sigma^*\sigma \right) \left[1 + \frac{N\sigma^*\sigma}{(E^- - \mathrm{i}\,b)(E^- - j - a)} \right] \tag{10.2.18}$$

$$= N\left[\frac{1}{\lambda^2} - \frac{1}{(E^- - \mathrm{i}\,b)(E^- - j - a)} \right] ,$$

and are left with an ordinary twofold integral over the commuting variables a, b,

$$\overline{Z(E^-, j)} \;=\; N \int \frac{da\,db}{2\pi} \exp\left\{ -\frac{N}{2\lambda^2}(a^2 + b^2) \right\} \left(\frac{E^- - \mathrm{i}\,b}{E^- - j - a} \right)^N \tag{10.2.19}$$

$$\times \left[\frac{1}{\lambda^2} - \frac{1}{(E^- - \mathrm{i}\,b)(E^- - j - a)} \right] .$$

Differentiating according to (10.2.4), we proceed to the averaged Green function

$$\overline{G(E^-)} \;=\; \int \frac{da\,db}{2\pi} \exp\left[-\frac{N}{2\lambda^2}(a^2 + b^2) \right] \left(\frac{E^- - \mathrm{i}\,b}{E^- - a} \right)^N \tag{10.2.20}$$

$$\times \frac{1}{E^- - a} \left\{ \frac{N}{\lambda^2} - \frac{N+1}{(E^- - \mathrm{i}\,b)(E^- - a)} \right\} .$$

10.2.4 Two Remaining Saddle-Point Integrals

Each of the two terms in the curly bracket above yields a double integral which can in fact be seen as a product of separate single integrals; each of the latter can be done by the saddle-point method in the limit of large N. I take on the one over a first and consider

$$\int_{-\infty}^{\infty} da\, A(a) \mathrm{e}^{-Nf(a)} \approx \sum \sqrt{2\pi/Nf''(\hat{a})} A(\hat{a}) \mathrm{e}^{-Nf(\hat{a})} \tag{10.2.21}$$

where $A(a)$ is independent of N,

$$f(a) = a^2/2\lambda^2 + \ln(E^- - a), \tag{10.2.22}$$

and $\hat{a}$ is the highest saddle of the (modulus of the) integrand for which the path of steepest descent (and thus constant phase of the integrand) in the complex a plane can be continuously deformed back into the original path of integration without crossing any singularity; should there be several such saddles, degenerate in height, and one has to sum their contributions. The saddle-point equation $f'(a) = 0$ is quadratic and has the two solutions

$$a_\pm = \frac{E^-}{2}\left(1 \pm \sqrt{1 - (\tfrac{2\lambda}{E^-})^2}\right) \tag{10.2.23}$$

for which the square root is assigned a branch cut in the complex energy plane along the real axis from -2λ to $+2\lambda$.

(1) $(\frac{E}{2})^2 < \lambda^2$: The small imaginary part of E^- requires, for $E > 0$, evaluating the square root on the lower lip of the cut such that upon dropping $\mathrm{Im}E^-$, we have the saddle points $a_\pm = E/2 \mp \mathrm{i}\{\lambda^2 - (E/2)^2\}^{1/2}$. They are mutual complex conjugates and lie in the complex a plane on the circle of radius λ around the origin. The integrand in (10.2.20) vanishes for $a \to \pm\infty$ on the real axis and has no singularities at finite a except for a pole at $a = E^-$, right below the real axis. The original path of integration along the real axis can be deformed so as to climb along a path of constant phase of the integrand over the saddle at a_-, without crossing the said pole; a path of constant phase from $-\infty$ to $+\infty$ over the saddle at a_+ also exists but cannot be deformed back to the original path without crossing the pole. Thus, only the saddle at a_- contributes.

(2) $(\frac{E}{2})^2 > \lambda^2$: Expanding the square root for $\epsilon \to 0^+$ we have $a_\pm = (E/2)(1 \pm \sqrt{1 - (2\lambda/E)^2}) - \mathrm{i}(\epsilon/2)(1 \pm 1/\sqrt{1 - (2\lambda/E)^2})$. Again, the saddle point a_+ lies below the real axis while a_- lies above such that only the latter can contribute. Thus, for all cases we conclude that

$$\int_{-\infty}^{\infty} da\, A(a)\mathrm{e}^{-Nf(a)} \approx \sqrt{\frac{2\pi}{Nf''(a_-)}}\, A(a_-)\mathrm{e}^{-Nf(a_-)}, \tag{10.2.24}$$

where the phase of the square root in the denominator is defined by the steepest descent direction of integration across the saddle point.

A little surprise is waiting for us in the integrals over b in (10.2.20),

$$\int_{-\infty}^{\infty} db\, B(b)\mathrm{e}^{-Ng(b)} \qquad \text{with} \qquad g(b) = b^2/2\lambda^2 - \ln(E^- - \mathrm{i}b). \tag{10.2.25}$$

Now $b = -\mathrm{i}E$ is a zero of the integrand rather than the location of a singularity. Saddles are encountered for $\mathrm{i}b_\pm = a_\pm = (E^-/2)(1 \pm \sqrt{1 - (2\lambda/E^-)^2})$. The saddle points $b_\pm$ are obtained by a clockwise $(\pi/2)$-rotation of the $a_\pm$ about the origin in the complex a plane. Two cases again arise:

(1) $(E/2)^2 < \lambda^2$: The saddles lie symmetrically to the imaginary axis of the complex a-plane at $b_\pm = -\mathrm{i}E/2 \mp \sqrt{\lambda^2 - (E/2)^2}$. They are degenerate in height since $\mathrm{Reg}(b_+) = \mathrm{Reg}(b_-)$. Since the integrand has no singularity at any finite point b, the path of integration can be deformed so as to climb uphill from $-\infty$ to b_+, then descend to the zero at $-\mathrm{i}E$, climb again to b_-, and finally descend toward $+\infty$. That deformed path may be chosen as one of constant phase, except at the zero of the integrand where no phase can be defined, such that, the phase jumps upon passing through the exceptional point.

(2) $(E/2)^2 > \lambda^2$: In that case both saddle points are purely imaginary, $b_\pm = -\mathrm{i}(E/2)\big(1 \pm \sqrt{1 - (2\lambda/E)^2}\big)$, and b_+ is further away from the origin than b_-; see Fig. 10.1. The saddle at b_+ is higher than that at b_- since $\mathrm{Re}g(b_+) < \mathrm{Re}g(b_-)$, and one might thus naively expect only b_+ to be relevant. But surprisingly, b_+ does not enter at all, as the following closer inspection reveals. The phase of the integrand, i.e., $\mathrm{Im}g(x+\mathrm{i}y) = xy/\lambda^2 + \arctan[x/(E+y)]$, vanishes along the imaginary axis. That path of the vanishing phase passes through both saddles and the zero of the integrand; along it the integrand has a minimum at b_- and a maximum at b_+. Another path of constant vanishing phase goes through each saddle; along the one through b_-, the modulus of the integrand has a maximum at b_-; that latter path extends to $\pm\infty + \mathrm{i}0$ and makes b_- the only contributing saddle. No excursion from b_- to b_+ can be steered toward $+\infty + \mathrm{i}0$ without climbing higher after reaching b_+, except for the inconsequential excursion returning to b_- along the imaginary axis. To summarize the discussion of the b-integral, I note that

$$\int_{-\infty}^{\infty} db\, B(b)\mathrm{e}^{-Ng(b)} \approx \begin{cases} \displaystyle\sum_{\hat{b}=b_+,b_-} \sqrt{\frac{2\pi}{Ng''(\hat{b})}}\, B(\hat{b})\mathrm{e}^{-Ng(\hat{b})} & \text{for } \left(\frac{E}{2}\right)^2 < \lambda^2 \\[2ex] \displaystyle\sqrt{\frac{2\pi}{Ng''(b_-)}}\, B(b_-)\mathrm{e}^{-Ng(b_-)} & \text{for } \left(\frac{E}{2}\right)^2 > \lambda^2 . \end{cases} \tag{10.2.26}$$

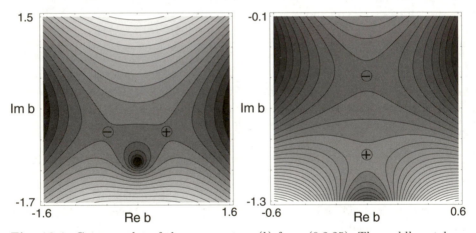

Fig. 10.1. Contour plot of the exponent $-g(b)$ from (9.2.25). The saddles at $b_\pm$ are marked as $\oplus$ and $\ominus$. Height is indicated in shades of grey, ranging from black (low) to white (high). Left for case (1), i.e., $(E/2)^2 < \lambda^2$, with $E = 1$, $\lambda = 1/\sqrt{2}$. The two saddles are of equal height. The path of integration climbs up the valley from the left through b_- and descends to the zero of $\exp(-Ng(b))$ on the imaginary axis, then again uphill to b_+, and therefrom down the valley toward $+\infty$. Right for case (2), i.e., $(E/2)^2 > \lambda^2$, with $E = \sqrt{2} + 0.1$, $\lambda = 1/\sqrt{2}$. We recognize that the saddle at b_+ lies higher than that at b_- and also see the logarithmic "hole" below b_+. As described in the text the path of integration climbs up the valley from the left toward the saddle at b_- and descends on the other side into the valley to the right

Combining the a- and b-integrals, we first treat the case labelled (2) above, i.e., $(E/2)^2 > \lambda^2$. Since only the saddles at a_- and b_- contribute, we get

$$\overline{G(E^-)} \approx \frac{1}{2\pi(E^- - a_-)} \left\{ \frac{N}{\lambda^2} - \frac{N+1}{(E^- - a_-)(E^- - ib_-)} \right\}$$

$$\times \sqrt{\frac{(2\pi)^2}{N^2 f''(a_-) g''(b_-)}} \, \exp\left[-N\big(f(a_-) + g(b_-)\big) \right] . \qquad (10.2.27)$$

However, due to $a_- = ib_-$ we have $f(a_-) + g(b_-) = 0$ such that the exponential factor equals unity. Similarly, $f''(a_-) = g''(b_-) = \lambda^{-2} - (E^- - a_-)^{-2}$ whereupon the curly bracket above cancels as $N^{-1}\big(f''(a_-)g''(b_-)\big)^{-1/2}\{\ldots\} = 1 + \mathcal{O}(1/N)$. But since corrections of relative order $1/N$ are already made in the saddle-point approximation, $\overline{G(E^-)} = (E^- - a_-)^{-1} \approx \big\{(E^-/2)\big(1 + \sqrt{1 - (2\lambda/E^-)^2}\big)\big\}^{-1}$. Setting $\epsilon = 0$, we finally get the purely real result $\overline{G(E^-)} \approx \big\{(E/2)\big(1 + \sqrt{1 - (2\lambda/E)^2}\big)\big\}^{-1}$. Thus, the average density vanishes, $\overline{\varrho(E)} \approx 0$ for $(E/2)^2 > \lambda^2$.

In the hope of a finite mean density of levels, we finally turn to case (1), $(E/2)^2 < \lambda^2$. According to the above findings, we confront contributions from the two pairs of saddles $\{a_-, b_-\}$ and $\{a_-, b_+\}$. The first pair is dealt with in full analogy with the discussion of the corresponding pair $\{a_-, b_-\}$ under case (2) and furnishes $\overline{G(E^-)}$ with $\big(E/2 - i\sqrt{\lambda^2 - (E/2)^2}\big)^{-1}$. The pair a_-, b_+ is incapable of providing anything of asymptotic weight: The argument of the exponential in (10.2.27) is replaced with $-N\big(f(a_-) + g(b_+)\big) = -N\big(a_-^2/(2\lambda^2) + \ln a_+ - a_+^2/(2\lambda^2) - \ln a_-\big)$ which is purely imaginary due to $a_- = a_+^*$, such that the exponential remains unimodular however large N. Collecting all other N-dependent factors according to the model of (10.2.27), we get $N^{-1}\{\ldots\} = N^{-1}\{N/\lambda^2 - (N+1)/(a_+ a_-)\} = -1/N\lambda^2$, i.e., an overall weight of order $1/N$. Indeed then, only the pair $\{a_-, b_-\}$ contributes in the limit $N \to \infty$, just as in case (2).

It remains to collect the limiting form of the averaged Green function,

$$\lim_{N\to\infty} \overline{G(E^-)} = \begin{cases} \dfrac{1}{(E/2)(1+\sqrt{1-(2\lambda/E)^2}} & \text{for} \quad (\tfrac{E}{2})^2 > \lambda^2 \\[2mm] \dfrac{1}{E/2 - i\sqrt{\lambda^2 - (E/2)^2}} & \text{for} \quad (\tfrac{E}{2})^2 < \lambda^2 , \end{cases} \qquad (10.2.28)$$

and, after taking the imaginary part, *Wigner's* semicircle law,

$$\lim_{N\to\infty} \overline{\varrho(E)} = \begin{cases} 0 & \text{for} \quad (\tfrac{E}{2})^2 > \lambda^2 \\[2mm] \dfrac{1}{\pi\lambda^2}\sqrt{\lambda^2 - (E/2)^2} & \text{for} \quad (\tfrac{E}{2})^2 < \lambda^2 . \end{cases} \qquad (10.2.29)$$

Needless to say, the edges of the semicircular level density at $E = \pm 2\lambda$ are rounded off at finite N, as may be studied by scrutinizing next-to-leading-order corrections. The importance of the semicircle law is somewhat indirect: It must be scaled out of the energy axis before the predictions of random-matrix theory for spectral fluctuations take on their universal form to which generic dynamical systems are so amazingly faithful.

10.3 Superalgebra

10.3.1 Motivation and Generators of Grassmann Algebras

A compaction of our notation is indicated before we can generalize the above discussion of the mean density of levels $\overline{\varrho(E)}$ to two-point correlation functions like $\overline{\varrho(E)\varrho(E')}$. We need to generalize some basic notions of algebra to accommodate both commuting and anticommuting variables as components of vectors and elements of matrices.

We had previously (see Sect. 4.12) introduced Grassmann variables $\eta_1, \ldots, \eta_N$ with the properties $\eta_i\eta_j + \eta_j\eta_i = 0$ and shall stick to these as the "generators of a 2^N-dimensional Grassmann algebra". Whenever convenient, we shall, as already done in the previous section, complement these N generators by N more such, $\eta_1^*, \ldots, \eta_N^*$, so as to simply have a 2^{2N}-dimensional Grassmann algebra. An internal structure is provided to the latter by defining complex conjugation as $(\eta_i)^* = \eta_i^*$ and $(\eta_i^*)^* = -\eta_i$.

10.3.2 Supervectors, Supermatrices

We shall have to handle $(2M)$-component vectors with M commuting components $S_1, \ldots, S_M$ and M anticommuting components $\chi_1, \ldots, \chi_M$,

$$\Phi = \begin{pmatrix} S \\ \chi \end{pmatrix}, \quad S = \begin{pmatrix} S_1 \\ \vdots \\ S_M \end{pmatrix}, \quad \chi = \begin{pmatrix} \chi_1 \\ \vdots \\ \chi_M \end{pmatrix}. \tag{10.3.1}$$

Quite intentionally, I have denoted the components χ_i with a Greek letter differing from the one reserved for the generators η_i since it will be necessary to allow the χ_i to be odd functions of the generators like, e.g., $\chi_i = \sum_j c_{ij}\eta_j + \sum_{ijkl} c_{ijkl}\eta_j\eta_k\eta_l \ldots$ with commuting coefficients c. Similarly, the commuting character of the S_i does not prevent the appearance of η's but restricts the S_i to be even in the η's, like $S_i = z_i + \sum_{ijk} z_{ijk}\eta_j\eta_k + \ldots$. If a verbal distinction between commuting numbers without and with additions even in the η's is indicated I shall call the former numerical; an addition even in the Grassmannians is a nilpotent quantity;[1] the coefficients c and z in the above characterizations of anticommuting χ's and commuting S's are meant to be numerical. In somewhat loose but intuitive jargon, I shall sometimes refer to commuting variables as bosonic and anticommuting ones as fermionic. At any rate, a vector of the structure (10.3.1) is called a supervector.

If the need ever arises, one can generalize the definition of a supervector such as to allow for the number of elements of the bosonic and fermionic subvectors to be different, M_B for S and M_F for χ. Moreover, M_F need not coincide with the number N of generators of the Grassmann algebra. Unless explicitly stated otherwise, I shall in the following take $M_B = M_F = M$.

---

[1] A nilpotent quantity has vanishing powers with integer exponents upward of some smallest positive one.

Supermatrices are employed to relate supervectors linearly. We introduce such $2M \times 2M$ matrices as

$$F = \begin{pmatrix} a & \sigma \\ \rho & b \end{pmatrix}, \qquad F\Phi = \begin{pmatrix} a & \sigma \\ \rho & b \end{pmatrix} \begin{pmatrix} S \\ \chi \end{pmatrix} = \begin{pmatrix} aS + \sigma\chi \\ \rho S + b\chi \end{pmatrix}. \tag{10.3.2}$$

For $F\Phi$ to have the same structure as Φ, where the first M components are bosonic and the last M ones are fermionic, of the four $M \times M$ blocks in F, the matrices a and b must have bosonic elements while σ and ρ must have fermionic elements. It is customary to speak of a as the Bose–Bose block or simply the Bose block, of b as the Fermi–Fermi or simply the Fermi block, of σ as the Bose–Fermi block, and of ρ as the Fermi–Bose block. For products of supermatrices, the usual rules of matrix multiplication must hold. (If $M_B \neq M_F$, a is $M_B \times M_B$, b is $M_F \times M_F$, while the two blocks with fermionic elements are rectangular matrices, σ an $M_B \times M_F$ one and ρ an $M_F \times M_B$ one.)

We need to define transposition for supermatrices. Since it is desirable to retain the usual rule $(F_1 F_2)^T = F_2^T F_1^T$, we must modify the familiar rule for transposing purely bosonic matrices to

$$F^T = \begin{pmatrix} a & \sigma \\ \rho & b \end{pmatrix}^T = \begin{pmatrix} \tilde{a} & \tilde{\rho} \\ -\tilde{\sigma} & \tilde{b} \end{pmatrix}; \tag{10.3.3}$$

As done previously in this book, I reserve the tilde to denote the familiar matrix transposition and shall use the superscript T for transposition of supermatrices that includes the unfamiliar minus sign in the Fermi–Bose block. A further price for retaining the usual rule for transposition of products is that double transposition is not the identity operation,

$$(F^T)^T = \begin{pmatrix} a & -\sigma \\ -\rho & b \end{pmatrix} \neq F. \tag{10.3.4}$$

A useful extension of the notion of a trace to supermatrices is the supertrace

$$\mathrm{Str} F = \mathrm{Tr} a - \mathrm{Tr} b = \sum_{m=1}^{M} (a_{mm} - b_{mm}). \tag{10.3.5}$$

The minus sign in front of the trace of the Fermi block is beneficial by bringing about the usual cyclic invariance for the supertrace,

$$\mathrm{Str} F_1 F_2 = \mathrm{Str} F_2 F_1, \tag{10.3.6}$$

since $\mathrm{Str} F_1 F_2 = \mathrm{Tr}(a_1 a_2 + \sigma_1 \rho_2 - \rho_1 \sigma_2 - b_1 b_2) = \mathrm{Tr}(a_1 a_2 + \sigma_1 \rho_2 + \sigma_2 \rho_1 - b_1 b_2)$ is obviously symmetric in the labels 1 and 2. It follows that $\mathrm{Str} MFM^{-1} = \mathrm{Str} F$ where M is another supermatrix and M^{-1} is its inverse.

A most welcome consequence of the cyclic invariance of the supertrace is the following identity for the logarithm of a product,

$$\mathrm{Str} \ln FG = \mathrm{Str} \ln F + \mathrm{Str} \ln G \tag{10.3.7}$$

whose proof parallels that of the corresponding identity for ordinary matrices. Sketching that proof provides an opportunity to recall the definition of the logarithm as a series,

$$\ln F = \ln \left[1 + (F - 1) \right] = - \sum_{n=1}^{\infty} \frac{(-1)^n}{n} (F - 1)^n , \qquad (10.3.8)$$

up to analytic continuation for F outside the range of convergence. To display unity as the reference matrix about which the expansion goes I momentarily introduce $F = 1 + f$, $G = 1 + F^{-1}g$ such that $FG = 1 + f + g$. Thus, the identity to be proven may be written as

$$\mathrm{Str}\ln(1 + f + xg) \stackrel{?}{=} \mathrm{Str}\ln(1 + f) + \mathrm{Str}\ln[1 + x(1 + f)^{-1}g] \qquad (10.3.9)$$

where x is a real parameter which can obviously be sneaked in without gain or loss of generality, but with profit in convenience. Expanding the x-dependent logarithms about the unit matrix, we transform our identity to

$$- \sum_{n=1}^{\infty} \frac{(-1)^n}{n} \mathrm{Str}(f + xg)^n \stackrel{?}{=} \mathrm{Str}\ln(1 + f) - \sum_{n=1}^{\infty} \frac{(-1)^n}{n} \mathrm{Str}\left(\frac{x}{1 + f}g\right)^n .$$
$$(10.3.10)$$

When differentiating with respect to x, we benefit from the cyclic invariance of the trace and rid ourselves of the factors $(1/n)$ in the nth terms of the expansions, such that the expansions take the forms of geometric series,

$$\frac{1}{1 + f + xg} g \stackrel{?}{=} \frac{1}{1 + (1 + f)^{-1}xg} \frac{1}{1 + f} g . \qquad (10.3.11)$$

But now the question mark can be dropped since the two sides do equal one another due to $(ab)^{-1} = b^{-1}a^{-1}$. We can also wave good-bye to the question mark in (10.3.9); the functions of x on the two sides have coinciding derivatives for all x and take the same value for $x = 0$. Needless to say, our proof of (10.3.7) applies to traces of logarithms of products of ordinary matrices as well.

10.3.3 Superdeterminants

The definition of a superdeterminant $\mathrm{Sdet}F$ to be given presently is meant to preserve three familiar properties of usual determinants: (1) a supermatrix F and its transpose F^T should have the same superdeterminant, $\mathrm{Sdet}F = \mathrm{Sdet}F^T$, (2) the superdeterminant of a product should be the product of the superdeterminants of the factor supermatrices, $\mathrm{Sdet}F_1 F_2 = \mathrm{Sdet}F_1 \mathrm{Sdet}F_2$; and (3) for a supermatrix, the logarithm of its superdeterminant should equal the supertrace of its logarithm, $\ln \mathrm{Sdet}F = \mathrm{Str}\ln F$.

We shall find these requirements met when the superdeterminant is defined as the ratio of usual determinants

$$\mathrm{Sdet}F = \det(a - \sigma b^{-1}\rho)/\det b \quad \text{if} \quad \det b_0 \neq 0 \qquad (10.3.12)$$

where b_0 denotes the numerical part of the Fermi block b. If $\det b_0 = 0$, the superdeterminant of F does not exist. Its inverse may, however, be given a meaning,

$$(\mathrm{Sdet}\, F)^{-1} = \det(b - \rho a^{-1}\sigma)/\det a \qquad \text{if} \qquad \det a_0 \neq 0, \tag{10.3.13}$$

provided the numerical part a_0 of the Bose block is nonsingular. If both a_0 and b_0 are nonsingular, the compatibility of the two definitions must be proven.

No work is required to that end if F is block diagonal with $\sigma = \rho = 0$, so that the superdeterminant is the ratio of the determinants of the Bose and Fermi blocks,

$$\mathrm{Sdet}\, F = \frac{\det a}{\det b} \qquad \text{for} \qquad \sigma = \rho = 0, \tag{10.3.14}$$

a case of some importance for applications.

The compatibility proof for general F is a good opportunity to shape up for less trivial adventures to follow later. Under the assumption mentioned, we may rewrite (10.3.12) as $\mathrm{Sdet}\, F = \det(1 - a^{-1}\sigma b^{-1}\rho)\det a/\det b$ and (10.3.13) as $(\mathrm{Sdet}\, F)^{-1} = \det(1 - b^{-1}\rho a^{-1}\sigma)\det b/\det a$ such that we need to show the equality of the two usual $N \times N$ determinants $\det(1 - a^{-1}\sigma b^{-1}\rho)$ and $\left(\det(1 - b^{-1}\rho a^{-1}\sigma)\right)^{-1}$ or, equivalently, the equality of $\mathrm{Tr}\ln(1 - a^{-1}\sigma b^{-1}\rho)$ with $-\mathrm{Tr}\ln(1 - b^{-1}\rho a^{-1}\sigma)$. To that end, we expand both logarithms and need to verify that

$$\mathrm{Tr}(a^{-1}\sigma b^{-1}\rho)^n = -\mathrm{Tr}(b^{-1}\rho a^{-1}\sigma)^n. \tag{10.3.15}$$

But this can be checked by rewriting the l.h.s. in two steps. The first relies on the cyclic invariance of the usual trace, l.h.s. $= \mathrm{Tr}\sigma[b^{-1}\rho(a^{-1}\sigma b^{-1}\rho)^{n-1}a^{-1}]$. No less easy even though a little unfamiliar is the second step. Again under the protection of the trace operation, I want to change the order of the factors σ and $[\dots]$; but since both factor matrices are now fermionic, a minus sign results, $\mathrm{Tr}\sigma[\dots] = -\mathrm{Tr}[\dots]\sigma$, simply because the matrix elements of σ and $[\dots]$ anticommute; the fermionic character of $[\dots]$ follows from the fact that the latter matrix involves an odd number of fermionic factors ρ and σ; with the minus sign generated by shifting σ to the right, we have indeed arrived at the equality (10.3.15) and thus at the compatibility of (10.3.12) and (10.3.13).

The proof of the equality of the superdeterminants of F and F^T,

$$\mathrm{Sdet}\, F = \mathrm{Sdet}\, F^T \tag{10.3.16}$$

is a nice aside left to the reader. I shall, however, pause to show that indeed

$$\ln \mathrm{Sdet}\, F = \mathrm{Str}\ln F. \tag{10.3.17}$$

To that end, I write the general supermatrix F as a product,

$$F = \begin{pmatrix} a & \sigma \\ \rho & b \end{pmatrix} = \begin{pmatrix} a & 0 \\ 0 & b \end{pmatrix} \begin{pmatrix} 1 & a^{-1}\sigma \\ b^{-1}\rho & 1 \end{pmatrix}, \tag{10.3.18}$$

and use (10.3.7) for the supertrace of its logarithm,

$$\operatorname{Str}\ln F = \operatorname{Str}\ln\begin{pmatrix} a & 0 \\ 0 & b \end{pmatrix} + \operatorname{Str}\ln\begin{pmatrix} 1 & a^{-1}\sigma \\ b^{-1}\rho & 1 \end{pmatrix}. \tag{10.3.19}$$

The first of the supertraces appearing on the r.h.s. is simply evaluated as

$$\operatorname{Str}\ln\begin{pmatrix} a & 0 \\ 0 & b \end{pmatrix} = \operatorname{Str}\begin{pmatrix} \ln a & 0 \\ 0 & \ln b \end{pmatrix} = \operatorname{Tr}\ln a - \operatorname{Tr}\ln b = \ln\frac{\det a}{\det b}, \tag{10.3.20}$$

whereas for the second, we must invoke once more the definition of the logarithm as a series; momentarily abbreviating as $\nu = a^{-1}\sigma, \mu = b^{-1}\rho$,

$$\operatorname{Str}\ln\begin{pmatrix} 1 & \nu \\ \mu & 1 \end{pmatrix} = \sum_{n=1}^{\infty}\frac{(-1)^{n-1}}{n}\operatorname{Str}\begin{pmatrix} 0 & \nu \\ \mu & 0 \end{pmatrix}^n.$$

But only even-order terms of the series have nonvanishing supertraces such that

$$= -\sum_{n=1}^{\infty}\frac{1}{2n}\operatorname{Str}\begin{pmatrix} (\nu\mu)^n & 0 \\ 0 & (\mu\nu)^n \end{pmatrix} = -\sum_{n=1}^{\infty}\frac{1}{2n}\left[\operatorname{Tr}(\nu\mu)^n - \operatorname{Tr}(\mu\nu)^n\right]$$

$$= -\sum_{n=1}^{\infty}\frac{1}{n}\operatorname{Tr}(\nu\mu)^n = \operatorname{Tr}\ln(1-\nu\mu) = \ln\det(1-\nu\mu). \tag{10.3.21}$$

Putting (10.3.20, 21) into (10.3.19), we get $\operatorname{Str}\ln F = \ln\left(\det(a-\sigma b^{-1}\rho)/\det b\right) = \ln\operatorname{Sdet}F$, the desired result.

Finally, the product rule

$$\operatorname{Sdet}(F_1 F_2) = (\operatorname{Sdet}F_1)(\operatorname{Sdet}F_2) \tag{10.3.22}$$

follows from the above as $\ln\operatorname{Sdet}F_1 F_2 = \operatorname{Str}\ln F_1 F_2 = \operatorname{Str}\ln F_1 + \operatorname{Str}\ln F_2 = \ln\operatorname{Sdet}F_1 + \ln\operatorname{Sdet}F_2 = \ln\left[\operatorname{Sdet}F_1\operatorname{Sdet}F_2\right]$.

10.3.4 Complex Scalar Product, Hermitian and Unitary Supermatrices

As we have already seen when deriving the semicircle law in the previous section, it is sometimes desirable to associate complex conjugates with Grassmann variables. Our present goal of setting up a suitable superalgebraic framework means that we need to define complex conjugation of supervectors. Well then, we shall work with

$$\Phi = \begin{pmatrix} S \\ \chi \end{pmatrix}, \qquad \Phi^* = \begin{pmatrix} S^* \\ \chi^* \end{pmatrix}, \qquad (\Phi^*)^* = \begin{pmatrix} S \\ -\chi \end{pmatrix}, \tag{10.3.23}$$

whereby we simply double the number of generators of the Grassmann algebra formed by the Fermionic components to $2M_F$. As already done in Sect. 10.2, the definition of complex conjugation is completed by requiring that $(\chi_i^*)^* = -\chi_i$, $(\chi_i\chi_j)^* = \chi_i^*\chi_j^*$, and $(\chi_i + \chi_j)^* = \chi_i^* + \chi_j^*$. It follows that $\chi_i^*\chi_i$ is real. However, $\chi_i + \chi_i^*$ is not real since its complex conjugate is $-\chi_i + \chi_i^*$.

We shall employ the scalar product

$$(\Phi_1, \Phi_2) = \sum_{i=1}^{M} (S_{1i}^* S_{2i} + \chi_{1i}^* \chi_{2i}) \tag{10.3.24}$$

which enjoys the familiar property $(\Phi_1, \Phi_2)^* = (\Phi_2, \Phi_1)$.

The Hermitian conjugate or "adjoint" of a supermatrix F is obtained by taking both the complex conjugate and the transpose,

$$F^\dagger = (F^T)^* = (F^*)^T ; \tag{10.3.25}$$

this is useful because of $(F\Phi_1, \Phi_2) = (\Phi_1, F^\dagger \Phi_2)$. We should note that in contrast to twofold transposition, twofold Hermitian conjugation is the identity operation,

$$F^{\dagger\dagger} = \begin{pmatrix} a & \sigma \\ \rho & b \end{pmatrix}^{\dagger\dagger} = \begin{pmatrix} a^\dagger & \tilde{\rho}^* \\ -\tilde{\sigma}^* & b^\dagger \end{pmatrix}^\dagger = \begin{pmatrix} a & -\sigma^{**} \\ -\rho^{**} & b \end{pmatrix} = F . \tag{10.3.26}$$

Incidentally, the foregoing implies that for the Bose–Fermi and Fermi–Bose blocks, we can maintain the usual definition of Hermitian conjugation, $\sigma^\dagger = \tilde{\sigma}^*$, which implies $\sigma^{\dagger\dagger} = -\sigma$.

A Hermitian supermatrix can be defined as usual as a supermatrix equalling its own adjoint; it has the structure

$$H = H^\dagger = \begin{pmatrix} a & \rho^| \\ \rho & b \end{pmatrix} \qquad \text{where} \qquad a^\dagger = a, \ b^\dagger = b . \tag{10.3.27}$$

Unitary supermatrices preserve the scalar product of supervectors,

$$(U\Phi_1, U\Phi_2) = (\Phi_1, \Phi_2), \qquad UU^\dagger = U^\dagger U = 1 . \tag{10.3.28}$$

10.3.5 Diagonalizing Supermatrices

Special attention is due to Hermitian supermatrices. The eigenvalue problem $H\Phi = h\Phi$ yields coupled homogeneous equations for the bosonic and fermionic components S and χ or, after eliminating of one of them,

$$(a - h - \rho^\dagger \frac{1}{b-h} \rho) S = 0 \tag{10.3.29}$$

$$(b - h - \rho \frac{1}{a-h} \rho^\dagger) \chi = 0 . \tag{10.3.30}$$

The special structure of these equations forbids the appearance of too many eigenvalues: the first makes sense only if the numerical part of $b - h$ has a non-vanishing determinant and thus provides M_B eigenvalues for the Bose block; similarly, the second requires $\det(a - h)_0 \neq 0$ and gives M_F eigenvalues for the Fermi block. To reveal this structure explicitly it is well to consider the simplest case $M_B = M_F = 1$ for which (10.3.29) may be rewritten as the quadratic equation $h^2 - (a+b)h + ab - \rho^*\rho = 0$; one of the two solutions, $h_{\text{Bghost}} = b - \rho^*\rho/(a-b)$, must be discarded since $h_{\text{Bghost}} - b$ has a vanishing numerical part, i.e., is nilpotent. Similarly, (10.3.30) has the spurious solution $h_{\text{Fghost}} = a + \rho\rho^*/(a-b)$ such that we are left with one eigenvalue each for the Bose and Fermi blocks,

$$h_B = a + \rho^*\rho/(a-b), \quad h_F = b + \rho^*\rho/(a-b), \quad \text{for } M_B = M_F = 1. \quad (10.3.31)$$

Needless to say, no eigenvalue exists if coincidentally $a = b$ and $\rho^*\rho \neq 0$. An eigenvalue is called positive if its numerical part is positive.

Momentarily staying with $M_B = M_F = 1$, we may wonder which unitary supermatrix diagonalizes H; it is $H = U\text{diag}(h_B, h_F)U^{-1}$ with

$$U = \begin{pmatrix} 1 + \eta\eta^*/2 & \eta \\ \eta^* & 1 - \eta\eta^*/2 \end{pmatrix}, \quad U^{-1} = \begin{pmatrix} 1 + \eta\eta^*/2 & -\eta \\ -\eta^* & 1 - \eta\eta^*/2 \end{pmatrix} \quad (10.3.32)$$

where nilpotency $\eta = -\rho^*/(a-b)$, $\eta^* = +\rho/(a-b)$, and $U^{-1} = U^\dagger$. Obviously again, for $\rho^* \neq 0$, nilpotency of $a - b$ would preclude diagonalizability.

Our expressions for the eigenvalues h_B and h_F, the matrices U and U^{-1}, and for η and η^* remain valid when b is continued to arbitrary complex values; they are in fact most often used for imaginary b. Even though H then ceases to be Hermitian and U to be unitary, U still diagonalizes H. However, unitarity of U can be restored by redefining complex conjugation for Grassmannians such that η and η^* remain complex conjugates with b an arbitrary complex number (See Problems 10.3 and 10.4).

10.4 Superintegrals

10.4.1 Some Bookkeeping for Ordinary Gaussian Integrals

To properly build up the notion of superintegrals as integrals over both commuting and anticommuting variables, it is well to recall the familiar ordinary Gaussian integral

$$\int \left(\prod_{i=1}^{M} \frac{d^2 S_i}{\pi} \right) \exp \left(- \sum_{i,j=1}^{M} S_i^* a_{ij} S_j \right) = \frac{1}{\det a}, \quad (10.4.1)$$

where the range of integration is the full complex plane for each of the M complex integration variables S_i and the $M \times M$ matrix a must have a positive real part for convergence. The differential $d^2 S_i$ is meant as $d\text{Re}S_i d\text{Im}S_i$. For maximal convenience in changing integration variables, it is sometimes advantageous to think of the differential volume element as an ordered antisymmetric product, the so-called wedge product,[2] $d^2 S_i = d\text{Re}S_i \wedge d\text{Im}S_i = -d\text{Im}S_i \wedge d\text{Re}S_i = (2\text{i})^{-1}dS_i^* \wedge$

---

[2] Readers not previously familiar with the wedge product of differentials of commuting variables have every right to be momentarily confused. They have the choice of either spending a quiet hour with, e.g., Ref. [362], or simply ignoring the following remark and convincing themselves of the correctness of (10.4.2) in some other way, most simply by expanding the integrand in powers of c.

$dS_i = -(2\mathrm{i})^{-1}dS_i \wedge dS_i^*$; the variables S_i and S_i^*, as well as their differentials dS_i and dS_i^*, may then, with some caution, be regarded as independent such that the complex S_i-plane is tiled with area elements $dS_i^* \wedge dS_i = -dS_i \wedge dS_i^*$. One bonus of this bookkeeping device lies in the possibility of changing integration variables in the sense of independent analytic continuation in both S_i and S_i^*, so as to give up the kinship between S_i and S_i^* as mutual complex conjugates. The simplest such change shifts one, but not the other, of these variables by a constant complex number. Thus, e.g.,

$$\int d^2S\, e^{-S^*(S+c)} = \pi\,.$$ (10.4.2)

For any change of the integration variables, the artifice of the wedge product readily yields the familiar transformation of the differential volume element,

$$\prod_{i=1}^{M} d^2 S_i = \det \frac{\partial(S^*,S)}{\partial(z^*,z)} \prod_{i=1}^{M} d^2 z_i = J\!\left(\frac{S^*,S}{z^*,z}\right) \prod_{i=1}^{M} d^2 z_i\,.$$ (10.4.3)

In particular, for a linear transformation which keeps the S_i^* unchanged but reshuffles the S_i,

$$z_i = \sum_j a_{ij} S_j \quad \Longrightarrow \quad J\!\left(\frac{S}{z}\right) = \frac{1}{\det a}\,.$$ (10.4.4)

In fact, (10.4.1) can be seen as an application of (10.4.4): With the latter transformation introduced into the Gaussian integral, we may read S_i^* and z_i as complex conjugates and have $\left((2\pi\mathrm{i})^{-1}\int dS^* \wedge dz \exp(-S^*z)\right)^M = \left(\pi^{-1}\int d^2S\, e^{-S^*S}\right)^M = 1$.

With these remarks in mind, we may change the integration variables in (10.4.1) as $S_i \to S_i - \sum_j (a^{-1})_{ij} z_j$, $S_i^* \to S_i^* - \sum_j z_j^* (a^{-1})_{ji}$ to get

$$(\det a) \int \left(\prod_{i=1}^{M} \frac{d^2 S_i}{\pi}\right) \exp\left(-\sum_{i,j=1}^{M} S_i^* a_{ij} S_j\right) \exp\left[\sum_{i=1}^{M}(S_i^* z_i + z_i^* S_i)\right]$$

$$\equiv \left\langle\!\left\langle \exp\left\{\sum_{i=1}^{M}(S_i^* z_i + z_i^* S_i)\right\}\right\rangle\!\right\rangle = \exp\left[\sum_{i,j=1}^{M} z_i^* (a^{-1})_{ij} z_j\right].$$ (10.4.5)

The angular brackets $\langle\!\langle\ldots\rangle\!\rangle$ in the foregoing identity suggest a change of perspective, hopefully not too upsetting to the reader;[3] they are a shorthand for the average of $(\ldots)$ w.r.t. the normalized Gaussian distribution appearing in the first member. In fact, we may read the quantity in (10.4.5) as the moment generating function for that Gaussian. A nice application is Wick's theorem, which we find

---

[3] Truly daring readers might even enjoy contemplating the kinship of (10.4.5) as well as its superanalytic generalizations below (see (10.4.10, 17)) with the Hubbard–Stratonovich transformation.

by differentiating w.r.t. the auxiliary variables $\{z_i, z_i^*\}$ and subsequently setting these to zero; the simplest cases read

$$\langle S_i S_j^* \rangle = (a^{-1})_{ij} \tag{10.4.6}$$
$$\langle S_i S_j S_k^* S_l^* \rangle = \langle S_i S_k^* \rangle \langle S_j S_l^* \rangle + \langle S_i S_l^* \rangle \langle S_j S_k^* \rangle$$
$$= (a^{-1})_{ik}(a^{-1})_{jl} + (a^{-1})_{il}(a^{-1})_{jk},$$

and the general case is offered to the interested reader as Problem 10.5.

10.4.2 Recalling Grassmann Integrals

By now, we have often used the Grassmann analogue (4.12.1) of (10.4.1),

$$\int \left(\prod_{n=1}^{M} d\chi_n^* d\chi_n \right) \exp\left(- \sum_{ij}^{1...M} \chi_i^* b_{ij} \chi_j \right) = \det b, \tag{10.4.7}$$

which holds for any $M \times M$ matrix b and, like all Grassmann integrals, does not require boundaries to become definite. To tune in to what is to follow, I propose once more to prove that integral representation of $\det b$, in a way differing from the original one of Sect. 4.12. Changing integration variables as

$$\rho_i = \sum_j b_{ij} \chi_j \tag{10.4.8}$$

while holding on to the χ^*'s and employing the Jacobian (4.12.17),

$$J(\tfrac{\chi}{\rho}) = \det b \quad \Longrightarrow \quad \prod_i d\chi_i^* d\chi_i = \det b \prod_i d\chi_i^* d\rho_i, \tag{10.4.9}$$

we see the Gaussian integral (10.4.7) yielding the determinant $\det b$, up to the inconsequential factor $\left[\int d\chi^* d\rho \exp(-\chi^* \rho) \right]^M = \left[\int d\chi^* d\rho (1 - \chi^* \rho) \right]^M = 1$.

The fruit of another change of integration variables in (10.4.7), $\chi_i \to \chi_i - \sum_j (b^{-1})_{ij} \eta_j$, $\chi_i^* \to \chi_i^* - \sum_j \eta_j^* (b^{-1})_{ji}$ with constant Grassmannians η_i, η_i^*, is worthy of respectful consideration,

$$(\det b)^{-1} \int \left(\prod_{n=1}^{M} d\chi_n^* d\chi_n \right) \exp\left[- \sum_{ij}^{1...M} \chi_i^* b_{ij} \chi_j + \sum_i^{1...M} (\chi_i^* \eta_i + \eta_i^* \chi_i) \right]$$
$$\equiv \left\langle \exp\left[\sum_{i=1}^{M} (\chi_i^* \eta_i + \eta_i^* \chi_i) \right] \right\rangle = \exp\left[\sum_{ij}^{1...M} \eta_i^* (b^{-1})_{ij} \eta_j \right]. \tag{10.4.10}$$

We confront the Grassmannian analogue of the moment generating function (10.4.5) that invites pushing the analogy to ordinary Gaussian integrals toward "means" like $\langle \chi_i \chi_j^* \rangle$. By differentiating (10.4.10) w.r.t. to some of the η's and η^*'s and setting all of these "auxiliary" parameters to zero thereafter, we get

$$\langle \chi_i \rangle = \langle \chi_i^* \rangle = 0, \qquad \langle \chi_i \chi_j^* \rangle = (b^{-1})_{ij} = \frac{\partial}{\partial b_{ji}} \ln \det b; \tag{10.4.11}$$

the latter identity is, of course, an old friend from determinantology (see, e.g., (7.4.5)) and may show that the "probabilistic" perspective in which I am indulging here does have useful implications, its admittedly frivolous appearance notwithstanding. Upon taking more derivatives, we encounter the Grassmannian version of Wick's theorem, e.g.,

$$\langle \chi_i \chi_j \chi_k^* \chi_l^* \rangle = -\langle \chi_i \chi_k^* \rangle \langle \chi_j \chi_l^* \rangle + \langle \chi_i \chi_l^* \rangle \langle \chi_j \chi_k^* \rangle . \tag{10.4.12}$$

It is well to note that the different powers of $\det a$ and $\det b$ in the Jacobians (10.4.4, 9) and thus in the Gaussian integrals (10.4.1, 7) determine the starting point of superanalysis in random-matrix theory.

10.4.3 Gaussian Superintegrals

The most important, albeit not most general, Gaussian superintegral is just the product of the Gaussian integral representations for $\det b$ and $1/\det a$,

$$\frac{\det b}{\det a} = \left[\mathrm{Sdet} \begin{pmatrix} a & 0 \\ 0 & b \end{pmatrix} \right]^{-1} \tag{10.4.13}$$

$$= \int \left(\prod_{i=1}^{M} \frac{d^2 S_i}{\pi} d\eta_i^* d\eta_i \right) \exp \left[-\sum_{ij} \left(S_i^* a_{ij} S_j + \eta_i^* b_{ij} \eta_j \right) \right],$$

which I hurry to compact with the help of $\Phi = \binom{S}{\eta}$ and $F = \binom{a\,0}{0\,b}$ as

$$\int d\Phi^* d\Phi \exp(-\Phi^\dagger F \Phi) = (\mathrm{Sdet} F)^{-1} . \tag{10.4.14}$$

We have already used this in Sect. 10.2 for the generating function Z in (10.2.3, 7). There, as again below, we profit from having lifted the matrices a, b to an exponent which makes subsequent averages over these matrices easy to implement, provided that their elements are independent Gaussian random numbers. We should not forget the existence condition $a + a^\dagger > 0$.

Once and for all, I have introduced here the differential volume element for the integration over a supervector Φ,

$$d\Phi^* d\Phi = \prod_{i=1}^{M} \frac{d^2 S_i}{\pi} d\eta_i^* d\eta_i = \prod_{i=1}^{M} \frac{dS_i^* \wedge dS_i}{2\pi\mathrm{i}} d\eta_i^* d\eta_i . \tag{10.4.15}$$

If ever needed, one may allow for the numbers M_B and M_F of bosonic and fermionic components of Φ to be different.

We need to talk about transformations of integration variables. As long as we stick to separate transformations for bosonic and fermionic variables (i.e., forbid any mixing of the two types) and further restrict ourselves to linear reshufflings of the fermionic variables, we just assemble the respective Jacobians (10.4.4, 9) as

$$J\left(\frac{S^*, S, \eta^*, \eta}{z^*, z, \chi^*, \chi} \right) = \det \left(\frac{\partial (S^*, S)}{\partial (z^*, z)} \right) \det \left(\frac{\partial (\eta^*, \eta)}{\partial (\chi^*, \chi)} \right)^{-1}$$

$$= \mathrm{Sdet} \left(\frac{\partial (S^*, S, \eta^*, \eta)}{\partial (z^*, z, \chi^*, \chi)} \right) . \tag{10.4.16}$$

The final form of the Jacobian, often called the Berezinian, also holds for transformations mixing bosonic and fermionic variables; of course, only such transformations are to be admitted which lead to M_B new bosonic and M_F fermionic variables; in particular, the new bosonic variables may have additive pieces even in the old fermionic ones. For a proof of the general validity of the last expression for the Berezinian, I refer the reader to *Berezin*'s book[365].

The compact version (10.4.14) of our superintegral is actually more general than the original, (10.4.13), inasmuch as it does not require the supermatrix F to be block diagonal with vanishing Bose–Fermi and Fermi–Bose blocks. To see this, we simply replace the integration variables Φ by $\Psi = F\Phi$ while leaving the Φ^* unchanged. The Berezinian of that transformation is $J\left(\frac{\Phi}{\Psi}\right) = (\mathrm{Sdet}F)^{-1}$; in the remaining integral $\int d\Phi^* d\Psi \exp(-\Phi^\dagger \Psi)$ we may, in the sense of analytic continuation already alluded to in the previous subsections, reidentify Ψ with the complex conjugate of Φ, whereupon the integral takes the form (10.4.13) with $F = 1$ and is therefore equal to unity. Of course, the Bose–Bose block F_{BB} of F must obey $F_{BB} + F_{BB}^\dagger > 0$ for the integral (10.4.14) to exist.

It is both interesting and useful to extend Wick's theorem to superanalysis, i.e., to "moments" of a "normalized Gaussian distribution" $(\mathrm{Sdet}\,F)\exp(-\Phi^\dagger F\Phi)$. In analogy with our above reasoning for the bosonic and fermionic moment-generating functions (10.4.5, 10), we avail ourselves of a superanalytic one by changing integration variables in (10.4.14) as $\Phi_i \to \Phi_i - \sum_{j=1}^{2M}(F^{-1})_{ij}\Psi_j$, $\Phi_i^* \to \Phi_i^* - \sum_{j=1}^{2M}\Psi_j^*(F^{-1})_{ji}$ to get

$$(\mathrm{Sdet}F)\int d\Phi^* d\Phi \, \exp(-\Phi^\dagger F\Phi)\, \exp\left(\Phi^\dagger\Psi + \Psi^\dagger\Phi\right)$$
$$\equiv \left\langle \exp\left(\Phi^\dagger\Psi + \Psi^\dagger\Phi\right)\right\rangle = \exp(\Psi^\dagger F^{-1}\Psi)\,. \qquad (10.4.17)$$

By differentiating w.r.t. the auxiliary variables $\{\Psi_i, \Psi_i^*\}$, we obtain the superanalytic generalization of (10.4.6, 12),

$$\langle \Phi_i\Phi_j^*\Phi_k\Phi_l^*\rangle = \langle\Phi_i\Phi_j^*\rangle\langle\Phi_k\Phi_l^*\rangle + (-1)^?\langle\Phi_i\Phi_l^*\rangle\langle\Phi_k\Phi_j^*\rangle$$
$$= (F^{-1})_{ij}(F^{-1})_{kl} + (F^{-1})_{il}(F^{-1})_{kj}\,; \qquad (10.4.18)$$

the unspecified sign factor is -1 if an odd number of commutations of fermionic variables is involved in establishing the order of indices $ilkj$ from $ijkl$ and $+1$ otherwise. Needless to say, that identity encompasses the purely bosonic (10.4.6) and the purely fermionic (10.4.12) as special cases.

10.4.4 Some Properties of General Superintegrals

One more issue comes up with changes of integration variables in integrals like

$$I = \int_R d^{M_B}S \int d^{M_F}\chi f(S,\chi)\,, \qquad (10.4.19)$$

where $d^{M_B}S = dS_{M_B}\ldots dS_2 dS_1$, $\quad d^{M_F}\chi = d\chi_{M_F}\ldots d\chi_2 d\chi_1$ and where the M_B bosonic integration variables S_i, as well as their range of integration R, are purely numerical, i.e., contain no even nilpotent admixtures. According to (4.12.7,8), I is well defined provided that the ordinary integral remaining converges, once the Grassmannian one is done. We may, however, think of a change of the integration variables which brings in new bosonic variables with nilpotent admixtures even in the Grassmannians χ_i, formally $S = S(z, \chi)$. The question arises as to what the original integration range R turns into and what is to be understood by the new bosonic part of the differential volume element.

Berezin realized that consistency with the previously established rule (10.4.16) of changing integration variables requires us simply to ignore the idempotent admixtures of the z_i everywhere in the integrand, in $d^{M_B}S$, as well as in the new integration range, provided that the function $f(S, \chi)$ and all of its derivatives vanish on the boundary of the original range R. Referring the interested reader to [364] for the discussion of the general case, I confine myself here to sketching the consistency proof for $M_B = 1, M_F = 2$. Then, the integral (10.4.19) takes the form

$$I = \int_a^b dS \int d\chi_2 d\chi_1 f(S, \chi) \tag{10.4.20}$$

where S, a, b are purely numerical. The transformation

$$S = z + Z(z)\chi_1\chi_2 \tag{10.4.21}$$

has the Jacobian $dS/dz = 1 + Z'(z)\chi_1\chi_2$. Accepting Berezin's rule that the new boundaries should still be a, b, we need to find the conditions for the integral

$$\tilde{I} = \int_a^b dz \int d\chi_2 d\chi_1 \frac{dS}{dz} f(z + Z(z)\chi_1\chi_2, \chi) \tag{10.4.22}$$

to equal I in (10.4.20). Expanding the integrand, we get $f(z + Z(z)\chi_1\chi_2, \chi) = f(z, \chi) + \frac{\partial f(z,0)}{\partial z} Z(z)\chi_1\chi_2$, and thus

$$\begin{aligned}
\tilde{I} &= \int_a^b dz \int d\chi_2 d\chi_1 \frac{dS}{dz} \left[f(z, \chi) + \frac{\partial f(z,0)}{\partial z} Z(z)\chi_1\chi_2 \right] \\
&= I + \int_a^b dz \frac{\partial f(z,0) Z(z)}{\partial z} \\
&= I + f(b,0)Z(b) - f(a,0)Z(a) .
\end{aligned}$$

Obviously, now, $\tilde{I} = I$ provided $f = 0$ at the boundaries $S = a, b$.

More caution is indicated when the integrand does not vanish at the boundaries of the range R, as the following example shows. Let the integrand $f(S, \chi)$ in (10.4.20) be $g(S + \chi_1\chi_2) = g(S) + g'(S)\chi_1\chi_2$. The integral comes out as $I = g(b) - g(a)$. If one insists in shifting the bosonic integration variable to

$z = S + \chi_1\chi_2$, one should rewrite the integral by introducing unit step functions,

$$
\begin{aligned}
I &= \int_{-\infty}^{\infty} dS \int d\chi_2 d\chi_1 g(S + \chi_1\chi_2)\Theta(b - S)\Theta(S - a) \\
&= \int_{-\infty}^{\infty} dz \int d\chi_2 d\chi_1 g(z)\Theta(b - z + \chi_1\chi_2)\Theta(z - a - \chi_1\chi_2) \\
&= \int_{-\infty}^{\infty} dz \int d\chi_2 d\chi_1 g(z)[\Theta(b-z) + \delta(b-z)\chi_1\chi_2][\Theta(z-a) - \delta(z-a)\chi_1\chi_2] \\
&= g(b) - g(a) \, .
\end{aligned}
\tag{10.4.23}
$$

Fortunately, we shall be concerned mostly with integrands that vanish, together with all of their derivatives, at the boundary of the integration range R and in that benign case, one enjoys the further rules

$$
\int_R d^{M_B} S \int d^{M_F} \chi \frac{\partial}{\partial S_i}(fg) = 0 \, ,
$$
$$
\int_R d^{M_B} S \int d^{M_F} \chi \frac{\partial}{\partial \chi_i}(fg) = 0 \, ,
\tag{10.4.24}
$$

which are useful for integration by parts.

10.4.5 Integrals over Supermatrices, Parisi–Sourlas–Efetov–Wegner Theorem

We shall have ample opportunity to deal with integrals over manifolds of non-Hermitian supermatrices

$$
Q = \begin{pmatrix} a & \rho^* \\ \rho & ib \end{pmatrix}
\tag{10.4.25}
$$

with real ordinary numbers a, b and a pair of Grassmannians ρ, ρ^*. In particular, for integrands of the form $f(\mathrm{Str}Q, \mathrm{Str}Q^2)$ that vanish for $a \to \infty, b \to \infty$, we shall need the identity

$$
I = \int dQ \, f(\mathrm{Str}\, Q, \mathrm{Str}\, Q^2) = f(0,0) \qquad \text{where} \qquad dQ = \frac{dadb}{2\pi} d\rho^* d\rho \, .
\tag{10.4.26}
$$

This is mostly referred to as the *Parisi–Sourlas–Efetov–Wegner* (PSEW) theorem, in acknowledgment of its historical emergence [4] [370–374]. An important special case is

$$
\int dQ \, \exp\left(-\frac{1}{2}\mathrm{Str}\, Q^2\right) = 1 \, ,
\tag{10.4.27}
$$

and the correctness of that latter identity is obvious from $\mathrm{Str}Q^2 = a^2 + b^2 + 2\rho^*\rho$. An immediate generalization is $\int dQ \, \exp\left(-c\,\mathrm{Str}\, Q^2\right) = 1$ with an arbitrary

---

[4] Useful generalizations, in particular to higher dimensions, and a watertight proof can be found in Ref. [374].

positive parameter c. To prove the general case (10.4.26), we need to recall from Sect. 10.3.5 that Q can be diagonalized as $Q = U\text{diag}(q_B, iq_F)U^{-1}$ with U given in (10.3.32), but $\eta = -\rho^*/(a - ib), \eta^* = \rho/(a - ib)$. The eigenvalues read

$$q_B = a + \frac{\rho^*\rho}{a - ib}, \qquad iq_F = ib + \frac{\rho^*\rho}{a - ib}. \qquad (10.4.28)$$

To do the superintegral in (10.4.26), it is convenient to change integration variables from a, b, ρ, ρ^* to q_B, q_F and the Grassmannian angles η, η^*. Evaluating the Berezinian for that transformation, we get the measure

$$dQ = \frac{dq_B dq_F}{2\pi(q_B - iq_F)^2} d\eta^* d\eta. \qquad (10.4.29)$$

The integral in search thus takes the form

$$I = \int \frac{dq_B dq_F}{2\pi(q_B - iq_F)^2} d\eta^* d\eta f(q_B - iq_F, q_B^2 + q_F^2) \qquad (10.4.30)$$

which is discomfortingly undefined: the bosonic integral diverges due to the pole in the measure while the Grassmannian integral vanishes. The following trick [369, 374] helps to give a meaning to the unspeakable $0 \cdot \infty$. One sneaks the factor $\exp(-c\text{Str}\,Q^2)$ into the integrand and considers the c-dependence of the integral

$$\begin{aligned} I(c) &= \int dQ\, f(\text{Str}\,Q, \text{Str}\,Q^2) \exp(-c\,\text{Str}\,Q^2) \qquad (10.4.31) \\ &= \int \frac{dq_B dq_F}{2\pi(q_B - iq_F)^2} d\eta^* d\eta f(q_B - iq_F, q_B^2 + q_F^2) \exp\{-c(q_B^2 + q_F^2)\} \end{aligned}$$

which taken at face value, is as undefined for any value of c as for $c = 0$. However, the situation improves for the derivative $dI(c)/dc$ if we differentiate under the integral. Due to $q_B^2 + q_F^2 = (q_B + iq_F)(q_B - iq_F)$, the pole of the integrand at the origin of the $q_B - q_F$-plane now cancels such that the still vanishing Grassmann integral enforces $dI(c)/dc = 0$. We conclude that $I(c)$ itself is independent of c and may thus be evaluated at, say, $c \to \infty$. The latter limit is accessible after the transformation $Q \to Q/\sqrt{2c}$. Since the Berezinian of that latter transformation equals unity,

$$\begin{aligned} I &= \lim_{c\to\infty} \int dQ\, f[\text{Str}\,Q/\sqrt{2c}, \text{Str}\,Q^2/(2c)] \exp\left(-\frac{1}{2}\text{Str}\,Q^2\right) \\ &= f(0,0) \int dQ \exp\left(-\frac{1}{2}\text{Str}\,Q^2\right) = f(0,0). \qquad (10.4.32) \end{aligned}$$

10.5 The Semicircle Law Revisited

To illustrate the conciseness of the superanalytic formulation, I propose to resketch the derivation of the semicircle law using the newly established language.

The average generating function (10.2.7) can be written as the superintegral

$$\overline{Z(E^-, j)} = \int d\Phi^* d\Phi \, \overline{\exp\left(-i\Phi^\dagger F\Phi\right)} = \overline{\text{Sdet}^{-1}F} \,,$$
$$F = \hat{E} \otimes 1_N - \hat{1} \otimes H - \hat{J} \otimes 1_N \qquad (10.5.1)$$

over the 2N-component supervector $\Phi = \binom{S}{\eta}$ with $d\Phi^* d\Phi = \prod_{i=1}^N \frac{d^2 S_i}{\pi} d\eta_i^* d\eta_i$. The three summands in the $2N \times 2N$ matrix F have been written as direct products of 2×2 and $N \times N$ matrices, a hat distinguishes the former,

$$\hat{E} = E^- \begin{pmatrix} 1 & 0 \\ 0 & 1 \end{pmatrix} = E^- \hat{1}, \quad \hat{J} = \begin{pmatrix} j & 0 \\ 0 & 0 \end{pmatrix}. \qquad (10.5.2)$$

The GUE average (10.2.11) reads

$$\overline{\exp\{i\Phi^\dagger \hat{1} \otimes H\Phi\}} = \exp\left\{-\frac{1}{2}\overline{(\Phi^\dagger \hat{1} \otimes H\Phi)^2}\right\} = \exp\left\{-\frac{N\lambda^2}{2}\text{Str}\,\tilde{Q}^2\right\},$$
$$\tilde{Q} = \frac{1}{N}\begin{pmatrix} \sum_m z_m z_m^* & \sum_m z_m \eta_m^* \\ \sum_m \eta_m z_m^* & \sum_m \eta_m \eta_m^* \end{pmatrix} \qquad (10.5.3)$$

and yields

$$\overline{Z(E^-, j)} = \int d\Phi^* d\Phi \, \exp\left\{-N\text{Str}\left[i(\hat{E} - \hat{J})\tilde{Q} + \frac{\lambda^2}{2}\tilde{Q}^2\right]\right\}. \qquad (10.5.4)$$

This demands a Hubbard-Stratonovich transformation for us to come back to a Gaussian superintegral over the supervector Φ. The resulting representation (10.2.14) of the average generating function can be seen as a fourfold superintegral over the elements of a 2×2 supermatrix

$$\hat{Q} = \begin{pmatrix} a & \sigma^* \\ \sigma & ib \end{pmatrix}, \quad \text{Str}\,\hat{Q}^2 = a^2 + b^2 + 2\sigma^*\sigma, \quad d\hat{Q} = \frac{dadb}{2\pi} d\sigma^* d\sigma,$$
$$\qquad (10.5.5)$$

and takes the form

$$\overline{Z(E^-, j)} = \int d\hat{Q} \, e^{-\frac{N}{2\lambda^2}\text{Str}\,\hat{Q}^2}\left(\int d\phi^* d\phi \, e^{-i\phi^\dagger(\hat{E} - \hat{Q} - \hat{J})\phi}\right)^N$$
$$= \int d\hat{Q} \, e^{-\frac{N}{2\lambda^2}\text{Str}\,\hat{Q}^2}\text{Sdet}^{-N}(\hat{E} - \hat{Q} - \hat{J}) \qquad (10.5.6)$$
$$= \int d\hat{Q} \, \exp\left\{-N\text{Str}\left[\frac{1}{2\lambda^2}\hat{Q}^2 + \ln(\hat{E} - \hat{Q} - \hat{J})\right]\right\}.$$

Here the original $4N$-fold superintegral over the components of the supervector Φ and their conjugates could be replaced by the Nth power of a fourfold integral over the two-component supervector $\phi = \binom{z}{\eta}$. Needless to say, all matrices appearing in (10.5.6) are 2×2.

While in Sect. 10.2 we proceeded by doing the two Grassmann integrals over the skew elements of $\hat{Q}$ rigorously and reserved a saddle-point approximation for the integral over the diagonal elements, here we shall treat all variables equally and propose a saddle-point approximation for the whole superintegral

$$\overline{Z(E^-, j)} = \int d\hat{Q}\, e^{-NA(\hat{Q}, \hat{j})},$$

$$A(\hat{Q}, \hat{J}) = \mathrm{Str}\left[\frac{1}{2\lambda^2}\hat{Q}^2 + \ln\left(\hat{E} - \hat{Q} - \hat{J}\right)\right]. \tag{10.5.7}$$

Before embarking on that adventure, it is well to pause by remarking that the foregoing integral must equal unity if we set $\hat{J} = 0$,

$$\overline{Z(E^-, 0)} = \int d\hat{Q}\, e^{-NA(\hat{Q}, 0)} = 1, \tag{10.5.8}$$

since unity is indeed the value of a block diagonal superdeterminant whose Bose and Fermi blocks are identical, cf (10.2.3) or (10.4.13). This is relevant since due to (10.2.4) we eventually want to set j and thus the matrix $\hat{J}$ equal to zero,

$$\overline{G(E^-)} = \int d\hat{Q}\, \mathrm{Str}\left(\frac{\hat{J}}{j}\frac{1}{\hat{E} - \hat{Q}}\right) e^{-NA(\hat{Q}, 0)}. \tag{10.5.9}$$

Now, the spirit of the saddle-point approximation, we may appeal to the assumed largeness of N and pull the preexponential factor out of the integral and evaluate it at the saddle-point value $\hat{Q}_0$ of the supermatrix $\hat{Q}$. Then, the remaining superintegral yields unity, and we are left with

$$\overline{G(E^-)} = \mathrm{Str}\left(\frac{\hat{J}}{j}\frac{1}{\hat{E} - \hat{Q}_0}\right). \tag{10.5.10}$$

To find $\hat{Q}_0$, we must solve the saddle-point equation

$$\delta A(\hat{Q}, 0) = \mathrm{Str}\,\delta\hat{Q}\left(\frac{\hat{Q}}{\lambda^2} - \frac{1}{\hat{E} - \hat{Q}}\right) = 0 \quad\Longrightarrow\quad \frac{\hat{Q}_0}{\lambda^2} = \frac{1}{\hat{E} - \hat{Q}_0}. \tag{10.5.11}$$

Assuming $\hat{Q}_0$ diagonalizable, we may start looking for diagonal solutions; there is indeed no undue loss of generality in that since all other solutions are accessible from diagonal ones by unitary transformations. For diagonal saddles, however, the saddle-point equation becomes an ordinary quadratic equation and yields two solutions (cf (10.2.23))

$$Q_\pm = (E^-/2)\left(1 \pm \sqrt{1 - (2\lambda/E^-)^2}\right). \tag{10.5.12}$$

Precisely as in Sect. 10.2 we must evaluate the square root in $Q_\pm = a_\pm = ib_\pm$ on the lower lip of the cut along the real axis from -2λ to $+2\lambda$ in the complex energy plane. Concentrating immediately on $(E/2)^2 < \lambda^2$, i.e., case (1) of Sect. 10.2.4 $Q_\pm = a_\pm = ib_\pm = E/2 \mp i\sqrt{\lambda^2 - (E/2)^2}$.

Four possibilities arise for the diagonal 2×2 matrix in search,

$$\begin{pmatrix} Q_+ & 0 \\ 0 & Q_+ \end{pmatrix}, \quad \begin{pmatrix} Q_+ & 0 \\ 0 & Q_- \end{pmatrix}, \quad \begin{pmatrix} Q_- & 0 \\ 0 & Q_+ \end{pmatrix}, \quad \begin{pmatrix} Q_- & 0 \\ 0 & Q_- \end{pmatrix}. \tag{10.5.13}$$

According to (10.5.10, 11), each of the first two would contribute Q_+/λ^2, and the last two Q_-/λ^2 to the average Green function. Therefore, the first two must be discarded right away since the imaginary part of $\text{Im}\overline{G(E_-)} = \pi\varrho(E)$ must be positive. This corresponds with our discarding the saddle a_+ in Sect. 10.2.4, arguing that the original contour of integration could not be continuously deformed so as to pass through that saddle without crossing a singularity. The question remains whether both of the last two contribute or only one of them and then which. To get the answer, we scrutinize the Gaussian superintegral over the fluctuations around the saddle. The second variation $\delta^2 A(\hat{Q},0)$ of the action, i.e., the second-order term in $\delta\hat{Q}$ of $A(\hat{Q}_0 + \delta\hat{Q},0) - A(\hat{Q}_0,0)$, reads

$$\delta^2 A(\hat{Q},0) = \frac{1}{2}\,\text{Str}\left[\frac{(\delta\hat{Q})^2}{\lambda^2} - \frac{(\delta\hat{Q}\hat{Q}_0)^2}{\lambda^4}\right] \qquad (10.5.14)$$

and must be evaluated for the two candidates for $\hat{Q}_0$. Denoting by $\delta^2 A_{--}$ the result for $\text{diag}(Q_-, Q_-)$ and by $\delta^2 A_{-+}$ that for $\text{diag}(Q_-, Q_+)$ (for simplicity at $E = 0$),

$$\delta^2 A_{--} = \frac{1}{\lambda^2}\,\text{Str}\,(\delta\hat{Q})^2 = \frac{1}{\lambda^2}(\delta a^2 + \delta b^2 + \delta\sigma^*\delta\sigma)\,,$$

$$\delta^2 A_{-+} = \frac{1}{\lambda^2}(\delta a^2 + \delta b^2)\,. \qquad (10.5.15)$$

Obviously now, the second saddle does not contribute at all (to leading order in N) since $\int d\delta\hat{Q}\exp(-N\delta^2 A_{+-}) = 0$, such that we arrive at

$$\overline{G(E^-)} = \frac{1}{\lambda^2}\left(E^-/2 + i\sqrt{\lambda^2 - (E^-/2)^2}\right) \qquad \text{for} \qquad (E/2)^2 < \lambda^2\,,$$
$$(10.5.16)$$

the result already established by a less outlandish calculation in Sect. 10.2; see (10.2.28). The reader is invited to go through the analogous reasoning for $(E/2)^2 > \lambda^2$. At any rate, the semicircle law follows immediately by taking the imaginary part as before.

Incidentally, it is quite remarkable that the nonscalar saddle $\text{diag}(Q_-, Q_+)$ does not contribute since it would come with a whole manifold of companions: Indeed, a $\hat{Q}$ solving the saddle-point equation gives rise to further solutions $U\hat{Q}U^\dagger$ with any unitary transformation U, provided that $\hat{Q} \neq U\hat{Q}U^\dagger$, i.e., provided that $\hat{Q}$ is not proportional to the unit matrix. As a final remark on the saddle $\text{diag}\,(Q_-, Q_+)$, I mention that it corresponds to the saddle $\{a_-, b_+\}$ of Sect. 10.2.4 which there, too, did not contribute in leading order, due to the vanishing of the preexponential factor at the saddle.

Should the previously uninitiated suffer from a little dizziness due to our use of the saddle-point approximation for a superintegral, cure might come from Problem 10.7.

10.6 The Two-Point Function of the Gaussian Unitary Ensemble

We proceed to calculate Dyson's two-point cluster function for the Gaussian unitary ensemble and show that it equals its counterpart for the circular unitary ensemble (see Sect. 4.13),

$$
Y_{\text{GUE}}(e) = Y_{\text{CUE}}(e) = \left(\frac{\sin \pi e}{\pi e} \right)^2 .
\tag{10.6.1}
$$

As in Sect. 10.2, I shall start with the Green function $G(\mathcal{E}) = N^{-1}\text{Tr}(\mathcal{E} - H)$ but now must consider the GUE average of the product of two such, $\overline{G(\mathcal{E}_1)G(\mathcal{E}_2)}$. We shall see that average behaves rather differently for large N, depending on whether the two complex energies lie on the same side or on different sides of the real axis. In the former case, the average product tends to the product of averages,

$$
\overline{G(E_1 + i\delta)G(E_2 + i\delta)} \xrightarrow{N \to \infty} \overline{G}(E_1 + i\delta)\, \overline{G}(E_2 + i\delta) ,
\tag{10.6.2}
$$

while in the latter case $\overline{G(E_1 + i\delta)G(E_2 - i\delta)}$ tends, apart from normalization, to a function of the single variable

$$
e = (E_1 - E_2)N\overline{\varrho}\left(\frac{E_1 + E_2}{2} \right) ,
\tag{10.6.3}
$$

i.e., the energy difference measured in units of the mean level spacing at the center energy. Anticipating the validity of the factorization (10.6.2) and employing (10.2.2), we can immediately check the connected density-density correlator, $\overline{\Delta\varrho(E_1)\Delta\varrho(E_2)} = \overline{\big(\varrho(E_1) - \overline{\varrho}(E_1)\big)\big(\varrho(E_2) - \overline{\varrho}(E_2)\big)}$, to be given by the real part of the connected version $\overline{\Delta G(E_1 - i0^+)\Delta G(E_2 + i0^+)}$ of the averaged product of two Green functions,

$$
\overline{\Delta\varrho(E_1)\Delta\varrho(E_2)} = \frac{1}{2\pi^2} \text{Re}\, \overline{\Delta G(E_1 - i0^+)\Delta G(E_2 + i0^+)} .
\tag{10.6.4}
$$

10.6.1 The Generating Function

The product of two Green functions can be obtained from the generating function

$$
\begin{aligned}
Z(E_1^-, E_2^+, E_3^-, E_4^+) &= \frac{\det(E_3^- - H)\det(E_4^+ - H)}{\det(E_1^- - H)\det(E_2^+ - H)} \\
&= (-1)^N \frac{\det \text{i}(E_3^- - H)\det \text{i}(E_4^+ - H)}{\det \text{i}(E_1^- - H)\det(-\text{i})(E_2^+ - H)}
\end{aligned}
\tag{10.6.5}
$$

by differentiating as

$$
G(E_1^-)G(E_2^+) = \frac{1}{N^2} \frac{\partial^2 Z(E_1^-, E_2^+, E_3^-, E_4^+)}{\partial E_1 \partial E_2} \bigg|_{E_1 = E_3, E_2 = E_4} .
\tag{10.6.6}
$$

This is analogous to (10.2.3, 4), and the proof parallels that in (10.2.5): Each differentiation brings about one factor G and the determinants cancel pairwise once we set $E_1 = E_3, E_2 = E_4$. Silly as it may appear to have sneaked various factors i in the last member of (10.6.5) (since they obviously cancel against the factor $(-1)^N$), such precaution pays, similarly as in Sect. 10.2, once we employ the Gaussian superintegral (10.4.13) to represent quotients of determinants,

$$Z = (-1)^N \int \left(\prod_{i=1}^{N} \frac{d^2 z_{1i}}{\pi} d\eta_{1i}^* d\eta_{1i} \right) \exp i \left(z_1^\dagger (H - E_1^-) z_1 + \eta_1^\dagger (H - E_3^-) \eta_1 \right) \quad (10.6.7)$$

$$\times \int \left(\prod_{i=1}^{N} \frac{d^2 z_{2i}}{\pi} d\eta_{2i}^* d\eta_{2i} \right) \exp i \left[-z_2^\dagger (H - E_2^+) z_2 + \eta_2^\dagger (H - E_4^+) \eta_2 \right],$$

with bosonic vectors z_1, z_2 and fermionic vectors η_1, η_2, all having N components. Indeed, the i's now conspire to ensure convergence of the bosonic Gaussian integrals, just as in (10.2.6).

To save space, we introduce the $(4N)$-component supervector

$$\Phi = \begin{pmatrix} z_1 \\ z_2 \\ \eta_1 \\ \eta_2 \end{pmatrix} = \begin{pmatrix} \Phi_1 \\ \Phi_2 \\ \Phi_3 \\ \Phi_4 \end{pmatrix} \quad (10.6.8)$$

such that the differential volume element $d\Phi^* d\Phi$ comprises those in the preceding integral, (10.6.7). Writing the energy variables with their infinitesimal imaginary offsets as $E_\alpha + i(-1)^\alpha 0^+$, $\alpha = 1, 2, 3, 4$ and employing the symbols L_α where $L_1 = L_3 = L_4 = -L_2 = 1$ to inconspicuously accommodate the funny looking convergence ensurer $L_2 = -1$ in the exponent, we can write the generating function in the compact form

$$Z = (-1)^N \int d\Phi^* d\Phi \exp \left\{ i \sum_{\alpha=1}^{4} \Phi_\alpha^\dagger L_\alpha [H - E_\alpha - i(-1)^\alpha 0^+] \Phi_\alpha \right\}. \quad (10.6.9)$$

All is set now for the average over the GUE; the superhocus-pocus has served to move the random Hamiltonian matrix H upstairs into an exponent. Recalling the variances (10.2.9) and the Gaussian average of exponentials (10.2.10), in the spirit of (10.2.11),

$$\overline{\exp \left(i \sum_{\alpha=1}^{4} \Phi_\alpha^\dagger L_\alpha H \Phi_\alpha \right)} = \exp \left(-\frac{1}{2} \overline{\left(\sum \Phi_\alpha^\dagger L_\alpha H \Phi_\alpha \right)^2} \right)$$

$$= \exp \left[-\frac{1}{2} \overline{\left(z_1^\dagger H z_1 - z_2^\dagger H z_2 + \eta_1^\dagger H \eta_1 + \eta_2^\dagger H \eta_2 \right)^2} \right] \quad (10.6.10)$$

$$= \exp \left(-\frac{\lambda^2}{2N} \sum_{\alpha,\beta} L_\alpha L_\beta \sum_{m,n} \Phi_{\alpha m}^* \Phi_{\alpha n} \Phi_{\beta n}^* \Phi_{\beta m} \right).$$

Once at work down-sizing bulky expressions, why not continue with 4×4 matrices

$$L = \mathrm{diag}(1, -1, 1, 1), \quad \hat{E} = \mathrm{diag}(E_1^-, E_2^+, E_3^-, E_4^+),$$

$$\tilde{Q}_{\alpha\beta} = \frac{1}{N} \sum_{m=1}^{N} \Phi_{\alpha m} \Phi_{\beta m}^* \tag{10.6.11}$$

and write the averaged generating function analogously to (10.5.4) as

$$\overline{Z}(\hat{E}) = (-1)^N \int d\Phi^* \, d\Phi \exp\left\{ N \mathrm{Str}\left[-i\hat{E}\tilde{Q}L - \frac{\lambda^2}{2}(\tilde{Q}L)^2 \right] \right\}. \tag{10.6.12}$$

The supermatrix $\tilde{Q} = \tilde{Q}^\dagger$ is Hermitian and has a non-negative Bose–Bose block.

It remains to do the $8N$-fold superintegral in (10.6.12) which, unfortunately, contains a quartic expression in the integration variables of the exponent. Taking a short breath, it will be good to realize that at the corresponding stage of the calculation of the mean Green function in Sect. 10.2, we had to employ the Hubbard–Stratonovich transformation to return to Gaussian integrals.

10.6.2 Unitary Versus Hyperbolic Symmetry

Had we placed all energy arguments $\mathcal{E}_\alpha$ of the generating function on one and the same side of the real axis, like $\mathcal{E}_\alpha = E^-$, we would have arrived at (10.6.12) with all four $L_\alpha = 1$, i.e., $L = \hat{1}$; then the integrand would be invariant under $\tilde{Q} \to U\tilde{Q}U^\dagger$ with an arbitrary unitary supermatrix since $\mathrm{Str}\, UMU^\dagger = \mathrm{Str} M$. I strongly recommend that the reader consider that case and run through all steps of Sect. 10.5 to check that everything goes through, the doubling of the dimension of the matrices $\tilde{Q}, \hat{Q}$ notwithstanding. Precisely, because of dimension doubling, the saddle-point integral gives the average product $\overline{G(E_1^-)G(E_2^-)}$ as the product of averages $\overline{G(E_1^-)}\ \overline{G(E_2^-)}$, as claimed in (10.6.2) .

The case is much less trivial when the energy arguments lie on both sides of the real axis, $\mathcal{E}_\alpha = E_\alpha + i(-1)^\alpha 0^+$. Then the exponential in the superintegral (10.6.12) would be invariant under the transformation

$$\tilde{Q} \to T\tilde{Q}T^\dagger \tag{10.6.13}$$

where a 4×4 supermatrix T satisfies

$$T^\dagger L T = L \tag{10.6.14}$$

if $\hat{E}$ were proportional to the 4×4 unit matrix, $\hat{E} \to E\hat{1}$. Of course, we need to allow for $E_1 = E_3 \neq E_2 = E_4$ such that the "hyperbolic" alias "pseudo-unitary" symmetry (10.6.13, 14) does not hold rigorously. But since the cumulant function in search decays to zero on an energy scale of the order of the mean level spacing which for the normalization of H chosen is of the order $1/N$, the relevant difference of the two energies in question is very small, $E_1 - E_2 = \mathcal{O}(1/N)$, and the hyperbolic symmetry does indeed hold to leading order in N.

The name "hyperbolic" symmetry should be commented on. If L were the unit matrix, the restriction (10.6.14) would render T unitary and the ensuing unitary symmetry of the exponent (10.6.12) would be the one pertaining to the disconnected part $\overline{G(E_1)\,G(E_2)}$ of the two-point function, as discussed above. It is precisely the single negative element in the Bose block of $L = \mathrm{diag}(1, -1, 1, 1)$ that replaces the unitary symmetry with the pseudounitary alias hyperbolic one we need to deal with when after the cumulant function. I shall refer to supermatrices T with the property $T^\dagger LT = L$ as pseudounitary. The reader is invited to check that these matrices form a group (see Problem 10.9).

For use later, I propose to show that the $4{\times}4$ supermatrix $\tilde{Q}L$ in the integrand of (10.6.12) can be diagonalized by a pseudounitary matrix. That diagonalizability arises due to the Hermiticity $\tilde{Q} = \tilde{Q}^\dagger$ and the positivity of the Bose block of $\tilde{Q}$.

Since L is diagonal and has the 2×2 unit matrix in the Fermi block, we may save space and labor and still retain the essence of the proof by first restricting ourselves to the toy problem of ordinary 2×2 matrices. For these, it is easy to see that the manifold $\mathcal{M}_2$ of Hermitian matrices $Q = Q^\dagger$ for which QL with $L = \mathrm{diag}\,(1, -1)$ is pseudounitarily diagonalizable is

$$\mathcal{M}_2: \qquad Q = Q^\dagger = \begin{pmatrix} a & c^* \\ c & b \end{pmatrix}, \qquad |c|^2 < \left(\frac{a+b}{2}\right)^2. \tag{10.6.15}$$

Nondiagonal matrices of the form QL are not Hermitian but will in general still be diagonalizable by a similarity transformation that we readily find as

$$QL = \begin{pmatrix} a & -c^* \\ c & -b \end{pmatrix} = S \begin{pmatrix} q_+ & 0 \\ 0 & q_- \end{pmatrix} S^{-1},$$

$$q_\pm = \tfrac{a-b}{2} \pm \sigma\sqrt{\left(\tfrac{a+b}{2}\right) - |c|^2},$$

$$S = \begin{bmatrix} c^* & \tfrac{a+b}{2} - \sigma\sqrt{(\tfrac{a+b}{2})^2 - |c|^2} \\ \tfrac{a+b}{2} - \sigma\sqrt{(\tfrac{a+b}{2})^2 - |c|^2} & c \end{bmatrix}, \tag{10.6.16}$$

$$\sigma = \mathrm{sign}(a+b),$$

$$\det S = -2\big[(\tfrac{a+b}{2})^2 - |c|^2\big] + |a+b|\sqrt{(\tfrac{a+b}{2})^2 - |c|^2}.$$

Note that the inequality in (10.6.15) secures positivity of the radicand in the above square roots and thus the reality of the off-diagonal elements $S_{12} = S_{21}$ of the matrix S. The two eigenvalues $q_\pm$ of QL are real and different from one another, while $\det S > 0$. Direct calculation yields $S^\dagger LSL = \binom{1\,0}{0\,1}\det L$. But since $\det S > 0$, we can renormalize as $T = S/\sqrt{\det S}$ such that T is indeed pseudounitary, $T^\dagger LT = L$. Positive matrices Q fall in our manifold $\mathcal{M}_2$: Positivity of Q requires $a > 0, b > 0, ab > |c|^2$ and is compatible with $|c|^2 < (a+b)^2/4$ since $ab \le (a+b)^2/4 = (a-b)^2/4 + ab$.

On the other hand, the case of negative radicand would still leave QL diagonalizable by the above S, with complex eigenvalues $q_\pm$; but since the off-diagonal elements $S_{12} = S_{21}$ also become complex, pseudounitarity is lost. Finally, a vanishing radicand yields $\det S = 0$ and altogether precludes diagonalizability.

To proceed from the toy problem to the manifold $\mathcal{M}_4$ of 4×4 supermatrices diagonalizable by pseudounitary transformations, we take two steps. The first is to let $Q = Q^\dagger$ be 4×4 and block diagonal with the Bose block from $\mathcal{M}_2$. Then, the toy argument goes through practically unchanged where T s also block diagonal; the toy problem teaches us that the Bose block of T is pseudounitary, $T_{BB}^\dagger L_{BB} T_{BB} = L_{BB}$; the Fermi block T_{FF} is unitary since the Hermitian 2×2 matrix $(QL)_{FF} = Q_{FF} L_{FF} = Q_{FF}$ can be diagonalized by a unitary 2×2 matrix. Then, the 4×4 matrix T itself is pseudounitary in the sense $T^\dagger L T = L$.

The second step of elevation amounts to showing that a 4×4 matrix QL can be brought to block diagonal form by a pseudounitary matrix T'. It is indeed easy to see that the job required is done by

$$T' = \begin{pmatrix} \sqrt{1 - \gamma^\dagger \gamma k} & \gamma^\dagger \\ -\gamma k & \sqrt{1 - \gamma k \gamma^\dagger} \end{pmatrix}, \tag{10.6.17}$$

$$T'^\dagger = \begin{pmatrix} \sqrt{1 - k\gamma^\dagger \gamma} & -k\gamma^\dagger \\ \gamma & \sqrt{1 - \gamma k \gamma^\dagger} \end{pmatrix}. \tag{10.6.18}$$

Here, I have given the typographically less voluminous name k to the 2×2 block $T_{BB} = \mathrm{diag}(1, -1) \equiv k$ and called γ and $\gamma^\dagger$ a pair of 2×2 matrices with four independent anticommuting entries each, viz., γ_{ij} and $(\gamma^\dagger)_{ij} = \gamma_{ji}^*$, such that γ_{ji}^* is complex conjugate to γ_{ji}. There are eight independent parameters in T' and $T'^\dagger$; these are all Grassmannians and just right in number to make the equations $(T'^\dagger QLT')_{BF} = (T'^\dagger QLT')_{FB} = 0$ generically solvable. Pseudounitarity, on the other hand, is built into (10.6.17, 18). When block diagonality is thus achieved, the Bose and Fermi blocks and their eigenvalues will change only by acquiring nilpotent additions; such additions cannot foul up the positivity of the Bose block since positivity of an eigenvalue is a property of its numerical part.

Further preparation for our intended use of the above results is to be undertaken. To ensure convergence of certain Gaussian integrals, we shall need to analytically continue the 4×4 matrices Q such that their Fermi blocks become anti-Hermitian rather than Hermitian, in analogy with what we had to do in (10.5.5) (where we put $b \to ib$). Then, we can still employ the matrices T' and $T'^\dagger$ to bring QL to block diagonal form. If we stick to the conventional definition of complex conjugation for Grassmannians, the pseudounitarity of T' is lost. However, according to the model of the discussion in the concluding paragraph of Sect. 10.3.5, we may restore pseudounitarity by appropriately redefining complex conjugation of Grassmannians; in the following, we shall tacitly imagine that such restoration is implemented together with the continuation of Q_{BB} to anti-Hermiticity.

Finally, we turn to the Hubbard–Stratonovich transformation needed to turn the integral over the supervectors Φ, Φ^* in (10.6.12) into a Gaussian. We cannot uncritically employ the usual formula which would involve an unrestricted integral over an auxiliary 4×4 supermatrix. The correct procedure was pioneered in a rather different context by *Schäfer* and *Wegner* [375], later adapted to superanalytic needs by *Efetov* [372], and comprehensively discussed in Ref. [376]; it amounts to restricting the integration range for the auxiliary supermatrix to the manifold $\mathcal{M}_4$. To explain the essence of the idea, it suffices once more to consider ordinary 2×2 matrices and prove the integral identity

$$\exp\left[-\tfrac{1}{2}\mathrm{tr}(\tilde{Q}L)^2\right] = -\int_{\mathcal{M}_2} dQ \exp\left[-\tfrac{1}{2}\mathrm{tr}(QL)^2 + \mathrm{i}\,\mathrm{tr}QL\tilde{Q}L\right] \qquad (10.6.19)$$

for matrices $\tilde{Q} \in \mathcal{M}_2$; as usual, the integration measure is $dQ = da\,db\,d^2c/(2\pi^2)$. The minus sign signals that the integral is far from being a usual Gaussian. For the proof, we first bring $\tilde{Q}L$ to diagonal form by a suitable pseudounitary transformation, $T\tilde{Q}LT^{-1} = \mathrm{diag}(q_+, q_-)$. Subsequently, when T is absorbed in the integration variable Q, neither the integration range nor the measure is changed. The remainder of the proof is a straightforward calculation,

$$\int_{-\infty}^{\infty} da \int_{-\infty}^{\infty} db \int_{|c|^2 < (\frac{a+b}{2})^2} d^2c \, \exp\left[-\tfrac{1}{2}(a^2 + b^2 - 2|c|^2) + \mathrm{i}(aq_+ - bq_-)\right]$$

$$= \pi \int_{-\infty}^{\infty} da \int_{-\infty}^{\infty} db \left\{\exp\left[-(\tfrac{a-b}{2})^2\right] - \exp\left(-\tfrac{a^2+b^2}{2}\right)\right\} \mathrm{e}^{\mathrm{i}(aq_+ - bq_-)}. \qquad (10.6.20)$$

The first term in the curly bracket in the foregoing integrand depends only on the difference $a - b$ and thus might be expected to give rise to a term proportional to the delta function $\delta(q_+ - q_-)$; it can't since for $Q \in \mathcal{M}_2$, the two eigenvalues $q_\pm$ of QL necessarily differ from one another. Only the second term in the square bracket survives; it is negative and obviously yields $-2\pi^2 \exp[-\tfrac{1}{2}(q_+^2 + q_-^2)] = -2\pi^2 \exp[-\tfrac{1}{2}\mathrm{tr}(\tilde{Q}L)^2]$. The extension of the calculation to 4×4 supermatrices and an integral over the manifold $\mathcal{M}_4$ does not present difficulties and gives, after a rescaling suitable for our need in (10.6.12),

$$\exp\left[-\tfrac{N\lambda^2}{2}\mathrm{Str}(\tilde{Q}L)^2\right] = -\int_{\mathcal{M}_4} dQ \exp\left\{-N\mathrm{Str}\left[\tfrac{1}{2\lambda^2}(QL)^2 - \mathrm{i}QL\tilde{Q}L\right]\right\}. \qquad (10.6.21)$$

10.6.3 Efetov's Nonlinear Sigma Model

By inserting the Hubbard–Stratonovich transformation (10.6.21) in the averaged generating function (10.6.12), we can do the Gaussian integral over the supervector Φ. As in Sect. 10.5, we get

$$\overline{Z}(\hat{E}) = (-1)^{N+1} \int_{\mathcal{M}_4} dQ \exp\left\{-N\mathrm{Str}\left[\tfrac{1}{2\lambda^2}(QL)^2 + \ln\left(\mathrm{i}L\hat{E} - \mathrm{i}LQL\right)\right]\right\}. \qquad (10.6.22)$$

For the Q integral to converge in the Fermi block, it is imperative to generalize the previously encountered 2×2 matrices of the structure (10.5.5) to 4×4 such that the previous $Q_{FF} = \mathrm{i}b$ becomes an anti-Hermitian 2×2 matrix. No extra precaution is necessary for the 2×2 Bose block Q_{BB}, since $Q \in \mathcal{M}_4$ implies

$Q_{BB} \in \mathcal{M}_2$. Both of the foregoing facts can be seen at work in the following inequality:

$$
\begin{aligned}
\mathrm{Str}(QL)^2 &= \mathrm{Tr}\left[\left\{Q_{BB}\begin{pmatrix}1 & 0 \\ 0 & -1\end{pmatrix}\right\}^2 - Q_{FF}^2\right] \quad\quad (10.6.23) \\
&= (Q_{11} - Q_{22})^2 + 2\det Q_{BB} + \mathrm{Tr}\, Q_{FF}Q_{FF}^{\dagger} \\
&= a^2 + b^2 - 2|c|^2 + \mathrm{Tr}\, Q_{FF}Q_{FF}^{\dagger} \\
&= 2\left[\left(\tfrac{a-b}{2}\right)^2 + \left(\tfrac{a+b}{2}\right)^2 - |c|^2\right] + \mathrm{Tr}\, Q_{FF}Q_{FF}^{\dagger} > 0,
\end{aligned}
$$

which in turn does indeed ensure convergence for the integral in (10.6.22).

To achieve minor embellishments in (10.6.22), we invoke $\mathrm{Str}\ln(iL\hat{E} - iLQL) = \ln\mathrm{Sdet}(L\hat{E} - LQL) = \ln\mathrm{Sdet}\,L + \ln\mathrm{Sdet}(\hat{E} - QL)$ and $\mathrm{Sdet}\,L = -1$ and even use QL rather than Q as integration variables. Moreover, an overall minus sign (that in the Hubbard–Stratonovich transformation (10.6.21)) is swept into the integration measure and will receive attention only in the final expression for dQ at the end of the present subsection. For now we confront

$$
\overline{Z}(\hat{E}) = \int_{\mathcal{M}_4 L} dQ \exp\left\{-N\mathrm{Str}\left[\frac{1}{2\lambda^2}Q^2 + \ln(\hat{E} - Q)\right]\right\}. \quad\quad (10.6.24)
$$

The very appearance of this integral invites a saddle-point approximation. In contrast to previous appearances of that approximation, a whole manifold of saddle points is lurking around the corner, so we ought to proceed with care.

First, we are interested in $E_1 - E_2 = \mathcal{O}(1/N)$ and make such smallness manifest by writing $\mathrm{Str}\ln(\hat{E} - Q) = \mathrm{Str}\ln(\tilde{E} - Q) + \mathrm{Str}(\hat{E} - \tilde{E})(\tilde{E} - Q)^{-1}$ where $\tilde{E} = \mathrm{diag}(E^-, E^+, E^-, E^+)$ and the energy E is suitably chosen such that $\hat{E} - \tilde{E} = \mathcal{O}(1/N)$. The diagonal matrix $\tilde{E}$ differs from $E\hat{1}$ only by the various imaginary infinitesimals; wherever these latter are dispensable, I shall indulge in the shorthand $\tilde{E} \to E\hat{1} \to E$. With these conventions,

$$
\begin{aligned}
\overline{Z}(\hat{E}) &= \int dQ \exp[-NA(Q)]\exp\left[-N\mathrm{Str}(\hat{E} - \tilde{E})\frac{1}{\tilde{E} - Q}\right] \\
A(Q) &= \mathrm{Str}\left[\frac{1}{2\lambda^2}Q^2 + \ln(\tilde{E} - Q)\right] \quad\quad (10.6.25)
\end{aligned}
$$

where the first exponential enjoys the pseudounitary symmetry discussed in the previous subsection; the second exponential, on the other hand, is effectively independent of N and thus immaterial for the saddle-point equation

$$
\frac{Q}{\lambda^2} = \frac{1}{E - Q} \quad\quad (10.6.26)
$$

which is identical in appearance with that found for the Green function $\overline{G}(E^-)$, (10.5.11). In analogy to that previous case, we may start looking for diagonal solutions and eventually extend to nondiagonal ones by pseudounitary transformations. But while we were previously led to a "scalar" saddle, i.e., a matrix proportional to the unit matrix, a nonscalar saddle is waiting for us now. To find a suitable solution Q^s of (10.6.26), it is well to take maximum benefit from what we already know; therefore, we look at the two-point function that results from taking derivatives w.r.t. E_1 and E_2 and then setting $E_1 = E_3, E_2 = E_4, E = (E_1 + E_2)/2$,

$$\overline{G(E_1^-)G(E_2^+)} = \int dQ \exp[-NA(Q)] \tag{10.6.27}$$

$$\times \frac{Q_{11}}{\lambda^2} \frac{Q_{22}}{\lambda^2} \exp\left\{ -N\mathrm{Str}\left[(\hat{E} - \tilde{E})(\tilde{E} - Q)^{-1}\right]\right\}.$$

The factors Q_{11}/λ^2 and Q_{22}/λ^2 correspond to $G(E_1^-)$ and $G(E_2^+)$, respectively, such that the principal suspect of being the relevant saddle is[5]

$$Q^s = E/2 + i\Lambda\sqrt{\lambda^2 - (E/2)^2}, \tag{10.6.28}$$

$$\Lambda = \mathrm{diag}(1, -1, 1, -1); \tag{10.6.29}$$

indeed, the two positive entries correspond to the saddle previously found for the advanced Green function $\overline{G}(E^-)$, and the two negative ones would have resulted had we chosen to evaluate the retarded Green function $\overline{G}(E^+)$; moreover, it is reasonable that the Fermi block in Λ should equal the Bose block since we might as well have generated the two-point function by taking derivatives w.r.t. E_3 and E_4.

To elevate the principal suspect (10.6.28, 29) to the doubtless exclusive relevant diagonal saddle, we must secure solid evidence, rather than rely on the intuitive reasoning just put forth. After all, there are sixteen possibilities for assigning a sign to the square root term in the four diagonal elements. I shall come back to that task at the end of this subsection.

Inasmuch as the saddle (10.6.28, 29) is not proportional to the unit matrix, it gives rise to a whole manifold of saddle points since it is accompanied by the following other solutions of the saddle-point equation:

$$Q^s(T) = TQ^sT^{-1} = E/2 + iT\Lambda T^{-1}\sqrt{\lambda^2 - (E/2)^2}, \tag{10.6.30}$$

where T is any pseudounitary matrix. Accounting for all of these saddle points amounts to an integration to which we must turn now.

It is customary to speak of a "zero-dimensional nonlinear supermatrix sigma model" when left with the average generating function in the form of an integral over $T\Lambda T^{-1}$; these supermatrices form a manifold characterized by

$$Q_0 \equiv T\Lambda T^{-1}, \qquad Q_0^2 = 1, \qquad Q_0^\dagger = LQ_0L, \qquad \mathrm{Str}\, Q_0 = \mathrm{Str}\, \Lambda = 0. \tag{10.6.31}$$

The present model is zero-dimensional, inasmuch as we do not encounter a supermatrix-valued field with spatial variations that attaches a different supermatrix to every point in real space; while such fields are encountered in the theory of disordered media, the random-matrix ensembles of Wigner and Dyson spare us the complications of field theory [366]. Incidentally, such "complications" are not overwhelming, as we shall see in Sect. 10.9.

---

[5] In specifying the saddle (10.6.28), we have immediately restricted ourselves to the energy interval $|E/2| < \lambda$ to which the spectrum is confined by the semicircle law.

The manifold of distinct saddle points is not quite as voluminous as the group of pseudounitary supermatrices T (which is commonly called $U(1,1/2)$). For instance, $Q_0 = T\Lambda T^{-1}$ is unchanged if a given T is multiplied from the right by any diagonal unitary 4×4 matrix. The latter matrices form the subgroup $U(1) \times U(1) \times U(1) \times U(1)$, and the integration range in question can thus be reduced to the coset space $U(1,1/2)/U(1) \times U(1) \times U(1) \times U(1)$. Actually, the group of pseudounitary matrices commuting with Λ is larger than just the one of the diagonal matrices such that the integration range must be further reduced.

To that end, it is convenient to reorder our 4×4 matrices. We agreed on

$$Q = \begin{pmatrix} Q_{BB} & Q_{BF} \\ Q_{FB} & Q_{FF} \end{pmatrix} \quad \text{and} \quad \hat{E} = \text{diag}(E_1^-, E_2^+, E_3^-, E_4^+) \qquad (10.6.32)$$

with 2×2 blocks Q_{BB} etc.; that convention is sometimes referred to as the "Bose–Fermi" notation. Now, we permute the four rows and columns according to $1234 \to 1324$ such that

$$\begin{aligned}
\tilde{E} &\to \text{diag}(E^-, E^-, E^+, E^+) \\
\hat{E} &\to \text{diag}(E_1^-, E_3^-, E_2^+, E_4^+) \\
\Lambda &\to \text{diag}(1, 1, -1, -1) \\
L &\to \text{diag}(1, 1, -1, 1) \qquad\qquad\qquad\qquad\qquad (10.6.33)
\end{aligned}$$

but shall retain the previous names $\tilde{E}, \hat{E}, \Lambda, L, Q$. Again, the matrices Q can be written as 2×2 matrices where all four elements are 2×2 supermatrices; the upper left block then refers to "advanced" Green functions where $\text{Im}E < 0$, and the lower right one to "retarded" functions where $\text{Im}E > 0$; therefore, the reordering in question is referred to as the "advanced-retarded" notation. Whenever suitable, we shall write Λ and L as 2×2 matrices with 2×2 elements as well,

$$\Lambda = \begin{pmatrix} 1 & 0 \\ 0 & -1 \end{pmatrix}, \qquad L = \begin{pmatrix} 1 & 0 \\ 0 & -k \end{pmatrix}, \qquad k = \begin{pmatrix} 1 & 0 \\ 0 & -1 \end{pmatrix}. \qquad (10.6.34)$$

With that reordering done, we may imagine the Q block diagonalized by a suitable pseudounitary matrix T_0,

$$Q = T_0 \begin{pmatrix} Q_a & 0 \\ 0 & Q_r \end{pmatrix} T_0^{-1}, \qquad\qquad\qquad\qquad (10.6.35)$$

where Q_a, Q_r are not yet diagonal. Of the sixteen independent parameters in Q (yes, sixteen indeed, since we started from a Hermitian 4×4 matrix; neither the analytic continuation to an anti-Hermitian Fermi block nor the subsequent multiplication with L could change that number), eight are accommodated in the blocks Q_a, Q_r and eight in the matrix T_0 which brings about the block diagonalization. We shall see in Subsect. 10.6.4 that the integral over blocks Q_a, Q_r is taken care of by the saddle-point approximation and that the matrices T_0 in effect span the saddle-point manifold. Then, it is also clear which group, larger than $U(1) \times U(1) \times U(1) \times U(1)$, leaves $\Lambda = \text{diag}(1, 1, -1, -1)$ invariant: It is $U(1/1) \times U(1/1)$, formed by block diagonal matrices with one unitary and one pseudounitary 2×2 supermatrix along the diagonal; all such matrices indeed commute with Λ. Consequently, the integration manifold for T_0 is the eight-parameter coset space $U(1, 1/2)/U(1/1) \times U(1/1)$ remaining when the group of pseudounitary 4×4 matrices is deprived of the subgroup $U(1/1) \times U(1/1)$.

We may parametrize the block diagonalizing matrix as[6]

$$T_0 = \begin{pmatrix} \sqrt{1 + \Gamma^\dagger k \Gamma} & \Gamma^\dagger k \\ \Gamma & \sqrt{1 + \Gamma \Gamma^\dagger k} \end{pmatrix} \tag{10.6.36}$$

such that the eight parameters in T_0 are located within a 2×2 supermatrix Γ and its adjoint $\Gamma^\dagger$. We easily verify that

$$T_0^\dagger = \begin{pmatrix} \sqrt{1 + \Gamma^\dagger k \Gamma} & \Gamma^\dagger \\ k \Gamma & \sqrt{1 + k \Gamma \Gamma^\dagger} \end{pmatrix}, \quad T_0^{-1} = T_0 \mid_{\Gamma \to -\Gamma} \tag{10.6.37}$$

as well as pseudounitarity, $T_0^\dagger L T_0 = L$. It is worth noting that with our new ordering of rows and columns, the supermatrix rule of adjunction reigns only within the supermatrix Γ. An important benefit of the parametrization (10.6.36, 37) is that all four 2×2 blocks within T_0 can be diagonalized simultaneously with the help of a unitary 2×2 supermatrix U and a pseudounitary one V, i.e., $U^\dagger U = 1$ and $V^\dagger k V = k$. Indeed, if such a pair yields

$$\Gamma = V t U^{-1} \quad \text{with} \quad t = \begin{pmatrix} t_B & 0 \\ 0 & t_F \end{pmatrix} \tag{10.6.38}$$

we immediately have

$$\Gamma^\dagger k = U t^* k V^{-1}, \qquad \Gamma^\dagger k \Gamma = U t^* k t U^{-1}, \qquad \Gamma \Gamma^\dagger k = V t^* t V^{-1}. \tag{10.6.39}$$

Moreover, T_0 takes the simple form

$$T_0 = \begin{pmatrix} U & 0 \\ 0 & V \end{pmatrix} \begin{pmatrix} \sqrt{1 + k t^* t} & k t^* \\ t & \sqrt{1 + k t^* t} \end{pmatrix} \begin{pmatrix} U^{-1} & 0 \\ 0 & V^{-1} \end{pmatrix}. \tag{10.6.40}$$

We may conclude that the eigenvalue t_F is bounded in modulus, $|t_F|^2 \leq 1$, whereas t_B is unbounded. Thus, our saddle-point manifold $U(1, 1/2)/U(1/1) \times U(1/1)$ is revealed as noncompact. To go through with parameter counting, we note that the complex eigenvalues accommodate four, whereas the remaining four sit in the diagonalizing 2×2 supermatrices U and V. These matrices can be chosen as

$$U = \begin{pmatrix} \sqrt{1 + \rho \rho^*} & \rho \\ \rho^* & \sqrt{1 + \rho^* \rho} \end{pmatrix} = \begin{pmatrix} 1 + \rho \rho^*/2 & \rho \\ \rho^* & 1 + \rho^* \rho/2 \end{pmatrix} = \exp \begin{pmatrix} 0 & \rho \\ \rho^* & 0 \end{pmatrix} \tag{10.6.41}$$

$$V = \begin{pmatrix} \sqrt{1 - \sigma \sigma^*} & i\sigma \\ i\sigma^* & \sqrt{1 - \sigma^* \sigma} \end{pmatrix} = \begin{pmatrix} 1 - \sigma \sigma^*/2 & i\sigma \\ i\sigma^* & 1 - \sigma^* \sigma/2 \end{pmatrix} = \exp \begin{pmatrix} 0 & i\sigma \\ i\sigma^* & 0 \end{pmatrix}$$

---

[6] This is similar in structure to T' in (10.6.17); note that we have reordered rows and columns in between.

with Grassmannians $\rho, \rho^*, \sigma, \sigma^*$; the unitarity $U^\dagger U = 1$ and the pseudounitarity $V^\dagger k V = k$ are readily checked.

Anticipating Subsect. 10.6.5, we note for instant use the integration measure

$$
\begin{aligned}
dQ &= dQ_a dQ_r d\mu(T_0) \frac{(q_{aB} - q_{rB})^2 (q_{aF} - q_{rF})^2}{(q_{aB} - q_{rF})^2 (q_{aF} - q_{rB})^2} \\
&\equiv dQ_a dQ_r d\mu(T_0)\, F(Q_a, Q_r) \, ;
\end{aligned}
\tag{10.6.42}
$$

here, dQ_a, dQ_r denote the product of differentials of the respective four matrix elements which may be thought of as replaced by the product of the (bosonic) eigenvalue differentials $dq_{aB} dq_{aF} dq_{rB} dq_{rF}$ and of the differentials of the Grassmannian "angles" in the diagonalizing matrices and the Berezinian for the diagonalization; finally, $d\mu(T_0)$ is the integration measure of the coset space spanned by the T_0 characterized above,

$$
d\mu(T_0) = \frac{d^2 t_B d^2 t_F d\rho^* d\rho d i\sigma^* d i\sigma}{\pi^2 (|t_B|^2 + |t_F|^2)^2} \, .
\tag{10.6.43}
$$

Finally, we can return to the question why we could disregard all but one of the diagonal saddles, viz., (10.6.28, 29). To that end, we consider fluctuations δQ around the diagonal saddles. These can be separated into "soft modes" that remain within a symmetry-induced manifold around a given diagonal saddle and "hard modes" transverse to that manifold. Only the hard modes can be integrated over by the saddle-point approximation and should be checked for their contribution to the two-point function. We can identify these hard modes from (10.6.35) as

$$
\delta_\perp Q = T \begin{pmatrix} \delta Q_a & 0 \\ 0 & \delta Q_r \end{pmatrix} T^{-1}
\tag{10.6.44}
$$

where T is any of the pseudounitary transformations that lead from a diagonal saddle-point matrix to any block diagonal matrix through (10.6.35). Checking the contributions of those hard modes for all sixteen diagonal saddles is more work than is advisable to display here [376]. Therefore, I resort to a more effective, albeit somewhat frivolous, procedure and indiscriminately admit arbitrary fluctuations

$$
\delta Q = \begin{pmatrix} \delta a & \delta\sigma^\dagger \\ \delta\sigma & \delta ib \end{pmatrix} \qquad \text{(Bose–Fermi notation)}
\tag{10.6.45}
$$

with 2×2 blocks δa etc.. As in Sect. 10.5, we may inspect the second variation $\delta^2 A$ of the action $A(Q) = \mathrm{Str}\left[Q/(2\lambda^2) + \ln(\tilde{E} - Q) \right]$; note that the action $A(Q)$ in (10.6.25) is the same function of Q as the action $A(Q, J = 0)$ in (10.5.7); thus, the second variation is still given by (10.5.14). Here, we insert the suspects, $\mathrm{diag}\,(B_a, B_r, F_a, F_r)$ with $B_{a,r} = \pm 1$, $F_{a,r} = \pm 1$, of which that where $B_a = F_a = 1$, $B_r = F_r = -1$ is our principal one. In evaluating the supertrace of the present 4×4 supermatrices, we may, as in Sect. 10.5, choose $E = 0$, just to avoid unnecessarily lengthy formulae, and thus get

$$
\begin{aligned}
\lambda^2 \delta^2 A(B_a, B_r, F_a, F_r) &= (\delta a_{aa})^2 + (\delta a_{rr})^2 + (1 + B_a B_r)\delta a_{ar}^* \delta a_{ar} \quad (10.6.46)\\
&\quad + (\delta b_{aa})^2 + (\delta b_{rr})^2 + (1 + F_a F_r)\delta b_{ar}^* \delta b_{ar}\\
&\quad + (1 + B_a F_a)\delta\sigma_{aa}^* \delta\sigma_{aa} + (1 + B_r F_r)\delta\sigma_{rr}^* \delta\sigma_{rr}\\
&\quad + (1 + B_a F_r)\delta\sigma_{ra}^* \delta\sigma_{ra} + (1 + B_r F_a)\delta\sigma_{ar}^* \delta\sigma_{ar} \, .
\end{aligned}
$$

To keep the further effort within reasonable limits, I declare the two Bose entries, $B_a = -B_r = 1$, nonnegotiable and appeal to our previous experience: In Sects. 10.2.4 and 10.5 the saddles $\mathrm{diag}(a_+, ib_\pm) = \mathrm{diag}(Q_+, Q_\pm)$ could not contribute to $\overline{G}(E^-)$ (just as $\mathrm{diag}(a_-, ib_\pm) = \mathrm{diag}(Q_-, Q_\pm)$ not to $\overline{G}(E^+)$) since the original path of integration could not be deformed so as to pass through them without crossing a singularity of the integrand. It remains to check on the four possibilities $F_a = \pm 1, F_r = \pm 1$, of which $F_a = -F_r = 1$ corresponds to our principal suspect. The four cases give rise to the following four second variations:

(i) $F = B = \mathrm{diag}\,(1, -1)$,

$$
\lambda^2 \delta^2 A_{+-} = (\delta a_{aa})^2 + (\delta a_{rr})^2 + (\delta b_{aa})^2 + (\delta b_{rr})^2 + 2\delta\sigma_{aa}^* \delta\sigma_{aa} + 2\delta\sigma_{rr}^* \delta\sigma_{rr} \, ,
$$

(ii) $F = -B = \mathrm{diag}\,(-1, 1)$,

$$
\lambda^2 \delta^2 A_{-+} = (\delta a_{aa})^2 + (\delta a_{rr})^2 + (\delta b_{aa})^2 + (\delta b_{rr})^2 + 2\delta\sigma_{ra}^* \delta\sigma_{ra} + 2\delta\sigma_{ar}^* \delta\sigma_{ar} \, ,
$$

(iii) $F = 1$,

$$
\lambda^2 \delta^2 A_{++} = \delta a_{aa}^2 + \delta a_{rr}^2 + \delta b_{aa}^2 + \delta b_{rr}^2 + 2\delta b_{ar}^2 + 2\delta\sigma_{aa}^* \delta\sigma_{aa} + 2\delta\sigma_{ra}^* \delta\sigma_{ra} \, ,
$$

(iv) $F = -1$,

$$
\lambda^2 \delta^2 A_{--} = \delta a_{aa}^2 + \delta a_{rr}^2 + \delta b_{aa}^2 + \delta b_{rr}^2 + 2\delta b_{ar}^2 + 2\delta\sigma_{rr}^* \delta\sigma_{rr} + 2\delta\sigma_{ar}^* \delta\sigma_{ar} \, .
$$

When attempting to do the Gaussian integrals over the fluctuations δQ with the (negative of the) foregoing quadratic forms in the exponents, we must pay a price for not having restricted δQ to hard modes: Each missing commuting variable in a quadratic form gives rise to a factor ∞, and each missing Grassmann variable to a factor 0. Thus, cases (i,ii) formally have the weight $(\infty \times 0)^2$, whereas cases (iii,iv) have the "smaller" weight $\infty \times 0^2$. We shall presently convince ourselves that each undefined product $(\infty \times 0)$ in these weights actually has the value 1 such that cases (iii,iv) do not contribute at all while case (i), i.e. the principal suspect, and case (ii) give identical contributions. To see two diagonal saddles with vanishing contributions justifies having discarded them above. But to find case (ii) as of equal weight as our principal suspect is a worry we must ease immediately.

The diagonal saddle from case (ii) lies on the manifold generated pseudo-unitarily from the principal suspect and therefore is automatically accounted for when integrating over the said manifold. To check on that claim, we must specify the pseudounitary matrix relating the two saddles. It is easy to verify that

$$T = \begin{pmatrix} 1 & 0 & 0 & 0 \\ 0 & 0 & 0 & -1 \\ 0 & 0 & 1 & 0 \\ 0 & 1 & 0 & 0 \end{pmatrix} \qquad \text{(advanced-retarded notation)} \qquad (10.6.47)$$

is indeed of the structure (10.6.36) with $\Gamma = \begin{pmatrix} 0 & 0 \\ 0 & 1 \end{pmatrix}$ and yields (note that the advanced-retarded notation is used) $T \operatorname{diag}(1, 1, -1, -1) T^{-1} = \operatorname{diag}(1, -1, -1, 1)$.

Now, the finite value unity of the seemingly undefined products $\infty \times 0$ is checked at leisure in a way reminiscent of our "proof" of the Parisi–Sourlas–Efetov–Wegner theorem in Sect. 10.4.5: We sneak a parameter $c < 1$ into those quadratic terms of the second variation (10.6.46) which would otherwise vanish for a given diagonal saddle in the fashion $(1 + B_a B_r) \delta a_{ar}^* \delta a_{ar} \to (1 + c B_a B_r) \delta a_{ar}^* \delta a_{ar}$. Then, the Gaussian superintegrals for cases (i,ii) equal unity, independently of c; cases (iii,iv), however, similarly yield vanishing superintegrals; the auxiliary parameter c can eventually approach unity.

10.6.4 Implementing the Zero-Dimensional Sigma Model

Having put the integration measure dQ and the integration range $U(1, 1/2)/U(1/1) \times U(1/1)$ at our disposal, now we can do the integrals in the two-point function (10.6.27). We employ the reordered matrices and write

$$\overline{G(E_1^-) G(E_2^+)} = \int dQ_a \exp\left\{ -N \operatorname{Str}\left[\frac{Q_a^2}{2\lambda^2} + \ln(E^- - Q_a) \right] \right\} \qquad (10.6.48)$$

$$\times \int dQ_r \exp\left\{ -N \operatorname{Str}\left[\frac{Q_r^2}{2\lambda^2} + \ln(E^+ - Q_r) \right] \right\} F(Q_a, Q_r)$$

$$\times \int d\mu(T_0) \exp\left[N \operatorname{Str}(\tilde{E} - \hat{E})(\tilde{E} - Q)^{-1} \right] \frac{Q_{1BB}}{\lambda^2} \frac{Q_{2BB}}{\lambda^2}.$$

Here, Q_{1BB}, Q_{2BB} are the Bose–Bose entries of the 2×2 supermatrices Q_a, Q_r. The exponentials in the first and second lines involve only Q_a and Q_r. Both of these exponentials are $U(1/1)$ invariant such that we confront, with respect to both Q_a and Q_r, a saddle-point integral analogous to that encountered for the average Green function in Sects. 10.2 and 10.5. In particular, (10.5.7,8) apply (with $J = 0$), since the prefactor $F(Q_a, Q_r)$ equals unity at the saddle point; indeed, at the saddle (10.6.28), yields

$$\left. \begin{aligned} q_{aB}^s = q_{aF}^s = \frac{E_1}{2} + i\sqrt{\lambda^2 - (\frac{E_1}{2})^2} = \lambda^2 \overline{G}(E_1^-) \\ q_{rB}^s = q_{rF}^s = \frac{E_2}{2} - i\sqrt{\lambda^2 - (\frac{E_2}{2})^2} = \lambda^2 \overline{G}(E_2^+) \end{aligned} \right\} \implies F^s(Q_a, Q_r) = 1 \, (10.6.49)$$

Denoting by

$$\begin{aligned} Q^s(T_0) = T_0 Q^s T_0^{-1} &= E/2 + i T_0 \Lambda T_0^{-1} \sqrt{\lambda^2 - (E/2)^2} \\ &= E/2 + i\pi \lambda^2 \overline{\varrho}(E) \, T_0 \Lambda T_0^{-1} \end{aligned} \qquad (10.6.50)$$

the general member of the saddle-point manifold, we arrive at the integral

$$\overline{G(E_1^-)G(E_2^+)} \approx \int d\mu(T_0) \exp\left[N\mathrm{Str}(\tilde{E} - \hat{E})\frac{Q^s(T_0)}{\lambda^2}\right]$$

$$\times \frac{[Q^s(T_0)]_{1BB}}{\lambda^2} \frac{[Q^s(T_0)]_{2BB}}{\lambda^2} \tag{10.6.51}$$

which constitutes, as already announced above, Efetov's zero-dimensional super-matrix sigma model. Now, the exponential has an argument of order unity, as $N \to \infty$, due to $E - \hat{E} = \mathcal{O}(1/N)$, and this is why we cannot invoke any further large-N approximation in doing the integral over T_0. The approximate-equality sign indicates the loss of a correction of relative weight $1/N$ in the saddle-point approximation which is finally implemented here; the reader is asked to keep that loss in mind when the equality sign is used again in the following.

Labor is saved by recalling from (10.6.4) that we are really after the connected two-point function $\overline{\Delta G(E_1^-)\Delta G(E_2^+)} = \overline{G(E_1^-)G(E_2^+)} - \overline{G}(E_1^-)\overline{G}(E_2^+)$. When subtracting the product of averages $\overline{G}(E_1^-)\overline{G}(E_2^+)$ from both sides of (10.6.44) and (10.6.47), we can pull that product into the integrands such that it is the term $\frac{Q_{1BB}^s}{\lambda^2}\frac{Q_{2BB}^s}{\lambda^2}$ from which the subtraction is made. Such manipulation is justified by the identity

$$1 = \overline{Z(E_1^-, E_2^+, E_1^-, E_2^+)}$$

$$= \int dQ_a \exp\left\{-N\mathrm{Str}\left[\frac{Q_a^2}{2\lambda^2} + \ln(E^- - Q_a)\right]\right\}$$

$$\times \int dQ_r \exp\left\{-N\mathrm{Str}\left[\frac{Q_r^2}{2\lambda^2} + \ln(E^+ - Q_r)\right]\right\}F(Q_a, Q_r)$$

$$\times \int d\mu(T_0) \exp\left[N\mathrm{Str}(\tilde{E} - \hat{E})(E - Q)^{-1}\right] \tag{10.6.52}$$

which is obvious from the definition (10.6.5) of the generating function. That identity remains intact, to leading order in N, when the integrations over Q_a and Q_r are done in the saddle-point approximation,

$$\int d\mu(T_0) \exp\left[N\mathrm{Str}(\tilde{E} - \hat{E})\frac{Q^s(T_0)}{\lambda^2}\right]\Bigg|_{E_3=E_1, E_2=E_4} \approx 1, \tag{10.6.53}$$

whereupon the connected two-point function can be written as

$$\overline{\Delta G(E_1^-)\Delta G(E_2^+)} = \int d\mu(T_0) \exp\left[N\mathrm{Str}(\tilde{E} - \hat{E})\frac{Q^s(T_0)}{\lambda^2}\right] \tag{10.6.54}$$

$$\times \left\{\frac{[Q^s(T_0)]_{1BB}}{\lambda^2} \frac{[Q^s(T_0)]_{2BB}}{\lambda^2} - \overline{G}(E_1^-)\overline{G}(E_2^+)\right\}.$$

Now, I propose to reveal explicitly the dependence of the integrand on the integration variables $t_B, t_F, \rho, \rho^*, \sigma, \sigma^*$. First turning to the supertrace in the exponent, we realize, with the help of (10.6.46), that it must be proportional to $E_1^- - E_2^+$ since both Λ and the unit matrix have vanishing supertraces. To fully evaluate that supertrace we insert the parametrization (10.6.40, 41) of T_0

and thus get $\mathrm{Str}[(\tilde{E}-\hat{E})\frac{Q^{\mathrm{s}}(T_0)}{\lambda^2}] = i\pi\overline{\varrho}(E)\mathrm{Str}(\tilde{E}-\hat{E})T_0\Lambda T_0^{-1} = -i\pi\overline{\varrho}(E)\big[(E_1^- - E_2^+)/2\big]\mathrm{Str}\Lambda T_0\Lambda T_0^{-1} = -i2\pi\overline{\varrho}(E)(E_1^- - E_2^+)(|t_B|^2 + |t_F|^2)$. At this point, the familiar spacing variable e from (10.6.3) comes into play with a negative imaginary infinitesimal attached to it, and the exponential in (10.6.50) takes the form

$$\exp\left[N\mathrm{Str}(\tilde{E}-\hat{E})\frac{Q^{\mathrm{s}}(T_0)}{\lambda^2}\right] = \exp\left[-i2\pi e^-\big(|t_B|^2 + |t_F|^2\big)\right]. \tag{10.6.55}$$

The absence of the Grassmannians $\rho, \rho^*, \sigma, \sigma^*$ from the exponential (10.6.55) is noteworthy since it allows us to discard everything in the curly bracket in (10.6.55) except the term proportional to the product $\rho\rho^*\sigma\sigma^*$. To find that term, once more we invoke the parametrization (10.6.40, 41) of T_0 and get

$$\begin{aligned}
Q_{1BB}/\lambda^2 &= \overline{G}(E_1^-) + i2\pi\overline{\varrho}(E_1)\big(Uktt^*U^{-1}\big)_{BB}, \\
Q_{2BB}/\lambda^2 &= \overline{G}(E_2^+) - i2\pi\overline{\varrho}(E_2)\big(Vktt^*V^{-1}\big)_{BB}, \\
\big(Uktt^*U^{-1}\big)_{BB} &= |t_B|^2 + \big(|t_B|^2 + |t_F|^2\big)\rho\rho^*, \\
\big(Vktt^*V^{-1}\big)_{BB} &= |t_B|^2 - \big(|t_B|^2 + |t_F|^2\big)\sigma\sigma^*.
\end{aligned} \tag{10.6.56}$$

Now, we see that the desired term in the prefactor in (10.6.55) is proportional to $\big(|t_B|^2 + |t_F|^2\big)^2$ which factor cancels against the denominator in the integration measure (10.6.43). By doing the fourfold Grassmann integral, we get

$$\overline{\Delta G(E_1^-)\Delta G(E_2^+)} = (2\pi)^2\overline{\varrho}(E_1)\overline{\varrho}(E_2)\int_{|t_F|\leq 1}\frac{d^2t_F}{\pi}\frac{d^2t_B}{\pi}$$
$$\times \exp\left[-i2\pi\big(|t_B|^2 + |t_F|^2\big)e^-\right]. \tag{10.6.57}$$

Herein, t_B ranges over the whole complex plane and this is why the negative imaginary infinitesimal addition to the spacing variable had to be taken along; the restriction of t_F to the unit disc around the origin of the complex t_F plane comes with the parametrization (10.6.40) and was already noted above. With these remarks in mind, we evaluate the integrals and, after taking the real part according to (10.6.4), find the two-point cluster function of the GUE,

$$\frac{\overline{\Delta\varrho(E_1)\Delta\varrho(E_2)}}{\overline{\varrho}(E_1)\overline{\varrho}(E_2)} = \delta(e) - \left(\frac{\sin\pi e}{\pi e}\right)^2 = \delta(e) - Y_{GUE}(e), \tag{10.6.58}$$

which equals that of the circular unitary ensemble derived in 4.13.3.

10.6.5 Integration Measure of the Nonlinear Sigma Model

To fulfill the promise of deriving the integration measure (10.6.42, 43) for the saddle-point manifold, we start with some warm-up exercises:

(i) The first is to return to an integral $\int d\Phi^* d\Phi f(\Phi, \Phi^*)$ over a supervector $\Phi = \binom{S}{\eta}$ and change the integration variables to a new complex conjugate pair of supervectors Φ', Φ'^* according to $\Phi = \Phi(\Phi')$. We had convinced ourselves of $d\Phi^* d\Phi = d\Phi'^* d\Phi' J(\frac{\Phi^*}{\Phi'^*})J(\frac{\Phi}{\Phi'})$ with the Berezinian

$$J(\frac{\Phi}{\Phi'}) = \mathrm{Sdet}\left(\frac{\partial}{\partial\Phi_i'}\Phi_j\right) \equiv \mathrm{Sdet}\, D_{ij}^T = \mathrm{Sdet}\, D_{ij}; \tag{10.6.59}$$

see (10.4.16). The supermatrix of derivatives $D^T = \frac{\partial}{\partial \Phi'}\Phi$ also determines the transformation of the elementary squared length in the space of supervectors, $(\delta\Phi, \delta\Phi) = \sum_i \delta\Phi_i^* \delta\Phi_i$. Let $\delta\Phi_i = \sum_j \delta\Phi_j' D_{ji}^T = \sum_i D_{ij}\delta\Phi_j'$, where transposition is, of course, meant in the supermatrix sense to account for the anticommutativity of a fermionic D_{ij} with a fermionic $\delta\Phi_j'$. Then, the squared length element reads

$$(\delta\Phi, \delta\Phi) = (D\delta\Phi', D\delta\Phi') = (\delta\Phi', D^\dagger D\delta\Phi') \equiv (\delta\Phi', G\delta\Phi'). \tag{10.6.60}$$

The "metric" supermatrix $G = D^\dagger D$ occurring herein has as its superdeterminant the square of the Berezinian under discussion,

$$\text{Sdet}\, G = \text{Sdet}\, D^\dagger D = |\text{Sdet}\, D|^2 = J\left(\frac{\Phi^*, \Phi}{\Phi'^*, \Phi'}\right) = \left| J\left(\frac{\Phi}{\Phi'}\right) \right|^2. \tag{10.6.61}$$

Thus, we may say that the Berezinian is related to the metric in superspace.

(ii) In a second warm-up, we exploit the connection between metric and Berezinian for an integral over a Hermitian 2×2 supermatrix,

$$
\begin{aligned}
I &= \int dH f(H), \\
H &= \begin{pmatrix} a & \rho^* \\ \rho & b \end{pmatrix} = U \begin{pmatrix} h_B & 0 \\ 0 & h_F \end{pmatrix} U^{-1} \\
dH &= \frac{1}{\pi} da\, db\, d\rho^*\, d\rho.
\end{aligned}
\tag{10.6.62}
$$

As new integration variables, we wish to employ the two eigenvalues h_B, h_F and the Fermionic "angles" η^*, η in the diagonalizing matrix U, given in (10.3.31, 32). A straightforward calculation yields the Berezinian (see Problem 10.11)

$$\text{Sdet}\, D = \text{Sdet}\, \frac{\partial(a, b, \rho^*, \rho)}{\partial(h_B, h_F, \eta^*, \eta)} = \frac{1}{(h_B - h_F)^2}. \tag{10.6.63}$$

A shortcut to that result involves the metric in the space of supermatrices engendered by the squared length element

$$
\begin{aligned}
(\delta H, \delta H) \equiv \text{Str}\, (\delta H)^2 &= \text{Str}\left(\delta U h U^{-1} + U \delta h U^{-1} + U h \delta U^{-1}\right)^2 \\
&= \text{Str}\left([U^{-1}\delta U, h] + \delta h\right)^2 \\
&= \text{Str}\left([U^{-1}\delta U, h]^2 + (\delta h)^2\right).
\end{aligned}
\tag{10.6.64}
$$

Now filling in U, U^{-1} from (10.3.32),

$$\text{Str}\, (\delta H)^2 = (\delta h_B)^2 - (\delta h_F)^2 + 2(h_B - h_F)^2 \delta\eta^* \delta\eta \tag{10.6.65}$$

$$
= (\delta h_B, \delta h_F, \delta\eta^*, \delta\eta)
\begin{pmatrix}
1 & 0 & 0 & 0 \\
0 & -1 & 0 & 0 \\
0 & 0 & 0 & (h_B - h_F)^2 \\
0 & 0 & -(h_B - h_F)^2 & 0
\end{pmatrix}
\begin{pmatrix}
\delta h_B \\
\delta h_F \\
\delta\eta^* \\
\delta\eta
\end{pmatrix}
$$

$$
= (\delta h_B, \delta h_F, \delta\eta^*, \delta\eta)\, G
\begin{pmatrix}
\delta h_B \\
\delta h_F \\
\delta\eta^* \\
\delta\eta
\end{pmatrix}.
$$

But since the integration measure in the present case is dH, rather than $d\Phi d\Phi^*$ as before, the superdeterminant of the metric matrix G in the above length element is the square of the Jacobian (10.6.63). The reader might appreciate the greater ease of the "metric way" of getting that Jacobian, compared to straightforwardly evaluating the superdeterminant of the derivative matrix D.

With so much lure laid out, now we go the metric way toward the principal goal of this subsection, derivation of the integration measure (10.6.42, 43). As in the second warm-up above, the problem is to change integration variables from the matrix elements of the 4×4 matrix Q to the eigenvalues $q_{aB}, q_{aF}, q_{rB}, q_{rF}$ and the twelve parameters in the diagonalizing matrix T. The latter number is twelve even though the group $U(1, 1/2)$ of pseudounitary matrices has sixteen parameters since, as remarked before, if some matrix T diagonalizes Q, $Q = TQT^{-1}$, so does any $T' = Te^{-i\phi}$ with ϕ a real diagonal matrix; thus the set of matrices relevant for the envisaged change of integration variables is the coset space $U(1, 1/2)/U(1) \times U(1) \times U(1) \times U(1)$. We had argued further that that set can be split into the eight-parameter coset space $U(1, 1/2)/U(1, 1) \times U(1, 1)$ of the block diagonalizing matrices T_0 relevant to the saddle-point manifold, see (10.6.35), and the four-parameter space of matrices T_1 diagonalizing the 2×2 blocks Q_a, Q_r (i.e., the space $U(1, 1) \times U(1, 1)/U(1) \times U(1) \times U(1) \times U(1)$),

$$Q = T_0 \begin{pmatrix} Q_a & 0 \\ 0 & Q_r \end{pmatrix} T_0^{-1} = T_0 T_1 q T_1^{-1} T_0^{-1} \tag{10.6.66}$$

and

$$T_1 = \begin{pmatrix} u & 0 \\ 0 & v \end{pmatrix}, \tag{10.6.67}$$

with unitary u and pseudounitary v, the latter both 2×2. To start our metric walk, we write the increment of $Q = TqT^{-1}$ as $\delta Q = [\delta T T^{-1}, Q] + T\delta q T^{-1}$ and develop a 16×16 metric matrix G in $(\delta Q, \delta Q) = (\delta Q', G\delta Q')$ where $\delta Q'$ is made up of the increments δq and $T^{-1}\delta T$. In analogy with (10.6.64),

$$\begin{aligned} \mathrm{Str}(\delta Q)^2 &= \mathrm{Str}\{[T^{-1}\delta T, q]^2 + (\delta q)^2\} \tag{10.6.68} \\ &= \sum_{i \neq j}^{1...4} (-k_i)(T^{-1}\delta T)_{ij} (q_i - q_j)^2 (T^{-1}\delta T)_{ji} + \sum_i^{1...4} k_i(\delta q_i)^2 \end{aligned}$$

where k_i is equal to $+1$ or -1 when the label i is, respectively, bosonic or fermionic. The latter sign factor stems, of course, from the definition of the supertrace. Obviously, then, the metric matrix G is block diagonal, a fully diagonal 4×4 block pertains to the eigenvalue increments δq_i, and a 12×12 block is for the $(T^{-1}\delta T)_{ij}$. The superdeterminant $\mathrm{Sdet}G$ factorizes accordingly, and the 4×4 block equals unity. Thus we need to worry only about the remaining 12×12 supermatrix which I choose to call G_T. We meet $M_B = 4$ bosonic labels and $M_B = 8$ fermionic labels in G_T since the eigenvalues are all bosonic.

Inasmuch as we may choose the twelve independent increments $(T^{-1}\delta T)_{ij}$ with $i \neq j$ as increments of the parameters in T, we immediately conclude from the second line in (10.6.64) that G_T is itself block diagonal, with six 2×2 blocks, since a given $(T^{-1}\delta T)_{ij}$ couples only to $(T^{-1}\delta T)_{ji}$. Thus, the metric matrix has the superdeterminant (see Problem 10.2) $\mathrm{Sdet}G_T = \prod_{i \neq j}^{1...4}(q_i - q_j)^{4k_ik_j}$, whose square root is the Berezinian in[7]

$$dQ = \left[\prod_{i \neq j}(q_i - q_j)^{2k_ik_j} \right] \left[\prod_{i \neq j}\left(T^{-1}dT\right)_{ij} \right] \left(\prod_i dq_i \right). \qquad (10.6.69)$$

As an intermediate result, we have the integration measure of the full coset space $U(1,1/2)/U(1) \times U(1) \times U(1) \times U(1)$

$$d\mu(T) = \prod_{i \neq j}\left(T^{-1}dT\right)_{ij}. \qquad (10.6.70)$$

Completely analogous reasoning yields an analogous result for the change of variables from matrix elements to eigenvalues and "angles" in the diagonalizing matrices u, v of the blocks $Q_\mathrm{a}, Q_\mathrm{r}$. By combining these results with (10.6.70) we get (10.6.42). Note that all intrablock eigenvalue differences are then absorbed in $dQ_\mathrm{a}, dQ_\mathrm{r}$ such that only interblock eigenvalue differences remain on display, together with the yet to be determined integration measure $d\mu(T_0)$ of the reduced coset space $U(1,1/2)/U(1,1) \times U(1,1)$.

When accounting for $T = T_0T_1$, yet another transformation of variables must be effected. It is well to keep in mind the restrictions on the pseudounitary matrices T_0 and T_1 which bring about the appropriate number of independent parameters. Pseudounitarity itself allows us to consider all diagonal elements as dependent on the off-diagonal ones: Indeed, $T^\dagger LT = L$ yields $|T_{ii}|^2L_{ii} = L_{ii} - \sum_{j(\neq i)}T_{ij}^\dagger L_{jj}T_{ji}$, and this fixes the moduli of the diagonal elements; the phases may be chosen arbitrarily anyway, as already mentioned. Moreover, the matrix T_1 needed to diagonalize the 2×2 blocks Q_a and Q_r can be chosen block diagonal itself and involves four fermionic parameters, two each for u and v; see (10.6.63); the matrix T_0 which brings Q to block diagonal form has diagonal 2×2 blocks T_0^{11} and T_0^{22} which depend on the two off-diagonal blocks T_0^{12} and T_0^{21}; that structure is obvious from the explicit form (10.6.36); the number of independent parameters in T_0 is thus eight, as stated previously; four each are bosonic and fermionic. With these preliminary remarks in mind, we go the metric way toward the Berezinian for our final transformation and consider

$$\mathrm{Str}\left[(T^{-1}\delta T)_\mathrm{ind}\right]^2$$

$$= \mathrm{Str}\left[\left\{ \begin{pmatrix} u^{-1}\delta u & 0 \\ 0 & v^{-1}\delta v \end{pmatrix} + \begin{pmatrix} u^{-1} & 0 \\ 0 & v^{-1} \end{pmatrix} T_0^{-1}\delta T_0 \begin{pmatrix} u & 0 \\ 0 & v \end{pmatrix} \right\}_\mathrm{ind} \right]^2 \qquad (10.6.71)$$

$$= \mathrm{str}\left\{ \left[(\delta uu^{-1})_\mathrm{ind}\right]^2 + \left[(\delta vv^{-1})_\mathrm{ind}\right]^2 + 2\left(T_0^{-1}\delta T_0\right)_\mathrm{ind}^{12}\left(T_0^{-1}\delta T_0\right)_\mathrm{ind}^{21} \right\}.$$

---

[7] The product of increments $(T^{-1}dT)_{ij}$ must be read as a wedge product, as far as the bosonic increments are concerned.

To attract the reader's attention to the change of the dimension of the super-matrices from four to two in the foregoing chain, I employ two slightly differing symbols for the supertrace; the subscript "ind" reminds us of the need to drop the diagonal elements of $\delta u u^{-1}$ and $\delta v v^{-1}$ and even the two diagonal blocks $(T_0^{-1}\delta T_0)^{11}$ and $(T_0^{-1}\delta T_0)^{22}$, to keep the correct number of independent incre-ments. Thus, it is obvious once more that the Berezinian for the envisaged change of variables (from the twelve independent increments in $T^{-1}\delta T$ to the indepen-dent increments in $u^{-1}\delta u$, $v^{-1}\delta v$, and $T_0^{-1}\delta T_0$) factorizes in a $4{\times}4$ part pertaining to the parameters in u and v and an 8×8 superdeterminant pertaining to $d\mu(T_0)$; the latter is determined by the last term in (10.6.71).

To finally nail down $d\mu(T_0)$, we inspect the last term in (10.6.71) and, using the structure (10.6.40) of T_0, note that

$$
\left(T_0^{-1}\delta T_0\right)^{12} = U\left\{\sqrt{1+k|t|^2}\left[(U^{-1}\delta U)kt^* + k\delta t^*\right]\right.
$$
$$
\left. -kt^*(V^{-1}\delta V)\sqrt{1+k|t|^2} - kt^*\delta\sqrt{1+k|t|^2}\right\}V^{-1},
$$
$$
\left(T_0^{-1}\delta T_0\right)^{21} = V\left\{\sqrt{1+k|t|^2}\left[(V^{-1}\delta V)t + \delta t\right]\right.
$$
$$
\left. -t(U^{-1}\delta U)\sqrt{1+k|t|^2} - t\delta\sqrt{1+k|t|^2}\right\}U^{-1}.
$$

The reader will recall that $t = \mathrm{diag}(t_B, t_F)$ is the diagonalized version of T_0 and thus harbors four real bosonic parameters. Two independent fermionic parameters live in U and V each, as is indeed obvious from the explicit form (10.6.41). As independent increments in the matrices $(U^{-1}\delta U)$ and $(V^{-1}\delta V)$, thus we may take their off-diagonal parts, just as was the case for the blocks $u^{-1}\delta u$ and $v^{-1}\delta v$ above. Before inserting their explicit forms in the above off-diagonal blocks of $T_0^{-1}\delta T_0$, it is well to simplify the latter a bit by introducing intermediate variables momentarily replacing t_B and t_F,

$$
s = \frac{t}{\sqrt{1+k|t|^2}} \quad,\text{ i.e., }\quad s_B = \frac{t_B}{\sqrt{1+|t_B|^2}}, \quad s_F = \frac{t_F}{\sqrt{1-|t_F|^2}}. \tag{10.6.72}
$$

Thus, these blocks take the form

$$
\left(T_0^{-1}\delta T_0\right)^{12} = U\sqrt{1+k|t|^2}\left\{(U^{-1}\delta U)ks^* - ks^*(V^{-1}\delta V) + k\delta s^*\right\}\sqrt{1+k|t|^2}\, V^{-1},
$$
$$
\left(T_0^{-1}\delta T_0\right)^{21} = V\sqrt{1+k|t|^2}\left\{-s(U^{-1}\delta U) + (V^{-1}\delta V)s + \delta s\right\}\sqrt{1+k|t|^2}\, U^{-1}.
$$

The square-root factors which surround the curly brackets here are diagonal 2×2 matrices with entries $S_B \equiv \sqrt{1+|t_B|^2}$ and $S_F \equiv \sqrt{1-|t_F|^2}$; they will eventually drop out of the Berezinian we are after since that Berezinian will turn out to be a block diagonal superdeterminant with a 4×4 Bose block and an equally large Fermi block both of which involve S_B and S_F only in a common and thus cancelling prefactor $(S_B S_F)^4$. To see this, we need to evaluate the last term in (10.6.71). An easy little calculation starting from (10.6.41) establishes

$$
(U^{-1}\delta U)_{\text{ind}} = \begin{pmatrix} 0 & \delta\rho \\ \delta\rho^* & 0 \end{pmatrix}, \quad (V^{-1}\delta V)_{\text{ind}} = \begin{pmatrix} 0 & i\delta\sigma \\ i\delta\sigma^* & 0 \end{pmatrix} \tag{10.6.73}
$$

and some more patient scribbling accounting for T_0, as in (10.6.40), yields

$$
\begin{aligned}
\mathrm{str}\,\left(T_0^{-1}\delta T_0\right)_{\mathrm{ind}}^{12}&\left(T_0^{-1}\delta T_0\right)_{\mathrm{ind}}^{21}\\
=\quad & S_B^4\delta s_B^*\delta s_B + S_B^2 S_F^2(s_F^*\delta\rho + s_B^*\delta i\sigma)(s_F\delta\rho^* - s_B\delta i\sigma^*) \qquad(10.6.74)\\
& + S_F^4\delta s_F^*\delta s_F + S_B^2 S_F^2(s_B^*\delta\rho^* + s_F^*\delta i\sigma^*)(s_B\delta\rho - s_F\delta i\sigma)\,.
\end{aligned}
$$

The promised block structure of the Berezinian and the cancelling of the factors S_B and S_F therein is now manifest. The Fermi block involves the matrix

$$
\begin{pmatrix}
0 & |s_F|^2 - |s_B|^2 & 0 & -2s_B s_F^* \\
|s_B|^2 - |s_F|^2 & 0 & -2s_B^* s_F & 0 \\
0 & 2s_B^* s_F & 0 & |s_F|^2 - |s_B|^2 \\
2s_B s_F^* & 0 & |s_B|^2 - |s_F|^2 & 0
\end{pmatrix}
\qquad(10.6.75)
$$

which contributes the factor $(|s_B|^2 + |s_F|^2)^{-2}$ to the Berezinian. Straightening out various factors 2 and π cavalierly left unnoticed along the way, we get

$$
d\mu(T_0) = \frac{1}{(|s_B|^2 + |s_F|^2)^2}\frac{d^2 s_B}{\pi}\frac{d^2 s_F}{\pi}d\rho^* d\rho d i\sigma^* d i\sigma\,. \qquad(10.6.76)
$$

An elementary final transformation dispatches the auxiliary variables s_B and s_F in favor of the eigenvalues t_B and t_F and yields the reward for our efforts and (hopefully, not too much) suffering,

$$
d\mu(T_0) = \frac{1}{(|t_B|^2 + |t_F|^2)^2}\frac{d^2 t_B}{\pi}\frac{d^2 t_F}{\pi}d\rho^* d\rho d i\sigma^* d i\sigma\,, \qquad(10.6.77)
$$

which was previously announced in (10.6.43) and used in Subsect. 10.6.4. Thus, the superanalytic derivation of the two-point cluster function is completed.

10.6.6 Back to the Generating Function

In Subsect. 10.6.4, we shifted our attention away from the generating function $\overline{Z}(\hat{E})$ toward the two-point function, an observable indicator of spectral fluctuations. The generating function is worthy of interest itself since it represents the nonlinear supermatrix sigma model most concisely and has the two-point function as a second derivative, according to (10.6.6). Therefore, we shall conclude this section by evaluating $\overline{Z}(\hat{E})$, picking up the thread after (10.6.25). After inserting the integration measure (10.6.42, 43) there, we obtain, in the fashion of (10.6.51),

$$
\begin{aligned}
\overline{Z}(\hat{E}) &= \int d\mu(T_0)\exp\left[N\mathrm{Str}(\tilde{E} - \hat{E})\frac{Q^{\mathrm{s}}(T_0)}{\lambda^2}\right]\\
&= \int d\mu(T_0)\exp\left[N\mathrm{Str}(\tilde{E} - \hat{E})\,T_0\,\overline{G}(\tilde{E})\,T_0^{-1}\right]\,. \qquad(10.6.78)
\end{aligned}
$$

The average Green function with the matrix argument $\tilde{E}$ is to be read as

$$
\begin{aligned}
\overline{G}(\tilde{E}) &= \mathrm{diag}\left(\overline{G}(E^-), \overline{G}(E^-), \overline{G}(E^+), \overline{G}(E^+)\right)\\
&= \mathcal{P}\int\frac{dE'\overline{\varrho}(E')}{E - E'} + i\pi\overline{\varrho}(E)\Lambda\,. \qquad(10.6.79)
\end{aligned}
$$

Note that the real part of $\overline{G}(\tilde{E})$ is proportional to the unit matrix.

As it stands in (10.6.78), the mean generating function still contains nonuniversal properties of the GUE. It does give rise to the universal cluster function, as described in Subsect. 10.6.4, and it is well to recall why that is so. As required by (10.6.6), we must differentiate w.r.t. E_1 and E_2 and set $E_1 = E_3, E_2 = E_4$. Only then does the exponent in (10.6.78) shed the real part of the average Green function since $\text{Str}(\tilde{E} - \hat{E}) = -\text{Str}\hat{E}$ and $\text{Str}\hat{E} = 0$ for $E_1 = E_3, E_2 = E_4$. By subtracting the squared mean Green function from the resulting mean two-point function to get the cumulant function and finally absorbing the mean level density $\overline{\varrho}(E)$ in the rescaled energy variable $e = (E_1 - E_2)N\overline{\varrho}(E)$ we arrive at the universal cluster function in (10.6.58).

We shall encounter the generating function (10.6.78) again in the next section for a non-Gaussian matrix ensemble.

10.7 Universality of Spectral Fluctuations: Non-Gaussian Ensembles

Hackenbroich and *Weidenmüller* presented an alternative derivation of the two-point function of ensembles of random matrices [377] which does not assume Gaussian statistics for the matrix elements H_{ij}. These ensembles are characterized by matrix densities of the form

$$P(H) \propto \exp[-N\text{Tr}\, V(H)], \tag{10.7.1}$$

constrained only such that moments exist. The appearance of the trace secures invariance of the ensemble w.r.t. the desired "canonical" transformations, be these unitary, orthogonal, or symplectic. Here, I shall confine myself to the unitary case of complex Hermitian matrices H. By choosing $V(H) = H^2$, we would be back to the Gaussian ensemble for which the mean density of levels is given by the semicircle law. For general $V(H)$, that latter law does not reign, but after rescaling the energy axis in the usual way, $e = EN\overline{\varrho}(E)$, we shall recover the same two-point function as for the GUE above. This will be no surprise for the reader who has studied level dynamics and appreciated, in particular, the arguments of Sect. 6.10. Universality of spectral fluctuations is at work here. Once we have ascertained the independence of the two-point function $Y(e)$ of the function $V(H)$ and invoke the ergodicity à la Pandey (see Sect. 4.13.2), we shall have gone a long way toward understanding why a single dynamical system with global chaos in its classical phase space has universal spectral spectral fluctuations.

As a preparation to the reasoning of Hackenbroich and Weidenmüller, we need to familiarize ourselves with delta functions of Grassmann variables.

10.7.1 Delta Functions of Grassmann Variables

For Grassmann as well as for ordinary variables, one defines the delta function by

$$f(\eta_0) = \int d\eta \, \delta(\eta - \eta_0) f(\eta) \, . \tag{10.7.2}$$

With $f(\eta) = f_0 + f_1 \eta$, we check the simple representations

$$\delta(\eta - \eta_0) = \frac{1}{i} \int d\sigma \, e^{i\sigma(\eta - \eta_0)} = \eta - \eta_0 \, . \tag{10.7.3}$$

Neither representation suggests drawing anything peaked, but who would want to draw graphs for functions of Grassmann variables anyway. A more serious comment is that $\delta(a\eta) = a\delta(\eta)$ for an arbitrary complex number a; in particular, $\delta(-\eta) = -\delta(\eta)$. The latter two properties are well worth highlighting since they contrast to $\delta(ax) = |a|^{-1}\delta(x)$ for real a. The first of the representations in (10.7.3) is obviously analogous to the familiar Fourier integral representation of the ordinary delta function, $\delta(x) = (2\pi)^{-1} \int dk \, \exp ikx$.

Thus equipped, we can proceed to delta functions of supermatrices. Starting with the simplest case, 2×2 supermatrices of the form

$$Q = \begin{pmatrix} a & \eta^* \\ \eta & ib \end{pmatrix} , \tag{10.7.4}$$

we define $\delta(Q - Q')$ as the product of delta functions of the four matrix elements

$$\delta(Q - \tilde{Q}) = \delta(a - a')\delta(b - b')\delta(\sigma^* - \sigma'^*)\delta(\sigma - \sigma') \tag{10.7.5}$$

and note the integral identities

$$\begin{aligned}
\delta(Q - Q') &= \frac{1}{2\pi} \int d\hat{Q} \, \exp\left[iN\mathrm{Str}\hat{Q}(Q - Q')\right] \\
f(Q') &= \int dQ d\hat{Q} \, \exp\left[iN\mathrm{Str}\hat{Q}(Q - Q')\right] f(Q) \, ;
\end{aligned} \tag{10.7.6}$$

all integration measures are defined here according to our convention $dQ = (dadb/2\pi)d\sigma^* d\sigma$; note that the factor N in the exponent is sneaked in at no cost, for the sake of later convenience. When generalizing to $2n \times 2n$ supermatrices, replacing a in (10.7.4) by an $n \times n$ Bose block etc., we still accompany each bosonic differential increment with a factor $1/\sqrt{2\pi}$ and therefore must replace the factor $(2\pi)^{-1}$ in the first of the identities (10.7.6) by $(2\pi)^{-n^2}$ to get the correct power of 2π required by the Fourier-integral representation of the ordinary delta function. The second of the identities (10.7.6) remains intact for any n.

10.7.2 Generating Function

We return to the generating function (10.6.9) for the two-point function and average over the non-Gaussian ensemble (10.7.1),

$$
\begin{aligned}
\overline{Z}(\hat{E}) &= \int dH P(H) Z(\hat{E}) \\
&= (-1)^N \int d\Phi^* d\Phi \int dH P(H) \exp[\mathrm{i}\,\Phi^\dagger L (H - \hat{E})\Phi]. \qquad (10.7.7)
\end{aligned}
$$

The admitted non-Gaussian character of $P(H)$ forbids us from resorting to the Hubbard–Stratonovich transformation employed in Sect. 10.6.1. We still enjoy the unitary invariance $P(H) = P(H')$ with $H' = UHU^\dagger$ and U an arbitrary unitary $N \times N$ matrix. Due to that invariance, the integrand of the foregoing supervector integral is, after the ensemble average, a "level-space scalar", i.e., it depends on the vector Φ and its adjoint only through

$$
\sum_{i=1}^{N} \Phi_{\alpha i} \Phi^*_{\beta i} L_\beta = N \tilde{Q}_{\alpha\beta} L_\beta ; \qquad (10.7.8)
$$

note that the 4×4 supermatrix $\tilde{Q}$ was already encountered in (10.6.11).

Instead of Hubbard–Stratonovich, now we use the second of the delta function identities (10.7.6) for 4×4 supermatrices and $f(Q) = 1$ to write[8]

$$
\begin{aligned}
\overline{Z}(\hat{E}) &= (-1)^N \int d\Phi^* d\Phi \int dQ d\hat{Q} \exp[\mathrm{i}N\mathrm{Str}\,\hat{Q}(Q - \tilde{Q}L)]\overline{\exp[\mathrm{i}\,\Phi^\dagger L(H - \hat{E})\Phi]} \\
&= (-1)^N \int dQ d\hat{Q} \exp[\mathrm{i}N\mathrm{Str}\,Q\hat{Q}] \int d\Phi^* d\Phi \overline{\exp[\mathrm{i}\,\Phi^\dagger L(H - \hat{E} - \hat{Q})\Phi]} \\
&= (-1)^N \int dQ d\hat{Q} \exp(\mathrm{i}N\mathrm{Str}\,Q\hat{Q}) \overline{\left(\mathrm{Sdet}\, L(H - \hat{E} - \hat{Q})\right)^{-1}} \\
&= \int dQ d\hat{Q} \exp(\mathrm{i}N\mathrm{Str}\,Q\hat{Q}) \overline{\left(\mathrm{Sdet}\, (H - \hat{E} - \hat{Q})\right)^{-1}} \\
&= \int dQ d\hat{Q} \exp(\mathrm{i}N\mathrm{Str}\,Q\hat{Q}) \overline{\exp\mathrm{Str}\,\mathrm{Tr}\ln(H - \hat{E} - \hat{Q})} . \qquad (10.7.9)
\end{aligned}
$$

Even though the reader should be at peace by now with compact notation, it may be well to spell out that the superdeterminant above is $4N \times 4N$ and that within it we must read H as the tensor product of the $N \times N$ random matrix H in level space with the 4×4 unit matrix in superspace, whereas conversely $\hat{E}$ and $\hat{Q}$ are 4×4 matrices in superspace and act like the $N \times N$ unit matrix in level space. In the last line of the foregoing equation, I have written the overall supertrace (which sums over $4N$ diagonal elements) as $\mathrm{Str}\,\mathrm{Tr}$, to reserve the symbol Str for the four-dimensional superspace, denoting by Tr the trace in level space.

---

[8] Our unscrupulous change of the order of integrations as well as the naive use of the saddle-point approximation for the N-fold energy integral later admittedly give a certain heuristic character to this section.

At this point, we may imagine the matrix H diagonalized in level space and do the ensemble average with the help of the joint density of levels

$$P(E) = \mathcal{N} \exp[-N \sum_i V(E_i)] \prod_{i<j} (E_i - E_j)^2 \qquad (10.7.10)$$

following from (10.7.1); the normalization constant $\mathcal{N}$ secures $\int d^N E P(E) = 1$. Then, the ensemble averaged exponential in (10.7.9) reads

$$\overline{\exp \operatorname{Str} \operatorname{Tr} \ln(H - \hat{E} - \hat{Q})} \qquad (10.7.11)$$

$$= \mathcal{N} \int d^N E \exp \left(2 \sum_{i<j} \ln |E_i - E_j| - N \sum_i V(E_i) - \sum_i \operatorname{Str} \ln (E_i - \hat{E} - \hat{Q}) \right).$$

Inasmuch as we are interested in large values of N, the foregoing N-fold integral over the energies E_i can be done in the saddle-point approximation. To that end, we simply observe that the third of the three terms in the exponent is smaller than the first two by one order in N such that it gives rise to a "prefactor" that can be evaluated at the saddle and then taken out of the integral. The remaining integral cancels against the normalization factor, $\mathcal{N}/\mathcal{N} = 1$, so that we are simply left with the saddle-point value of the prefactor,

$$\overline{\exp \operatorname{Str} \operatorname{Tr} \ln(H - \hat{E} - \hat{Q})} = \exp \left[- \sum_i \operatorname{Str} \ln (E_i^{\mathrm{sp}} - \hat{E} - \hat{Q}) \right]; \quad (10.7.12)$$

here E_i^{sp} denotes the saddle-point value of the ith integration variable, determined by the saddle-point equation

$$\sum_{j(\neq i)} \frac{2}{E_i^{\mathrm{sp}} - E_j^{\mathrm{sp}}} - NV'(E_i^{\mathrm{sp}}) = 0. \qquad (10.7.13)$$

We are done with the ensemble average and confront integrals over the supermatrices $Q, \hat{Q}$ for the average generating function

$$\overline{Z}(\hat{E}) = \int dQ d\hat{Q} \exp \left[iN \operatorname{Str} Q\hat{Q} - \sum_i \operatorname{Str} \ln (E_i^{\mathrm{sp}} - \hat{E} - \hat{Q}) \right]. \qquad (10.7.14)$$

Now, it is indicated to split the logarithm in the foregoing exponent into a leading-order term and a perturbation smaller by one order in N, as in the transition from (10.6.24) to (10.6.25), by again introducing the center energy $E = (E_1 + E_2)/2$ and the diagonal matrix $\tilde{E} = (E^-, E^-, E^+, E^+)$ as in (10.6.33), immediately taking $E_1 = E_3, E_2 = E_4$. The analogue of (10.6.25),

$$\overline{Z}(\hat{E}) = \int dQ d\hat{Q} \exp \left[iN \operatorname{Str} Q\hat{Q} - \sum_i \operatorname{Str} \ln (E_i^{\mathrm{sp}} - \tilde{E} - \hat{Q}) \right]$$

$$\times \exp \left[- \sum_i \operatorname{Str} (E_i^{\mathrm{sp}} - \tilde{E} - \hat{Q})^{-1} (\hat{E} - \tilde{E}) \right], \qquad (10.7.15)$$

has its first exponential invariant under the hyperbolic transformations discussed in Sect. 10.6.2. When invoking the saddle-point approximation for the $\hat{Q}$ integral, we must treat the second exponential above as a prefactor and thus involve the saddle-point equation

$$iQ = \frac{1}{N} \sum_i \frac{1}{\tilde{E} + \hat{Q}^{\text{sp}} - E_i^{\text{sp}}} \approx \int \frac{dE'\, \overline{\varrho}(E')}{\tilde{E} + \hat{Q}^{\text{sp}} - E'} = \overline{G}(\tilde{E} + \hat{Q}^{\text{sp}}) ; \quad (10.7.16)$$

it is in line with the leading-order treatment in N when we approximate here the sum by an integral with the mean level density as a weight and so involve the mean Greens function $\overline{G}$ on the right-hand side; at any rate, the foregoing equation implicitly fixes the saddle-point value $\hat{Q}^{\text{sp}}(Q, \tilde{E})$ as a function of the matrices Q and $\tilde{E}$. The Parisi–Sourlas–Efetov–Wegner theorem secures the integral over the fluctuations $\delta\hat{Q}$ around that saddle at unity so that we are left with the integral over the supermatrix Q

$$\overline{Z}(\hat{E}) = \int dQ \, \exp\left[iN\text{Str}\, Q\hat{Q}^{\text{sp}} - \sum_i \text{Str}\ln\left(E_i^{\text{sp}} - \tilde{E} - \hat{Q}^{\text{sp}}\right)\right]$$
$$\times \exp\left[- N\text{Str}\, \overline{G}(\tilde{E} + \hat{Q}^{\text{sp}})(\hat{E} - \tilde{E})\right] \quad (10.7.17)$$

which itself invites a final saddle-point approximation. Here again, the second exponential plays the role of a prefactor, and the saddle-point equation reads

$$i\hat{Q}^{\text{sp}} + iQ\frac{\partial \hat{Q}^{\text{sp}}}{\partial Q} - \overline{G}(\tilde{E} + \hat{Q}^{\text{sp}})\frac{\partial \hat{Q}^{\text{sp}}}{\partial Q} = 0 . \quad (10.7.18)$$

From this and the previous saddle-point equation (10.7.16), we conclude that

$$\hat{Q}^{\text{sp}}(Q^{\text{sp}}, \tilde{E}) = 0 \quad \Longrightarrow \quad iQ^{\text{sp}} = \overline{G}(\tilde{E}) . \quad (10.7.19)$$

But due to the hyperbolic invariance of the first exponential in (10.7.17), a whole manifold of saddles arises. We insert the integration measure (10.6.42, 43) and get, as in (10.6.78) and in the fashion of (10.6.51),

$$\overline{Z}(\hat{E}) = \int d\mu(T_0) \exp\left[\text{Str}(\tilde{E} - \hat{E})\, T_0\, \overline{G}(\tilde{E})\, T_0^{-1}\right] . \quad (10.7.20)$$

The mean Green function $\overline{G}(\tilde{E})$ is expressed in terms of the mean level density $\overline{\varrho}(E)$ as in (10.6.79), but the latter is of course ensemble specific, and in this case is not given by the semicircle law.

Thus recovering the generating function of the nonlinear sigma model, we see that the non-Gaussian random-matrix ensemble under consideration does indeed have the same two-point function as the GUE, save for the different mean density of levels. Nevertheless, the universal cluster function arises once that density is scaled away, as described in the discussion of the generating function (10.6.78) at the end of the preceding section.

10.8 Universal Spectral Fluctuations of Sparse Matrices

Another important ensemble of matrices, for which universality à la Wigner and Dyson has been demonstrated, is formed by random Hermitian matrices with independent and identically distributed entries. I shall not enter into a detailed discussion but rather refer the reader to the important contributions of *Khorunzhy, Khoruzhenko*, and *Pastur* [378] and *Anna* and *Zee* [379].

The early work by *Mirlin* and *Fyodorov* is closest to the spirit of this chapter [380]. The latter authors pioneered a superanalytic approach based on a functional generalization of the Hubbard–Stratonovich transformation. They found that the most important characteristic of the ensembles with independent elements is the mean number p of non-zero elements per row or column. More precisely, if the density of non-zero entries is an even function with finite moments and if p grows with the matrix size, as $p \sim N^\alpha$ with $0 < \alpha \leq 1$, both the mean and the higher cluster functions of the level density come out as for the Gaussian ensembles of the same symmetry (real symmetric, complex Hermitian, or quaternion real). The only "nontrivial" case thus appears to be one of *very sparse* matrices that have a *finite* mean number p of non-zero entries randomly placed among the about $N - p$ zeros per row, even when N grows indefinitely. The mean density of levels then deviates from the semicircular form, but still, one has the truly remarkable property that, as long as p exceeds a certain threshold value, $p > p_c$, the spectral correlations remain the same as for the Gaussian ensembles. Only when p is decreased below p_c, a kind of Anderson localization occurs, and the eigenvalues become decorrelated [381]. The numerical value of the threshold p_c is non universal, i.e., it depends on the density of non-zero entries and can be found by solving a certain integral equation. However, numerical experiments with sparse matrices show [382] that the value of p_c is typically smaller than two. It is indeed an impressive manifestation of universality à la Dyson and Wigner that two non-zero elements out of thousands or more vanishing ones suffice to produce the statistics familiar from the Gaussian (or circular) ensembles.

What is crucial here is the arbitrary position of the non-zero entries within the rows or columns. In the subsequent section, we shall see that quite different statistics arise if the non-zero elements are bound to be concentrated near the principal diagonal of the matrix.

10.9 Thick Wires, Banded Random Matrices, One-Dimensional Sigma Model

10.9.1 Banded Matrices Modelling Thick Wires

We have seen a link between quantum chaos and disordered systems in our discussion of quantum localization in Chap. 7. The disordered systems considered there were strictly one-dimensional and that restriction can now be eased in favor of the quasi one-dimensional behavior of a particle hopping from one lattice site to neighboring sites in a thick wire. I propose to return Anderson's tight-binding Hamiltonian of (7.2.1)

$$H_{mn} = T_m \delta_{mn} + W_r \delta_{m,n+r}, \tag{10.9.1}$$

where the indices refer to sites, T_m is a single-site potential, and W_r is a hopping amplitude.

Anderson's model can account for an electron moving in a thick wire of length L and cross-section S in the following way. We think of the wire as divided in L/l_{el} slices, whose length l_{el} is the mean free path l_{el}, i.e., the mean distance between scattering impurities. Each such slice provides $b \equiv k_F^2 S$ transverse "channel states." The moving electron can make intraslice transitions between different channel states and may also hop from one slice to one of the two neighboring sites. The site label in Anderson's Hamiltonian may be chosen such that it increases by b within one slice before proceeding to doing the same for the neighboring slice to the right. Very schematically, we may associate $r = \pm b$ with an interslice hop and $|r| < b$ with an intraslice transition. Thus, the total number of "sites" in the "lattice" is $N = bL/l_{el} = k_F^2 SL/l_{el}$, and the Hamiltonian matrix (10.9.1) has a band structure: Its elements H_{mn} are either strictly zero outside a band of width $2b$ around the diagonal, $H_{mn} = 0$ for $|m - n| > b$, or at least fall to zero sufficiently rapidly as the skewness $|m - n|/b$ grows large. To introduce disorder, we take the elements H_{mn} as independent random numbers with the joint density

$$P(H) = \mathcal{N} \prod_{i=1}^{N} \exp\left(-\frac{H_{ii}^2}{2J_{ii}}\right) \prod_{i<j} \exp\left(-\frac{|H_{ij}|^2}{2J_{ij}}\right) \tag{10.9.2}$$

where the variances J_{ij} suitably express the assumed bandedness. The exponential or the box distribution are examples convenient to work with,

$$J_{ij} = J(|i - j|) = \begin{cases} (\lambda^2/2b)\exp(-|i - j|/b) \\ (\lambda^2/2b)\Theta(b - |i - j|) \end{cases}, \tag{10.9.3}$$

where λ fixes the unit of energy; all subsequent explicit results will be specified for the exponential distribution. We face an ensemble of random Hermitian matrices *not* invariant under unitary transformations; nor do we allow for time-reversal invariance, inasmuch as the off-diagonal elements have equal variances for real and imaginary parts. Of course, the bandwidth b should be small, $b/N \to 0$ as $N \to \infty$, or else we would be back to the GUE.

A great deal more physics is accessible through the quasi one-dimensional character of this model, compared to the strictly one-dimensional model of Chap. 7. We cannot even take for granted localization of all eigenfunctions since the effectively long-range random hops of range $b = k_F^2 S \sim S$ make for competition between diffusion and quantum coherence. Localization will prevail only if the length L of the sample is larger than the localization length l while for short samples, i.e., $L < l$, classical diffusion is not hindered by quantum coherence.

It is well to support the foregoing argument by an order-of-magnitude estimate of the various quantities involved. As the particle hops about in the lattice, starting at some site, classical diffusion will cease to provide an appropriate description once its spatial uncertainty has reached the quantum localization length l. Such quantum cutoff of diffusion must surely happen if the time needed for diffusive exploration of the sample length L, the so-called Thouless time $\tau_c = L^2/D$, is large compared to the time scale $\hbar/\Delta$ on which the particle begins to resolve the level spacing Δ, i.e., to feel the discreteness of the spectrum. Assuming that the hopping particle is a fermion whose energy is near the Fermi energy $E_F = \hbar^2 k_F^2/2m$, we may estimate the level spacing as $\Delta = E_F/(k_F^3 LS)$ since $k_F^3 LS$ is the total number of single-particle states with energies between 0 and E_F. Similarly, the diffusion constant is roughly given by the elastic mean free path $l_{\rm el}$ and the Fermi velocity $\hbar k_F/m$ as $D = l_{\rm el}\hbar k_F/m$, such that the localization condition $L^2/D \gg \hbar/\Delta$ yields $L \gg l_{\rm el}(k_F^2 S) \equiv l$. Conversely, if the particle can diffusively explore the whole sample before resolving the level spacing, one has the case of a short sample in the sense $L \ll l_{\rm el}(k_F^2 S) \equiv l$. The length scale dividing short and long is just the localization length

$$l = l_{\rm el}(k_F^2 S) = b l_{\rm el}\,. \tag{10.9.4}$$

In quasi one-dimensional wires, $l \gg l_{\rm el}$. To spell out the important conclusion once more, diffusion prevails if the wire is short, whereas localization becomes effective in long wires. The detailed behavior of the eigenvectors of the banded matrices under study turns out to be governed by a scaling parameter depending on N and b as

$$x \propto N/b^2\,, \tag{10.9.5}$$

as was first found numerically *Casati, Molinary*, and *Izrailev* [383, 384] and later confirmed analytically by *Fyodorov* and *Mirlin* [385].

10.9.2 Inverse Participation Ratio and Localization Length

Although in presenting Lloyd's model in Chap. 7 we calculated the mean localization length l, we shall presently follow *Fyodorov* and *Mirlin* [385] and discuss quantum localization for the ensemble of banded random matrices (10.9.3) in terms of the so-called inverse participation ratio, a quantity intimately related to l. For the μth eigenvector of a matrix H, that quantity is defined as

$$P_\mu = \sum_{n=1}^{N} |\psi_{\mu n}|^4\,. \tag{10.9.6}$$

If the eigenvector is "extended" rather than localized, $|\psi_{\mu n}| \sim 1/\sqrt{N}$ such that $P_\mu \propto 1/N$ and $P_\mu \to 0$ as $N \to \infty$, but a finite value of P_μ results for an exponentially localized vector, $\psi_{\mu n} = (2l_\mu)^{-\frac{1}{2}} \exp(-|n - n_\mu|/2l_\mu)$. A one-line calculation indeed reveals that $P_\mu \approx 1/4l_\mu$ for $l \gg 1$ and $N \to \infty$.

To characterize a whole matrix H rather than one of its eigenvectors, we employ the spectral average

$$\langle P_\mu \rangle = \frac{1}{N\bar{\varrho}(E)\Delta E} \int_{E-\Delta E/2}^{E+\Delta E/2} dE' \sum_\mu P_\mu\, \delta(E' - E_\mu)\Bigg|_{\Delta E \to 0} \tag{10.9.7}$$

and to make sure that we don't pick up an exceptional matrix, we perform the disorder average, i.e., the average over the ensemble (10.9.3),

$$P \equiv \overline{\langle P_\mu \rangle} = \frac{1}{N\overline{\varrho}(E)} \sum_\mu \overline{P_\mu \delta(E - E_\mu)}. \tag{10.9.8}$$

The ensemble-averaged density of levels $N\varrho(E)$ appearing here must be evaluated for the present matrix ensemble. I shall not devote separate space to that endeavor but refer the reader to Refs. [386–389] where it is shown that Wigner's semicircle law holds if the bandwidth grows no more slowly than $b^2 \propto N$ for $N \to \infty$, the case of interest. In fact, we shall see below that the semicircle law must arise as a by-product of our discussion of localization.

In Chap. 7, we extracted the localization length l from the real part of the Green function $\mathrm{Tr}\,(\mathcal{E} - H)^{-1}$; the inverse participation ratio P, on the other hand, requires the product of two matrix elements of the resolvent, $G_{nn}(\mathcal{E}) = \langle n|(\mathcal{E} - H)^{-1}|n\rangle = \sum_\mu |\psi_n^\mu|^2 (\mathcal{E} - E_\mu)^{-1}$, one retarded and the other advanced,

$$\begin{aligned}
N\overline{\rho}(E)P &= \lim_{\epsilon \downarrow 0} \frac{\epsilon}{\pi} \sum_n \overline{G_{nn}(E - i\epsilon)G_{nn}(E + i\epsilon)} \\
&= \lim_{\epsilon \downarrow 0} \frac{\epsilon}{\pi} \sum_{n,\mu,\nu} \overline{\frac{|\psi_n^\mu|^2 |\psi_n^\nu|^2}{(E - E_\mu - i\epsilon)(E - E_\nu + i\epsilon)}} \\
&= \lim_{\epsilon \downarrow 0} \frac{\epsilon}{\pi} \sum_{n,\mu} \overline{\frac{|\psi_n^\mu|^4}{(E - E_\mu)^2 + \epsilon^2}} = \sum_{n,\mu} \overline{|\psi_n^\mu|^4 \,\delta(E - E_\mu)}.
\end{aligned} \tag{10.9.9}$$

The reader will appreciate that only diagonal terms in the foregoing double sum over eigenvectors survive the limit as ϵ goes to zero.

I should hurry to add that there is a deep reason for not settling on the average Green function but on average products thereof when attempting to understand transport properties of quasi one-dimensional wires in terms of random band matrices. It turns out that a full characterization requires not just one such quantity but a whole set of multilocal ones like

$$\sum_\mu \overline{|\psi_{n_1}^\mu|^{2q} |\psi_{n_2}^\mu|^{2q} \dots |\psi_{n_k}^\mu|^{2q} \delta(E - E_\mu)} \tag{10.9.10}$$

with positive integer exponents q. The reason is that finite samples are not self-averaging in their transport behavior, such that sample-to-sample fluctuations require a probability density for the inverse participation ratio rather than just a mean value. Even though the calculation of the whole set of moments is not much more difficult than that of just the lowest-order set [385], our P, I hold fast to pedagogic principles and present the simplest case first.

10.9.3 One-Dimensional Nonlinear Sigma Model

The expression (10.9.9) for the inverse participation ratio closely resembles the average product of two Green functions from which we had previously extracted the two-point correlator of the level density. However, an important difference to the two-point correlator $\overline{G(E_1^-)G(E_2^+)}$ is that we no longer look at representational independent traces $G(E^\pm) = \sum_n G_{nn}(E^\pm)$ but at individual matrix elements $G_{nn}(E^\pm)$ in the "position" representation where the index n refers to lattice sites. That difference is no obstacle at all for the superanalytic procedure from the ensemble average via Hubbard–Stratonovich to the sigma model; we just have to carry along the site index and shall thus be led to a separate supermatrix Q_i for each site, whereupon the sigma model becomes one-dimensional rather than zero-dimensional. The zero-dimensional model will still be contained as the limit $b \gg \sqrt{N}$.

We can even obtain the product $G_{nn}(E^-)G_{nn}(E^+)$ from a generating function, in analogy to (10.6.5, 6). Leaving the specification of an appropriate generating function to the reader as Problem 10.13, I proceed here to apply the moment-generating function (10.4.17) and write the superintegral

$$G_{nn}(E^-)G_{nn}(E^+) = (-1)^? \int d\Phi^* d\Phi \left(z_{1n}^* z_{1n} z_{2n}^* z_{2n}\right) \tag{10.9.11}$$

$$\times \exp\left[i \sum_{\alpha=1}^{4} \Phi_\alpha^\dagger L_\alpha \left(H - E_\alpha + i\Lambda_\alpha \epsilon\right)\Phi_\alpha\right],$$

not bothering about an overall sign since we know the final result must be positive. This differs from the generating function for the GUE (10.6.9) only by the appearance of the prefactor $z_{1n}^* z_{1n} z_{2n}^* z_{2n}$ and in that one and only one energy argument E appears; the positive infinitesimal ϵ remains on display here until eventually dealt with in the fashion of (10.9.9). Clearly, a $(4N)$-component supervector Φ like (10.6.8) is employed, but it is now convenient to proceed immediately to the "advanced-retarded" notation and order the components as

$$\Phi = \begin{pmatrix} z_1 \\ \eta_1 \\ z_2 \\ \eta_2 \end{pmatrix} = \begin{pmatrix} \Phi_1 \\ \Phi_2 \\ \Phi_3 \\ \Phi_4 \end{pmatrix} ; \tag{10.9.12}$$

then the diagonal 4×4 matrices $L = \mathrm{diag}(1, 1, -1, 1)$ and $\Lambda = \mathrm{diag}(1, 1, -1, -1)$ immediately take the form familiar from (10.6.33). Were the reader inclined to ponder for a moment about the similarity and the reason for the slight differences between (10.6.9) and (10.9.11), that moment would be well spent.

As in Sect. 10.6, we proceed to the Gaussian average. The disorder ensemble (10.9.3, 4) now gives $\overline{\exp(i \sum_{ij} H_{ij} u_{ij})} = \exp(-\frac{1}{2} \sum_{ij} J_{ij} u_{ij}^* u_{ij})$ where $u_{ij} = \sum_\alpha \Phi_{\alpha i}^* L_\alpha \Phi_{\alpha j}$. It is precisely due to the dependence of the variance J_{ij} on the site indices i, j that now we must deal with the "local" 4×4 supermatrix

$$\tilde{Q}_{\alpha\beta i} = \Phi_{\alpha i}\Phi_{\beta i}^* \tag{10.9.13}$$

rather than with the "global" one in (10.6.11) and have $\overline{\exp(i \sum_{ij} H_{ij} u_{ij})} = \exp(-\frac{1}{2} \sum_{ij} J_{ij} \mathrm{Str}\tilde{Q}_i L\tilde{Q}_j L)$. Therefore, instead of the average generating function (10.6.12), we get

$$\overline{G_{nn}(E^-)G_{nn}(E^+)} = (-1)^? \int d\Phi^* d\Phi \left(z_{1n}^* z_{1n} z_{2n}^* z_{2n} \right) \tag{10.9.14}$$

$$\times \exp \operatorname{Str} \left(-\frac{1}{2} \sum_{ij} J_{ij} \tilde{Q}_i L \tilde{Q}_j L - \mathrm{i}E \sum_i \tilde{Q}_i L - \epsilon \sum_i \tilde{Q}_i L \Lambda \right).$$

With the disorder average out of the way, we attack the integral over the supervector Φ which as usual enters the exponent quartically. The superanalytic Hubbard–Stratonovich transformation (10.6.21) must be modified to deal with the product $\tilde{Q}_i L \tilde{Q}_j L$; it is easy to see that we may use the generalization

$$\exp \left(-\frac{1}{2} \sum_{ij} J_{ij} \tilde{Q}_i L \tilde{Q}_j L \right) \tag{10.9.15}$$

$$= (-1)^N \int_{\mathcal{M}_4^N} d^N Q \exp \operatorname{Str} \left[-\frac{1}{2} \sum_{ij} (J^{-1})_{ij} Q_i L Q_j L - \mathrm{i} \sum_i Q_i L \tilde{Q}_i L \right]$$

in which N auxiliary supermatrices Q_i appear as integration variables, each to be integrated over its own manifold $\mathcal{M}_4$.

As a little aside, we ought to convince ourselves of the validity of the foregoing identity. Really no more than the temporary introduction of an ordinary orthogonal $N \times N$ matrix is required to diagonalize the real symmetric variance matrix, $O^T J O = \operatorname{diag}(j_1, j_2, \ldots j_N)$. Thus, the quadratic form in the $\tilde{Q}_i$ reads $\sum_{ij} J_{ij} \operatorname{Str} \tilde{Q}_i L \tilde{Q}_j L = \sum_k j_k \operatorname{Str} \tilde{q}_k L \tilde{q}_L$ with $\tilde{q}_k = \sum_i O_{ik} \tilde{Q}_i$. But if $\tilde{Q}_i L$ is pseudounitarily diagonalizable, so is $\tilde{q}_k L$ such that we may use (10.6.21) N times, $\exp[-\frac{1}{2} \sum_k j_k \operatorname{Str} (\tilde{q}_k L)^2] = \prod_k (-1)^N \int dq_k \exp \left[- \operatorname{Str} (\frac{1}{2j_k} (q_k L)^2 - \mathrm{i} q_k L \tilde{q}_k L) \right]$. Undoing the orthogonal transformation we confirm (10.9.15).

With the help of the superanalytic Hubbard–Stratonovich transformation, (10.9.15), our averaged product of Green functions reads

$$\overline{G_{nn}(E^-)G_{nn}(E^+)} = (-1)^? \int_{\mathcal{M}_4^N} d^N Q \int d\Phi^* d\Phi \left(z_{1n}^* z_{1n} z_{2n}^* z_{2n} \right) \tag{10.9.16}$$

$$\times \exp \operatorname{Str} \left[-\frac{1}{2} \sum_{ij} (J^{-1})_{ij} Q_i L Q_j L - \sum_i \operatorname{Str} \tilde{Q}_i L \left(\mathrm{i}E - \epsilon \Lambda - \mathrm{i} Q_i L \right) \right],$$

and now the Φ-integral has the desired Gaussian form. The superanalytic variant (10.4.18) of Wick's theorem provides the following generalization of (10.6.22):

$$\overline{G_{nn}(E^-)G_{nn}(E^+)} = (-1)^? \int_{\mathcal{M}_4^N} d^N Q \left[g_{\mathrm{aa}}^{BB}(n) g_{\mathrm{rr}}^{BB}(n) + g_{\mathrm{ar}}^{BB}(n) g_{\mathrm{ra}}^{BB}(n) \right]$$

$$\times \exp \left[-\frac{1}{2} \sum_{ij} (J^{-1})_{ij} \operatorname{Str} Q_i L Q_j L - \sum_i \operatorname{Str} \ln L \left(\mathrm{i}E - \epsilon \Lambda - \mathrm{i} Q_i L \right) \right];$$

$$\tag{10.9.17}$$

it may be well to spell out that each site but the nth contributes the usual factor $\operatorname{Sdet}^{-1}(\mathrm{i}E - \epsilon \Lambda - \mathrm{i} Q_i L)$ while for the nth, to which the prefactor $z_{1n}^* z_{1n} z_{2n}^* z_{2n}$

pertains, Wick's theorem (10.4.18) and the generating function (10.4.17) give two factors, i.e., the nth superdeterminant of the foregoing form and the combination of matrix elements of

$$g(n) = (E - Q_n + i\epsilon\Lambda)^{-1} \rightarrow (E - Q_n)^{-1} \tag{10.9.18}$$

in the prefactor; the imaginary infinitesimal will turn out to be dispensable in the 4×4 supermatrix $g(n)$ since Q_i will become effective with a finite imaginary part. Note that only bosonic elements of $g(n)$ arise in (10.9.17) due to the purely bosonic character of the quantity $z_{1n}^* z_{1n} z_{2n}^* z_{2n}$ "averaged" over; the indices "1" and "3" do indeed refer to the bosonic entries of the supervector (10.9.12) in the advanced-retarded notation now employed. The same minor embellishments as in our previous processing of (10.6.22) are in order now, i.e., to use $\mathrm{Sdet}\, L = -1$ and employ $Q_i L$ in place of Q_i as integration variables. At the same time, I choose to expand the logarithm to first order in the positive infinitesimal ϵ and thus get

$$\overline{G_{nn}(E^-)G_{nn}(E^+)} = (-1)^? \int_{(\mathcal{M}_4 L)^N} d^N Q \exp\left[-A(Q)\right] \tag{10.9.19}$$

$$\times f_n \exp[-i\epsilon \sum_i \mathrm{Str}\,(E - Q_i)^{-1}\Lambda]$$

where the "action" and the prefactor f_n are defined by

$$A(Q) = \frac{1}{2}\sum_{ij}\left[(J^{-1})_{ij} - J_0^{-1}\delta_{ij}\right]\mathrm{Str}\, Q_i Q_j + \sum_i \mathrm{Str}\left[\frac{Q_i^2}{2J_0} + \ln\left(E - Q_i\right)\right]$$

$$f_n = g_{\mathrm{aa}}^{BB}(n)g_{\mathrm{rr}}^{BB}(n) + g_{\mathrm{ar}}^{BB}(n)g_{\mathrm{ra}}^{BB}(n) \tag{10.9.20}$$

and the abbreviation

$$J_0 = \sum_{j=1}^N J_{ij} = \frac{\lambda^2}{2b}\frac{1 + e^{-1/b}}{1 - e^{-1/b}} \approx \lambda^2 . \tag{10.9.21}$$

Even though there is no explicit large factor decorating the action $A(Q)$, the summation over the N sites suggests that A is proportional to N. The double sum in the first term also shares that property, as becomes obvious when we consider the exponential form of J_{ij} given in (10.9.4) and its well known inverse [390]. We easily check

$$(J^{-1})_{ij} = \frac{2b\lambda^{-2}}{1 - e^{-2/b}}\left[(1 + e^{-2/b})\delta_{ij} - e^{-1/b}(\delta_{i,j+1} + \delta_{i,j-1})\right],$$

$$(J^{-1})_{ij} - J_0^{-1}\delta_{ij} = \frac{B}{2}\left(2\delta_{ij} - \delta_{i,j+1} - \delta_{i,j-1}\right), \tag{10.9.22}$$

$$B = \frac{4b\lambda^{-2}e^{-1/b}}{1 - e^{-2/b}} \approx \frac{2b^2}{\lambda^2} .$$

At any rate, inasmuch as we are concerned with $b \gg 1$, a saddle-point approximation appears in order for the Q-integrals. To prepare for that step, we write

$Q_i = Q_i^s + \delta Q_i$ and expand the action to second order in the increment δQ_i,

$$A(Q) = A(Q^s) + \delta A + \delta^2 A + \dots$$

$$\delta A = \mathrm{Str} \sum_i \delta Q_i \{ J_0^{-1} Q_i^s - (E - Q_i^s)^{-1} + \sum_j [(J^{-1})_{ij} - J_0^{-1}\delta_{ij}]Q_j^s \},$$

$$\delta^2 A = \frac{1}{2}\mathrm{Str} \sum_{ij} [(J^{-1})_{ij} - J_0^{-1}\delta_{ij}]\delta Q_i \delta Q_j, \tag{10.9.23}$$

$$+ \frac{1}{2}\mathrm{Str}\left[J_0^{-1}\delta Q_i^2 - (E - Q_i^s)^{-1}\delta Q_i (E - Q_i^s)^{-1}\delta Q_i \right].$$

Obviously, the saddle-point equation $\delta A = 0$ possesses homogeneous solutions $Q_i^s = Q^s$ determined by the familiar equation

$$J_0^{-1}Q^s = (E - Q^s)^{-1}, \tag{10.9.24}$$

and by reasoning precisely as in Sect. 10.6, we establish the relevant diagonal solution. Notational differences apart, we recover (10.6.28) and, for the manifold of nondiagonal saddles, (10.6.30). However, even though the diagonal saddle is homogeneous, we must allow for a separate pseudounitary transformation T_i at each site such that the manifold of saddles associated with the ith site reads

$$Q_i^s = E/2 - i\pi\overline{\varrho}(E)J_0 T_i \Lambda T_i^{-1}, \tag{10.9.25}$$

where $\overline{\varrho}(E) = (2\pi J_0)^{-1}(4J_0 - E^2)^{1/2}$ is the semicircular mean density of levels which, the reader will recall, was extracted from the analogue (10.5.11) of the homogeneous saddle-point equation (10.9.25) in Sect. 10.5.

It is appropriate to pause for a moment and appreciate that the homogeneous solution (10.9.24) can hardly be expected to be reliable when boundary effects come into play for $b \lesssim N$. Thus, we should expect $1 \ll b \ll N$ as the range of validity of our procedure from this point on.

As in (10.6.35), we employ pseudounitary matrices T_{0i} to block diagonalize the supermatrices Q_i,

$$Q_i = T_{0i} \begin{pmatrix} Q_{ai} & 0 \\ 0 & Q_{ri} \end{pmatrix} T_{0i}^{-1} \tag{10.9.26}$$

and from (10.6.42), we take the integration measure

$$d^N Q = \prod_{i=1}^{N} dQ_{ai} dQ_{ri} d\mu(T_{0i}) \frac{(q_{aBi} - q_{rBi})^2 (q_{aFi} - q_{rFi})^2}{(q_{aBi} - q_{rFi})^2 (q_{aFi} - q_{rBi})^2}$$

$$\equiv \prod_{i=1}^{N} dQ_{ai} dQ_{ri} d\mu(T_{0i}) \, F(Q_{ai}, Q_{ri}) \tag{10.9.27}$$

where $d\mu(T_{0i})$ is as in (10.6.43). The 2×2 blocks Q_{ai}, Q_{ri} at each site contain the hard modes whose action increasing fluctuations can be integrated out by the saddle-point approximation. Thus, in analogy to (10.6.51), we get

$$\overline{G_{nn}(E^-)G_{nn}(E^+)} = \int \left(\prod_i d\mu(T_{0i}) \right) f_n \tag{10.9.28}$$

$$\times \exp\left\{ -i\epsilon \sum_i \mathrm{Str}\, \frac{Q_i^s}{J_0} \Lambda - \frac{1}{2} \sum_{ij} [(J^{-1})_{ij} - J_0^{-1}\delta_{ij}]\,\mathrm{Str}\, Q_i^s Q_j^s \right\}$$

where according to (10.9.18, 26), we could replace $g(n)$ by Q_n^s/J_0. Each of the T_{0i} integrals is over the coset space $U(1,1/2)/U(1/1)U(1/1)$. A slightly more suggestive form of the foregoing expression arises when we realize from (10.9.25) that the supermatrices Q_i^s depend on the block diagonalizing matrices T_{0i} only through

$$Q(T_{0i}) \equiv T_{0i}\Lambda T_{0i}^{-1} \tag{10.9.29}$$

and that in the exponent in (10.9.28), we may replace Q_i^s by $-i\pi\overline{\varrho}(E)J_0 Q(T_{0i})$. As already remarked around (10.6.31), these latter dimensionless 4×4 supermatrices are the basic constituents of the nonlinear supermatrix sigma model obeying

$$Q(T_{0i})^2 = 1\,, \quad Q(T_{0i})^\dagger = LQ(T_{0i})L\,, \quad \mathrm{Str}\, Q(T_{0i}) = \mathrm{Str}\,\Lambda = 0\,. \tag{10.9.30}$$

Together with (10.9.22) they allow us to write

$$\overline{G_{nn}(E^-)G_{nn}(E^+)} = \int \left(\prod_i d\mu(T_{0i}) \right) f_n\, \mathrm{e}^{-S(\{T_{0i}\})}\,, \tag{10.9.31}$$

$$S(\{T_{0i}\}) = \frac{\gamma}{2} \sum_i \mathrm{Str}\, Q(T_{0i})Q(T_{0,i+1}) + \epsilon\pi\overline{\varrho} \sum_i \mathrm{Str}\, Q(T_{0i})\Lambda$$

$$= -\frac{\gamma}{4} \sum_i \mathrm{Str}\,[Q(T_{0i}) - Q(T_{0,i+1})]^2 + \epsilon\pi\overline{\varrho} \sum_i \mathrm{Str}\, Q(T_{0i})\Lambda\,,$$

$$\gamma = [\pi\overline{\varrho}(E)J_0]^2 B \propto 2b^2\,.$$

The dimensionless (but energy dependent!) parameter $\gamma \propto 2b^2 \gg 1$ defines a correlation length for the supermatrices $Q(T_{0i})$, as becomes especially transparent if we pass to a continuum description through $Q(T_{0i}) - Q(T_{0,i+1}) \to -\partial Q(x)/\partial x$ and $\sum_{i=1}^N \to \int_0^L dx$. Continuum description or discrete, we find ourselves up to our ears immersed in field theory when we have to implement the one-dimensional sigma model set up in the foregoing expression for the average product of two Green functions, a task a bit more involved than for the zero-dimensional model (cf. Sect. 10.6.4).

10.9.4 Implementing the One-Dimensional Sigma Model

Efetov and *Larkin* first showed that the continuum version of the one-dimensional model describes thick disordered wires [391]. These authors also demonstrated the localization of eigenfunctions in the limit of long wires and found the localization length $l = 2\gamma l_{\mathrm{el}}$. Here I shall keep following Ref. [385] where the full statistics of localization was treated.

Let us put our goal in view. The mean inverse participation ratio is given by (10.9.9, 34) as

$$
\begin{aligned}
P &= \frac{1}{N} \sum_{n=1}^{N} P_n\,, \\
P_n &= \lim_{\epsilon \to 0} \frac{\epsilon}{\pi \overline{\varrho}} \overline{G_{nn}(E - i\epsilon)G_{nn}(E + i\epsilon)} \qquad (10.9.32) \\
&= \lim_{\epsilon \to 0} \frac{\epsilon}{\pi \overline{\varrho}} \int \Big[\prod_i d\mu(T_{0i}) \Big] f_n\, e^{-S(\{T_{0i}\})}\,.
\end{aligned}
$$

The matrix $g(n)$ whose elements enter the prefactor f_n according to (10.9.20) is determined by (10.9.18, 27, 32) as

$$
g(n) = \frac{E}{2J_0} - i\pi\overline{\varrho}(E)Q(T_{0n})\,. \qquad (10.9.33)
$$

We proceed to do the superintegral in (10.9.32) recursively, working our way inward site per site. We start at the edge, $i = 1$, and imagine that $j - 1 \geq 1$ steps were already taken and yielded

$$
\begin{aligned}
Y_j(T_{0j}) &\equiv \prod_{i=1}^{j-1} \Big[\int d\mu(T_{0i})\, e^{\mathrm{Str}\big\{ -\frac{\gamma}{2} Q(T_{0i})Q(T_{0,i+1}) - \eta Q(T_{0i})\Lambda \big\}} \Big]\,, \\
\text{where}\quad \eta &= \pi\overline{\varrho}\epsilon\,. \qquad (10.9.34)
\end{aligned}
$$

We can continue until after $(n-1)$ steps we arrive with $Y_n(T_{0n})$ at the site whose contribution P_n to the inverse participation ratio P we focus on; that site requires extra care since it attaches the extra factor f_n to the integrand. Along the way we may rewrite the above integral as the recursion relation

$$
\begin{aligned}
Y_j(T_0) &= \int d\mu(T_0')\, E(T_0, T_0')\, Y_{j-1}(T_0')\,, \\
E(T_0, T_0') &= \exp\Big[-\frac{\gamma}{2}\mathrm{Str}\, Q(T_0)Q(T_0') - \eta\,\mathrm{Str}\, Q(T_0')\Lambda \Big]\,, \qquad (10.9.35) \\
Y_1 &= 1\,.
\end{aligned}
$$

But we may equally well work our way inward starting from the other end, at $i = N$ with $Y_N = 1$, and arrive after $N - n$ steps with the contribution $Y_{N-n}(T_{0n})$ to the final integral over T_{0n}. Combining all of the pieces, that last integral reads

$$
P_n = \lim_{\eta \to 0} \eta \int d\mu(T_0)\, Y_n(T_0) Y_{N-n}(T_0) \frac{f_n}{(\pi\overline{\varrho})^2} e^{-\eta\,\mathrm{Str}\, Q(T_0)\Lambda}\,. \qquad (10.9.36)
$$

Now, the parametrization (10.6.40, 41) of the matrix T_0 must be invoked, together with the integration measure (10.6.43). The recursion relation (10.9.35) then appears as no small hurdle to jump since the function Y_j might depend on all eight independent variables entering T_0, i.e., the Grassmannians $\rho, \rho^*, \sigma, \sigma^*$ and the two complex eigenvalues t_B, t_F. A decisive step ahead of us is to show that only the two moduli $|t_B|, |t_F|$ of these eight variables actually do enter. Fearlessly launching ourselves into the fight for that simplification, we scrutinize the kernel $E(T_0, T_0')$ in (10.9.35). Simply to check is $\mathrm{Str}\, Q(T_0')\Lambda = 4(|t_B'|^2 + |t_F'|^2)$. The evaluation of $\mathrm{Str}\, Q(T_0)Q(T_0')$ is a little harder but still quite straightforward. Gruesome expressions to walk away from with displeasure can be avoided by replacing the moduli of the eigenvalues $t_B = |t_B|e^{i\varphi_B}, t_F = |t_F|e^{i\varphi_F}$ of T_0 with

$$
\begin{aligned}
\lambda_B &= 1 + 2|t_B|^2 & \text{ranging in} & \quad 1 \le \lambda_B < \infty, \\
\lambda_F &= 1 - 2|t_F|^2 & \text{ranging in} & \quad -1 \le \lambda_F \le 1
\end{aligned}
\tag{10.9.37}
$$

and temporarily employing the abbreviations

$$
\mu_B = \sqrt{\lambda_B^2 - 1}, \quad \mu_F = \sqrt{1 - \lambda_F^2}, \tag{10.9.38}
$$

$$
i\delta\varphi = \frac{1}{2}(\rho'\rho^* - \rho\rho^{*\prime} + \sigma'\sigma^* - \sigma\sigma^{*\prime})
$$

$$
\tilde\rho = \rho' - \rho, \quad \tilde\rho^* = \rho^{*\prime} - \rho^*, \quad \tilde\sigma = \sigma' - \sigma, \quad \tilde\sigma^* = \sigma^{*\prime} - \sigma^*;
$$

$$
\tilde\varphi_B = \varphi_B' - \varphi_B - \delta\varphi, \quad \tilde\varphi_F = \varphi_F' - \varphi_F - \delta\varphi;
$$

here primed quantities refer to T_0' and thus to integration variables in the recursion relation (10.9.35). Thus equipped, we get

$$
\mathrm{Str}\, Q(T_0')\Lambda = 2(\lambda_B' - \lambda_F') \tag{10.9.39}
$$

and

$$
\begin{aligned}
\mathrm{Str}\, &Q(T_0)Q(T_0') \\
= \ & 2(\lambda_B\lambda_B' - \lambda_F\lambda_F') + (\lambda_B - \lambda_F)(\lambda_B' - \lambda_F')(\tilde\rho\tilde\rho^* - \tilde\sigma\tilde\sigma^*) \\
& -2\big(1 - \tfrac{1}{4}\tilde\rho\tilde\rho^*\tilde\sigma\tilde\sigma^*\big)\big(\mu_B\mu_B' \cos\tilde\varphi_B + \mu_F\mu_F' \cos\tilde\varphi_F\big) \\
& -(\tilde\rho\tilde\rho^* - \tilde\sigma\tilde\sigma^*)\big(\mu_B\mu_B' \cos\tilde\varphi_B - \mu_F\mu_F' \cos\tilde\varphi_F\big) \\
& +i\tilde\sigma\tilde\rho^*\big(\mu_B'\mu_F e^{-i\tilde\varphi_B} + \mu_F'\mu_B e^{i\tilde\varphi_F}\big)e^{-i(\varphi_B - \varphi_F)} \\
& -\tilde\rho i\tilde\sigma^*\big(\mu_B'\mu_F e^{i\tilde\varphi_B} + \mu_F'\mu_B e^{-i\tilde\varphi_F}\big)e^{i(\varphi_B - \varphi_F)} .
\end{aligned}
\tag{10.9.40}
$$

Now, we may seek a solution of the recursion relation (10.9.35) in the form $Y_k(\lambda_B, \lambda_F)$. Accounting for the transformation (10.9.37), we confront

$$
\begin{aligned}
Y_j(\lambda_B, \lambda_F) = \ & \int_1^\infty d\lambda_B' \int_{-1}^1 d\lambda_F' \frac{1}{(\lambda_B' - \lambda_F')^2} \int_0^{2\pi} \frac{d\varphi_B' d\varphi_F'}{(2\pi)^2} \int d\rho^{*\prime} d\rho \, d i\sigma^{*\prime} d i\sigma' \\
& \times \exp\big[-2\eta(\lambda_B' - \lambda_F') - (\gamma/2)\mathrm{Str}\, Q(T_0)Q(T_0')\big] Y_{j-1}(\lambda_B', \lambda_F') .
\end{aligned}
\tag{10.9.41}
$$

We immediately conclude that the nilpotent phase shifts $\delta\varphi$ in $\tilde\varphi_B, \tilde\varphi_F$ within the kernel can be eliminated by simply shifting the integration variables φ_B', φ_F' to $\tilde\varphi_B, \tilde\varphi_F$. Thereafter, the four primed Grassmannian integration variables appear only in the differences $\tilde\rho = \rho' - \rho$, etc., and may be replaced by the latter; due to $d\tilde\rho = d\rho'$, etc., the integration measure remains unchanged.

The pole in the integration measure at $\lambda_B' = \lambda_F' = 1$ deserves comment. It corresponds to the pole at $t_B' = t_F' = 0$ for the original parametrization (10.6.43) and needs to be cancelled if the above integral is to exist. Searching for such cancellation, we inspect the phase-averaged exponential

$$\int_0^{2\pi} \frac{d\tilde\varphi_B d\tilde\varphi'_F}{(2\pi)^2} \exp\left[-(\gamma/2)\mathrm{Str}\, Q(T_0)Q(T'_0)\right] \tag{10.9.42}$$

$$= \mathrm{e}^{-\gamma(\lambda_B\lambda'_B - \lambda_F\lambda'_F)}\left[E_0 + E_1\tilde\rho i\tilde\sigma^* - E_1^* i\tilde\sigma\tilde\rho^* + E_2(\tilde\rho\tilde\rho^* - \tilde\sigma\tilde\sigma^*) + E_3\tilde\rho\tilde\rho^*\tilde\sigma\tilde\sigma^*\right]$$

where the four quantities E_i are functions of $\lambda_B, \lambda_F, \lambda'_B, \lambda'_F$ and read

$$E_0 = I_0(\gamma\mu'_B\mu_B)I_0(\gamma\mu'_F\mu_F), \tag{10.9.43}$$

$$E_1 = \frac{\gamma}{2}\left(\mu'_B\mu_F I_1(\gamma\mu'_B\mu_B)I_0(\gamma\mu'_F\mu_F) + (B\leftrightarrow F)\right)\mathrm{e}^{-\mathrm{i}(\varphi_B - \varphi_F)},$$

$$E_2 = \frac{\gamma}{2}\Big\{-(\lambda_B - \lambda_F)(\lambda'_B - \lambda'_F)E_0$$

$$\qquad + \left[\mu_B\mu'_B I_1(\gamma\mu'_B\mu_B)I_0(\gamma\mu'_F\mu_F) - (B\leftrightarrow F)\right]\Big\},$$

$$E_3 = \frac{\gamma^2}{2}(\lambda_B - \lambda_F)(\lambda'_B - \lambda'_F)\Big\{(\lambda'_B\lambda_B + \lambda'_F\lambda_F)E_0$$

$$\qquad - \left[\mu_B\mu'_B I_1(\gamma\mu'_B\mu_B)I_0(\gamma\mu'_F\mu_F) - (B\leftrightarrow F)\right]\Big\}$$

$$\equiv (\lambda_B - \lambda_F)(\lambda'_B - \lambda'_F)L(\lambda'_B, \lambda'_F|\lambda_B, \lambda_F).$$

Cancellation indeed! Due to the definition of μ_B, μ_F in (10.9.38) and the small-argument behavior of the Bessel function I_1, the three functions E_1, E_2, E_3 all approach zero linearly as $\lambda'_B, \lambda'_F \to 1$ and thus give rise to converging integrals over λ'_B, λ'_F in (10.9.41). We can wave good-bye for good to the terms with E_1, E_2 since these are annulled by the subsequent fourfold Grassmann integral. The term with E_3 does survive since it comes with the maximal Grassmann monomial which integrates to unity, $\int d\tilde\rho^* d\tilde\rho d\tilde\sigma^* d\tilde\sigma\, \tilde\rho\tilde\rho^*\tilde\sigma\tilde\sigma^* = 1$. Finally, the term with E_0 does not vanish at the boundary $\lambda'_B = \lambda'_F = 1$ and thus engenders a diverging bosonic integral; it is multiplied by the vanishing Grassmann integral $\int d\tilde\rho^* d\rho d\sigma^* d\sigma = 0$, however, and the formal product $\infty \times 0$ is assigned the value $\exp[-\gamma(\lambda_B - \lambda_F)]Y_{j-1}(1,1)$ by the Parisi–Sourlas–Efetov–Wegner theorem of Sect. 10.4.5. Thus, the recursion relation takes the form

$$Y_j(\lambda_B, \lambda_F) = \mathrm{e}^{-\gamma(\lambda_B - \lambda_F)}Y_{j-1}(1,1) \tag{10.9.44}$$

$$\qquad + \int_1^\infty d\lambda'_B \int_{-1}^1 d\lambda'_F \frac{\lambda_B - \lambda_F}{\lambda'_B - \lambda'_F}L(\lambda'_B, \lambda'_F|\lambda_B, \lambda_F)$$

$$\qquad \times \mathrm{e}^{-2\eta(\lambda'_B - \lambda'_F)}Y_{j-1}(\lambda'_B, \lambda'_F).$$

Before trying to solve for Y_j, it is well to pull the expression (10.9.36) for the inverse participation ratio P_n to the level reached for the function Y_j, i.e., to do the phase and Grassmann integrals. To that end, we should write the quantity $f_n/(\pi\bar\rho)^2$ more explicitly. Recalling (10.9.20, 20), we easily find

$$f_n/(\pi\bar\rho)^2 = -2\rho^*\rho\sigma^*\sigma\lambda_B(\lambda_B - \lambda_F) + \dots, \tag{10.9.45}$$

where the dots refer to submaximal Grassmann monomials annulled by the subsequent integration, analogous to the fate of the functions E_1, E_2 in the recursion relation. The desired reformulation of (10.9.36) is

$$P_n = \lim_{\eta \to 0} 2\eta \int_1^\infty \int_{-1}^1 \frac{d\lambda_B d\lambda_F \lambda_B}{\lambda_B - \lambda_F} e^{-2\eta(\lambda_B - \lambda_F)} Y_{N-n}(\lambda_B, \lambda_F) Y_n(\lambda_B, \lambda_F) \qquad (10.9.46)$$

and suggests the next step ahead by its very appearance: The integral must be proportional to $1/\eta$ for the limit $\eta \to 0$ to exist. Such singular behavior can be contributed only by large values of λ_B. But in that range, $\lambda_B - \lambda_F \to \lambda_B$, and we are led to suspect that $Y_n(\lambda_B, \lambda_F)$ becomes independent of λ_F. The ansatz

$$Y_n(\lambda_B, \lambda_F)|_{\lambda_B \gg \lambda_F} = y_n(2\eta\lambda_B) \equiv y_n(z), \qquad y_1 = 1 \qquad (10.9.47)$$

will in fact turn out to be consistent and immediately embellishes P_n to

$$P_n = 2 \int_0^\infty dz e^{-z} y_{N-n}(z) y_n(z). \qquad (10.9.48)$$

Now the recursion relation (10.9.44) must be updated to the ansatz (10.9.47) in the large-λ_B limit. As a first simplification, we can dispense with the exponentially small boundary term, and thus

$$y_j(2\eta\lambda_B) = \int_1^\infty d\lambda_B' \frac{\lambda_B e^{-2\eta\lambda_B'}}{\lambda_B'} \int_{-1}^1 d\lambda_F' L(\lambda_B', \lambda_F'|\lambda_B, \lambda_F) \, y_{j-1}(2\eta\lambda_B') \,. \qquad (10.9.49)$$

The integral over λ_F' becomes explicitly doable once we appreciate that $\mu_B \mu_B' \approx \lambda_B \lambda_B' - \frac{1}{2}(\frac{\lambda_B'}{\lambda_B} + \frac{\lambda_B}{\lambda_B'}) \gg 1$ and replace the Bessel functions with large arguments by their asymptotic forms expansions $I_n(z) = (2\pi z)^{-1/2} e^z (1 - \frac{4n^2-1}{8z} \dots)$. Then, the kernel $L(\lambda_B', \lambda_F'|\lambda_B, \lambda_F)$ takes the asymptotic form

$$
\begin{aligned}
L(\lambda_B', \lambda_F'|\lambda_B, \lambda_F) \;=\; & \frac{\gamma^{3/2}}{2\sqrt{2\pi\lambda_B\lambda_B'}} \exp\left[-\frac{\gamma}{2}\left(\frac{\lambda_B'}{\lambda_B} + \frac{\lambda_B}{\lambda_B'}\right) + \gamma\lambda_F'\lambda_F\right] \\
& \times \left\{\left[\lambda_F'\lambda_F + \frac{1}{2\gamma} + \frac{1}{2}\left(\frac{\lambda_B'}{\lambda_B} + \frac{\lambda_B}{\lambda_B'}\right)\right] \right. \\
& \left. \times I_0(\gamma\mu_F'\mu_F) + \mu_F'\mu_F I_1(\gamma\mu_F'\mu_F)\right\}
\end{aligned}
\qquad (10.9.50)
$$

which we can subject to the λ_F'-integration required in (10.9.49). The integral identity (see Problem 10.14)

$$\int_{-1}^1 dx e^{ax} I_0(b\sqrt{1-x^2}) = 2\frac{\sinh\sqrt{a^2+b^2}}{\sqrt{a^2+b^2}} \qquad (10.9.51)$$

and its first derivatives w. r. t. the real parameters a, b yield

$$y_j(z) \;=\; \int_0^\infty dz' \sqrt{\frac{z}{z'^3}} e^{-z'} L_\gamma\left(\frac{z'}{z}\right) y_{j-1}(z') = \int_0^\infty du\, u^{-3/2} e^{-uz} L_\gamma(u) y_{j-1}(uz),$$

$$L_\gamma(u) \;=\; \sqrt{\frac{\gamma}{2\pi}} e^{-\frac{\gamma}{2}(u+\frac{1}{u})} \left[\cosh\gamma - \frac{\sinh\gamma}{2\gamma} + \frac{1}{2}\left(u+\frac{1}{u}\right)\sinh\gamma\right]. \qquad (10.9.52)$$

One arrow remains in our quiver and will be shot presently, the largeness of the parameter $\gamma \propto 2b^2 \gg 1$. The kernel $L_\gamma(u)$ therefore has a sharp peak of width $1/\sqrt{\gamma} \ll 1$ at $u = 1$. Promising to demonstrate consistency later, I assume that the function $y_j(uz)$ varies slowly in u across the width of the peak of $L_\gamma(u)$, whatever the value of z may be. Then, it is advisable to shift and rescale the integration variable as

$$u = 1 + \frac{\xi}{\sqrt{\gamma}} \tag{10.9.53}$$

such that the following expansions in powers of $1/\sqrt{\gamma}$ become available:

$$
\begin{aligned}
L_\gamma(u) &= \frac{1}{2}\sqrt{\frac{\gamma}{2\pi}}e^{\gamma-\frac{\gamma}{2}(u+\frac{1}{u})}\left[1 - \frac{1}{2\gamma} + \frac{1}{2}\left(u+\frac{1}{u}\right)\right] \\
&= \sqrt{\frac{\gamma}{2\pi}}e^{-\frac{\xi^2}{2}}\left[1 - \frac{1}{4\gamma} + \frac{\xi^2}{4\gamma} + \frac{\xi^3}{2\sqrt{\gamma}} - \frac{\xi^4}{2\gamma} + \frac{\xi^6}{8\gamma} + \mathcal{O}(\gamma^{-3/2})\right], \\
u^{-3/2} &= 1 - \frac{3\xi}{2\sqrt{\gamma}} + \frac{15\xi^2}{8\gamma} + \mathcal{O}(\gamma^{-3/2}), \\
e^{-zu} &= e^{-z}\left[1 - \frac{z\xi}{\sqrt{\gamma}} + \frac{z^2\xi^2}{2\gamma} + \mathcal{O}(\gamma^{-3/2})\right], \\
y_{j-1}(uz) &= y_{j-1}(z) + \frac{z\xi}{\sqrt{\gamma}}y'_{j-1}(z) + \frac{z^2\xi^2}{2\gamma}y''_{j-1}(z) + \mathcal{O}(\gamma^{-3/2}). \tag{10.9.54}
\end{aligned}
$$

Upon doing the ξ-integrals, we transform the recursive integral equation (10.9.52) into the recursive differential equation[9]

$$e^z y_j(z) = y_{j-1}(z) + \frac{z^2}{2\gamma}\left\{y_{j-1}(z) - 2y'_{j-1}(z) + y''_{j-1}(z)\right\}. \tag{10.9.55}$$

We should appreciate that the unit steps of the site index j are minute compared to the correlation length γ; thus, a continuum approximation according to $j \to 2\gamma\tau$ is indicated, with τ ranging in the interval $[0, N/(2\gamma) \equiv x]$. Finally, we anticipate that $y_j(z)$ will decay with growing z on a scale $\propto 1/\gamma$ and introduce the rescaled independent variable $y = 2\gamma z$,

$$y_j(z) = y_{2\gamma\tau}(y/2\gamma) \equiv Y(y, \tau) \quad \text{with} \quad y \in [0, \infty), \quad \tau \in [0, x = \frac{N}{2\gamma}]. \tag{10.9.56}$$

Such rescalings yield $e^{-z} = e^{-y/2\gamma} \to 1$, reveal the first two terms in the curly bracket in (10.9.55) as negligible, and turn the recursion relation into the partial differential equation

$$\frac{\partial}{\partial\tau}Y(y,\tau) = (y^2\frac{\partial^2}{\partial y^2} - y)Y(y,\tau) \equiv \hat{L}Y(y,\tau) \tag{10.9.57}$$

---

[9] Equation 72 on p. 3818 of Ref. [385] contains typos and should be read as our Eq. (10.9.55).

and the "initial" condition in (10.9.35, 51) into

$$Y(y, \tau = 0) = 1.$$ (10.9.58)

The very absence of all parameters from the foregoing initial-value problem demonstrates the consistency of the scaling assumptions made above, provided, of course, that a solution exists.

The (local) inverse participation ratio P_n last given in (10.9.48) can also be expressed in terms of the function $Y(y, \tau)$,

$$P_n \to \frac{1}{\gamma} \int_0^\infty dy\, Y(y, x - \tau) Y(y, \tau).$$

Upon averaging over sites according to (10.9.32), we express the mean participation ratio as

$$P = \frac{1}{N} \sum_{n=1}^N P_n = \frac{2}{N} \int_0^\infty dy \int_0^x d\tau\, Y(y, x - \tau) Y(y, \tau).$$ (10.9.59)

This is already an important asymptotic result, worthy of highlighting. We may read the prefactor $P_{\mathrm{GUE}} \equiv 2/N$ of the foregoing integral as the inverse participation ratio of the GUE and refer our P to that unit. The result,

$$\frac{P}{P_{\mathrm{GUE}}} = \int_0^\infty dy \int_0^x d\tau\, Y(y, x - \tau) Y(y, \tau) \equiv \beta(x),$$ (10.9.60)

depends only on the scaling parameter

$$x = N/2\gamma \propto N/b^2.$$ (10.9.61)

As already noted above, *Casati, Molinari,* and *Izrailev* [383] established such scaling through numerical work and even conjectured that $\beta(x) = 1 + x/3$. Such surprisingly simple behavior was indeed borne out by *Fyodorov's* and *Mirlin's* analysis of Ref. [385] which we are spreading out here. It will be convenient to go for the final goal via the Laplace transform

$$\tilde{\beta}(p) = \int_0^\infty dx\, e^{-px} \beta(x) = \int_0^\infty dy\, \tilde{Y}(y, p)^2.$$ (10.9.62)

Solutions of the above partial differential equation may be sought as compositions of eigenfunctions of the operator $\hat{L} = y^2 \frac{d^2}{dy^2} - y$. Such eigenfunctions which decay to zero for $y \to \infty$, are related to the modified Bessel functions $K_r(t)$ as $f_r(y) = 2\sqrt{y} K_r(2\sqrt{y})$ and come with the eigenvalues $(r^2 - 1)/4$. For imaginary indices $r = i\nu$, $\nu \in (0, \infty)$, the functions $f_{i\nu}(y)$ are mutually orthogonal and can be used as a basis in the sense of the Lebedev–Kontorovich transformation [392]. That transformation relates a function $F(x)$ where $0 < x < \infty$ to its transform $\tilde{F}(\nu)$ as

$$F(x) = \int_0^\infty d\nu\, K_{i\nu}(x) \tilde{F}(\nu),$$ (10.9.63)

$$\tilde{F}(\nu) = \frac{2}{\pi} \nu \sinh \pi\nu \int_0^\infty \frac{dx}{x} K_{i\nu}(x) F(x),$$

provided (1) that $F(x)$ is piecewise differentiable in $(0, \infty)$ and (2) thet there is some positive number ϵ such that

$$\int_0^\epsilon dx\, x^{-1}|F(x)\ln x| < \infty \qquad \text{and} \qquad \int_\epsilon^\infty dx\, x^{-1/2}|F(x)| < \infty. \tag{10.9.64}$$

A little thought shows that we cannot naively invoke that transformation to represent $Y(y, \tau)$ as a superposition of eigenfunctions $e^{i\frac{\nu^2+1}{4}\tau}2\sqrt{y}K_{i\nu}(2\sqrt{y})$ since the initial condition $Y(y, 0) = 1$ would yield the function $F(x) = 1/x$ which does not qualify as the member of a Lebedev–Kontorovich pair. There is a way out found by Fyodorov and Mirlin and easily followed. The differential operator $\hat{L}$ has one additional eigenfunction which decays to zero for $y \to \infty$ and pertains to the eigenvalue 0; indeed, $\hat{L}f_1(y) = 0$ where $f_1 = 2\sqrt{y}K_1(2\sqrt{y})$. By simply including that eigenfunction, we represent the function $Y(y, \tau)$ in search as

$$Y(y, \tau) = b_1 2\sqrt{y}K_1(2\sqrt{y}) + \int_0^\infty d\nu\, b(\nu)\, e^{-\frac{\nu^2+1}{4}\tau}\, 2\sqrt{y}K_{i\nu}(2\sqrt{y}) \tag{10.9.65}$$

and determine the expansion coefficients b_1 and $b(\nu)$ from the initial condition $Y(y, 0) = 1$. That condition reads

$$\frac{1 - b_1 2\sqrt{y}K_1(2\sqrt{y})}{2\sqrt{y}} = \int_0^\infty d\nu\, b(\nu)\, 2\sqrt{y}K_{i\nu}(2\sqrt{y}) \tag{10.9.66}$$

and reveals $b(\nu)$ as the would-be Lebedev–Kontorovich transform of the function $F(x) = \frac{1}{x} - b_1 K_1(x)$. Due to the small-argument behavior of the Bessel function, $K_1(x) \to \frac{1}{x} + \frac{x}{2}\ln\frac{x}{2}$ for $x \to 0$, $F(x) \to (1 - b_1)\frac{1}{x}) - b_1\frac{x}{2}\ln\frac{x}{2}$, and we conclude that the coefficient b_1 is uniquely determined as $b_1 = 1$. No other choice would meet the first of the conditions (10.9.64); the second of these is also met since for $b_1 = 1$, the large-x behavior $F(x) \to \frac{1}{x}$. Therefore, our function $Y(y, \tau)$ is determined up to quadratures, namely,

$$b(\nu) = \frac{2}{\pi}\nu \sinh \pi\nu \int_0^\infty \frac{dx}{x} K_{i\nu}(x)\left[\frac{1}{x} - K_1(x)\right] = \frac{2}{\pi}\frac{\nu \sinh\frac{\pi\nu}{2}}{1+\nu^2} \tag{10.9.67}$$

and the composition (10.9.65)

$$Y(y, \tau) = 2\sqrt{y}\left[K_1(2\sqrt{y}) + \frac{2}{\pi}\int_0^\infty d\nu\, \frac{\nu \sinh\frac{\pi\nu}{2}}{1+\nu^2}\, e^{-\frac{1+\nu^2}{4}\tau}K_{i\nu}(2\sqrt{y})\right]. \tag{10.9.68}$$

The final expression for the expansion coefficient in (10.9.65) as well as the following Laplace transform

$$\begin{aligned}\tilde{Y}(y, p) &= \frac{1}{p}\int_0^\infty du\, u\frac{(\mu+1) + u^2(\mu-1)}{(1+u^2)^2}\, J_{\mu-1}(2\sqrt{y}u) \\ &\equiv \frac{1}{p}\int_0^\infty du\, F(u)\, J_{\mu-1}(2\sqrt{y}u), \\ \mu &= \sqrt{4p+1}\end{aligned} \tag{10.9.69}$$

were found in Ref. [385] by some ingenious juggling with integrals involving trigonometric, hyperbolic, and Bessel functions. I shall sketch these interesting calculations below but propose to proceed to the final result first.

Inserting the Laplace transform $\tilde{Y}(y,p)$ into the scaling function $\tilde{\beta}(p)$ of (10.9.61), we obtain

$$p^2 \tilde{\beta}(p) = \int_0^\infty dy \left[\int_0^\infty du\, F(u)\, J_{\mu-1}(2\sqrt{yu}) \right]^2. \tag{10.9.70}$$

A simple change of the integration variable y to $t = 2\sqrt{y}$ and the orthogonality $\int_0^\infty dt\, t J_\nu(tu) J_\nu(tv) = \frac{1}{u}\delta(u-v)$ of the Bessel functions produce

$$p^2 \tilde{\beta}(p) = \frac{1}{2} \int_0^\infty du\, u^{-1} F(u)^2 = p + \frac{1}{3}. \tag{10.9.71}$$

Reverting to the Laplace transform gives the celebrated scaling function

$$\beta(x) = 1 + x/3. \tag{10.9.72}$$

It is customary to reexpress that result in terms of the localization length $l(N,\gamma) \propto 1/P$. Remembering that $x = N/2\gamma$, we find $l(N,\gamma)$ as the geometric mean of its limiting values for complete delocalization and full localization

$$\frac{1}{l(N,\gamma)} = \frac{1}{l(N,\infty)} + \frac{1}{l(\infty,\gamma)}. \tag{10.9.73}$$

Clearly, then, there is no threshold or critical value of x for the onset of localization but rather a continuous transition from the delocalization typical of GUE matrices to well pronounced localization, as the scaling parameter x grows from $\mathcal{O}(\frac{1}{N})$ to $\mathcal{O}(N)$.

10.10 Problems

10.1. Let ρ and σ be $M \times M$ matrices with anticommuting entries. Using $(\rho_{ij}^*)^* = -\rho_{ij}$, show that $\rho^{\dagger\dagger} = -\rho$, $\widetilde{\rho\sigma} = -\tilde{\sigma}\tilde{\rho}$, and $(\rho\sigma)^\dagger = -\sigma^\dagger \rho^\dagger$.

10.2. A common factor x in a bosonic row or column of a superdeterminant can be pulled out as the factor x multiplying the superdeterminant; in contrast, a factor y in a fermionic row or column as $1/y$. True?

10.3. Let ρ and ρ^* with $(\rho^*)^* = -\rho$ be a pair of Grassmannians and c an ordinary complex number. Define a new pair $\eta = c\rho$ and $\bar{\eta} = c\rho^* = (c/c^*)\eta^*$. Show that $(\overline{\bar{\eta}}) = -\eta$ such that one may call $\bar{\eta}$ a complex conjugate of η.

10.4. Show that the 2×2 supermatrix $H = \begin{pmatrix} a & \rho^\dagger \\ \rho & b \end{pmatrix}$ with real a and imaginary b, even though not Hermitian, is still diagonalized by (10.3.31, 32); give a redefinition of the complex conjugation of Grassmann variables that restores unitarity of the matrix U in (10.3.32). Convince yourself that no integral over the Grassmannians η and η^* is affected by such a redefinition of the complex conjugation.

10.5. Use the generating function (10.4.5) to prove Wick's theorem

$$\langle S_{i_1} \ldots S_{i_n} S_{j_1}^* \ldots S_{j_n}^* \rangle = \sum_{\mathcal{P}} (a^{-1})_{i_1, \mathcal{P} j_1} \ldots (a^{-1})_{i_n, \mathcal{P} j_n}$$

where the sum is over all $n!$ permutations of the $\{j\}$ and $\langle S_i S_j^* \rangle = (a^{-1})_{ij}$.

10.6. Why is the action $A(\hat{Q}, 0)$ from (10.5.7) invariant under unitary transformations of the 2×2 matrix $\hat{Q}$?

10.7. Show that the saddle-point approximation gives the Grassmann integral $\int d\eta_1 d\eta_2 \exp(-N\eta_2\eta_1)$ exactly. Generalize to arbitrary Gaussian integrals, Grassmann, ordinary, and super.

10.8. Convince yourself of $\overline{G(E^+)} = \overline{G(E^-)}^*$.

10.9. Show that pseudounitary $N \times N$ matrices form a group.

10.10. Show that pseudounitary matrices leave the scalar product $(\Phi, \hat{L}\Phi)$ invariant in the sense $(T\Phi, \hat{L}T\Phi) = (\Phi, \hat{L}\Phi)$.

10.11. Let $H = \begin{pmatrix} a & \rho^\dagger \\ \rho & b \end{pmatrix} = U \begin{pmatrix} h_B & 0 \\ 0 & h_F \end{pmatrix} U^{-1}$ be a Hermitian 2×2 supermatrix, and imagine a superintegral $\int dH f(H)$ where $dH = da\,db\,d\rho^*d\rho$. Change integration variables to the eigenvalues h_B, h_F and the Grassmannian "angles" η^*, η in the diagonalizing matrix U given in (10.3.31, 32). Show that the Berezinian reads $J(\frac{a,b,\rho^*,\rho}{h_B,h_F,\eta^*,\eta}) = (h_B - h_F)^{-2}$.

10.12. Convince yourself of the equivalence of the delta functions (10.6.1) and (10.6.3) for 2×2 supermatrices. Argue that the two i's in the integral representation could be replaced by any complex number. Why does the factor $(2\pi)^{-1}$ in (10.6.1) have to be dropped for 4×4 supermatrices?

10.13. Which generating function would have the product of two matrix elements of the resolvent as derivatives? Which point of Chap. 7 provides a clue? Write the generating function as a superintegral and give the expression provided by the one-dimensional nonlinear sigma model.

10.14. Even though possibly not aware of it, you know the integral identity (10.9.51). Consider the Fourier transform of a spherically symmetric function $F(|\boldsymbol{x}|)$ in three dimensions, $\int d^3x\, e^{i\boldsymbol{k}\cdot\boldsymbol{x}} F(|\boldsymbol{x}|) = \int_0^\infty dr\, r^2 F(r) f(kr)$ where

$$f(\rho) = \int_0^{2\pi} d\varphi \int_0^\pi d\theta \sin\theta\, e^{i\rho\cos\theta} = 4\pi \frac{\sin\rho}{\rho}$$

To save labor the polar axis was chosen parallel to the vector $\boldsymbol{k}$ here. Now choose axes less wisely and enjoy

$$\int_{-1}^1 dx\, e^{iax} J_0(b\sqrt{1-x^2}) = 2\frac{\sin\sqrt{a^2+b^2}}{\sqrt{a^2+b^2}}.$$

as the fruit of lesser wisdom. Analytic continuation to imaginary a, b produces (10.9.51).

References

Chapter 1

1. M.V. Berry, M. Tabor: Proc. R. Soc. Lond. **A356**, 375 (1977)
2. G.M. Zaslavskii, N.N. Filonenko: Sov. Phys. JETP, **8**, 317 (1974)
3. M.V. Berry, M. Tabor: Proc. R. Soc. Lond. **A349**, 101 (1976)
4. M.V. Berry: In G. Iooss, R.H. Helleman, R. Stora (eds.): Les Houches Session XXXVI 1981, *Chaotic Behavior of Deterministic Systems* (North-Holland, Amsterdam, 1983)
5. O. Bohigas, R. Haq, A. Pandey: In K. Böckhoff (ed.) *Nuclear Data for Science and Technology* (Reidel, Dordrecht, 1983)
6. G. Bohigas, M.-J. Giannoni: In *Mathematical and Computational Methods in Nuclear Physics*, Lecture Notes in Physics 209 (Springer, Berlin, Heidelberg 1984)
7. O. Bohigas, M.J. Giannoni, C. Schmit: Phys. Rev. Lett. **52**, 1 (1984)
8. T.A. Brody, J. Flores, J.B. French, P.A. Mello, A. Pandey, P.S.M. Wong: Rev. Mod. Phys. **53**, 385 (1981)
9. N. Rosenzweig, C.E. Porter: Phys. Rev. **120**, 1698 (1960)
10. H.S. Camarda, P.D. Georgopulos: Phys. Rev. Lett. **50**, 492 (1983)
11. M.R. Schroeder: J. Audio Eng. Soc. **35**, 307 (1987)
12. H.J. Stöckmann, J. Stein: Phys. Rev. Lett. **64**, 2215 (1990)
13. H. Alt, H.-D. Gräf, H.L. Harney, R. Hofferbert, H. Lengeler, A. Richter, P. Schart, H.A. Weidenmüller: Phys. Rev. Lett. **74**, 62 (1995)
14. H. Alt, H.-D. Gräf, R. Hofferbert, C. Rangacharyulu, H. Rehfeld, A. Richter, P. Schart, A. Wirzba: Phys. Rev. **E 54**, 2303 (1996)
15. H. Alt, C. Dembowski, H.-D. Gräf, R. Hofferbert, H. Rehfeld, A. Richter, R. Schuhmann, T. Weiland: Phys. Rev. Lett. **79**, 1026 (1997)
16. Th. Zimmermann, H. Köppel, L.S. Cederbaum, C. Persch, W. Demtröder: Phys. Rev. Lett. **61**, 3 (1988); G. Persch, E. Mehdizadeh, W. Demtröder, T. Zimmermann, L.S. Cederbaum: Ber. Bunsenges. Phys. Chem. **92**, 312 (1988)
17. H. Held, J. Schlichter, G. Raithel, H. Walther: Europhys. Lett. **43**, 392 (1998)
18. C. Ellegaard, T. Guhr, K. Lindemann, H.Q. Lorensen, J. Nygård, M. Oxborrow: Phys. Rev. Lett. **75**, 1546 (1995)
19. C. Ellegaard, T. Guhr, K. Lindemann, J. Nygård, M. Oxborrow: Phys. Rev. Lett. **77**, 4918 (1996)
20. A. Hoenig, D. Wintgen: Phys. Rev. **A39**, 5642 (1989)
21. M. Oxborrow, C. Ellegaard: In *Proceedings of the 3rd Experimental Chaos Conference* (Edinburgh 1995)

22. S. Deus, P.M. Koch, L. Sirko: Phys. Rev. E **52**, 1146 (1995)
23. O. Legrand, C. Schmit, D. Sornette: Europhys. Lett. **18**, 101 (1992)
24. U. Stoffregen, J. Stein, H.-J. Stöckmann, M. Kuś, F. Haake: Phys. Rev. Lett. **74**, 2666 (1995)
25. P. So, S.M. Anlage, E. Ott, R.N. Oerter: Phys. Rev. Lett. **74**, 2662 (1995)
26. M.L. Mehta: *Random Matrices and the Statistical Theory of Energy Levels* (Academic, New York 1967; 2nd edition 1991)
27. C.E. Porter (ed.): *Statistical Theory of Spectra* (Academic, New York 1965)
28. T. Guhr, A. Müller-Groeling, H.A. Weidenmüller: Physics Reports **299**, 192 (1998)
29. P. Pechukas: Phys. Rev. Lett. **51**, 943 (1983)
30. T. Yukawa: Phys. Rev. Lett. **54**, 1883 (1985)
31. T. Yukawa: Phys. Lett. **116A**, 227 (1986)
32. S. Fishman, D.R. Grempel, R.E. Prange: Phys. Rev. Lett. **49**, 509 (1982); Phys. Rev. **A29**, 1639 (1984)

Chapter 2

33. E.P. Wigner: *Group Theory and its Applications to the Quantum Mechanics of Atomic Spectra* (Academic, New York 1959)
34. C.E. Porter (ed.): *Statistical Theories of Spectra* (Academic, New York 1965)
35. M.L. Mehta: *Random Matrices and the Statistical Theory of Spectra* (Academic, New York 1967; 2nd edition 1991)
36. D. Delande, J.C. Gay: Phys. Rev. Lett. **57**, 2006 (1986)
37. G. Wunner, U. Woelck, I. Zech, G. Zeller, T. Ertl, F. Geyer, W. Schweitzer, H. Ruder: Phys. Rev. Lett. **57**, 3261 (1986)
38. T.H. Seligman, J.J.T. Verbaarschot: Phys. Lett. **108A**, 183 (1985)
39. G. Casati, B.V. Chirikov, D.L. Shepelyansky, I. Guarneri: Phys. Rep. **154**, 77 (1987)
40. G. Casati, B.V. Chirikov, F.M. Izrailev, J. Ford: In G. Casati, J. Ford (eds.): *Stochastic Behavior in Classical and Quantum Hamiltonian Systems* Lecture Notes in Physics, Vol. 93 (Springer, Berlin, Heidelberg 1979)
41. B.V. Chirikov: preprint no. 267, Inst. Nucl. Physics Novosibirsk (1969); Phys. Rep. **52**, 263 (1979)
42. F. Haake, M. Kuś, R. Scharf: Z. Physik **B65**, 381 (1987);
43. M. Kuś, R. Scharf, F. Haake: Z. Physic **B66**, 129 (1987)
44. F. Haake, M. Kuś, R. Scharf: In *Fundamentals of Quantum Optics II,* ed. by F. Ehlotzky, Lecture Notes in Physics Vol. 282 (Springer, Berlin, Heidelberg 1987)
45. R. Scharf, B. Dietz, M. Kuś, F. Haake, M.V. Berry: Europhys. Lett. **5**, 383 (1988)
46. F. Dyson: J. Math. Phys. **3**, 1199 (1962)
47. M. Zirnbauer: private communication and to be published

Chapter 3

48. C.E. Porter (ed.): *Statistical Theories of Spectra* (Academic, New York 1965)
49. M.L. Mehta: *Random Matrices and the Statistical Theory of Spectra* (Academic, New York 1967; 2nd edition 1991)

50. J. von Neumann, E.P. Wigner: Phys. Z. **30**, 467 (1929)
51. R. Scharf, B. Dietz, M. Kuś, F. Haake, M.V. Berry: Europhys. Lett. **5**, 383 (1988)
52. O. Bohigas, R.U. Haq, A. Pandey: In K.H. Böchhoff (ed.): *Nuclear Data for Science and Technology* (Reidel, Dordrecht, 1983)
53. Th. Zimmermann, H. Köppel, L.S. Cederbaum, C. Persch, W. Demtröder: Phys. Rev. Lett. **61**, 3 (1988); G. Persch, E. Mehdizadeh, W. Demtröder, T. Zimmermann, L.S. Cederbaum: Ber. Bunsenges. Phys. Chem. **92**, 312 (1988)
54. H. Held, J. Schlichter, G. Raithel, H. Walther: Europhys. Lett. **43**, 392 (1998)
55. M.R. Schroeder: J. Audio Eng. Soc. **35**, 307 (1987)
56. H.-J. Stöckmann, J. Stein: Phys. Rev. Lett. 64, 2215 (1990)
57. H. Alt, H.-D. Gräf, H.L. Harney, R. Hofferbert, H. Lengeler, A. Richter, P. Schart, H.A. Weidenmüller: Phys. Rev. Lett. **74**, 62 (1995)
58. H. Alt, H.-D. Gräf, R. Hofferbert, C. Rangacharyulu, H. Rehfeld, A. Richter, P. Schart, A. Wirzba: Phys. Rev. **E 54**, 2303 (1996)
59. H. Alt, C. Dembowski, H.-D. Gräf, R. Hofferbert, H. Rehfeld, A. Richter, R. Schuhmann, T. Weiland: Phys. Rev. Lett. **79**, 1026 (1997)
60. A. Richter: In D.A. Hejhal, J. Friedman, M.C. Gutzwiller, A.M. Odlyzko (eds.): *Emerging Applications of Number Theory* IMA volume 109, p. 109 (Springer, New York, 1998)
61. C. Ellegaard, T. Guhr, K. Lindemann, H.Q. Lorensen, J. Nygård, M. Oxborrow: Phys. Rev. Lett. **75**, 1546 (1995)
62. C. Ellegaard, T. Guhr, K. Lindemann, J. Nygård, M. Oxborrow: Phys. Rev. Lett. **77**, 4918 (1996)
63. U. Stoffregen, J. Stein, H.-J. Stöckmann, M. Kuś, F. Haake: Phys. Rev. Lett. **74**, 2666 (1995)
64. P. So, S.M. Anlage, E. Ott, R.N. Oerter: Phys. Rev. Lett. **74**, 2662 (1995)
65. H.-J. Stöckmann: *Quantum Chaos, An Introduction* (Cambridge University Press, Cambridge, 1999)
66. M.V. Berry: In G. Iooss, R.H.G. Helleman, R. Stora (eds.): Les Houches, Session XXXVI, 1981, *Chaotic Behavior of Deterministic Systems* (North-Holland, Amsterdam, 1983)
67. T.A. Brody, J. Floris, J.B. French, P.A. Mello, A. Pandey, S.S.M. Wong: Rev. Mod. Phys. **53**, 385 (1981) Appendix B

Chapter 4

68. M.V. Berry, M. Tabor: Proc. R. Soc. Lond. **A356**, 375 (1977)
69. O. Bohigas, M.J. Giannoni, C. Schmit: Phys. Rev. Lett. **52**, 1 (1984); J. de Phys. Lett. **45**, 1015 (1984)
70. S.W. McDonald, A.N. Kaufman: Phys.Rev. Lett. **42**, 1189 (1979)
71. G. Casati, F. Valz-Gris, I. Guarneri: Lett. Nouvo Cimento **28**, 279 (1980)
72. M.V. Berry: Ann. Phys. (USA) **131**, 163 (1981)
73. M.L. Mehta: *Random Matrices and the Statistical Theory of Energy Levels* (Academic, New York 1967; 2nd edition 1991)
74. C.E. Porter (ed.): *Statistical Theories of Spectra* (Academic, New York 1965)
75. T. Guhr, A. Müller-Groeling, H.A. Weidenmüller: Phys. Rep. **299**, 189 (1998)
76. E.P. Wigner: Proc. 4th Can. Math. Congr., Toronto, 1959, p. 174
77. M. Kuś, J. Mostowski, F. Haake: J. Phys. A: Math. Gen. **21**, L 1073–1077 (1988)

78. F. Haake, K. Zyczkowski: Phys. Rev. **A42**, 1013 (1990)
79. M. Feingold, A. Peres: Phys. Rev. **A34**, 591 (1986)
80. B. Mehlig, K. Müller, B. Eckhardt: Phys. Rev. **E59**, 5272 (1999)
81. C.E. Porter, R.G. Thomas: Phys. Rev. **104**, 483 (1956)
82. T.A. Brody, J. Floris, J.B. French, P.A. Mello, A. Pandey, S.S.M. Wong: Rev. Mod. Phys. **53**, 385 (1981)
83. M.V. Berry: J. Phys. **A10**, 2083 (1977)
84. V.N. Prigodin, : Phys. Rev. Lett. **75**, 2392 (1995)
85. H.-J. Stöckmann: *Quantum Chaos, An Introduction* (Cambridge University Press, Cambridge, 1999)
86. A. Pandey: Ann. Phys. **119**, 170–191 (1979)
87. M.L. Mehta: Commun. Math. Phys. **20**, 245 (1971)
88. F.J. Dyson: J. Math. Phys. (N.Y.) **3**, 166 (1962)
89. F.J. Dyson: Commun. Math. Phys. **19**, 235 (1970)
90. F.J. Dyson: J. Math. Phys. **3**, No. 1, 140, 157, 166 (1962)
91. B. Dietz, F. Haake: Z. Phys. B**80**, 153 (1990)
92. F.J. Dyson: Commun. Math. Phys. **47**, 171–183 (1976)
93. B. Riemann: Monatsberichte d. Preuss. Akad. d. Wissensch., Berlin 671 (1859)
94. H.M. Edwards: *Riemann's Zeta Function* (Academic, New York, 1974)
95. H.L. Montgomery: Proc. Symp. Pure Math. **24**, 181 (1973)
96. A.M. Odlyzko: Math. of Comp. **48**, 273 (1987)
97. M.V. Berry, J.P. Keating: In J.P. Keating, I.V. Lerner *Supersymmetry and Trace Formulae: Chaos and Disorder* (Plenum, New York, 1998)
98. C. Itzykson, J.-B. Zuber: *Quantum Field Theory* (McGraw-Hill, New York 1890)
99. F.A. Berezin: *Method of Second Quantization* (Academic, New York 1966)
100. F.A. Berezin: *Introduction to Superanalysis* (Reidel, Dordrecht, 1987)
101. K. Efetov: *Supersymmetry in Disorder and Chaos* (Cambridge University Press, Cambridge, 1997)
102. I. Newton: *Universal Arithmetic* (1707); see e.g., W.W. Rouse Ball: *A Short Account of the History of Mathematics*, 4th edition (1908) or www.maths.tcd.ie/pub/HistMath/People/Newton/RouseBall/RB_Newton.html
103. A. Mostowski, M. Stark: *Introduction to Higher Algebra* (Pergamon, Oxford 1964)
104. M. Marden: *Geometry of Polynomials* (American Mathematical Society, Providence, 1966)
105. E. Bogomolny, O. Bohigas, P. Leboeuf: Phys. Rev. Lett. **68**,2726 (1992); J. Math. Phys. **85**, 639 (1996)
106. M. Kuś, F. Haake, B. Eckhardt: Z. Physik **B92**, 221 (1993)
107. F. Haake, M. Kuś, H.-J. Sommers, H. Schomerus, K. Życkowski, J. Phys. A: Math. Gen. **29**, 3641 (1996)
108. M.V. Berry, J.P. Keating: J. Phys. **A23**, 4839 (1990)
109. E. Bogomolny: Comments At. Mol. Phys. **25**, 67 (1990); Nonlinearity **5**, 805 (1992)
110. H.-J. Sommers, F. Haake, J. Weber: J. Phys. A: Math. Gen. **31**, 4395 (1998)
111. F. Haake, H.-J. Sommers, J. Weber: J. Phys. A: Math. Gen. **32**, 6903 (1999)
112. R.E. Hartwig, M.E. Fischer: Arch. Rat. Mech: Anal. **32**, 190 (1969)
113. U. Smilansky: Physica **D109**, 153 (1997)

Chapter 5

114. I.C. Percival: Adv. Chem: Phys. **36**, 1 (1977)
115. M. Brack, R.K. Bhaduri: *Semiclassical Physics* (Addison-Wesley, Reading, Mass., 1997)
116. V.P. Maslov: *Théorie des Perturbations et Méthodes Asymptotiques* (Dunod, Paris, 1972)
117. V.P. Maslov, M.V. Fedoriuk: *Semiclassical Approximation in Quantum Mechanics* (Reidel, Boston, 1981)
118. J.B. Delos: Adv. Chem. Phys. **65**, 161 (1986)
119. J.-P. Eckmann, R. Sénéor: Arch. Rational Mech. Anal. **61**, 153 (1976)
120. R.G. Littlejohn, J.m. Robbins: Phys. Rev. **A 36**, 2953 (1987)
121. N. Rosenzweig, C.E. Porter: Phys. Rev. **120**, 1698 (1960)
122. M.V. Berry, M. Robnik: J. Phys. A: Math. Gen. **17**, 2413 (1984)
123. M.V. Berry, M. Tabor: Proc. R. Soc. Lond. **A349**, 101 (1976)
124. M.V. Berry, M. Tabor: Proc. R. Soc. Lond. **A356**, 375 (1977)
125. E.B. Bogomolny, U. Gerland, C. Schmit: Phys. Rev. **E59**, R1315 (1999)

Chapter 6

126. P. Pechukas: Phys. Rev. Lett. **51**, 943 (1983)
127. T. Yukawa: Phys. Rev. Lett. **54**, 1883 (1985)
128. T. Yukawa: Phys. Lett. **116A**, 227 (1986)
129. F. Haake, M. Kuś, R. Scharf: Z. Phys. **B65**, 381 (1987)
130. K. Nakamura, H.J. Mikeska: Phys. Rev. **A**35, 5294 (1987)
131. M. Kuś: Europhys. Lett. **5**, 1 (1988)
132. O. Bohigas, M.J. Giannoni, C. Schmit: Phys. Rev. Lett. **52**, 1 (1984)
133. F. Haake, M. Kuś: Europhys. Lett. **6**, 579 (1988)
134. F. Dyson: J. Math. Phys. **3**, 140 (1962); see also M.L. Mehta: *Random Matrices and the Statistical Theory of Energy Levels* (Academic, New York 1967; 2nd edition 1991)
135. P. Gaspard, S.A. Rice, H.J. Mikeska, K. Nakamura: Phys. Rev. **A42**, 4015 (1990)
136. F. Calogero, C. Marchioro: J. Math. Phys. **15**, 1425 (1974)
137. B. Sutherland: Phys. Rev. **A5**, 1372 (1972)
138. J. Moser: Adv. Math. **16**, 1 (1975)
139. S. Wojciechowski: Phys. Lett. **111A**, 101 (1985)
140. P.D. Lax: Comm. Pure Appl. Math. **21**, 467 (1968)
141. L.D. Landau, E.M. Lifshitz: *Quantum Mechanics, Non-relativistic Theory*, Vol. 3 of *Course of Theoretical Physics* (Pergamon Press, London,1959)
142. P.A. Braun: Rev. Mod. Phys. **65**, 115 (1993)
143. P.A. Braun, S. Gnutzmann, F. Haake, M. Kuś, K. Życzkowski: to be published
144. M.R. Zirnbauer: In I.V. Lerner, J.P. Keating, D.E. Khmelnitskii (eds.): *Supersymmetry and Trace Formulae; Chaos and Disorder* (Kluwer Academic/Plenum, New York, 1999)
145. E.B. Bogomolny, B. Georgeot, M.-J. Giannoni, C. Schmit: Phys. Reports **291**, 219 (1997)
146. J.P. Keating, F. Mezzadri, Nonlinearity **13**, 747 (2000)

147. B. Dietz: Dissertation, Essen (1991)
148. B. Dietz, F. Haake, Europhys. Lett. **9**, 1 (1989)
149. H. Frahm, H.J. Mikeska: Z. Phys. **B65**, 249 (1986)
150. M. Abramowitz, I. Stegun: *Handbook of Mathematical Functions* (Dover, New York 1965)
151. F. Haake, G. Lenz: Europhys. Lett. **13**, 577 (1990)
152. D.J. Thouless: Physics Reports **13C**, 93 (1974)
153. P. Gaspard, S.A. Rice, H.J. Mikeska, K. Nakamura: Phys. Rev. **A42**, 4015 (1990)
154. D. Saher, F. Haake, P. Gaspard: Phys. Rev. **A44**, 7841 (1991)
155. B.D. Simons, A. Hashimoto, M. Courtney, D. Kleppner, B.L. Altshuler: Phys. Rev. Lett. **71**, 2899 (1993)
156. M. Kollmann, J. Stein, U. Stoffregen, H.-J. Stöckmann, B. Eckhardt: Phys. Rev. **E49**, R1 (1994)
157. D. Braun, E. Hofstetter, A. MacKinnon, G. Montambaux: Phys. Rev. **B55**, 7557 (1997)
158. D. Braun, G. Montambaux: Phys. Rev. **B50**, 7776 (1994)
159. J. Zakrzewski, D. Delande: Phys. Rev. **E47**, 1650 (1993); ibid. 1665 (1993); D. Delande, J. Zakrzewski: J. Phys. Soc. Japan **63**, Suppl. A, 101 (1994)
160. F. von Oppen: Phys. Rev. Lett. **73**,798 (1994); Phys. Rev. **E51**, 2647 (1995)
161. Y.V. Fyodorov, H.-J. Sommers: Phys. Rev. **E51**, R2719 (1995); Z. Physik **B 99**, 123 (1995)
162. S. Iida, H.-J. Sommers: Phys. Rev. **E49**, 2513 (1994)
163. E. Akkermans, G. Montambaux: Phys. Rev. Lett. **68**, 642 (1992)
164. Y.V. Fyodorov, A.D. Mirlin: Phys. Rev. **B51**, 13403 (1995)
165. M. Sieber, H. Primack, U. Smilanski, I. Ussishkin, H. Schanz: J. Phys, **A28**, 5041 (1995)
166. H.-J. Stöckmann: *Quantum Chaos, An Introduction* (Cambridge University Press, Cambridge, 1999)
167. M. Barth, U. Kuhl, H.-J. Stöckmann: Phys. Rev. Lett. **82**, 2026 (1999)
168. M.V. Berry: J. Phys. **A10**, 2083 (1977)
169. V.N. Prigodin, : Phys. Rev. Lett. **75**, 2392 (1995)
170. X. Yang, J. Burgdörfer: Phys. Rev. **A46**, 2295 (1992)
171. B.D. Simons, B.L. Altshuler: Phys. Rev. Lett. **70**, 4063 (1993); Phys. Rev. **B48**, 5422 (1993)
172. J. Zakrzewski: Z. Physik **B98**, 273 (1995)
173. I. Guarneri, K. Życzkowski, J. Zakrzewski, L. Molinari, G. Casati: Phys. Rev. **E52**, 2220 (1995)
174. G. Lenz, F. Haake: Phys. Rev. Lett. **65**, 2325 (1990)
175. G. Lenz: Dissertation (Essen, 1992);
176. H. Risken: *The Fokker Planck Equation*, Springer Ser. Synergetics, Vol. 18 (Springer, Berlin, Heidelberg 1984)
177. A. Pandey, M.L. Mehta: Comm. Math. Phys. **87**, 449 (1983)
178. M.L. Mehta, A. Pandey: J. Phys. A**16**, 2655 (1983)
179. J.B. French, U.K.B. Kota, A. Pandey, S. Tomsovic: Ann. Phys. (N.Y.) **181**, 198 (1988)
180. A. Pandey: In T.H. Seligman, H. Nishioka (eds.): *Quantum Chaos and Statistical Nuclear Physics* (Springer, Berlin, Heidelberg 1986)
181. F.J. Dyson: J. Math. Phys. **13**, 90 (1972)
182. L.A. Pastur: Th. Math. Phys. **10**, 67 (1972)
183. L.A. Bunimovich: Funct. Anal. Appl. **8**, 73 (1974)

184. A. Pandey, D. Forster, F. Haake: unpublished
185. R. Scharf: J. Phys. A**21**, 4130 (1988)
186. R. Scharf: J. Phys. A**21**, 2007 (1988)
187. G. Lenz, F. Haake: Phys. Rev. Lett. **67**, 1 (1991)
188. E. Caurier, B. Grammaticos, A. Ramani: J. Phys. A**23**, 4903 (1990)

Chapter 7

189. S. Fishman, D.R. Grempel. R.E. Prange: Phys. Rev. Lett. **49**, 509 (1982); Phys. Rev. A**29**, 1639 (1984)
190. D.L. Shepelyansky: Phys. Rev. Lett. **56**, 677 (1986)
191. G. Casati, J. Ford, I. Guarneri, F. Vivaldi: Phys. Rev. A**34**, 1413 (1986)
192. Shan-Jin Chang, Kang-Jie Shi: Phys. Rev. A**34**, 7 (1986)
193. M. Feingold, S. Fishman, D.R. Grempel, R.E. Prange: Phys. Rev. B**31**, 6852 (1985)
194. M.V. Berry, S. Klein: Eur. J. Phys. **18**, 222 (1997)
195. T. Dittrich, U. Smilansky: Nonlinearity **4**, 59 (1991);ibid. **4**, 85 (1991)
196. N. Argaman, Y. Imry, U. Smilansky: Phys. Rev. B**47**, 4440 (1993)
197. K.B. Efetov: *Supersymmetry in Disorder and Chaos* (Cambridge University Press, Cambridge, 1997)
198. Y. Imry: *Introduction to Mesoscopic Physics* (Oxford University Press, Oxford, 1997)
199. J.E. Bayfield, P.M. Koch: Phys. Rev. **33**, 258 (1974)
200. E.J. Galvez, B.E. Sauer, L. Moorman,P.M. Koch, D. Richards: Phys. Rev. Lett. **61**, 2011 (1988)
201. J.E. Bayfield, G. Casati, I. Guarneri, D.W. Sokol: Phys. Rev. Lett. **63**, 364 (1989)
202. P.M. Koch: Physica D**83**, 178 (1995)
203. P.M. Koch, K.A.H. van Leeuwen: Physics Reports **255**, 289 (1995)
204. M.R.W. Bellermann, P.M. Koch, D.R. Mariani, D. Richards: Phys. Rev. Lett. **76**, 892 (1996)
205. O. Benson, G. Raithel, H. Walther: In G. Casati, B. Chirikov: *Quantum Chaos* (Cambridge University Press, Cambridge, 1995)
206. R. Bluemel, W.P. Reinhardt: *Chaos in Atomic Physics* (Cambridge University Press, Cambridge, 1997)
207. G. Casati, B. Chirikov: *Quantum Chaos* (Cambridge University Press, Cambridge, 1995)
208. R. Graham, M. Schlautmann, P. Zoller: Phys. Rev. **A 45**, R19 (1992)
209. F.L. Moore, J.C. Robinson, C. Bharucha, P.E. Williams, M.G. Raizen: Phys. Rev. Lett. **73**, 3974 (1994)
210. J.C. Robinson, C. Bharucha, F.L. Moore, R. Jahnke, G.A. Georgakis, Q. Niu,, M.G. Raizen, B Sundaram: Phys. Rev. Lett. **74**, 3963 (1995)
211. F.L. Moore, J.C. Robinson, C. Bharucha, B Sundaram, M.G. Raizen: Phys. Rev. Lett. **75**, 4598 (1995)
212. H.-J. Stöckmann: *Quantum Chaos, An Introduction* (Cambridge University Press, Cambridge, 1999)
213. B. Kramer, A. MacKinnon: Rep. Prog. Phys. **56**, 1469 (1993)
214. P.W. Anderson: Phys. Rev. **109**, 1492 (1958); Rev. Mod. Phys. **50**, 191 (1978)
215. D.J. Thouless: In *Session XXXI, 1979, Ill-Condensed Matter* R. Balian, R. Maynard, G. Thoulouse (eds.): Les Houches, (North-Holland, Amsterdam, 1979)

216. K. Ishii: Prog. Th. Phys., Suppl. **53**, 77 (1973)
217. H. Furstenberg: Trans. Ann. Math. Soc. **108**, 377 (1963)
218. A. Crisanti, G. Paladin, A. Vulpiani: *Products of Random Matrices* (Springer, Berlin, 1993)
219. B.V. Chirikov: preprint no. 367, Inst. Nucl. Physics Novosibirsk (1969); Phys. Rep. **52**, 263 (1979)
220. H. Weyl: Math. Ann. **77**, 313 (1916)
221. P. Lloyd: J. Phys. **C2**, 1717 (1969)
222. D.J. Thouless: J. Phys. **C5**, 77 (1972)
223. S.F. Edwards, P.W. Anderson: J. Phys. **F5**, 965 (1975)
224. B.V. Chirikov, F.M. Izrailev, D.L. Shepelyansky: Sov. Sci. Rev. **2C**, 209 (1981)
225. D.L. Shepelyansky: Phys. Rev. Lett. **56**, 677 (1986)
226. F. Haake, M. Kuś, J. Mostowski, R. Scharf: In F. Haake, L.M. Narducci, D.F. Walls (eds.): *Coherence, Cooperation, and Fluctuations* (Cambridge University Press, Cambridge, 1986)
227. F. Haake, M. Kuś, R. Scharf: Z. Phys. **B65**, 381 (1987)
228. M. Kuś, R. Scharf, F. Haake: Z. Phys. **B66**, 129 (1987)
229. R. Scharf, B. Dietz, M. Kuś, F. Haake, M.V. Berry: Europhys. Lett. **5**, 383 (1988)
230. F.T. Arecchi, E. Courtens, G. Gilmore, H. Thomas: Phys. Rev. **A6**, 2211 (1972)
231. R. Glauber, F. Haake: Phys. Rev. **A13**, 357 (1976)
232. K. Zyczkowski: J. Phys. **A23**, 4427 (1990)
233. A. Peres: *Quantum Theory: Concepts and Methods* (Kluwer Academic, New York, 1995)
234. F. Haake, D.L. Shepelyansky: Europhys. Lett. **5**, 671 (1988)

Chapter 8

235. A. Einstein: Verh. Dt. Phys. Ges. **19**, 82 (1917)
236. M.C. Gutzwiller: *Chaos in Classical and Quantum Mechanics* (Springer, New York, 1990)
237. M.C. Gutzwiller: J. Math. Phys. **12**, 343 (1971)
238. P. Cvitanovic: *Classical and Quantum Chaos. Cyclist Treatise* www.nbi.dk/ChaosBook/
239. M. Brack, R.K. Bhaduri: *Semiclassical Physics* (Addison-Wesley, Reading, Mass., 1997)
240. L.S. Schulman: *Techniques and Applications of Path Integration* (Wiley, New York, 1981)
241. G. Junker, H. Leschke: Physica **56D**, 135 (1992)
242. M. Tabor: Physica **6D**,195 (1983)
243. J.P. Keating: Nonlinearity **4**, 309 (1991) (Addison-Wesley, Reading, Mass., 1950; 2nd ed. 1980)
244. R. Courant, D. Hilbert: *Methoden der Mathematischen Physik* (Springer, Berlin; 3. Aufl. 1968); R. Bellmann: *Introduction to Matrix Analysis*, 2nd. ed. (McGraw-Hill, New York, 1970)
245. S. Levit, K. Möhring, U. Smilansky, T. Dreyfus: Ann. Phys. **114**, 223 (1978); S. Levit, U. Smilansky: Ann. Phys. **103**, 198 (1977); ibid. **108**, 165 (1977); K. Möhring, S. Levit, U. Smilansky: Ann. Phys. **127**, 198 (1980)
246. M. Morse: *Variational Analysis* (Wiley, New York, 1973)

247. E.T. Copson: *Asymptotic Expansions* (Cambridge University Press, Cambridge, 1965)

248. H.P. Baltes, E.R. Hilf: *Spectra of Finite Systems* (BI Wissenschaftsverlag, Mannheim, 1976)

249. R. Balian, F. Bloch: Ann. Phys. (USA) **63**, 592 (1971)

250. M.V. Berry, C.J. Howls: Proc. Roy. Soc. Lond. **A447**, 527 (1994)

251. K. Stewartson, R.T. Waechter: Proc. Camb. Phil. Soc. **69**, 353 (1971)

252. H. Primack, U. Smilansky: J. Phys. **A31**, 6253 (1998); H. Primack, U. Smilansky: Phys. Rev. Lett. **74**,4831 (1995); H. Primack: Ph.D. Thesis, The Weizman Institute of Science, Rehovot (19997)

253. P. Gaspard: Prog. Th. Phys. Supp. **116**, 59 (1994); P. Gaspard, D. Alonso: Phys. Rev. **A47**, R3468 (1993); D. Alonso, P. Gaspard: Chaos **3**, 601 (1993)

254. T. Prosen: Phys. Lett. **A223**, 323 (1997)

255. N.L. Balazs, A. Voros: Phys. Rep. **143**, 109 (1986)

256. T. Kottos, U. Smilansky: Ann. Phys. **274**, 76 (1999); T. Kottos, U. Smilansky:chao-dyn/9906008; H. Schanz, U. Smilansky: *Proceedings of the Australian Summer School on Quantum Chaos and Mesoscopics*, Canberra Australia, January 1999

257. M.V. Berry, J.P. Keating: J. Phys. **A23**, 4839 (1990)

258. J.P. Keating: Proc. R. Soc. Lond. **A436**, 99 (1992)

259. M.V. Berry, J.P. Keating: Proc. R. Soc. Lond. **A437**, 151 (1992)

260. B. Eckhardt, E. Aurell: Europhys. Lett. **9**, 509 (1989)

261. E.B. Bogomolny: Nonlinearity **5**, 805 (1992)

262. E. Doron, U. Smilansky: Nonlinearity **5**, 1055 (1995); H. Schanz, U. Smilansky: Chaos, Solitons & Fractals **5**, 1289 (1995); U. Smilansky: In E. Akkermans, G. Montambaux, J.L. Pichard: *Proceedings of the 1994 Summer School on "Mesoscopic Quantum Physics"* (North Holland, Amsterdam, 1991) ; B. Dietz, U. Smilansky: Chaos **3**, 581 (1993); C. Rouvinez, U. Smilansky: J. Phys. **A28**, 77 (1994)

263. E.G. Vergini: J. Phys. **A33**, 4709 (2000); E.G. Vergini, G.G. Carlo: J. Phys. **A33**, 4717 (2000)

264. R.G. Littlejohn: J. Math. Phys. **31**, 2952 (1990); J. Stat. Phys. **68**, 7 (1992)

265. I.C. Percival: Adv. Chem. Phys. **36**, 1 (1977)

266. S.C. Creagh, J.M. Robbins, R.G. Littlejohn: Phys. Rev. **A42**, 1907 (1990)

267. J.N. Mather: Comm. Math. Phys. **94**, 141 (1984)

268. J. Robbins: Nonlinearity **4**, 343 (1991)

269. J. Moser: Ergod. Theor.Dynam. Sys. **6**, 401 (1986)

270. P.A. Boasman, U. Smilansky: J. Phys. **A27**, 1373 (1994)

271. J.H. Hannay, A.M. Ozorio de Almeida: J. Phys. **A17**, 3429 (1984)

272. M.V. Berry: Proc. R. Soc. London **A400**, 229 (1985)

273. S. Fishman, J.P. Keating: J. Phys. **A 31**, L313 (1998)

274. P. Cvitanović, B. Eckhardt: J. Phys. **A24**, L237 (1991); Phys. Rev. Lett. **63**, 823 (1989); Nonlinearity **6**, 277 (1993)

275. E.B. Bogomolny, J.P. Keating: Phys. Rev. Lett. **77**, 1472 (1996)

276. O. Agam, B.L. Altshuler, A.V. Andreev: Phys. Rev. Lett. **75**, 4389 (1995)

277. P. Gaspard: *Chaos, Scattering and Statistical Mechanics* (Cambridge University Press, Cambridge, 1998)

278. E.B. Bogomolny, J.P. Keating: Private communication and to be published

279. E.B. Bogomolny, F. Leyvraz, C. Schmit: Commun. Math. Phys. **176**, 577 (1996); W. Luo, P. Sarnac: Commun. Math. Phys. **161**, 419 (1994)

280. E. Bogomolny, C. Schmit: Nonlinearity **6**, 523 (1993)

281. J.P. Keating, M. Sieber: Proc. R. Soc. London **A447**, 413 (1994)
282. R. Aurich, C. Matthies, M. Sieber, F. Steiner: Phys. Rev. Lett. **68**, 1629 (1992)
283. A.V. Andreev, B.L. Altshuler: Phys. Rev. Lett. **75**, 902 (1995)
284. M. Abramowitz, I. Stegun: *Handbook of Mathematical Functions* (Dover, New York 1965)
285. A.J. Lichtenberg, M.A. Liebermann: *Regular and Chaotic Dynamics*, Applied Mathematical Sciences, Vol. 38, 2nd ed. (Springer, New York, 1992)
286. M.V. Berry, M. Robnik: J. Phys. **A17**, 2413 (1984)
287. K.R. Meyer: Trans. Am. Math. Soc. **149**, 95 (1970)
288. H. Schomerus: J. Phys. **A31**, 4167 (1998)
289. A.M. Ozorio de Almeida, J.H. Hannay: J. Phys. **A20**, 5873 (1987)
290. S. Tomsovic, M. Grinberg, D. Ullmo: Phys. Rev. Lett. **75**, 4346 (1995)
291. D. Ullmo, M. Grinberg, S. Tomsovic: Phys. Rev. **E54**, 136 (1996)
292. M. Sieber: J. Phys. **A29**, 4715 (1996)
293. H. Schomerus, M. Sieber: J. Phys. **A30**, 4537 (1997)
294. H. Schomerus: Europhys. Lett. **38**, 423 (1997)
295. M. Sieber, H. Schomerus: J. Phys. **A31**, 165 (1998)
296. T. Poston, I.N. Stewart: *Catastrophe Theory and its Applications* (Pitman, London, 1978)
297. M. Kuś, F. Haake, D. Delande: Phys. Rev. Lett. **71**, 2167 (1993)
298. H. Schomerus, F. Haake: Phys. Rev. Lett. **79**, 1022 (1997)
299. M.V. Berry, J.P. Keating, S.D. Prado: J. Phys. **A31**, L243 (1998)
300. C. Manderfeld, H. Schomerus: chao-dyn/9911024
301. M.V. Berry, J.P. Keating, H. Schomerus: submitted to Proc. R. Soc. Lond. (Jan. 2000)
302. M.V. Berry: In: G. Casati, I. Guarneri and U. Smilansky (eds): *Proceedings of the International School of Physics "Enrico Fermi" Course CXLIII* G. Casati, I. Guarneri and U. Smilansky (Ios Press, Amsterdam, 2000)
303. M. Khodas, S. Fishman: Phys. Rev. Lett. **84**, 2837 (2000), chaodyn 9910040
304. J. Weber, F. Haake, P. Šeba: Phys. Rev. Lett. **85**, 3620 (2000), and chaodyn 0001013
305. C.W. Gardiner: *Quanntum Noise* (Springer, Berlin, 1991)
306. C. Manderfeld, J. Weber, F. Haake, P.A. Braun: to be published
307. D. Weiss, M.L. Roukes, A. Menschig, P. Grambow, K. von Klitzing, G. Weimann: Phys. Rev. Lett. **66**, 2790 (1991) ; D. Weiss, K. Richter, A. Menschig, R. Bergmann, H. Schweizer, K. von Klitzing, G. Weimann: Phys. Rev. Lett. **70**, 4118 (1993)
308. R. Fleischmann, T. Geisel, R. Ketzmerick: Phys. Rev. Lett. **68**, 1367 (1992)
309. G. Hackenbroich, F. von Oppen: Europhys. Lett. **29**, 151 (1995); Z. Physik **B97**, 157 (1995); K. Richter: Europhys. Lett. **29**, 7 (1995)
310. T.M. Fromhold, L. Eaves, F.W. Sheard, M.L. Leadbeater, T.J. Foster, P.C. Main: Phys. Rev. Lett. **72**, 2608 (1994); G. Mueller, G.S. Boebinger, H. Mathur, L.N. Pfeiffer, K.W. West: Phys. Rev. Lett. **75**, 2875 (1995); E.E. Narimanov, A.D. Stone: Phys. Rev. **B57**, 9807 (1998); E.E. Narimanov, A.D. Stone: Physica **D131**, 221 (1999); E.E. Narimanov, A.D. Stone, G.S. Boebinger: Phys. Rev. Lett. **80**, 4024 (1998)
311. A. Mekis, J.U. Nöckel, G. Chen, A.D. Stone, R.K. Chang: Phys. Rev. Lett. **75**, 2682 (1995); J.U. Nöckel, A.D. Stone: Nature **385**,47 (1997); C. Gmachl, F. Capasso, E.E. Narimanov, J.U. Nöckel, D.A. Stone, J. Faist, D.L. Sivco, A.Y. Cho: Science **280**, 1493 (1998); E.E. Narimanov, G. Hackenbroich, P. Jacquod, A.D. Stone: Phys. Rev. Lett. **83**, 4991 (1999)

312. M. Brack, S.M. Reimann, M. Sieber: Phys. Rev. Lett. **79**, 1817 (1997)
313. J. Main, G. Wunner: Phys. Rev. **A55**, 1743 (1997)
314. M. Fließer, A. Csordás, P. Szépfalusy, R. Graham: Phys. Rev. **A56**, R2533 (1997); ibid. A56, 4879 (1997); M. Fließer, R. Graham: Phys. Rev. **A59**, R27 (1999); Physica **D131**, 141 (1999); M. Fließer: Dissertation, Essen (2000)

Chapter 9

315. R. Graham, F. Haake: *Quantum Statistics in Optics and Solid State Physics,* Springer Tracts in Modern Physics, Vol. 66 (Springer, Berlin, Heidelberg 1973)
316. H. Spohn: Rev. Mod. Phys. **53**, 569 (1980)
317. F. Haake: Z. Phys. **B48**, 31 (1982)
318. F. Haake, R. Reibold: Phys. Rev. **A32**, 2462 (1985)
319. L.P. Kadanoff, G. Baym: *Quantum Statistical Mechanics* (Benjamin, New York 1962)
320. R. Bonifacio, P. Schwendimann, F. Haake: Phys. Rev. **A4**, 302, 854 (1971)
321. M. Gross, S. Haroche: Physics Reports **93**, 301 (1982)
322. L.D. Landau, E.M. Lifschitz: *Course of Theoretical Physics,* Vol. 5, Statistical Physics (Pergamon, London 1952)
323. A.J. Leggett: Prog. Th. Phys. Suppl.**69**, 80 (1980)
324. A.O. Caldeira, A.J. Leggett: Ann. Phys. (NY) **149**, 374 (1983)
325. F. Haake, D.F. Walls: Phys. Rev. **A36**, 730 (1987)
326. W.H. Zurek: Phys. Rev. **D24**, 1516 (1981); **D26**, 1862 (1982); Physics Today **44**, 36 (1991); Prog. Th. Phys. **89**, 281 (1993)
327. M. Brune, E. Hagley, J. Dreyer, X. Maître, A. Maali, C. Wunderlin, J.M. Raimond, S. Haroche: Phys. Rev. Lett. **77**, 4887 (1996); S. Haroche: Physics Today **51**, 36 (1998)
328. R. Grobe, F. Haake: Z. Phys. **B68**, 503 (1987) and Lect. Notes Phys. **282**, 267 (1987);
329. R.R. Puri, G.S. Agarwal: Phys. Rev. **A33**, 3610 (1986)
330. S.M. Barnett, P.L. Knight: Phys. Rev. **A33**, 2444 (1986)
331. P.A. Braun, D. Braun, F. Haake, J. Weber: Euro. Phys. Journal **D2**, 165 (1998)
332. P.A. Braun, D. Braun, F. Haake: Euro. Phys. Journal **D3**, 1(1998)
333. D. Braun, P.A. Braun, F. Haake: Physica **D131**, 265 (1999)
334. D. Braun, P.A. Braun, F. Haake: Opt. Comm.**179**, 195 (2000); and in P. Blanchard, D. Giulini, E. Joos, C. Kiefer, I.O. Stamatescu (eds.), *Decoherence: Theoretical, Experimental, and Conceptual Problems* (Springer, Berlin, 2000)
335. D. Braun: Chaos **9**, 760 (1999)
336. R. Grobe, F. Haake, H.-J. Sommers: Phys. Rev. Lett. **61**, 1899 (1988)
337. L.E. Reichl, Z.-Y. Chen, M. Millonas: Phys. Rev. Lett. **63**, 2013 (1989)
338. J. Ginibre: J. Math. Phys. **6**, 440 (1965)
339. M.L. Mehta: *Random Matrices and the Statistical Theory of Energy Levels* (Academic, New York 1967; 2nd ed. 1991)
340. M. Abramowitz, I.A. Stegun: *Handbook of Mathematical Functions* (Dover, New York 1970)
341. N. Lehmann, H.-J. Sommers: Phys. Rev. Lett. **67**, 941 (1991)
342. M. Kuś, F. Haake, D. Zaitsev, A. Huckleberry: J. Phys. **A 30**, 8635 (1997)
343. D. Zaitsev, M. Kuś, A. Huckleberry, F. Haake: J. Geometry and Physics, to appear

344. H. Flanders: *Differential Forms with Applications to the Physical Sciences* (Dover, New York, 1989)
345. R. Grobe, F. Haake: Phys. Rev. Lett. **62**, 2889 (1989)
346. R. Grobe: Ph.D. Thesis, Essen (1989)
347. L. Onsager: Phys. Rev. **37**, 405 (1931) and Phys. Rev. **38**, 2265 (1931)
348. G.S. Agarwal: Z. Phys. **258**, 409 (1973);
349. H.J. Carmichael, D.F. Walls: Z. Phys. **B23**, 299 (1976)
350. G.M. Zaslavski: Phys. Lett. **69A**, 145 (1978)
351. G.M. Zaslavski, Kh.-R.Ya. Rachko: Sov. Phys. JETP **49**, 1039 (1979)
352. T. Dittrich, R. Graham: Ann. Phys. **200**, 363 (1990)
353. F. Haake: J. Mod. Optics **47**, 2883 (2000)
354. G.S. Agarwal, R.R. Puri, R.P. Singh: Phys. Rev. **56**, 2249 (1997)
355. F. Waldner, D.R. Barberis, H. Yamazaki: Phys. Rev. **A31**, 420 (1985)
356. F. Haake, G. Lenz, R. Puri: J. Mod. Optics **37**, 155 (1990)
357. F.T. Arecchi, E. Courtens, G. Gilmore, H. Thomas: Phys. Rev. **A6**, 2211 (1972)
358. R. Glauber, F. Haake: Phys. Rev. **A13**, 357 (1976)
359. R.J. Glauber: Phys. Rev. **130**, 2529 (1963); **131**, 2766 (1963)
360. D.F. Walls, G. Milburn: Phys. Rev. **A31**, 2403 (1985)
361. P. Gaspard, S.A. Rice, H.J. Mikeska, K. Nakamura: Phys. Rev. **A42**, 4015 (1990)

Chapter 10

362. H. Flanders: *Differential Forms with Applications to the Physical Sciences* (Dover, New York, 1989)
363. C. Itzykson, J.-B. Zuber: *Quantum Field Theory* (McGraw Hill, New York 1890)
364. F.A. Berezin: *Method of Second Quantization* (Academic, New York, 1966)
365. F.A. Berezin: *Introduction to Superanalysis* (Reidel, Dordrecht, 1987)
366. K.B. Efetov: *Supersymmetry in Disorder and Chaos* (Cambridge University Press, Cambridge, 1997)
367. T. Guhr, A. Müller-Groeling, H.A. Weidenmüller: Phys. Rep. **299**, 192 (1998)
368. H.J. Sommers: Unpublished lectures (Essen, 1997)
369. Y.V. Fyodorov: in E. Akkermans, G. Montambaux, J.-L. Pichard, J. Zinn-Justin (eds): *Les Houches, Session LXI, 1994, Mesoscopic Quantum Physics* (Elsevier, Amsterdam 1995)
370. G. Parisi, N. Sourlas: Phys. Rev. Lett. **43**, 744 (1979)
371. A.J. McKane: Phys. Lett. **A 76**, 33 (1980)
372. K.B. Efetov: Adv. Phys. **32**, 53 (1983)
373. F. Wegner: Z. Physik **49**, 297 (1983)
374. F. Constantinescu, F. de Groote: J. Math. Phys. **30**, 981 (1989)
375. L. Schäfer, F. Wegner: Z. Phys. **B 38**, 113 (1968)
376. J.J.M. Verbaarschot, H.A. Weidenmüller, M.R. Zirnbauer: Phys. Rep. **129**, 367 (1985)
377. G. Hackenbroich, H.A. Weidenmüller: Phys. Rev. Lett. **74** 4118 (1995)
378. A.M. Khoruzhy, B.A. Khoruzhenko, L.A. Pastur: J. Phys. **A 28**, L31 (1995)
379. J.D. Anna, A. Zee: Phys. Rev. **E 53**, 1399 (1996)
380. A.D. Mirlin, Y.V. Fyodorov: J. Phys. **A 24**, 2273 (1991)
381. Y.V. Fyodorov, A.D. Mirlin: Phys. Rev. Lett. **67**, 2049 (1991)
382. S.N. Evangelou: J.Stat. Phys. **69**, 361 (1992)

383. G. Casati, L. Molinari, F. Izrailev: Phys. Rev. Lett. **64**, 16 (1990); J. Phys. **A24**, 4755 (1991)
384. S.N. Evangelou, E.N. Economou: Phys. Lett. **A151**, 345 (1991)
385. Y.V. Fyodorov, A.D. Mirlin: Int. J. Mod. Phys. **B8**, 3795 (1994)
386. M. Kuś, M. Lewenstein, F. Haake: Phys. Rev. **A44**, 2800 (1991)
387. L. Bogachev, S. Molchanov, L.A. Pastur: Mat. Zametki **50**, 31 (1991); S. Molchanov, L.A. Pastur, A. Khorunzhy: Theor. Math. Phys. **73**, 1094 (1992)
388. M. Feingold: Europhys. Lett. **71**, 97 (1992)
389. G. Casati, V. Girko: Rand. Oper. Stoch. Eq. **1**, 1 (1993)
390. M. Kac in *Statistical Physics: Phase Transitions and Superfluidity*, Vol. 1 (Gordon and Breach, New York, 1991)
391. K.B. Efetov, A.I. Larkin: Sov.Phys. JETP **58**, 444 (1983)
392. O.I. Marichev: *Handbook on Integral Transforms of Higher Transcendental Functions* (Ellis Horwood, New York, 1983)

Index

edes-Druck, Berlin
G, Würzburg